OXFORD QUICK REFERENCE

A Dictionary of
Construction, Surveying, and Civil Engineering

Professor Christopher Gorse is Director of the Leeds Sustainability Institute at Leeds Beckett University, where he holds a Chair in Construction and Project Management. He is currently the Chair of the Association of Researchers in Construction Management (ARCOM), Chair of the International Conference for Sustainable Ecological Engineering Design for Society (SEEDS), and a founding member of the Building Performance Network. Chris is a Chartered Builder and Engineering Professors' Council Member, holding Principal Investigator positions for major construction, environment, and energy research projects. Chris has written extensively on buildings, construction law, education, management, energy efficiency, digital innovation, and sustainability.

Dr David Johnston is a Professor of Building Performance Evaluation within Leeds Sustainability Institute. He has over twenty five years' experience of applied and theoretical research and consultancy in low-carbon housing and is a leading expert in coheating testing and building performance evaluation. He has led and managed numerous field trial projects in both new and existing dwellings, involving detailed in-use monitoring of energy consumption, the analysis of occupant behaviour, and detailed evaluations of the fabric and services performance of domestic buildings. He is an expert member of the British Standards Institution B/540/9 and CEN Technical Committee 89 Working Group 13. He is also the co-author of numerous technical reports and peer-reviewed publications.

Dr Martin Pritchard (BEng Hons, PGCHE, APDRAS, PhD, SFHEA, CEng FICE) is a Reader in Civil Engineering at Leeds Beckett University. He is a civil engineer and has conducted extensive international work on water purification systems and developed sustainable approaches to infrastructure works in developing countries. He also has a particular interest in geotechnics and computer modelling.

(⊕) SEE WEB LINKS

For recommended web links for this title, visit www.oxfordreference.com/page/constr when you see this sign.

The most authoritative and up-to-date reference books for both students and the general reader.

Accounting
Agriculture and Land
   Management
Animal Behaviour
Archaeology
Architecture
Art and Artists
Art Terms
Arthurian Literature and Legend
Astronomy
Bible
Biology
Biomedicine
British History
British Place-Names
Business and Management
Chemical Engineering
Chemistry
Christian Art and Architecture
Christian Church
Classical Literature
Computer Science
Construction, Surveying, and
   Civil Engineering
Cosmology
Countries of the World
Critical Theory
Dance
Dentistry
Ecology
Economics
Education
Electronics and Electrical
   Engineering
English Etymology
English Grammar
English Idioms
English Language
English Literature
English Surnames
Environment and Conservation
Everyday Grammar
Film Studies
Finance and Banking
Foreign Words and Phrases
Forensic Science
Fowler's Concise Modern
   English Usage
Geography
Geology and Earth Sciences

Hinduism
Human Geography
Humorous Quotations
Irish History
Islam
Journalism
Kings and Queens of Britain
Law
Law Enforcement
Linguistics
Literary Terms
London Place-Names
Marketing
Mathematics
Mechanical Engineering
Media and Communication
Medical
Modern Poetry
Modern Slang
Music
Musical Terms
Nursing
Philosophy
Physics
Plant Sciences
Plays
Political Quotations
Politics and International
   Relations
Popes
Proverbs
Psychology
Quotations
Quotations by Subject
Rhyming
Rhyming Slang
Saints
Science
Scottish History
Shakespeare
Social Work and Social Care
Sociology
Statistics
Synonyms and Antonyms
Weather
Weights, Measures,
   and Units
Word Origins
World Mythology
Zoology

Many of these titles are also available online at
www.oxfordreference.com

A Dictionary of

Construction, Surveying, and Civil Engineering

SECOND EDITION

CHRISTOPHER GORSE, DAVID JOHNSTON,
AND MARTIN PRITCHARD

Great Clarendon Street, Oxford, OX2 6DP,
United Kingdom

Oxford University Press is a department of the University of Oxford.
It furthers the University's objective of excellence in research, scholarship,
and education by publishing worldwide. Oxford is a registered trade mark of
Oxford University Press in the UK and in certain other countries

First edition 2012
Second edition 2020

Impression: 3

Published in the United States of America by Oxford University Press
198 Madison Avenue, New York, NY 10016, United States of America

British Library Cataloguing in Publication Data
Data available

Library of Congress Control Number: 2019945685

ISBN 978-0-19-883248-5

Printed and bound in Great Britain by
Clays Ltd, Elcograf S.p.A.

Contents

Preface

This dictionary aims to provide a comprehensive and up-to-date reference of construction, surveying, and civil engineering terms within one publication. The entries address the needs of those studying on built environment courses or engaged in professional practice as well as providing much useful information for those with a general interest in architecture or construction. Builders, trade suppliers, and contractors will find the dictionary a useful desktop manual that can be easily and quickly used to check unfamiliar terminology.

Due to the size and nature of the industry, it is difficult for even the most knowledgeable professionals to stay abreast of emerging issues and retain an encyclopedia of construction terms at the front of their mind. In such cases, the book acts as a reminder, giving reassurance of meanings and confidence in communication. Built environment professionals, including architects, building surveyors, building services engineers, construction managers, civil engineers, electrical engineers, facilities managers, mechanical engineers, and quantity surveyors should find the dictionary useful. It offers a quick and easy guide to terms and information that underpin the environment in which we live, both built and natural. In covering these terms, the authors have delved into the areas of building physics and science, and their application to design, structures, materials, and practices that inform construction and engineering. Issues associated with geology, geography, climate, and the natural environment within which buildings and structures are accommodated are also covered. The policy and legal frameworks that provide governance and procedure to the built environment are included, as well as the issues related to the professional organizations that hold influential positions within the industry.

The importance of the sustainability agenda and the related fields of building, engineering, and surveying mean that more people are taking an interest in construction and need to familiarize themselves quickly with the language and terminology used. The dictionary offers a good point of reference for those operating within the field and those just interested in learning more about the built environment and engineering. As a major consumer of the world's natural resources, the built environment and the impact that it has on the environment are high on the government agenda, with all policymakers ensuring that it is given due consideration. It is expected that the book, terms, and references will continue to grow and be in greater demand as the degree of importance associated with the built environment increases.

Acknowledgements

Our understanding of construction, surveying, and civil engineering has developed over the years, benefiting from the experience and knowledge of many professionals and colleagues. Some have made direct contributions to the dictionary, while others have been influential in shaping our understanding of the field. We would specifically like to thank all who have contributed to the pool of knowledge and those who we have had the pleasure of collaborating with on our journey.

Australia

Dr Dominic Ahiaga-Dagbui
 Dekin University
Dr Nicholas Chileshe
 University of South Australia
Martin Loosemore
 University of New South Wales
Prof. David Thorpe
 University of Southern Queensland

Belgium

Dr Geert Bauwens
 Katholieke Universiteit Leuven
Prof. Arnold Janssens
 University of Ghent
Prof. Gustaaf Roels
 KU Leuven
Prof. Dirk Saelens
 KU Leuven

China

Dr Wei Pan
 University of Hong Kong

Croatia

Prof. Anita Ceric
 University of Zagreb

Denmark

Dr Mikkel Kragh
 Danish Architecture Center

Finland

Prof. Jari Porras
 Lappeenranta-Lahti University of
 Technology

France

Dr Jean Carassus
 École des Ponts ParisTech
Prof. Jean-Phillipe Georges
 University of Lorraine
Prof. Éric Rondeau
 University of Lorraine

Germany

Prof. Olaf Droegehorn
 Harz University of Applied Sciences

Netherlands

Dr Christian Stuck
 Saxion University of Applied Sciences
Prof Jan de Wit
 Saxion University of Applied Sciences

Pakistan

Prof. Sher Jamal Khan
 National University of Sciences and
 Technology

Qatar

Amr Metwally
 Hamad Medical Corporation

Republic of Ireland

Dr Cormac Flood
 Coady Architects, Dublin
Prof. Lloyd Scott
 Technological University Dublin
Dr John Spillane
 University of Limerick

Russia

Dr Alexandra Klimova
ITMO University

Spain

Dr Aitor Erkoreka
University of the Basque Country

South Africa

Prof. Fidelis Emuze
Central University of Technology,
Free State
Prof. John Smallwood
Nelson Mandela University

Sweden

Prof. Karl Andersson
Luleå University of Technology
Prof. Christine Raisanen
Chalmers University of Technology

UK

Dr Emmanuel Abogye-Nimo
University of Brighton
Dr Ash Ahmed
Leeds Beckett University
Prof. Akintoye Akintola
Leeds Beckett University
Dr David Allinson
Loughborough University
Dr Philip Ashton
University of Brighton
Paul Barge
Addleshaw Goodard
Mike Bates
PL Projects
Dr Colin Booth
University of the West of England
Dr Adrian Bown
Leeds Beckett University
Prof. David Boyd
Birmingham City University
Matthew Brooke-Peat
Leeds Beckett University
Prof. Paul Chan ARCOM
University of Manchester
Dr Udeaja Chika
University of Salford
Dr Vivien Chow
Loughborough University

Dr Alexander Copping
University of Bath
Tom Craven
Leeds Beckett University
Prof. Andrew Dainty
Loughborough University
Prof. Mohammad Dastbaz
University of Suffolk
Ian Dickinson
Leeds Beckett University
Prof. Mohammed Dulaimi
Leeds Beckett University
Prof. Charles Egbu
President Chartered Institute
of Building
Dr Cliff Elwell
University College London
Barry Falconer
Leeds Beckett University
David Farmer
Leeds Beckett University
Dr Richard Fitton
University of Salford
Dr Martin Fletcher
Leeds Beckett University
Dr Fiona Fylan
Leeds Beckett University
Phillip Garrison
Leeds Beckett University
Dr Kassim Gidado
University of Brighton
Prof. Jacqui Glass
University College London
Dr Barry Gledson
Northumbria University
Dr David Glew
Leeds Beckett University
Ellen Glover
Leeds Beckett University
Prof. Jan Godsell
University of Warwick
Prof. Jack Goulding
University of Wolverhampton
Martin Green
Leeds Beckett University
Prof. David Greenwood
Northumbria University
Prof. Rajat Gupta
Oxford Brookes University

Tahira Hamid
  Leeds Beckett University
Dr Adam Hardy
  Leeds Beckett University
Dr Anthony Higham
  University of Salford
David Highfield
  Leeds Sustainability Institute.
Dr Paul Hirst
  Leeds Beckett University
Dr Jeffrey Hobday
  Rolls Royce
Prof. Will Hughes
  University Reading
Stewart Jones
  Leeds Sustainability Institute
Hadi Kazemi
  Leeds Beckett University
Dr Andrew King
  Nottingham Trent University
Dr Louise King
  University of the West of England
Dr Alfred Leung
  Leeds Beckett University
Prof. Kevin Lomas
  Loughborough University
Prof. Robert Lowe
  University College London
Dr Shu-Ling Lu
  University of Reading
Dr Michael McCarney
  Glasgow Caledonian University
Dr Patrick Manu
  University of Manchester
Dr Wilfred Masuwa Matipa
  Liverpool John Moores University
Dr Angela Maye-Banbury
  Sheffield Hallam University
Samantha Mepham
  Rider Levett Bucknall
Jennifer Muston
  Leeds Sustainability Institute
Dr Christopher Neilson
  University of Manchester
Prof. Alan Newall
  Leeds Beckett University
Killian Ngong
  Leeds Beckett University
Prof. Catherine Noakes
  University of Leeds

Dr Alex Opoku
  University College London
Dr Mohamed Osmani
  Loughborough University
Dr Alice Owen
  University of Leeds
Dr Noel Painting
  University of Brighton
Dr James Parker
  Leeds Beckett University
Prof. Parneet Paul
  Leeds Beckett University
Dr Poorang Piroozfar
  University of Brighton
Dr Francesco Pomponi
  Edinburgh Napier University
Prof. David Proverbs
  Birmingham City University
Dr Ani Raiden
  Nottingham Trent University
Karl Redmond
  Rider Levett Bucknall
Prof. Andrew Ross
  Liverpool John Moores University
Josie Rothera
  Leeds Beckett University
Dr Ajayi Saheed
  Leeds Beckett University
Dr Libby Schweber
  University of Reading
Dominic Miles Shenton
  Leeds Beckett University
Dr Fred Sherratt
  Anglia Ruskin University
Dr Shariful Shikder
  Leeds Beckett University
Prof. Gary Shuckford
  Leeds Sustainability Institute
Prof. Peter Skipworth
  Leeds Sustainability Institute
Anthony Smith
  Leeds Beckett University
Dr Melanie Smith
  Leeds Beckett University
Dr Simon Smith
  University of Edinburgh
Dr Robby Soetanto
  Loughborough University
Prof. Paul Stephenson
  Sheffield Hallam University

Prof. Ian Strange
  Leeds Beckett University
Prof. John Sturges
  Leeds Sustainability Institute
Prof. Andrew Swan
  Leeds Beckett University
Prof. William Swan
  University of Salford
Felix Thomas
  Leeds Beckett University
Dr Craig Thomson
  Glasgow Caledonian University
Prof. Tony Thorpe
  Loughborough University
Dr Niraj Thurairajah
  Northumbria University
Dr Apollo Tutesigensi
  University of Leeds
Michael White
  Leeds Beckett University

Paul Whitehead
  PL Projects
Prof. Jennifer Whyte
  Imperial College London
Mark Wilson
  Leeds Beckett University
Dr Hannah Wood
  University of Brighton
Dr Sam Zulu
  Leeds Beckett University

USA

Prof. Patricia Aloise-Young
  Colorado State University
Prof. Richard Cozzens
  Southern Utah University
Prof. Jeremy Farner
  Weber State University
Prof. Peter Young
  Colorado State University

abacus In classical architecture, the flat uppermost slab at the top of the *capital of a column.

abandonment of work To leave site and refuse to continue work. Must entail a complete stoppage of work with a clear intention not to continue or complete the work. The term is often used in arbitration and adjudication and marks a point where reference to legal proceedings can commence.

abatement of action An interruption of legal proceedings. A party, normally the defendant, serves an application of abatement giving reasons why proceedings should not continue.

abeyance 1. A state of suspension, temporarily inactive, cessation or put to one side for a period of time. **2.** When the ownership of a property has not yet been decided or the condition of a property is without an owner.

abioseston Non-living particulate matter found floating in watercourses.

abiotic Non-living chemical and physical matter (not biological). **Abiotic components** are parts of the environment; water, sunlight, oxygen, temperature, soil, and climate.

ablution fitting A large communal sanitary fitting used to perform ritual washing.

Abney level A hand-held instrument to measure vertical angles. *See also* CLINOMETER.

above grade A higher elevation than ground level. *See also* ABOVE GROUND LEVEL.

above ground level (US above grade) 1. The height above ground level. **2.** Work undertaken on a building after the building has come out of the ground.

Abram's law A rule that states the strength of concrete (or mortar) is inversely related to the water/cement (w/c) ratio. Provided there is sufficient water to ensure that the *hydration reaction occurs, the lower the water/cement ratio, the higher the strength of the concrete/mortar.

abrasion The wearing away of one material against another by *friction.

abrasion resistance The ability of a finish to withstand wear.

abrasive The use of friction/roughness to clean, grind, or polish a surface.

abrupt wave An unexpected increase in flow caused by a sudden change in flow conditions.

ABS (acrylonitrile butadiene styrene) A polymeric material composed of acrylonitrile, butadiene, and styrene to give a good balance of mechanical properties, used commonly as lavatory seats.

abscissa The horizontal or x-coordinate within the *Cartesian coordinate system.

abseil survey An inspection undertaken on a tall building/structure by a qualified person suspended via a rope.

absolute Not depending on anything else.

absolute assignment The irrevocable transfer of the entire debt or all legal rights.

absolute zero The lowest possible temperature. Defined as zero degrees *Kelvin, which is equivalent to −273.15 degrees on the Celsius scale.

absorbent The ability to absorb, i.e. to soak up a liquid.

absorption The process of absorbing a liquid.

absorption rate (absorption coefficient of water) The rate of water absorption, ingress by capillary action, into a material when exposed to a water medium over a period of time. The method for determining the water absorption property is carried out in accordance with BS EN 772-11. After drying to constant mass, a face of the masonry material is immersed in water for a specific period of time, both short- and long-term and the increase in mass is usually determined. The water absorption of the face of the unit exposed is measured, after drying to constant mass. The minimum immersion time is usually one hour, as specified in BS EN 771-4. The coefficient of water absorption is normally stated at 10, 30, and 90 minutes for masonry materials (bricks and blocks).

absorption refrigerator A refrigerator that operates using the absorption cycle rather than the vapour compression cycle. In such systems, a heat source, usually waste heat, is absorbed and used to evaporate the refrigerant rather than using a compressor to compress the refrigerant. Quiet in operation and capable of being driven by solar energy.

absorptive form *See* PERMANENT FORMWORK.

absorptivity (absorptive power) A property of a material that determines the amount of incident heat, light, or sound energy that is absorbed by the material.

ABT *See* ASSOCIATION OF BUILDING TECHNICIANS.

abut To adjoin; be next to; touch.

abutment 1. A solid structure, usually a pier or wall, which provides support to an arch, bridge, or vault. It enables the loads from the structure to be transmitted to the *foundations. **2.** The point where a roof slope intersects a wall that extends above the roof slope. **3.** The intersection between two building elements. **4.** In dam construction, the sides of the valley against which the dam is constructed.

abutment flashing A *flashing at an *abutment, usually made from lead, although copper and mortar have been used.

abutment piece (US) A *sill or *sole plate.

abutment wall The wall that extends beyond an *abutment.

ACA Form of Building Agreement The Association of Consultant Architects construction contract.

ACAS (Advisory, Conciliation, and Arbitration Service) An independent publicly funded organization founded in 1975 that acts to prevent and resolve employment disputes. It also provides professional advice and training.

accelerated weathering (artificial ageing) Normally a cyclic process that simulates adverse climatic conditions (such as temperature, ultraviolet radiation, and moisture), which is used to assess the durability of materials.

acceleration A measure of the rate at which *velocity is changing, i.e. a change in speed over time.

acceleration of work Agreement to complete the work in a shorter time. Within general law, the architect, contract administrator, or client has no power to instruct the contractor to complete the work in a shorter period of time than the expressed completion date, unless there is a provision (term) within the contract that allows for the acceleration of works.

accelerator An admixture that enhances early strength (hardening) but the long-term strength remains unaffected. Regularly used in cold weather when urgent repair work is needed. Calcium chloride is the most common accelerator, however, it reduces corrosion protection.

accelerometer An instrument for measuring acceleration.

acceptance An agreement, based on an offer, which forms a contract. An offer and acceptance are used in contract law to determine whether a contract exists. For a contract to exist, there needs to be an offer, acceptance of that offer, and consideration (the thing of value at the centre of the contract). Acceptance can be made orally, in writing, or by conduct. Acceptance must be unqualified.

accepted programme The schedule of works that has been agreed, under the contract terms, to be used to schedule activities and resources. The *Engineering and Construction Contract, previously the New Engineering Contract, makes reference to the accepted programme.

accepted risk Risks accepted by the client or employer which can influence the works but are beyond the contractor's control or scope of works. Term used in GC (General Contract)/Works Contract.

access 1. The method of gaining entry to a building, a room, a site, or services. **2.** The right or permission to use something (access to documents).

access chamber *See* INSPECTION CHAMBER.

access cover (access eye, inspection cover) The removable cover on an *inspection chamber.

access floor (raised floor) A floor that is suspended above the structural floor. Usually consists of removable panels supported on a metal grid that is raised off the floor by adjustable pedestals or battens. The space between the two floors can be used to route various services such as data, telephone, power, lighting, heating and cooling pipes, and ventilation.

access hole An opening, usually large enough for a person, that enables access to be gained to an area, an installation, or services.

accessories 1. Components not included as part of the original product. Often used to enhance or modify the performance of the original product.
2. Components that are used to connect various building elements or services together.

accessory box (accessory enclosure) A box that is used to store accessories.

access panel A small door, panel, piece of wood, metal, or plastic that is flush with the surrounding area but can be removed or opened to provide access to the space or object (such as a shut-off valve) behind it.

access point 1. (drains) An opening that is sealed or covered, or any point that allows entry into the *drainage system (surface or *foul water) for inspection, *rodding, flushing, cleaning, and removing blockages. Purpose-made access points can be fitted at the end of a drainage run directly into the side of the *pipe or from a junction fitting. **2.** (Wi-Fi) A hardware device such as a wireless router which allows a wireless device to connect to a network to transmit and receive data. An access point connects users to other users and other wireless local area networks (WLANs).

Access to Neighbouring Land Act 1992 Act designed to address problems that arise when entry to a neighbour's land is required to carry out work. The Act applies to basic preservation works only and would include maintenance and repair of buildings and property that cannot be reasonably undertaken within the boundary of the property. Application is made to the courts. The courts can only award access if it does not interfere with the enjoyment of an owner's land, nor should it cause hardship.

access to works The contractual right to have access to the place where the work is to be undertaken. The employer has to give the contractor access to the site otherwise it would be impossible for him/her to carry out the work. It is not a breach of contract where access is prevented by third parties over whom the employer has no control.

accident An identifiable and specific event that was not anticipated and was unexpected. Generally, the fact that something was an accident is not an adequate defence to an action brought in tort.

accidental error A mistake happening by chance that would influence accuracy; for instance, taking an incorrect reading without realizing the mistake.

acclivity An upward slope or incline.

accommodation works Activities that are undertaken or required to maintain structures, equipment, or land that is the property or under the control of a statutory undertaker; such work may include: bridges, fences, and gates to protect railways, substations, etc. Where there is a statutory obligation to protect such works it must be undertaken.

accord and satisfaction The purchase of a release from an *obligation. In *contract law an obligation arising under the contract may be purchased by means of consideration, without having to undertake the obligation. The accord is the agreement and the satisfaction represents the consideration taken into account to discharge the obligation and makes the agreement valid.

accordion door A door comprising a number of sections that fold together like the bellows of an accordion when opened.

accreditation A certification process by which an organization is awarded a standard of competency and credibility.

Accredited Construction Details (ACDs) A set of generic standardized construction details, developed in conjunction with industry, that have been created to support the industry in England and Wales to meet the energy efficiency requirements of Part L of the *Building Regulations. The details focus on *thermal bridging and *airtightness and incorporate a series of checklists that are designed to be used to demonstrate regulatory compliance. A separate set of ACDs have been devised for Scotland.

accrued rights The rights and duties of the parties following a contractor's determination of employment under the contract (remedies of either party). Such rights also give powers to contract administrators to issue instructions, in relation to work already completed, e.g. for defective work.

accumulation The increase of something over time.

ACDs *See* Accredited Construction Details.

acequia An irrigation canal/ditch used in Spain and the American Southwest.

acetone A flammable liquid from the family of ketones.

acetylene A hydrocarbon substance from the alkynes family.

acid A liquid with a pH value of less than 7.0.

acid deposition The process whereby acidic gases and particles that are contained within the Earth's atmosphere are deposited on land and water. This occurs via either *dry deposition or *wet deposition.

acidic A chemical compound whose properties are typical of an *acid.

acid rain A common term used to describe *wet deposition where the acidic gases and particles contained within the Earth's atmosphere are deposited on land, water, and surfaces via rain.

acknowledgement of service The formal step, by way of a letter, of responding to a claim. The letter would state that a defendant intends to dispute part or all of the claim, state the particulars of the claim, and record whether the court's jurisdiction is accepted. The procedure is governed by the Civil Procedure Rules.

acoustic board A board used on walls, floors, and ceilings that is designed to improve *sound absorption or *sound insulation.

acoustic clip A clip that is attached to the floor joists to provide support for an acoustic floor. Used to reduce *impact sound transmission on floating timber floors.

acoustic construction Any type of construction that improves the *acoustics of the construction. For instance, reduces *sound transmission, improves *sound insulation, or increases *sound absorption.

acoustic finish (acoustic finishing) A finish that reduces sound energy by absorption.

acoustic floor A floor that is designed to reduce sound transmission.

acoustic insulation Fabric, board, paints, or other material used to reduce the passage of sound.

acoustic lining A lining material that is designed to absorb noise and reduce *sound transmission. Used on pipes and ductwork.

acoustic plaster Plaster that has high sound absorption properties.

acoustics 1. A branch of physics that relates to the study of *sound generation, sound transmission, absorption, and reflection. It addresses waves in gases, liquids, or solids and includes topics that deal with vibration, sound, ultrasound and infrasound. **2.** In construction, the effect that sound has on a room or space. The internal characteristics of a room, place, or area that denote how well sound is carried and can be heard (ability to produce, control, receive, and effects produced).

acoustic tile A tile manufactured from *acoustic board that is usually used on ceilings to improve *sound absorption or *sound insulation.

acre An area of land equal to 4840 square yards (0.405 hectares).

acrylated rubber paints Special type of paints that are suitable for internal and external applications exposed to chemical attack, or wet and humid conditions. May be applied to metal or masonry by either brushing or spraying.

acrylic A compound derived from acrylic acid.

acrylic paints A type of paint in an acrylic polymer emulsion renowned for its quick drying properties.

acrylic resin *See* POLYMETHYLMETHACRYLATE.

acrylic sealants A type of *sealant derived from acrylic acid.

acrylic sheet A polymer material derived from acrylic acid, usually transparent.

acrylonitrile butadiene styrene *See* ABS.

action 1. The physical operation of something. **2.** A civil legal proceeding brought by one party against another. **3.** An **action (F)** is the term used in *Eurocodes to define a set of forces or loads applied to a structure or the ground. An **indirect action** is a set of deformations or accelerations caused by, for example, temperature and moisture changes. An action can be favourable (stabilizing) or unfavourable (destabilizing) and would have some *partial factor associated with it to make it into a *design action. The **effect of actions (E)** represents the resulting internal forces, moments, stresses and strains in the structure or ground. Actions can also be divided into a number of categories, for example:

- **permanent action (G)** Acts throughout a given period; hence time is irrelevant (e.g. the load from the structure upon the ground).
- **variable action (Q)** Acts for a set period of time (e.g. from wind or snow).
- **accidental action (A)** Acts for a short duration but which could have significant magnitude (e.g. impact of a vehicle on a bridge pier).
- **seismic action (AE)** Arises from motion of the ground caused by an earthquake.

active earth pressure (p$_a$) The horizontal pressure that is exerted on a retaining wall by the mass of soil it retains. Active pressure develops when the wall moves away from the retained mass and the soil expands sufficiently to mobilize its shear strength. The minimum horizontal stress (p$_a$) occurs when the failure strength of the soil is fully mobilized, represented by the minor principal stress, and is calculated by multiplying the *coefficient of active earth pressure (K$_a$) by the vertical stress at the point of consideration.

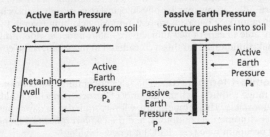

Active earth pressure (p$_a$) and passive earth pressure (Pp)

active fire protection A fire protection system that uses electrical and/or mechanical equipment to detect or suppress fires, such as *fire detectors, *alarms, *sprinkler systems, or *spray systems. They require regular maintenance to ensure that they are fully operational.

active leaf The most frequently used opening leaf on a double door. When closed, it is held in position by latching to the *inactive leaf.

active solar heating A heating system that utilizes electrical and/or mechanical equipment, such as solar collectors, pumps, and fans to collect solar energy from the sun to heat water.

active traffic management (ATM) A system that uses MIDAS (*see* MOTORWAY INCIDENT DETECTION AND AUTOMATIC SIGNALLING) to control the speed limit and hard-shoulder running on *smart motorways.

activity An item of work or task that is needed to complete a project. Used in *Critical Path Method scheduling. An **activity arrow** is used to represent an activity graphically. The time that is expected to be taken to complete an activity is known as the **activity duration**. An **activity schedule** is a list of all of the activities that are required to be carried out to complete a project.

Act of God A legal term meaning an occurrence caused by natural forces where foresight and prudence could not reasonably anticipate it and guard against it. The term is not generally used in construction contracts, *force majeure* is the term usually used to cover such occurrences.

Act of Parliament A statute that sets out the law, the legal will of Parliament in written form. The Acts provide the broad principles, with the detail of the law covered by regulations in the form of a statutory instrument.

actuator A mechanical device that is used to convert electrical, hydraulic, or pneumatic energy into mechanical energy. Used in flow control valves, meters, motors, pumps, switches, and relays.

acute angle An angle that is greater than zero degrees but less than 90°. In an **acute angle triangle** all three angles are acute.

adaptive comfort An approach to thermal comfort that is based upon the premise that when occupants are not thermally comfortable, they will adapt to attain comfort. Adaptations may be physiological (e.g. sweating), psychological (e.g. acclimatization), or behavioural (e.g. removing clothing layers). By adapting, occupants are able to experience subjective comfort under a wider range of environmental conditions than those considered under steady-state comfort metrics.

adaptive comfort temperature The *air temperature that centres a temperature range within which the majority of the occupants in a building will feel thermally comfortable. This temperature will vary from day to day depending upon the external environmental conditions, and the comfort range is dependent on building and occupant type (e.g. ±3°C for a new dwelling).

adaptor A device that enables the characteristics of one component to be matched to another component. Used for joining *pipework, *ductwork, and electrical cables.

addendum bills Bills of quantities produced to alter or modify parts of the original bills. Problems often occur where additional documents are used to modify the original bills as the addendum will refer to the original document, but the original document does not make reference to the addendum. If not cross-referenced properly such documents can be problematic. Addendums may be used to reduce the lowest tender so that it is within the client's budget or to cater for minor alterations.

additional work Work not identified or work over and above that specified in the contract. In standard building contracts additional work is dealt with under the various terms associated with *variations. *See also* EXTRA.

additive A substance that is added to another to alter (normally improve) properties or performance, e.g. an *admixture to concrete.

addressable system (intelligent fire alarm) A fire alarm system incorporating detectors that are capable of sending signals back to the main control unit. Such a system can be used to determine the location of a fire and activate the fire protection system.

adds Quantities of work, of a similar nature, that are added together under the same description. Adds are brought together before deducts are taken off, e.g. quantities of brickwork added together before the wall openings, windows and doors, are taken off (deducts).

adhesion A process whereby the molecules of two separate substances cling together to form a bond.

adhesive (glue) A substance used to bond two objects or materials together.

ad hoc Something that is undertaken only when necessary for the specific purpose in question and is not planned in advance.

ad idem A Latin expression that means at the same point, agreed, or of the same mind. Parties that have agreed to a contract are said to be *ad idem* when agreement has been reached on the contract terms.

adjacent Adjoining, next to, near, or close to.

adjoining property A building that has a common boundary or is attached to another property.

adjudication Method used to resolve disputes. An adjudicator is empowered by contract or statute to settle a dispute. A party to a construction contract has the right to refer a dispute to an adjudicator, where the parties to the contract have failed to resolve their contractual differences. The adjudicator must act independently, hearing the evidence and facts from both sides. Adjudicators are often lawyers or experts within the industry, but can be any person that the parties agree to. Adjudication was introduced as a quick and cost effective method of resolving disputes, when compared to *arbitration or litigation. The Housing Grant Construction Regeneration Act 1996 states that provisions for adjudication must be included within construction contracts. Where contracts do not comply with the Act, the Scheme for Construction Contracts will apply, thus adjudication is forced into contracts even if the construction contract does not have provisions to include it. The Construction Industry Council (CIC) procedures and the Technology and Construction Court (TeCSA) rules are often adopted as part of the contractual machinery associated with adjudication.

adjustable steel prop (ASP) A steel support that is adjustable along its length. Used to provide vertical support to floors and temporary beams.

administrative charges Costs associated with off-site secretarial, coordination, and management provisions. With off-site resources it may be difficult to identify an exact figure, for one contract, so a percentage for the *overheads, which includes administration, insurance, and banking charges are included in the *preliminaries.

administrative receiver A person or group appointed to act on behalf of the debenture holders to manage the company as a going concern, often before the company goes into liquidation.

admissibility of evidence The degree of consideration that can be given to evidence. Information, hearsay, or opinion of an expert that a tribunal finds useful in coming to its decision. Where issues are irrelevant or considered to be biased, overstated, or glorified, the tribunal may be restricted in the use of such evidence.

a

It is more common in criminal cases for evidence to be restricted for fear that the nature of the evidence will result in undue prejudice.

admixture A chemical product which is added to *concrete (<5% by mass) during mixing or by additional mixing, prior to placing, such that the normal concrete properties are changed, e.g. the speed of *strength development, improvement of *permeability, protection from fungicidal attack, improvement in *workability.

((⊕)) SEE WEB LINKS

- Home page of the UK Cement Admixtures Association (CAA), a trade association of UK admixture manufacture, providing information on types, current standards, and publication downloads.

adobe A building material made from water, sand, and soil containing clay that is dried in the sun. Straw or other organic fibrous materials may be added to reduce cracking, aid drying, and to help bind the material together. An **adobe brick**, or mud brick, is made by placing adobe into a mould.

adsorption The accumulation of a substance (gas or liquid) on a surface of another substance (liquid or solid), forming a molecular film.

advance (advance payment) Payment or part payment made before work is undertaken. Some standard forms of contract have provisions for advance payment. To be binding, the amount agreed and the date for payment must be inserted in the *appendix of the contract, together with the times and amounts of repayment. *Bonds are normally required for advanced payments.

advances on account A provision (term) used in construction contracts which refers to periodic payments during the progress of the work, normally at monthly intervals. The housing Grants Construction and Regeneration Act 1996 now requires that all construction contracts make provision for periodic payments. Periodic payments may be covered under the provisions for *interim certificates.

adverse possession Where the title of a person's land transfers to another without compensation, often called squatter's rights. Title of land may be acquired under the Limitations Act 1980. If a person occupies or remains in possession of land without rent or payment for a period of twelve years for private land or 30 years for Crown land, and makes uses of the land which are inconsistent with the owner's enjoyment or intended use, then the land may be acquired.

adverse weather conditions Weather conditions that restrict or preclude work from being undertaken on site, such as high winds and heavy rain.

advertisement for bids *See* SELECTED TENDER.

adze A carpenter's tool used for shaping wood. It has an axe-like head, long handle, and a thin blade perpendicular to the handle.

aerated concrete *See* AIRCRETE.

aeration The introduction of air (or gas) into a substance (e.g. water, concrete, sewage).

aerial Relating to the air, e.g. an **aerial photograph** can be taken from an aeroplane to produce maps and scale drawings of the ground.

aerobic Any treatment, activity, or process that requires the presence of air (oxygen).

aerochlorination Treatment of wastewater (e.g. *sewage) by a mixture of compressed air and chlorine gas.

aerodynamics The study of airflow around objects and the movement of objects through air, with particular reference to the forces experienced by an object as a result of the movement of air.

aerogel A range of synthetic solid translucent *hygroscopic materials that have a very low thermal conductivity and very low density. As their construction minimizes conductive and convective heat losses, they are very good thermal insulators. They are also excellent acoustic insulators. As their name suggests, they are manufactured by supercritically drying a gel to remove all of the liquid and replacing the liquid with a gas. Alumina, carbon, chromia, silica, vanadium, and zinc ferrite have all been used to produce aerogels.

aerotriangulation The use of angles and distances to orient and align overlapping *aerial photographs.

affidavit A factual or written statement or evidence made under oath. Documents sworn in under an affidavit are called exhibits.

affirmation of contract Where there is a breach of contract, the innocent party may choose to continue with the contract and affirm the contract, treating it as still being in force.

affirmative action (positive action) Steps taken by an employer to help or encourage groups of people who are under-represented in an activity or type or work, have different needs, or are at a disadvantage in some way to access work or training opportunities. 'Positive action' is the term commonly used within European countries. 'Affirmative action' tends to be used in other countries.

AFNOR (*Association Française de Normalisation*) The French national standards organisation.

a fortiori argument Latin term used in law to show that the case being presented falls well within the scope of something that has gone before, thus there is even greater reason that the situation under examination should be treated with at least the same consideration. The situation can be treated as *a maiore ad minus* e.g. if the situation applies to a large group then it should also hold true for a smaller group or if 8 m³ of concrete is contained in a dumper then 6 m³ would also fit into the same dumper. Alternatively, but less common, *a minore ad maius* argument can be used, which means from smaller to bigger, e.g. if two packs of bricks overloaded the scaffolding, then six packs of the same type of brick would also overload it. All of these arguments rely on the first assumption holding true, thus if two packs of bricks would not overload the scaffolding then there would be no evidence to check the second part of the statement. Where illogical arguments are used to justify a claim, it falls under the term *petitio principii.*

A-frame building A triangular-shaped building with a steeply sloping roof that extends to ground level.

African mahogany Common name for genus *Khaya* of the *meliaceae* family and genus *Afzelia* of the *fabaceae* (legumes) family trees.

after-flush The water remaining in a cistern after the toilet has been flushed. This water provides the water seal between flushing.

aftershock A small earthquake, normally one of many, that occurs after a large earthquake.

aftertack (after-tack) A defect in a paint film where the film remains tacky and sticky over an extended period of time.

ageing *See* PRECIPITATION HARDENING.

agency An organization that acts for and on the behalf of others. For instance, a construction recruitment agency would recruit staff for various construction companies.

agenda-driven research Focuses directly on addressing a particular industrial or social need. Sometimes this work is undertaken at national research laboratories (*see* TRANSPORT RESEARCH LABORATORY), rather than at universities.

agent 1. A person who has the authority to act on behalf of others. **2.** *See* SITE AGENT. **3.** A substance that changes the characteristics of another substance when added to it.

agglomerate To collect together in a round (ball) mass. **Agglomerates** are formed from volcanic fragments, consisting of various rock types, sizes, and shapes, fused together.

aggregate Granular material used in concrete, road construction, mortars, and plaster. Aggregates can originate from natural rocks, gravel and sands (e.g. limestone and sandstone), formed from artificial sources (via a thermal process), or be recycled inorganic construction waste. From June 2004 aggregate size is detailed by a lower (d) and upper (D) sieve sizes (expressed as d/D) in accordance with the appropriate European Standards: e.g. BS EN 12620 'aggregate for concrete'; BS EN 13043 'aggregates for bituminous mixtures and surface treatments for roads, airfields, and other trafficked areas'; BS EN 13285 'unbound mixtures'; and BS EN 13139 'aggregates for mortar'.

(((⊕))) SEE WEB LINKS
• Official website where the appropriate BS EN standards can be purchased/downloaded.

aggregate cement ratio Weight of aggregate divided by the weight of cement.

aggregation Bringing components, elements, data, etc. together.

aggregation wastewater The process of coming together/clustering of disinfection products and by-products. It facilitates in settling and subsequent removal of products from water in order to treat/purify it.

aggressive Exhibiting attacking or hostile behaviour, which can lead to damage.

aggressive environments Locations where acid rain, corrosive gas/liquids, chemical/industrial pollutants, etc. occur or are present. These elements can cause damage to materials such as concrete, steel, and timber.

agreement 1. The sense of a meeting of minds between two people. **2.** An essential part of a valid contract.

Agreement for Minor Building Works Standard form of contract, appropriate for use on projects up to a value of £90,000.

agreement to negotiate Agreement to make a contract. Although parties can agree to negotiate rates and make a *contract, due to the uncertainty that such practices bring, English law does not recognize a 'contract to negotiate' as a contract.

Agrément Certificate A certificate issued by the *British Board of Agrément that contains information on a product's durability, installation, and compliance with Building Regulations. Only issued once the product has successfully been assessed.

ahead of schedule (advance) An activity that is ahead of schedule is one that has been completed prior to the date recorded in the *programme or *activity schedule.

air-admittance valve A device that allows air to be admitted into sanitary pipework to equalize the pressure and maintain the water seal in the trap of the appliance.

air balancing The process of adjusting the air flows through the supply and extract grilles in ventilation or air-conditioning systems to provide the required air flow rates.

air barrier The part of the building envelope that has been designed and constructed to resist *air leakage from a conditioned to an unconditioned space. Air barriers can be constructed from a wide range of materials that are impervious to air, such as wet plaster and polythene sheeting. Materials that have been connected together to form an air barrier are termed an **air barrier system**.

air-blast cleaning High-pressure cleaning operation using air alone. Pneumatic hoses are often used to remove rubbish, dust, and debris that has become trapped between *steel reinforcement and *formwork. Formwork should be cleaned before any concrete is poured.

airborne laser scanning a measurement system where pulses of light are emitted from an instrument mounted on an aircraft and directed to the ground in a scanning pattern. The airborne laser scanner measures only the line-of-site vector from the laser scanner aperture to a point on the Earth's surface.

airborne noise Noise or sound that is admitted into the atmosphere. Typical sources of airborne sound include human voices, radios, musical instruments, and audio-visual equipment. *See also* AIRBORNE SOUND, IMPACT SOUND.

airborne sound Noise that propagates through air. *See also* AIRBORNE NOISE

air brick (ventilation brick) A brick containing a series of holes that run through the brick to allow ventilation. Usually built into a wall to ventilate rooms or a space under a ground floor level. Available in a range of different sizes.

air brush A small, air-operated spray gun that is used to produce a fine spray of paint.

air change The replacement of air within a space with new air, usually from outside.

air change rate The rate at which replacement air enters a space divided by the volume of that space. Expressed in air changes per hour (ac/h). Used to measure ventilation.

air compressor A machine that compresses (pressurizes) air, typically to power pneumatic tools such as drills, jackhammers, pumps, etc.

air conditioner (air conditioning) A mechanical device used to control temperature, relative humidity, and air quality within a space. Usually comprises an evaporator, a cooling coil, a compressor, and a condenser. Also commonly used to describe any system where refrigeration is included to provide cooling. A wide range of **air-conditioning systems** are available and they can be categorized according to their function:

- **Full air-conditioning systems**—provide all of the ventilation and fresh air requirements of the building. Temperature and humidity can be controlled within predetermined limits. The air supply is filtered, heated, or cooled to the required temperature, and is humidified or dehumidified to maintain acceptable levels of relative humidity.
- **Close control air-conditioning systems**—similar to full systems, but are capable of maintaining temperature and relative humidity within close control limits, typically within $\pm$ 0.5 °C and $\pm$ 5% RH.
- **Comfort cooling systems**—use refrigeration to reduce internal temperatures when required to produce more acceptable conditions for occupants. Ventilation and fresh air are not always supplied from a comfort cooling system, and may need to be provided separately.

aircrete A low-density porous material extensively utilized in the construction industry, usually in block form. The air content (*porosity) is typically between 60% and 85%. It is produced by mixing cement and/or pulverized fuel ash (PFA), lime, sand, water, and aluminium powder. The final process involves autoclaving for approximately 10 hours at high temperature and pressure. Aircrete (Ahmed and Kamau, 2017) accounted for a third of all concrete blocks in the UK. Such blocks are suitable as vertical load-bearing elements and provide the thermal insulation expected from typical UK wall construction. They may also be used as non-load-bearing outer leaves of masonry walls, external walls, and walls below ground level, where adequate care is essential to ensure their durability and protection from effects of the environment. The lightweight porous structure creates an effective moisture barrier with desirable thermal insulation properties. Their porous cellular structure and durability make them a recognized alternative in most below-ground situations. Aircrete originated from Scandinavia in the 1950s as an alternative to building with timber. Aircrete is also known commercially as **AAC (Autoclaved Aerated Concrete)**, **Celcon**, **Durox**, **Thermalite**, and **Topblock**. Aircrete is normally categorized as low, medium, and high density as depicted in the table below.

Classification and Physical Properties of Aircrete Blocks

Aircrete density	Compressive strength (N/mm²)	Density (kg/m³)	Thermal conductivity (W/mK)
Low	2.0 – 3.5	450	0.09 – 0.11
Medium	4.0 – 4.5	620	0.15 – 0.17
High	7.0 – 8.5	750	0.19 – 0.20

( ) SEE WEB LINKS

• Home page of the Aircrete Products Association, providing information regarding aircrete technology, its applications, and advantages.

air curtain A mechanical device that produces a narrow jet of high-velocity air that is directed down and across an opening. Commonly used on exterior doors, loading platforms, and refrigeration display cases to prevent air from one side of the opening mixing with the air from the other.

air diffuser An outlet or *grille designed to distribute air within a space. Commonly located in a *suspended ceiling.

air distribution The movement of air from one space to another.

air drain An empty space left around the external *perimeter of a *foundation to prevent moisture from the surrounding ground causing *damp.

air-dried timber *See* AIR-SEASONING.

air eliminator An automatic device used in plumbing to release air from the pipework.

air-entrained concrete A type of *concrete with purposely incorporated minute air bubbles to improve its freeze–thaw resistance.

air-entraining agent An *admixture which causes minute air bubbles to be incorporated into concrete, used to improve workability and frost resistance.

air exfiltration The uncontrolled exchange of air out of a building through a wide range of *air leakage paths in the building *envelope.

airfield soil classification A soil classification system for engineering purposes developed by A. Casagrande in the early 1940s, which became known as the **Unified Soil Classification System** in 1952. It described a standard system for classifying mineral and organo-mineral soils for engineering use based on the particle size, liquid limit, and plasticity values.

air flow exponent One of the output metrics obtained from undertaking a *fan pressurization test that is used to characterize how the air flows through the various adventitious cracks and gaps that exist in the building fabric. It should range from 0.5 to 1.0, with figures approaching 0.5 representing fully developed turbulent flow and figures approaching 1.0 representing more laminar flow. Turbulent flow is associated with air flow through a series of large apertures, whilst laminar flow is associated with air flow through a multitude of tiny gaps and cracks in the fabric.

air gap A gap between two components in a construction that is filled with air.

air grille (air grating) A framework of bars used to cover an air outlet that enables air to pass through.

air-handling equipment The equipment used to move air from one conditioned space to another in an air-conditioning system. Usually comprises a series of supply and extract fans and terminal units.

air-handling luminaire A *luminaire that is designed to allow air to be extracted through it. Used to reduce the heat gains from lighting and reduce the cooling load on the air-conditioning system.

air-handling unit (AHU, air handler) A packaged item of air-conditioning equipment that treats the incoming air prior to it being distributed around the building. Comprises a fan, combinations of heating and cooling coils, filters, humidifiers, and control dampers. Usually located on the roof with *refrigeration units and *boilers located nearby.

air heater *See* UNIT HEATER.

air house *See* AIR-INFLATED STRUCTURE.

air infiltration The uncontrolled exchange of air into a building through a wide range of air leakage paths in the building *envelope.

air-inflated structure A structure that has a double membrane supported by a series of tubes that are inflated using high-pressure air. **Air-supported structures** have a single membrane that is supported by low-pressure air. Both types of structures can be rapidly erected or dismantled and are mainly used as temporary structures.

air leakage The uncontrolled fortuitous exchange of air both into (*infiltration) and out of (*exfiltration) a building *envelope, space, or component through cracks, discontinuities, and other unintentional openings. It is driven by the wind and the stack effect. Typical air leakage points occur at: cracks, gaps, and joints in the structure; plasterboard dry-lining; windows, doors, and their surrounds; joist penetrations of external walls; boundary/wall junction; internal partition walls; loft service entries, ducts, and electrical components; permanent ventilators; loft hatches and skirting boards.

air leakage audit An inspection of a building that is carried out to identify the main *air leakage paths and any areas where there may be discontinuities in the *air barrier. Can be undertaken during construction or after the building is completed.

air leakage index The volume flow of air either into or out of a space, per square metre of building *envelope area, at a given pressure difference between the inside and the outside of the building (in the UK 50 Pa is used). For air leakage index, the area of the lowest floor is included in the building envelope area if it is not ground supported. Expressed in $m^3/(h.m^2)$. *See also* AIR LEAKAGE RATE.

air leakage path The path that air takes when leaking into or out of a building envelope or component.

air leakage point The point at which air leaks into or out of a building envelope or component.

air leakage rate The rate at which air leaks into or out of a building *envelope, space, or component, per unit volume, at a given pressure difference across the building envelope, space, or component (in the UK 50 Pa is used). Traditionally, expressed in ac/h, however, nowadays $m^3/(h.m^2)$ is more commonly used as it takes into consideration the effects of shape and size. *See also* AIR LEAKAGE INDEX.

airless spraying A method of painting that involves forcing paint through a nozzle at high pressure. Results in less overspray than *compressed air spraying.

air lift pump A deep (up to 200 m) *well pump, which is located below the *groundwater table. *Compressed air, from the surface, is pumped down the well where it bubbles into a large diameter pipe, lifting a *slurry mix to the surface. A key element to this pump is that there are no mechanical moving parts in the well, which would otherwise be abraded by the slurry causing maintenance issues.

air monitoring Sampling and measuring the quality (normally the level of pollutants) of the atmosphere/ambient air at regular intervals.

air permeability The volume flow of air either into or out of a space, per square metre of building *envelope area, at a given pressure difference between the inside and the outside of the building (in the UK 50 Pa is used). For air permeability, the area of the ground floor is included in the building envelope area. Expressed in $m^3/(h.m^2)$.

air permeability test A test that is undertaken to determine the air permeability of a building envelope, space, or component.

airport A location where non-military aircraft (i.e. only civil aeroplanes, helicopters, and airships) take off and land, equipped with a runway and facilities for handling passengers and cargo.

air pulse test (pulse technique) A test that is used to measure the *air leakage and *airtightness of a building. It is an alternative test method to the widely used *fan pressurization test. The technique involves subjecting the test building to a known volume change. This known volume change is achieved by releasing a very short (~1.5 seconds-long) pulse of compressed air into the test building, resulting in a pulse in the internal pressure. This pulse is then used to quantify the airtightness of the building. It operates at much lower pressures (3 to 10 Pa) than the fan pressurization technique.

air pulse test unit Equipment used to undertake an *air pulse test. It comprises the main air tank, a smaller reference pressure tank, a compressor which is used to pressurize the main air tank, a solenoid valve attached to each tank to control the opening and closing of the tanks, a nozzle, two pressure transducers, and a control box.

air quality The composition of *ambient air with respect to the *pollution levels.

air receiver (air vessel) A pressure vessel used in compressed air applications to store the compressed air and to equalize the pressure in the system.

air-release valve (pet cock) A device that is used to release trapped air from pipework or a fitting.

air-seasoned timber *See* AIR-SEASONING.

air-seasoning (air-dried timber, air-drying) The process of removing moisture from timber before it is used and put into service by drying in air. Two main methods of seasoning are air-seasoning (natural seasoning) and kiln-seasoning (artificial seasoning). In a living tree, the weight of water in the tree's vessel cells will frequently be greater than the dry weight of the tree itself. Seasoning is the removal of most of this moisture, and the stabilization of the moisture content before putting the timber into service. The primary aim of seasoning is to render the timber as dimensionally stable as possible. This ensures that once put into service as flooring, furniture, doors, etc., movement will be negligible. Seasoning involves removing most but not all of the moisture from the timber, and when this is undertaken other advantages accrue. Most wood-rotting fungi can grow in timber only if the moisture content is above 20–22%. Drying of timber occurs because of differences in vapour pressure from the centre of a piece of wood outwards. As the surface layers dry, the vapour pressure in these layers falls below the vapour pressure in the wetter wood further in, and a vapour pressure gradient is built up which results in the movement of moisture from centre to surface; further drying is dependent on maintaining this vapour pressure gradient. The steeper the gradient, the more rapidly the drying (seasoning) progresses, but in practice, too steep a gradient can cause the wood to split. For large-sized timbers, a combination of the two methods is often used. Air-seasoning is still practised in some countries where the cost of kiln-drying is high (particularly in the developing world), and it is still used to at least partially dry timbers of large cross-sectional area. Where such timber would take more than 4–5 weeks to kiln-dry, it is often air-dried to a moisture content of 25–30% first. In the UK, hardwood planks are usually air-dried for 18–24 months before being kiln-dried.

air shaft A vertical shaft within a building or tunnel that is used for ventilation.

air source heat pump (ASHP) A *heat pump that extracts heat from the outside ambient air and transfers it to the inside of a building. In summer, it can be reversed so excess heat from inside the building can be dumped to the outside air.

air space (air cavity) A space between two elements that is filled with air. Commonly known as a *cavity in masonry cavity construction.

air temperature (dry bulb temperature) The temperature of the air measured by a *thermometer that is shielded from moisture and radiation.

air terminal unit A mechanical device used in air-conditioning and *ventilation systems to supply and extract air from a space. *Fan coil units, *induction units, and *variable air volume units are all examples of air terminal units.

air termination network (air termination system) The part of a lightning protection system that is designed to intercept a lightning strike. Comprises a series of vertical rods that are placed on a roof or equipment, known as **air terminals**.

air test *See* FAN PRESSURIZATION TEST.

airtight inspection cover An *inspection cover that incorporates an airtight seal.

airtightness Used to describe the air leakage characteristics of a building. Frequently expressed in terms of a whole building leakage rate at an artificially induced pressure (in the UK 50 Pa is used). The smaller the air leakage rate at a given pressure difference across the building envelope, the greater the airtightness. The airtightness of a building determines the uncontrolled background ventilation or *leakage rate of a building which, together with ventilation caused by occupancy (opening windows, etc.), makes up the total ventilation rate for the building.

airtightness layer A layer within the building envelope, space, or component that resists air leakage. This may or may not be the same layer as the *air barrier.

airtightness test *See* FAN PRESSURIZATION TEST.

air-to-air heat-transmission coefficient *See* THERMAL TRANSMITTANCE.

air treatment The process of treating the air within an air-conditioning system. This may involve filtering the air to remove contaminants, and/or heating, cooling, humidifying, or dehumidifying the air. *See also* AIR CONDITIONER.

air valve A *valve that controls the flow of air into or out of a system.

air vent An opening that allows air to flow from one space to another.

air void An enclosed space filled with air.

air washer A mechanical device used to cool, clean, and dehumidify air. No longer used in air-conditioning systems due to health concerns.

airy stress function A function used to solve elastic analysis of plane stress distribution.

alarm Noise, created by a bell, gong, buzzer, horn, electric signal, or speaker to indicate a problem or danger where immediate action is required. A **fire alarm** is used to alert occupants of a building or a building site to the risk of a fire and requires the action of all persons evacuating the building (or building site) to go to an area that has been designated safe.

albic A *soil horizon from which clay and free iron oxides have been removed or iron oxides have been separated.

alcove A recessed or partially enclosed portion of a room.

aldrin A chlorinated pesticide that was used against soil-dwelling pests and for wood protection. It is a derivative of naphthalene ($C_{12}H_8C_{l6}$), which is closely related to dieldrin. It is no longer used because it is *toxic to animals and humans. It is a compound monitored in drinking water.

algae Mainly aquatic, *eukaryotic, photosynthetic organisms (e.g. seaweed). Once considered as a plant but now classified independently because they do not have true leaves, roots, or stems.

algorithm 20

algorithm A step-by-step problem-solving procedure, used particularly for solving mathematical problems in a finite number of steps.

aliasing An effect caused when an image or signal is sampled at a too-low frequency, causing jagged distortions in curves, diagonal lines in computer graphics, and static distortion in digital sound.

alidade A sighting device (telescope and attached parts) used on a *plane table for angular measurements.

alien enemy A person whose state (nationality) is at war with the land in which they reside. Such people may be allowed to leave the country or may be interned within the country. The term is occasionally used in construction contracts.

alignment The correct positioning of different components relative to something else, e.g. *setting out a road or railway on the ground.

aliquot To divide something equally without leaving a remainder.

alkali A substance (e.g. soluble salts) capable of neutralising an *acid. Common examples of alkali salts are sodium hydroxide, potassium hydroxide, calcium carbonate, and magnesium hydroxide. **Alkaline** conditions occur in soils or water when the pH is greater than 7.0.

alkali-aggregate reaction A chemical reaction between the aggregates in concrete and cement resulting in damage to the material.

alkali-resistant glass fibre A type of fibre utilized in concrete for protection against alkali reaction.

alkali-resistant paint Fast-drying paint for metals, concrete, and masonry which provides a tough, flexible film with excellent resistance to moisture, salt water, and alkali.

alkali-resistant primer A protective sealer used in masonry against stains and blistering caused by salts in the plaster.

alkyd paint A type of paint based on an oil-modified polyester resin (alkyd) which provides a surface that is highly resistant to wear.

alkyd plastics A type of polymer based on esters.

alkyd resin A synthetic resin used in paints.

all-in aggregate A mixture of coarse and fine aggregates.

all-in contract A *design and build or turnkey contract (*see* TURNKEY PROJECT), where the design and construction of the building is included in one contract.

allotropy A chemical element that can occur in more than one form **(allotrope)**. Allotropes have identical chemical composition but the *atoms are arranged differently by chemical bonds. *Diamond and graphite are two allotropes of *carbon. The carbon atoms in diamonds are arranged in a tetrahedral lattice, whereas in graphite they are arranged in a *hexagonal lattice.

allowable To permit or be acceptable. Allowable *bearing capacity is an acceptable (factored) pressure the ground can withstand from a foundation.

alloy A metallic material comprising two or more elements that can be either a metal or non-metal. Any metal that is not 100% pure is classified as an alloy. Examples of alloys are *steel, *stainless steel, galvanized steel, *cast iron (all based on iron), *bronze, *brass (both based on copper), and *solder (a mixture of lead and tin). The primary objective of alloys is to obtain a metal which comprises a mixture of the main attributes of each constituent.

alloy steel A *ferrous metal containing carbon and several other alloying elements.

all risks The list of risks that the *JCT 98 (Joint Contracts Tribunal) form of contract identifies as needing works insurance.

alluvium (alluvial, alluvial soil) Sediments (earth, silt, sand, gravel, etc.) deposited by flowing water (rivers and streams). Materials, sediment remaining on land that has been exposed to flooding or where a river previously flowed.

alms house A sheltered house or building for people without satisfactory income or support.

alpine Relating to high mountains (the Alps).

alteration or amendment of contract Alterations made to the contract that are written into the contract, or otherwise agreed, before the contract is executed (signed by the parties). Standard forms of contract are regularly updated to keep in line with court decisions and new statutes. The standard forms can also be amended to include specific requirements to suit either of the parties, so long as they are included prior to the contract being agreed and signed.

alternate bay construction A traditional method of concreting that uses a 'two-stage' process. In the first stage, the area to be concreted is divided into bays using formwork. Concrete is then poured into each alternative bay. As soon as is possible, the formwork between the bays is then removed. In the second stage, concrete is poured up against the existing concrete to fill the remaining bays. It is, however, becoming more common for concrete to be laid in a continuous slab.

alternate lengths work The underpinning of foundations that is carried out in small sections to ensure that support for the building remains in place at all time. To avoid settlement, one section of foundation should be underpinned, with the adjacent section of foundation remaining in place providing support until the new concrete foundation has gained sufficient strength. Thus, new underpinning is carried out in alternate sections with adjacent existing sections of foundation providing temporary support.

alternate proposal (US) An alteration or variation to the contract different from that described in the tender or bid documents.

alternating current (AC) An *electric current that reverses direction in a circuit at a constant *frequency. If the frequency is 60 Hertz, the direction in the flow of the current changes 60 times per second.

altitude The height of an object above a known *datum, normally sea level.

alum ($Al_2(SO_4)_3$, aluminium sulphate) A metal salt coagulant used in water purification treatment plants, which reacts with the alkalinity of the water to produce an insoluble metal hydroxide *floc.

aluminium A non-ferrous metal/alloy, chemical formula Al. Aluminium has the merit of being light (density 2,700 kg/m^3 as compared to 7,900 kg/m^3 for steel), ductile, and easily rolled into sheets and thin strips, and extruded into complex sections. It resists corrosion, especially if it is anodized. However, it has only one-third of the stiffness of steel, and it melts at the relatively low temperature of 550–600°C (in comparison to approximately 1500°C for steels). Its performance in fire is therefore not good, and so it cannot be used structurally in building. **Aluminium alloys** are easily formed due to its high ductility; this is evidenced by the thin aluminium foil sheet into which the relatively pure material may be rolled. In construction aluminium is used externally for window frames and structural glazing systems, roofing, and claddings, flashings, rainwater goods, etc. It is used internally for ceilings, panelling, ducting, light fittings, vapour barriers (as aluminium foil), architectural hardware, walkways, handrails, etc. Aluminium can be alloyed with lithium to further enhance its properties. These materials have lower densities (2500–2600 kg/m^3) in comparison with other metals, especially ferrous alloys, high specific moduli (elastic modulus-specific gravity ratios), and excellent fatigue and low temperature toughness properties. Aluminium–lithium alloys are mainly utilized in the aerospace industry.

aluminium cable An aluminium/aluminium alloy conductor material utilized for the transmission and distribution of electrical power. Since the late 1990s aluminium has replaced copper conductors for these applications, and is the standard material for electrical conductors. These conductor designs have consistently provided a superior combination of strength and conductivity for distribution and transmission applications. Aluminium and aluminium alloy conductor materials are excellent choices for wire and cable products in various electrical applications because of aluminium's excellent physical and electrical properties. Aluminium is light-weight—about a third as heavy as copper; it is an excellent conductor of heat and electricity, an excellent reflector of heat and light, it is highly resistant to corrosion, strong, and flexible, and can be made stronger or more flexible by alloying and/or heat treatments.

aluminium foil Aluminium sheets with a typical thickness of less than 0.2 mm. Typically laminated to other materials, e.g. plaster ceiling boards.

aluminium paint A special paint for various applications particularly where a bright reflective finish is required; especially for hot pipes.

aluminium primer An aluminium-based *pigment used particularly on timber surfaces.

aluminium roofing Sheet roofing material used to provide the protective finish or cladding to an aluminium roof.

aluminium sulphate *See* ALUM.

aluminium windows External windows manufactured from extruded sections of *aluminium that are either screwed or mechanically cleated together. They are usually anodized, powder-coated, or organically coated. Advantages are: durability, narrow frame sections, require little maintenance, and have good resistance to corrosion. Disadvantages are: aluminium is a good conductor of heat so frames need to incorporate a *thermal break, more expensive than steel, and they have a high embodied energy content.

aluminium-zinc coating Anti-corrosion treatment of metal-coated steel. An aluminium-zinc alloy coating applied to steel provides enhanced corrosion protection and can improve the lifetime of steel by up to four times that of galvanized steel under the same conditions.

ambient The immediate surrounding area. The **ambient conditions**, i.e. air temperature, wind speed, and humidity, can be defined for a given location.

ambiguity Something which is unclear, or has a degree of uncertainty or has more than one meaning. The standard forms of contract *ACA, *JCT 98, and IFC 98 have their own provisions to deal with discrepancies and ambiguities. Each contract has different interpretations.

amendment Change or correction to something, usually to documentation or a contract. Updates and revisions to standard forms of contract are often published and attached to the contract so that the contract encompasses changes in statutes and case law.

American bond *See* ENGLISH GARDEN WALL BOND.

ammeter An instrument used to measure electric current (amps) in a circuit.

ammonia A hazardous gas at ambient temperature with a very strong odour. Chemical formula NH_3.

amorphous Without shape, form, or structure. *Polymer chains can be packed in two distinct formations within the solidified *thermoplastic. Amorphous packing is where the chains are arranged with no order and *crystalline packing occurs when the chains 'line up' to form crystallites. Crystalline regions are denser than the amorphous regions and the amount of each region, in the solidified polymer, is reflected by its properties e.g. producing high-, medium-, and low-density grades of polyethylene.

ampere (amp A) The unit of current. One ampere is the amount of electric current for one *coulomb of charge per second.

amplifier A device that changes (normally increases) the *amplitude of a *signal. A **sound amplifier** is a device that makes *sound louder.

amplitude The distance from the mean (mid-point) value to the *crest or *trough of a wave or *oscillating signal.

anaerobic Without needing oxygen. An **anaerobic organism** does not need oxygen for *metabolism.

analogue (analog) A system that is based on a continuous varying physical quantity, e.g. when the volume on a stereo system is turned up, the sound through the loud-speakers increases at a constant rate. Since the 1990s, the analogue system has been superseded by digital technology.

analogy A comparison made between two things, used for the purpose of explanation. One thing, which is less understood, is compared with something that is well understood.

analysis A thorough investigation of something in order to understand.

anchor 1. To fix firmly and securely in position. **2.** A mechanical device that is built into a structure to prevent a component from moving.

anchorage 1. The process of anchoring. **2.** In a suspension bridge, the part of the bridge where the suspension cables are connected.

anchor block A block of wood set within a masonry wall, in place of a brick, that provides a surface for connecting other wooden items.

anchor plate (floor plate) A plate attached to a component that enables other components to be connected to it.

ancient lights Windows or openings in walls that have access to natural light for a period of twenty years or more have a right of access to unobstructed light. Under the Prescription Act 1932 the light cannot be obstructed without the owner's permission.

ancient monument A site, feature, or building of historical or archaeological importance listed under the Ancient Monuments and Archaeological Areas Act 1979. It is an offence to undertake any work on listed monuments without consent.

anemometer An instrument that is used to measure wind speed and direction.

aneroid barometer An instrument for measuring atmospheric pressure.

angiosperms Flowering plants that produce seeds enclosed in fruit. Over 250,000 species of angiosperms exist, making them the most dominant type of plant.

angle bead (plaster bead) A perforated metal strip in the shape of an angle that is used to protect and reinforce the corner of a plaster or plasterboard wall. The perforations are used to nail the strip to the surrounding plaster or plasterboard and act as a *key when the bead is plastered over.

angle block (glue block) A small wooden block, usually triangular in shape, that is fastened into the interior angle between two surfaces to strengthen and stiffen the joint. The interior angle is usually 90°.

angle board A board cut at a predetermined angle. Used as a guide for cutting other boards at the same angle.

angle brace (angle tie) A *brace fixed across an interior angle in a frame to improve the frame's rigidity. The brace can be temporary or permanent.

angle cleat (clip anchor) A short L-shaped angle used to connect components to structural members, e.g. attaching *precast concrete cladding panels to the main structural frame, or *purlins to *roof trusses.

angle closer *See* CLOSER.

angled (angling) Set at an angle.

angled tee A T-shaped pipe fitting where the main inlet of the *tee is at a shallow angle.

angle fillet A thin strip inserted into the internal angle between two surfaces to cover the joint. Triangular in cross-section.

angle float A plasterer's smoothing tool that is shaped to enable the plaster to shape, smooth, and form internal corners.

angle gauge A template that has been cut or made to set out or check corners.

angle grinder (disc grinder) A hand-held power tool with a small *abrasive rotating disc used to cut or grind masonry, concrete, or steel.

angle joint A joint between two surfaces that results in a change of direction.

angle of dip *See* DIP.

angle of friction (ϕ, angle of internal friction, angle of shearing resistance) Ratio between the shear stress and normal stress at failure, originally proposed by Coulomb in 1773:

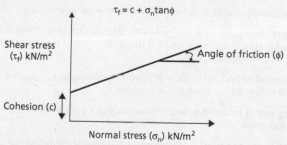

$$\tau_f = c + \sigma_n \tan\phi$$

Shear stress (τ_f) kN/m²

Angle of friction (ϕ)

Cohesion (c)

Normal stress (σ_n) kN/m²

Angle of friction: typical failure envelope of a sandy clay soil

angle of repose The steepest angle a granular heap of material would make to the horizontal when poured.

angle rafter *See* HIP RAFTER.

angle section (angle bar, angle iron) A steel L-shaped angle.

angle staff *See* ANGLE BEAD.

angle strut A structural member in the shape of an angle that acts in *compression.

angle tie *See* ANGLE BRACE.

angle tile (angular tile, arris tile) A tile in the shape of an angle that is used for tiling the corners of a wall, a *hip, or a *ridge.

angle trowel (twitcher) A plasterer's trowel that has been shaped to work with awkward edges and angles. Some trowels may have a 'V' cut out of them for external angles; others may have edges that are upturned to create internal angles and rounds.

angle valve A valve whose inlet is at 90° to the outlet. Used to control flow.

angstrom (A) A unit of length equal to 10^{-10} metres, used to measure wavelengths of electrometric radiation, i.e. very short wavelengths.

angular Relating to angles. **Angular distortion** is the relative rotation of a point, within a structure, about another point.

anhydrite A mineral also known as calcium sulphate commonly used as floor *screeding. It is anhydrous sulphate of lime, and differs from gypsum in not containing water.

anhydrous Containing no water.

anicut A *dam built in a river or stream for *irrigation purposes.

animal black A paint derived from animal products such as bones.

anion A negatively charge *ion, which is attracted to an *anode during *electrolysis.

anisotropic Having different properties in different directions. Timber is anisotropic because it is stronger loaded in the direction of the *grain rather than *perpendicular to the grain.

anneal (annealing) A heat treatment applied to materials, especially metals, to alter the chemistry, hence mechanical properties.

annealing point (glass) A specific temperature for a glass which corresponds to *zero residual stress.

annual ring (growth ring) A layer of wood that represents the annual growth in a tree's diameter.

annular In the shape of a ring.

annular nail A nail that contains a series of circular rings on the shaft. The rings improve grip, holding the nail firmly in position. Used for fixing plywood and other materials.

annunciator A device that can be used to visually or audibly indicate whether an event has taken place. For instance, these are used in *burglar alarm systems to indicate when an intrusion has taken place.

anode 1. The electrode in an electrochemical cell through which electric current flows. **2.** The positive terminal in an electrolytic cell. **3.** The negative terminal of a galvanic cell (battery). The electrode in a galvanic couple that experiences oxidation, or gives up electrons. *See also* SACRIFICIAL ANODE and ANODIC PROTECTION.

anodic protection (aP) A process utilizing electrolysis to improve the corrosion resistance of metals.

anodize (anodizing) A process using *electrolysis to coat a metal with a protective film.

anodized A metal which is corrosion-protected by *electrolysis.

anoxic A lack of oxygen.

ANSI American National Standards Institute.

antecedent drainage When a river has maintained its original direction of flow after *tectonic movement.

anthropogenic Activity undertaken by human beings that either has an influence on a phenomenon or results in a change. Of human origin or influence, often used alongside climate change* to indicate the human impact on the ecosystem and sustainability* of animals, plants, organisms, and other living matter.

anti-bandit glass A type of security glazing that is designed to be resistant to manual attack using a blunt instrument, such as a hammer. A wide range of glazing materials can be used, the most common being 11.3 mm 5-ply laminate and 11.5 mm 3-ply laminate. The specification for anti-bandit glazing is covered in BS5544: 1978 (1994).

anticipatory breach of contract Where it is known that a party will not perform its obligations to a contract. An example of such breaches would include a party stating that it will not carry out part or the whole of its work. The other party may act on the statement and sue for damages or repudiate the contract.

anti-climb paint A type of paint applied to pipes, fences, etc. to make it virtually impossible to climb (for intruders).

anticline An upward fold in *sedimentary rock, caused by *tectonic movement.

anti-condensation paint A type of paint used to reduce condensation.

anti-corrosive paints A paint utilized to protect iron and steel surfaces that contains a corrosive-resistant pigment (commonly *zinc-based) and a chemical- and moisture-resistant binder.

antidune A small ripple (sand hill), formed on a riverbed of a high flowing stream, which migrates upstream over a period of time.

antifreeze A liquid used to reduce the freezing point of liquids and as a heat-transfer medium.

anti-frost agent An additive used as frost protection.

anti-graffiti coating A coating that can be applied to a surface that either resists the application of graffiti or enables the graffiti to be removed easily. A wide variety of coatings are available, such as paints and resins.

anti-intruder chain link fencing A security fence constructed from galvanized or coated steel wire mesh, usually 1.8 m to 3 m in height.

anti-siphon pipe A pipe in a drainage system that emits air, preventing *induced siphonage.

anti-siphon trap A trap that maintains a water seal by preventing *siphonage.

anti-slip paint A tough, durable, high opacity floor coating providing anti-slip properties. Applied to concrete, masonry, tiles, wood, metal, and many other surfaces.

anti-static flooring (electrostatic discharge (ESD) flooring, static control flooring) A floor covering that is designed to reduce *static electricity, such as carpet, epoxy coatings, rubber tiles, or vinyl tiles.

anti-sun glass *See* SOLAR CONTROL GLAZING.

anti-vibration mounting (flexible mounting, resilient mounting) A mounting, usually in the form of a pad or mat, that is designed to prevent the noise from vibrating machinery being transmitted to the structure. Materials used include cork, felt, plastic, and rubber.

apartment A small dwelling housed in a *multistorey building. May also be called a flat or maisonette.

apex The tip or highest point of a building or structure.

aphotic zone The part of an ocean or lake that sunlight fails to reach.

apparent cohesion Fine-grained soil particles (e.g. silt and clay) being held together by capillary/surface water tension.

appeal A person or body has a right to make an application to a court or tribunal higher than that where the case was first considered for reconsideration of a decision made, where Acts of Parliament allow.

appendix An additional section of a document or book that normally contains additional or supporting information and is placed at the back of the document. In some standard contracts the appendix is an integral part of the contract and cannot operate unless the sections within the appendix are completed.

appliance ventilation duct *See* DUCTED FLUE.

application The appliance of a material to another material. For example, painting a coat of primer on the surface of a steel frame.

application rule In respect to the *Eurocodes, these are generally recognized rules (clauses that are not preceded by the letter P) used to satisfy the *principles which must be achieved.

applicator A tool used to place materials into their desired position. For example, a silicone gun is used to place silicone around windows to provide a sealant.

appraisal 1. An estimate of value. For a property it could be the current market value for insurance purposes or functional value for business purposes. **2.** Staff appraisals identify how a person has performed, whether existing goals have been met, whether he or she has established new business targets and areas of personal and *professional development.

apprentice A trainee who works under the stewardship of a skilled or trained person to learn a trade or profession for a period of time, described as an *apprenticeship. A person normally not yet certified as skilled or formally graduated from the institution of higher learning but seeking to learn more about the trades or professional position, then provided with an opportunity linked to an engineering, building, management, or construction-related organization where insights, experience, and knowledge about skills, trades, professions, and associated activities can be gained.

apprenticeship 1. Traditionally, this would describe the period of time served by an *apprentice to gain the required skills needed to become a qualified *tradesperson or professional. The written agreement that engages the apprentice and describes the training is called an indenture. **2.** A workplace-based learning system that generates opportunities for employment and training for people through work-based learning programmes which combine paid employment or work experience with on-the-job and off-the-job learning. National systems range from provision of vocational education and training through full-time schooling in vocational colleges and universities that offer teaching and training to a dual system based primarily on workplace-based apprenticeship learning. The dual system commonly includes part-time vocational learning and training in firms that accept apprentices. Apprenticeships are and have been considered a particularly powerful mechanism in providing a pathway to work for those who have no formal qualifications or work experience because of their disadvantaged background. They are now part of a package of education and training pathways that are available to all, regardless of background. Often the process provides the *apprentice with the necessary skills, experience, and knowledge of the various activities associated with the engineering, building, or construction processes with the view to gaining future employment within the same industry.

appropriate technology Technology that conforms to existing cultural, economic, environmental, and social conditions. Dr Schumacher is the person credited with the conception of the original term 'appropriate technology' in the early 1960s, which superseded his original term of 'intermediate technology'. The latter term portrayed a more second-class, less important status. However, the original concept of 'appropriate technology' for developing countries probably emanates from Gandhi, who advocated village-based technologies to help rural communities in India to become self-supporting.

appropriation of payments In the law of debtors and creditors, it is the application for a payment to pay a particular debt.

approval and satisfaction The act to state that the standard of processes or products are such that these meet specified conditions or a profession's requirements, or are perceived as being satisfactory. Contracts have provisions for a contract administrator or named professional to approve materials, workmanship, and operations. The nature and process associated with approval varies depending on the *standard form of contract.

approved document Qualified guidance, approved by a government body, giving detailed information on how the *Building Regulations can be accommodated within building design, construction, and maintenance.

Approved Installer Scheme Professional body and organization schemes to list those contractors that satisfy training and skill requirements to undertake work properly. Such schemes aim to remove or reduce the possibility of rogue traders and poor service.

approximate quantities The preliminary list of items that is difficult to measure and know exact quantities. The items are identified and listed in the preliminary *bills of quantities, and sometimes provisional costs or rates may be put against the work. When work is underway and can be measured, the approximate quantities are properly quantified and costed.

approximation An inexact number, cost, rate, or measurement that is useful to work with. Where there is insufficient information to produce an exact figure, it is necessary to use the information available to give a prediction. The symbol used to show that something is approximately equal to something is ≈.

appurtenant works (US) All additional works undertaken to complete a job even though these have not been identified or described in the bills or original quantities. Appurtenant means to belong to or be part of the whole.

apron 1. A horizontal piece of trim inserted underneath a window-sill on the inside. **2.** The tarmac or concrete area at an airport where aircraft stand when not in use.

apron flashing An L-shaped *flashing used to prevent water penetration at the junction between a vertical wall and a sloping roof. Commonly found on the lowest side of a chimney.

apron lining The horizontal facing used to cover the opening in the floor at a stair well.

aqueous Relating to or containing water.

aquifer A water-yielding strata. Permeable rock, sand, or gravel through which groundwater flows. The water can be extracted through a well dug down into the aquifer.

aramid Short for aromatic polyamide fibres that are a class of heat-resistant and strong synthetic fibres used primarily in aerospace and military applications. They are fibres in which the chain molecules are highly oriented along the fibre axis, allowing the strength of the bonds to be exploited.

arbitration A private legal process used to settle disputes. Arbitration is often adopted in favour of judicial proceedings, because the proceedings are conducted in a private forum, unless the parties agree the case is not available to the general public. Parties may prefer to have the case decided by an arbitrator or arbitrators that have technical or business expertise in the field of the dispute, since such expertise is not always available in the courts. There are often less grounds for appeal in arbitration, which means that once a decision is made it is easier to seek enforcement of an award. Rules of arbitration may be agreed to speed up the process, and although not always the case, arbitration is generally viewed as a quicker, more cost-effective way of settling disputes, when compared with going to court. However, arbitration is considered to be more costly and slower than *adjudication. Adjudication has many similarities to arbitration and has now been adopted by the industry as the main way of settling construction disputes.

arbitration agreement clause (agreement to refer) Provision or term contained within the contract that binds the parties to submit future disputes to arbitration.

arbitrator An impartial referee appointed by the parties concerned to make decisions on disputes. The jurisdiction (scope of work) and decision-making powers are agreed upon by the appointing parties. Arbitrators are often appointed because of their technical expertise and knowledge, and powers and procedures are often agreed within the arbitration rules *agreement and the Arbitration Act 1996.

arbitrator's award The award made by the arbitrator, which must deal with all matters referred, comply with any directions given in the submission, be final, have clear, certain meaning, and be consistent.

arc 1. A curve or a section of a circle. **2.** A luminous discharge caused by an electrical current bridging the gap between two electrodes in a circuit.

arcade An arched or sometimes covered passage or alley lined with shops.

arch A curved structure that is designed to span an opening and support weight. Many different types of arches exist, varying in shape and style.

Archaean (US Archean) Used to describe the earliest geological period/the oldest rock.

arch bar A curved steel or iron bar that supports the brickwork above a fireplace. Found in older properties in place of a *lintel.

arch brick 1. A wedge-shaped brick used to construct an arch. **2.** A very hard over-fired brick.

arch centre The temporary formwork used to support an arch under construction.

arch dam A curved water-retaining structure. The arch action is used in the horizontal plane to withstand water *pressure, which is transferred to the *abutments.

Archimedes' principle An object immersed in a fluid receives an upward force equating to the weight of fluid that is displaced. Lighter objects displace more fluid than their own weight and float; heavier objects displace less and sink.

Archimedes' screw A circular pipe containing a rotating helix. When the pipe is inclined at an angle of approximately 45° to the horizontal and the helix rotated, it can be used to lift water within the pipe. Typically used for transferring water from low-lying areas for irrigation purposes.

architect A professional who designs buildings in the UK, and is a member of *RIBA (Royal Institute of British Architects). In the UK the title 'registered architect' can only be used by qualified members of RIBA. Other countries have similar requirements. Traditionally the architect is associated with the development of the initial, conceptual, and final construction working drawing, normally based on the client's instructions or independently.

architect's appointment A rather dated standard form of agreement published by *RIBA (Royal Institute of British Architects) that contains provisions for the agreement of fees, services, and responsibilities. The form is occasionally used, although it has largely been replaced by other standard forms and procedure.

architect's instruction Written direction given to undertake work, normally issued where there is a change or variation to the *contract drawings. Depending on the profession of the person appointed as the client's representative or contract administrator, an *architect may not have the power to give instruction. Traditionally, the architect acted as the client's representative and contract administrator; however, it is quite common for other professions to undertake this role.

Architects' Registration Council (ARCUK) Operating under the Architects Act 1997, the organization was established to maintain a record of all 'registered architects'.

architectural drawings Drawings produced by the *architect that provide a detailed image of the proposed building. These normally form part of the *contract documents as used in the construction and building processes. The architectural drawings are subsequently used by the other members in the construction supply chain, such as the contractors and designers.

architectural ironmongery (architectural metalwork) Decorative or ornamental products made from iron or other metals, such as banisters, screens, and railings.

architectural sections Details that cut through the building to show how the components fit together within the building.

architecture The science, art, or field of study associated with the construction, detail, style, and aesthetics of buildings.

architrave The moulded trim that is used internally around door *sill and *jambs. Used to conceal the gap between the sill and jambs and the wall finish.

architrave bead A *stop bead used internally around door *sill and *jambs covered by the architrave.

architrave block A block-shaped trim used internally at the junction between the jamb and sill on a door.

architrave trunking An architrave with a removable or accessible cover that is designed to carry wiring.

arch stone (voussoir) A wedge-shaped stone used to construct an arch.

arc welding An inexpensive and widely used welding process, it involves an electric arc to melt the metals at the melting point. Sometimes the welding region is protected by an inert gas.

area The size of a surface confined within a boundary.

area way An area outside a basement, providing access, light, and air from outside.

arenaceous Sandy or having the appearance of sand.

argillaceous Clayey or having fine silt or clay particles.

argon An inert gas, chemical formula Ar. One of the Noble gases used in lightbulbs and to create an inert rich atmosphere for electric welding **(argon weld)**.

arid Little or no rainfall.

arithmetic check A mathematical check used in a long calculation. Certain parts of the calculation are recalculated by employing a different method and the answer is cross-checked.

arm The mechanical part of an excavator that is attached to the bucket.

armoured cable (US armored cable) An electrical cable that is designed to resist accidental breakage. Consists of a core of two to four plastic-sheathed insulated cables, which are encased in a flexible steel cable, and then sheathed in plastic. Used to transmit mains electricity underground.

array A group of photovoltaic modules that are connected together.

arris The sharp edge formed at an external corner where two surfaces meet. Commonly found in plasterwork and joinery.

arris fillet A triangular-shaped piece of wood that is nailed across the rafters and used to raise the roofing slates where they abut a chimney or wall. Similar to a *tilting fillet or eaves board.

arris gutter A V-shaped *gutter fixed to the *eaves of a building.

arris-wise (arris-ways) The diagonal laying of slates, tiles, bricks, or timber.

arrow A sign or marker to point the direction.

arrow diagram Project management technique for working out a logical arrangement of events, used to calculate the sequence of *activities, duration of the project, and *critical path. Circles or rectangles are used to represent events or activities and these are connected together by arrows. The duration, start, and finish times for each event are written within the activity boxes.

Art Deco A 1930s style of art and architecture, characterized by precise geometric shapes, bold colour, and outlines. The style was particularly prominent between 1925 and 1935 and was considered elegant, functional, and modern with its long, bold, and tall geometrical shapes, raised, exaggerated, or strong lines, and sweeping curves (unlike the more natural curves of *Art Nouveau). The Chrysler Building in New York is considered to be an example of Art Deco.

artefact An object made by a human, often associated with objects that have archaeological interest.

artery Main route in a road, rail, river, and drainage system.

artesian well A well where water flows out at the surface under *hydrostatic pressure, the water being drawn from a *confined aquifer.

articles of agreement Normally refers to the opening provisions within the contract; the core components of the contract.

articulated A joint that allows angular movement.

artificial Synthetic, not occurring naturally. Examples of artificial construction materials include: bricks, concrete, and plastics.

artificial ageing *See* ACCELERATED WEATHERING.

artificial aggregate Most lightweight aggregates with the exception of pumice.

artificial intelligence (AI) Computer systems that can perform intelligent human tasks, such as decision-making. Intelligence demonstrated by machines, in contrast to the natural intelligence displayed by humans and other animals. In some cases artificial intelligence is applied when a machine mimics cognitive functions that humans associate with other human minds, such as learning and problem-solving.

artificial lighting (artificial illumination) Lighting provided by artificial means, such as electric lighting, gas lighting, or an open fire.

artificial person A local authority, corporate body, or other legal entity that is recognized as a legal person capable of taking on rights and duties.

artificial seasoning An improved method of preventing bowing, cupping, twisting, or springing of boards or planks of lumber. During the latter portion of the timber-drying schedule, the boards or planks are passed between spaced pairs of rollers arranged to hold the lumber in a fixed plane. The lumber is therefore held in this plane while it is setting. Preferably the drying schedule includes intermittent exposure of the lumber to microwave radiation in an electronic kiln dryer.

artificial stone A synthetic product imitating stone. Various kinds have been used from the 19th century in building and for industrial purposes such as grindstones.

artists and tradespeople A contractual phrase used in early *JCT contracts to refer to parties employed by the client under a separate contract.

Art Nouveau A form of art and architecture that was most popular between 1890 and 1905. It is characterized by flowing forms and lines that are said to resemble natural forms such as flowers, leaves, and vines, and other organic forms. Some of the work of Antoni Gaudi, although very individual, can be said to be floral and natural Art-Nouveau style.

asbestos A naturally occurring heat- and fire-resistant fibre previously used extensively in construction, however, asbestos was outlawed in the UK in 1978 due to danger to human lungs.

asbestos encapsulation Process whereby asbestos is sprayed with a coating to minimize danger to human beings.

asbestos-free materials A material with 0% asbestos content, e.g. **asbestos-free board**.

asbestos removal Work undertaken to remove the banned construction material asbestos from buildings and equipment. The specialized work must be carried out by competent workers wearing protective clothing and breathing apparatus.

as-built drawing Drawings that show how a building has been constructed; the true position of service. Such information is essential for the safe maintenance and operation of the building.

as-built performance The actual performance of a building as constructed on site.

ascertain Within standard forms of contract, it is used to direct a person to find out for certain. Would preclude making a general assessment or making an estimate of quantities.

ASCII (American Standard Code for Information Interchange) A standard code that identifies letters, numbers, and certain symbols by code numbers for representing information on computer systems.

as-designed performance The performance of a building based upon the *design intent.

as-drawn wire Steel wire drawn through a die to harden it.

as-dug aggregate Quarried material delivered directly to site without any modification or treatment.

aseismic Not having or able to withstand earthquakes. Also denotes a fault where no earthquakes have occurred.

ashlar The stonework that is evenly dressed with 3 mm joints and provides the facings for buildings.

ashlaring 1. The short vertical members **(ashlar pieces)** that span between the *ceiling joist and the *rafter. **2.** A low *attic wall, usually 1 m in height and constructed in blockwork, that extends from attic floor to the ceiling. **3.** The process of bedding *ashlar in mortar.

ASHP *See* AIR SOURCE HEAT PUMP.

ASHRAE American Society of Heating, Refrigeration, and Air-Conditioning Engineers.

ASHVE American Society of Heating and Ventilation Engineers.

Asiatic closet (Eastern closet) A *water closet designed to be used in the squatting position, and comprises an elongated bowl that is positioned either at or just above the floor level. Rarely found in the UK.

ASP *See* ADJUSTABLE STEEL PROP.

aspect The orientation of a building on site.

aspect ratio The ratio between two dimensions, e.g. width to height ratio in terms of a building width to its height; a computer screen; or ratio of wing length to breadth in terms of an aircraft's wing.

asphalt A material comprised mainly of bitumen, sand, clay, and limestone. Used in road construction.

asphalt mixer Equipment used to heat and mix asphalt.

asphalt paver (asphalt plant) A self-propelled or towed wheeled or tracked machine used for laying asphalt in *flexible pavement construction.

asphalt roofing Normally, flat roofing made up of two or three coats of asphalt and binding materials.

asphalt soil stabilization The addition of asphalt to soil to improve its properties, i.e. reduce *permeability and increase *bearing capacity.

assembly The joining of the components that make up the final structure.

assembly gluing The gluing together of pre-assembled construction elements on site.

assent To agree or comply.

assessment of tender responses (vetting) Detailed assessment and evaluation of tenders received in order to advise the client.

asset Something of value, for example, within the built environment sector. Buildings, bridges, roads, sewage treatment plants, and power stations are all considered as assets.

asset management A systematic approach to operating, maintaining, upgrading, and disposing of *assets cost-effectively over their life cycle. The coordinated organization activity to realize value from those products, buildings, property, services, and any items that are owned by the organization. Realization of value will normally require a balancing of costs, risks, opportunities, and performance. Activity can also refer to the application of the elements of the asset management system. The term **activity** has a broad meaning and can include, for example, the approach, the planning, the plans, and their implementation (ISO 55000:2014). A related term is **facilities management**. In the engineering and technological environment, asset management can be considered as the process of developing, operating, safeguarding, and maintaining technological assets with the objective of maximizing net whole of life return.

assignment and subletting The transfer of a contract to another party. A party can transfer benefits to a contract, but cannot transfer a burden without express agreement from the other contracting parties.

Association of Building Technicians (ABT) A powerful *union of trade workers and technicians that was particularly prominent in the Victorian era. The association was formerly known as the Architects' and Surveyors' Assistants' Professional Union and as the Association of Architects, Surveyors, and Technical Assistants. The society amalgamated with the Amalgamated Union of Building Trade Workers (AUBTW) to form the Amalgamated Society of Woodworkers, Painters, and Builders (ASWPB).

(((⊕))) SEE WEB LINKS
• Trade Union Ancestors site with historical information.

as soon as possible Contractually, a direction given to use all means possible, giving due consideration of other work a party is committed to, to get a job done or materials delivered. When used in contracts or instructions, it has stronger connotations than in a reasonable time or forthwith.

asthenosphere A weak zone in the Earth's mantle.

ASTM (American Society for the Testing of Materials) A standards development organization that provides an independent source of technical standards for materials, products, systems, and services. The society was originally formed to address problems in standards in the railroad industry, which had previously been plagued with rail breaks and different standards.

((⊕)) SEE WEB LINKS
• Home page for the American Society for the Testing of Materials, including guides for design, manufacturing, and trade.

astragal 1. A small semicircular convex moulding. **2.** A vertical member attached to the *inactive or fixed leaf of a double door, which projects over the *active leaf when it is closed. Prevents the active leaf of the door being opened past the inactive leaf and seals the doors when they are in the closed position. **3.** (Scotland and US) A *glazing bar.

astronomy Scientific study of the sun, stars, and planets (the universe).

asymmetric Not symmetric, unequal.

atmometer An instrument used to measure the rate of evaporation.

atmosphere A gaseous region (air) that surrounds celestial bodies, such as the Earth.

atmospheric pressure The downward pressure caused by the weight of air above.

atom The smallest part of an element; made up of a dense nucleus containing both positively charged protons and electrically neutral neutrons, surrounded by negatively charged electrons.

atomic Relating to an atom or atoms.

at rest earth pressure The in-situ stress state of a soil where no horizontal or vertical strains occur. The ratio of horizontal stress (σ'_H) to vertical stress (σ'_v) is denoted by K_o, the coefficient of earth pressure at rest.

atrium A tall internal courtyard spanning a number of floors that has a glazed roof.

attachment of debts A procedure employed in the High Court where a judgement has been made against a debtor, and in order for that debtor to deliver sums of monies owed, third parties that owe money to the debtor can become part of the order to pay off the debts.

attendance Generally, the act of persons attending; however, different standard forms of contract can refer to general attendance as being plant, equipment, and temporary roads necessary to undertake works. Items described as special attendance are listed in the contract or *Standard Method of Measurement.

attenuation The reduction in sound level as a sound travels from one place to another. Measured in decibels.

attenuator A mechanical device installed within ductwork that is primarily designed to control fan noise. Comprises a circular or rectangular casing of acoustically absorbent material. It is normally located as close to the fan as possible; however, secondary units may be used where close control of noise is required, such as in auditoriums.

Atterberg limits (consistency limits) The physical properties of a fine-grained cohesive soil is normally directly related to its water content. There are four states in which the soil can exist, the boundaries between these states are known as the consistency limits.

	Four states	Consistency limits	Tests and definition of limits
Moisture content increases ↑	Liquid	↕ Liquid limit, LL	Empirical boundary, defined as the moisture content at which the soil is assumed to flow under its own weight. Measured using a *cone penetrometer or Casagrande cup (see CASAGRANDE, ARTHUR)
	Plastic	↕ Plastic limit, PL	Empirical boundary, defined as the moisture content at which a 3 mm diameter thread of soil can be rolled by hand without breaking up.
	Semi-Solid	↕ Shrinkage limit, SL	Moisture content below which no further reduction in volume will occur. Measured by slowly drying a sample out & periodically measuring its volume & mass.
	Solid		

Atterberg limits

attic (garret) A space in the roof that can be converted into a habitable room.

ATTMA The Air-Tightness Testing and Measurement Association.

auger A screw-shaped tool used to cut into wood or soil to produce a hole. The tools used to extract soil can be hand-held or mounted on tracked vehicles.

augmented reality A computer-generated image superimposed on the user's view of the real world.

aurora (Northern lights, Southern lights) A phenomenon caused by the interaction of atmospheric gases and solar particles producing streamers or bands of coloured light in the polar regions.

austenite A phase found in ferritic metals and alloys. Austenite is a *face-centred cubic (FCC) iron phase in iron and steel alloys with an FCC crystal structure.

austenizing A technique used to form the *austenite phase in a *ferrous metal alloy.

autobahn See MOTORWAY.

autoclave A pressurized device designed to heat aqueous solutions above their boiling point to achieve sterilization. The term 'autoclave' is also used to describe an industrial machine in which elevated temperature and pressure are used in processing materials, especially autoclaved aerated concrete (*see* AIRCRETE). The process can be used to sterilize equipment or materials.

autoclaved aerated concrete *See* AIRCRETE.

automatic controls Electrical or mechanical devices that are capable of operating items of equipment automatically according to predefined conditions.

automatic flushing cistern A cistern that flushes automatically, usually just after it has been used. Passive infrared sensors determine when the urinal or toilet has been used. Reduces water usage and ensures that the urinal or toilet is cleaned just after use.

automatic level An optical instrument used for *levelling. Prisms within the instrument ensure that a horizontal *collimation line will pass through the cross-hairs, provided that the instrument is approximately level.

autonomous car (driverless or self-drive car) A vehicle that is capable of monitoring the surrounding environment, using radar, computer vision, GPS, *odometry, etc., with little or no human input. *See also* SAE INTERNATIONAL STANDARD; SAE LEVELS OF AUTOMATION.

autoroute *See* MOTORWAY.

auto suppression system Part of an active fire protection system where detectors automatically activate the fire extinguishers to suppress the fire.

auxiliary equipment Back-up equipment, for example, a power supply that runs from its own generator rather than the National Grid.

available (availability) Standard forms of contracts use the term to mean contract drawings, bills of quantities, descriptions of works, and other documents that should be accessible where they are to be used. Documents would not be considered available if they were stored at a head office, if it was clear that their use was required on site. The use of *web portals for accessing documents could change how such documents are made available; terms within contracts will need to be adjusted to accommodate such access.

avalanche A sudden mass movement of snow down a mountainside.

average A number that represents the typical value in size of a collection of items being considered; calculated by dividing the sum of the items being considered by the number of items.

avulsion Sudden removal or loss of land by the action of water (a flood).

award *See* ARBITRATOR'S AWARD.

award of tender Notification by the *client or nominated body of acceptance of tender. Acceptance of the tender normally leads to the signing of the contract and binding agreement. Acceptance of the tender (offer) is normally a contract in itself;

however, the finalities of the agreement are often negotiated after the tender has been accepted.

awl A small, sharp-pointed tool used for marking hard surfaces or for punching small holes.

awning (terrace blind) An external covering that projects out from a building to provide protection from the sun and rain and is usually retractable.

axe 1. A tool used for chopping timber. **2.** A bricklayer's hammer used to cut and break bricks.

axed work Masonry (bricks and stonework) that is cut, formed, and shaped with a mason's hammer and chisel, and often used to form a brick or axed arch. When forming an axed arch, bricks are carefully shaped using a wooden template so that they are all the same size and fit together to form a neat brick arch. Joints in load-bearing brick arches are very fine, 3–4 mm, thus the bricks are cut precisely to form the arch.

(⊕) SEE WEB LINKS
• Water Recovery Group home page with information on brick arches used in the restoration of canals.

axial Relating to or forming an axis.

axial flow machine A pump or turbine with rotating blades parallel to the *axis about which they rotate. An **axial pump** (also known as a **propeller pump**) discharges fluid parallel to the pump shaft; an **axial turbine** has air/water flow parallel to its rotor axis.

axis 1. The vertical line and horizontal line (*plural* **axes**) on a graph, forming coordinates. **2.** A line on which an object would rotate. **3.** A line of symmetry around an object, e.g. the equator forms an axis around the centre of the Earth.

axisymmetric Symmetrical about an axis.

axonometric An image that is drawn slightly skewed so that more than one face of an object can be seen. With axonometric three-dimensional drawings, one of the axes is normally drawn vertically. Axonometric projections include *isometric, diametric, and trimetric.

(⊕) SEE WEB LINKS
• Demonstration of axonometric drawing

azimuth The horizontal angle, in degrees, from true north to a point directly beneath an observed object, e.g. the compass bearing of a star.

Bacillus coli An aerobic, rod-shaped, spore-producing bacterium that can be found in drinking water and may be harmful to human health.

backacter *See* BACKHOE.

back boiler A *boiler located at the back of a fireplace or stove. The hot water produced from the boiler can be used to provide a hot-water supply or for space heating.

back cutting Additional excavation that is required to form the correct formation level.

backdraught The rush of air through a door or window that is opened into an oxygen-starved room where a fire is burning. The fire uses all of the oxygen in the room to fuel the fire, and if a window or door is opened, air rushes into the oxygen-starved room providing fuel for the fire. Prior to the door being opened, the fire may have become so starved of oxygen that it is reduced to a smouldering fire; however, glowing embers can burst into flame once oxygen is present. When it is suspected that there is a fire in a room, the temperature of the door should be checked. If the door is hot, indicating that there is a fire behind it, the door should be left closed.

backdrop (backdrop connection, drop connection, US sewer chimney) A vertical pipe, located in an *inspection chamber, which allows a foul water drain to be connected to a sewer pipe at a lower level.

backer *See* BACKUP STRIP

backfall An adverse slope in a drain or gutter causing water to flow in the opposite direction than intended, which can lead to *ponding of water.

backfill (backfilling) Soil returned to an excavation, after a foundation has been cast or a pipe placed; typically compacted in layers.

back-flap hinge A *hinge with wide flaps—the flaps being the part of the hinge where the screw holes are located.

backflow Movement of liquid or gas in the opposite direction than intended. A **backflow valve** prevents such movement.

background (backing, base) The surface on which the first coat of plaster is applied.

background drawing A single drawing used to help coordinate and distribute building services. All subcontractors and designers should use the same background drawings and information to ensure that services can be properly coordinated without service runs clashing, e.g. two or more different service runs trying to occupy the same space. Where transparent service drawings are all laid over the top of the same background drawing, distribution problems and clashes can be detected and avoided. With digital information and extensive use of CAD, the use of overlays and background information is becoming more common. With any information management and coordination system, it is essential that the latest information and revisions are used.

background heating The process of providing constant low-temperature space heating to a space. Used to reduce the risk of condensation and to maintain a constant internal temperature.

background ventilation The uncontrollable air exchange between the inside and outside of a building through a wide range of *air leakage points and *air leakage paths in the building structure. Also known as *infiltration.

back gutter A *gutter that is installed on the 'back' or highest side of a chimney where it intersects the roof. Usually formed in lead, although other materials can be used.

backhoe (backacter, drag shovel, trench hoe) A wheeled tracker excavator (trademarked as a **JCB**), with buckets to its front and rear, each being independently connected to the tracker via a two-part hydraulic arm.

backing (base) 1. Any material installed behind a facing material to provide a base or support to the facing material, such as brickwork or concrete behind stonework. **2.** Any coat of paint or plaster that is not the final finishing coat.

backing bevel The V-shaped bevel found at the top of a hip rafter where the two roof slopes join.

back lintel A *lintel that is used to support the inner leaf of a wall.

back observation *See* BACKSIGHT.

back propping *See* SHORING.

back sawing A hand tool used for cutting, with a stiffened back section.

backset (set back) The distance from the leading edge of the door to the centre of the hole that is drilled for the spindle of the door handle or latch to be housed.

back shore (jack shore) An outer bottom support of a racking shore.

backsight A reading taken in surveying (e.g. levelling) to a reference location after the instrument has been repositioned. For example, a *foresight is taken before the level is moved to a staff that is visible from the new and old levelling station. The level is then set up at its new location and a backsight is taken.

backsplash *See* SPLASHBACK

backup strip (joint backing, backer) Compressible material, often foam strips, inserted to fill the majority of a movement joint so that the depth of sealant that is finally applied to seal the joint is limited. The backup strip usually has a finish that prevents a bond between the sealant and backup strip, allowing the sealant to stretch and compress within the movement joint without being inhibited by the filler strip.

backwater curve The longitudinal profile of a liquid surface in an open channel when the depth of water has been increased by an obstruction such as a dam or a weir.

bacteria bed A layer of sand which *effluent is passed through in sewage treatment plants. The action of air and *microorganisms in the bed help to break down the effluent.

baffle (baffle plate, baffler) A device that controls or restricts the flow of air, water, or light.

bag filter Air filter used in a balanced-flue gas fire.

bagged joint A *flat joint that has been finished using a coarse cloth, rather than a piece of wood or pointing tool. Usually used with handmade bricks.

bagging (bag rendering) The process of applying cement mortar to external brickwork using a brush or a hessian bag to fill in any gaps and joints. This results in a rough cement-like finish that can be painted over.

bag set Plaster or cement that has set in its bag, caused by storing the bag in damp conditions.

bailer A device that is used to remove sludge from the bottom of a mine.

balley The outer fortified walls surrounding a courtyard or keep; term associated with castles.

Bailey bridge A prefabricated lattice steel truss bridge designed for use by the military in the early 1940s. It can span up to 60 m and be assembled rapidly on site with basic hand tools.

bailment The legal process of placing goods with another person or organization on the condition that the goods will be returned. Goods are lent or deposited in the temporary custody of another. The bailment may be a simple gratuitous loan or for reward, as would be the case when something is hired. The holder is known as the **bailee**, the person who delivers or transfers the property is known as the **bailor**. For a bailment to be valid the holder (bailee) of the property must have physical control of the property and intent to possess it.

bainite A phase found in ferrous metals, for example, steel and cast iron. Bainite is comprised of ferrite and cementite.

Baker bell dolphin *See* DOLPHIN.

balance beam A timber beam that is pushed to open a lock gate when the water level is equal at either side.

balance box A counterbalance load that is located on a crane at the opposite side of the *jib.

balance bridge *See* BASCULE BRIDGE.

balanced construction Method used to ensure that there is the same amount of grain running in opposite directions in plywood. In three-ply timber, the thickness of the core timber would be twice that of either of the surface veneers. The grain of the two surface veneers would run in the same direction and the grain of core timber runs at 90 degrees to adjacent veneers, thus the amount of timber running in different directions is balanced.

balanced flue A flue where the air is drawn in and exhausted through separate compartments of the same flue. This is achieved using a concentric arrangement. Used in room-sealed appliances.

balanced sash A double-hung sash window that is balanced by weights hung on chains or ropes, or pre-tensioned springs. The balancing enables the sash to be opened smoothly and easily.

balanced step (dancing step, French flier) A step on a stairway that is slightly tapered at one end to produce a curve in a flight of stairs. Most tapered steps are very narrow on this inside edge allowing them to be accommodated by a central newel post; however, the wider tapered steps are easier to negotiate but require more room, often accommodated by an open stairwell.

balancing The process used when setting up *central heating systems and *commissioning them, where the inlet valve on each of the radiators is adjusted to ensure that the flow of hot water to each radiator is evenly distributed.

balancing valve A valve used for controlling the flow of a fluid through different legs or branches of a system to ensure that all outlets are balanced and receive equal fluid pressure.

balcony An open or covered platform, usually enclosed with a railing, that projects out from the external wall of a building.

baling A method of compressing waste to form bales to be either burnt or disposed of in landfill.

balk (US) *See* BAULK.

ballast 1. Stone or gravel used as a foundation material on roads and rail tracks. **2.** Unscreened gravel, containing small stones, sand, and grit, used to make concrete. **3.** Heavy weights that are used to stabilize or hold something down, e.g. weights that are carried in the hold of a cargo ship to increase its stability when it is empty. **4. ballast (choke)** An electronic device that controls the current through discharge lighting.

ball catch (bullet catch) A door fastener comprising a spring-loaded ball that is set within the mortise of the door. When the door is closed, the ball engages a striking plate on the jamb of the door that contains either a slight indentation or a hole that the ball sits within, thus keeping the door closed. The door is opened when applying enough force to the door so that the ball springs back out of the indentation or hole.

ball cock *See* BALL VALVE.

ball joint A connection, consisting of a rounded end that fits into a cup-shaped socket, which allows rotational movement.

balloon frame (balloon framing) A timber-frame construction consisting of long continuous vertical *studs that run from ground to eaves-level of a house. The timber-frame supporting walls extend the full height of the building, from the base plate to the roof plate. Each wall of the timber frame is normally prefabricated, typically in two-storey units, then simply erected and bolted to the other walls on site. Due to the units extending more than one storey, balloon framing can be a quicker method of construction than *platform frames which only extend one storey high. Platform frame units are easier to handle and transport, and because of this, they are often used in preference to balloon frames.

balloon grating (balloon wire) A large strainer placed over a rainwater outlet.

ball penetration test (Kelly ball) A field test used on freshly poured concrete to check the consistency for control purposes. The test consists of determining the depth to which a hemispherical ball, having a diameter of 150 mm and a weight 13.6 kg, will sink to under its own weight.

ball support A support that permits rotational movement but prevents displacements.

ball test A ball slightly smaller in diameter (i.e. 12 mm) than the pipe is passed down the pipe to determine if any restrictions in the pipe exist.

ball valve (ball cock, float valve) A valve in which the flow is regulated by the position of a moving ball. Typically found in toilet cisterns. When the cistern empties, the ball moves down with the water level and causes the valve to open. As the cistern fills, the ball rises to close the valve.

Baltimore truss A very strong *truss bridge; a subdivision of a *Pratt truss with additional bracing in the lower part of the bridge to help prevent *buckling and control *deflection. Typically it has been used for rail bridges.

baluster (banister) Edge guard to stairways, open edges of landings, roofs, and floors. Normally consists of a top-rail (coping or handrail), infill *spindles, or *balustrade that sit on the floor or bottom rail.

balustrade The vertical in-filling components of a structure that acts as a guard against an open edge of a roof, floor, landing, or stairway. A *baluster (edge guard) consists of vertical balusters, occasionally these may be supporting, but mostly they act as in-fill under the coping (top-rail or handrail).

band course *See* STRING COURSE.

banding *See* LIPPING.

band saw A power cutting device with a continuous vertically mounted blade.

band screen A mesh wire sieve, in the form of an endless moving belt, used at water and sewage treatment plants to remove solids so as to protect downstream equipment.

banister *See* BALUSTER.

bank 1. A raised area of land, e.g. the side of a waterway. **2.** The grouping of similar objects, e.g. a bank of lights.

banker (gauging board) Profile, often made of timber, to show the position of each brick course and mortar joint. Can be used to assist bricklayers to lay bricks, blocks, and stonework to a consistent height with consistent mortar joints.

bank guarantee A legally binding promise from a lending institution (bank) that the liabilities of a debtor will be met. The bank guarantee promises a sum of money to the beneficiary should something go wrong, e.g. if a buyer of goods experiences cash-flow problems, then the lending institution would still honour the transaction if a bank guarantee had been agreed.

banking Slope of a road surface.

bank material Soil, gravel, and rock before it is excavated or blasted from the ground.

bank of lifts A number of *lifts of material at the same location.

bankruptcy Legal declaration that a company or individual is unable to pay their creditors. The state procedure protects the debtor from demands of creditors and ensures the assets of the individual or company are equally distributed among the creditors. The procedure is laid down in Part IX of the Insolvency Act 1986.

((⊕)) SEE WEB LINKS
• The Insolvency Service

bank seating The end support of a bridge, i.e. a bridge abutment.

banksman (banksperson) A competent person who provides directions to a crane driver from where loads are attached or detached, because crane drivers, especially on tower cranes, do not always have a clear view of the loading and unloading areas.

bank storage Water that is held in the ground above the phreatic surface (*see* GROUNDWATER), from *run-off during heavy rains and flood periods.

banquette 1. A horizontal ledge. **2.** A footbridge.

Banwell Report In 1964 Sir Harold Banwell chaired a committee to provide a report on the state of the British construction industry and make recommendations. One of the main recommendations was the discouragement of open *tendering and an emphasis towards selective tendering.

bar 1. Unit of *pressure, equivalent to 100 kN/m^2. **2.** A solid length of material, usually steel, round, rectangular, or hexagonal in cross-section. **3.** Silt and sand deposited at a shallow or low *velocity part of a riverbed or the mouth of a river.

bar bender A machine that bends reinforcement bars to the desired shape before they are fixed in position and concrete is cast around them.

bar chain method An analysis method for *trusses. The deflected form of the truss is represented by a series of straight lines about the joints around which they rotate.

bar chair A special shaped reinforcement bar, which separates the top and bottom mats of reinforcement to provide the correct concrete cover.

bar chart (Gantt chart) A diagrammatic project management technique for illustrating the programme of works and resources associated with a given project or set of activities. A schedule or programme of activities plotted on a graph with time plotted on the x-axis. Each activity is represented by a solid or hatched bar, with the length of the bar extending for the proposed and actual duration of the activity, showing when the activity starts and finishes. With both estimated and actual durations shown on the same graph, any slippage of activities or changes to timing of events can be seen and tracked.

bar code (barcode) 1. A number given to steel reinforcement to identify shapes and characteristics of the reinforcement bar, also called **shape code. 2.** Series of lines or shapes that can be optically read by a scanner and correlated with data stored on a central processor. Most commonly used in supermarkets, but now is increasingly used to track parcels and materials in construction projects.

barcol hardness A hardness value obtained by measuring the resistance to penetration of a sharp, spring-loaded steel point. It is used to measure the degree of cure of a plastic.

bareface tenon The male part of a mortise-and-tenon timber joint with only one side of the tenon having a shoulder and the other side being a flush face.

bareface tongue Part of a timber joint made by rebating the end of timber on one side only so that the other side is flush. The rebate is often half the thickness of the original plank or sheet of wood. The thin section of tongue fits into a recess to provide a neat strong joint.

barge board (verge board, gable board) Diagonal piece of timber used to provide a decorative cover to the edge of the roof on a gable end, and serves the same function as the *fascia board. The barge board is either fixed under the edge of the roof tiles or slates, hiding the junction between the roof and the gable wall or is fixed to cover the side of the edge of roof tiles and wall. The barge board often meets the facia board and provides a neat continuation along the diagonal slope of the pitched roof.

barge course (verge course) Projecting brick coping or tiles next to the gable that protrude slightly.

barge flashing Flashing that extends over the gable tiles and barge board.

bar joist A flat truss structural member.

Barnes formula Used to calculate flow in sewers:

$$v = 107m^{0.7}\sqrt{i}$$

where v = velocity (ft/sec), m = hydraulic mean depth (in feet) and i = the slope of the sewer.

baroclinic An atmospheric condition in which density depends on both temperature and pressure; *cf.* *barotropic. A **baroclinic zone** is a region where a temperature gradient exists on a surface of constant pressure. Typically associated with mid-latitude and polar regions. Also applies to oceanic conditions.

barometer An instrument for measuring atmospheric pressure.

barotropic An atmospheric condition in which density depends only on pressure (independent of temperature) and vice versa; *cf* *baroclinic. A **barotropic weather system** is one in which temperature and pressure surfaces are coincident.

barrage A lower height dam placed across a watercourse (river or estuary) to control the water level.

barrel bolt A common type of door fastener comprising a metal rod set within a cylindrical case. When the door is closed, a knob attached to the metal rod is used to push it forwards by hand until it engages a separate smaller cylindrical case located on the jamb of the door.

barrel light Roof light with either curved glazing or reflector unit that resembles a barrel vault.

barrel nipple (US shoulder fitting) A pipe fitting comprising a short hollow cylinder with a slightly tapered thread at each end.

barrel roof A semicircular arch capable of spanning long distances.

barrette A large excavated pile, where the excavation is normally rectangular and undertaken by a grab.

barrier An obstacle obstructing access, e.g. a fence.

barrister A lawyer, being a member of one of the Four Inns of Court and who has been called to the Bar. The barrister is normally appointed by the client's solicitor once advocacy before a court is needed by the client. Barristers are engaged to provide specialist advice on points of law. There are an increasing number of barristers who specialize in construction law.

barrow run A temporary path, found on construction sites, that is used by loaded wheelbarrows to gain access across unstable and soft ground. Usually constructed from scaffold boards or wooden planks.

bar scale A line on a map, drawn to scale, to represent a set distance.

bar schedule (bending schedule) List of reinforcement needed, the size and shape (*see* BAR CODE) for each type of bar is specified and the quantities required are stated.

bar spacer A plastic object to position reinforcement in concrete, particularly in relation to maintaining the correct degree of cover for *formwork.

basal metabolism The amount of heat given off by a resting organism (e.g. a person) purely to maintain its basic functions in a comfortable environment.

basalt An extrusive *igneous rock; black to greyish-black in colour, aphanitic (fine-grained) and dense. Two principal types exist: olivine basalts (undersaturated in quartz), and tholeiitic basalts (quartz-saturated where mafic minerals are mainly pyroxenes). Typically used as an aggregate and in road construction.

basal till Material that is located between a moving glacier and the underlying bedrock, typically a mixture of rock fragments and boulders in a fine-grained sandy matrix.

bascule bridge A bridge in which the span is moveable, e.g. a drawbridge. It can have single or double spans, with each span counterbalanced to control the lowering and rising motions.

base 1. The lowest part of something that normally provides support, e.g. a foundation. **2.** A solvent in which the other compounds are held. **3.** A component of paint that is the material producing the required opacity (ability to mask colour of surface). The body of the paint may be increased by the addition of inert extenders such as silica, calcium carbonate, or barytes. **4.** A reference number used for counting digits, e.g. base 10 contains ten digits from zero to nine. **5.** A logarithm reference number; a number raised to a power (indicated by a superscript number), e.g. $10^2 = 100$, where 10 is the base. **6.** A substance that reacts with acids to form salts.

base bid (US) The initial bid against which an alternative bid is offered.

basecourse *See* BINDER COURSE.

base date Used in standard form contracts, such as *JCT (Joint Contract Tribunal), and IFC (Intermediate Form of Building Contract) to describe the date agreed and specified in the appendix by the parties. The base date is normally the same as the date of tender; other contracts may simply refer to the date of tender.

base exchange A water-softening process. *Hard water is passed through a tank, at mains pressure, containing *zeolite. The zeolite, a mineral reagent, absorbs the salts that harden the water. At regular intervals a salt solution is flushed over the zeolite to regenerate it.

base flood A flood with a 1 in 100 chance of occurring.

base flow The amount of flow in a river or stream emanating from groundwater, as opposed to run-off.

base gusset A stiff *rib that is located between a *base plate and steel column.

baseline A datum mark, period, event, or standard from which all other things can be measured and compared.

base load The capacity to produce something continuously at a certain level, e.g. the minimum amount of electricity from a power station.

base map An unsophisticated map on which data can be plotted. A base map tends to include boundaries, buildings, and roads with reference to geographic coordinates. An example of a base map is a topographic map.

basement An underground storey of a building.

baseplate 1. A steel plate located at the bottom of a column with holes predrilled for holding-down bolts, also known as the **stantion base**. **2.** The lower part of a theodolite used to level the instrument and fix it to the tripod.

base resistance (R_b) When the load is transmitted through the length of the pile to hard strata beneath, as defined in Eurocode 7. *See also* SHAFT RESISTANCE.

base shoe (shoe mould) A moulding, usually in the shape of a quarter circle, that is used to cover the junction between the skirting board and the floor.

base slab A concrete slab beneath a structure.

base station A *global positioning system (GPS) receiver positioned at a precise location.

basic prices *See* SCHEDULE OF PRICES.

basic rock A general term used for igneous rock with high concentrations of iron, magnesium, and calcium.

basin 1. A depression; in land that contains water, or at sea that contains sediment. **2.** An area drained by a river. **3. (washbasin)** A bathroom *sink used for face and hand washing. Usually made from *vitreous china.

basin mixer A *mixer tap on a washbasin.

basket weave pattern A type of pattern used when laying brick patios. The bricks are laid in pairs with each pair laid at a right angle to the surrounding pair.

bast fibres The fibres that come from inner bast tissue or bark of stems from dicotyledonous plants. Their function is to provide support to the plant. Retting is a process used to extract the fibres from the cellular and woody tissues, i.e. the plant stalks are rotted away from the fibres. Examples of the most common bast fibres are flax, hemp, and jute.

bat bolt *See* RAGBOLT.

batch A quantity of something that is produced in one operation. **Batching plant** is the equipment used to make the batch, e.g. a **concrete batching plant**.

batch box (measuring frame, gauge box) A box without a top and bottom (i.e. only sides), used for measuring the volume of the constituent of a mix. Once the box is full it is lifted away leaving the contents on the mixing platform. For example, aggregate, sand, and cement can be measured out in a **batching box** to form the correct proportions for a concrete mix.

bath A sanitary fitting where the body or part of the body is immersed in water for washing.

bathing waters Open waters (seawater or freshwater) having water quality, designated under an EC directive, suitable for public bathing (swimming).

bath mixer A *mixer tap on a bath.

batholith A major intrusion of igneous rock formed deep within the Earth's crust from slowly cooling magma. *See also* PLUTON; STOCK.

bath panel A board attached to the side or sides of a bath to conceal the pipework and the underneath of the bath.

bathroom A room containing a bath, a toilet, and a sink. A ¾ **bathroom** contains a shower, toilet, and a sink, while a ½ **bathroom** contains a toilet and a sink.

bathroom pod A bathroom where all of the plumbing, tiling, electrics, fixtures, and fittings have been *prefabricated in a factory ready for installation on site.

bathymetry The measurement of the depths of water, e.g. the depth of lakes, oceans, and seas.

batt 1. A rectangular blanket of insulation material. Used in cavity walls, stud walls, floors, and loft spaces. **2.** Short for battening.

batten A thin long strip of material (typically wood) used to strengthen or secure something.

battenboard A strong stable panel consisting of a *batten, *plywood, or laminated core with a wood *veneer face.

batten door (matchboarding door, ledged and braced door) A door constructed of boards in parallel rows secured by a *batten or *ledge across which all of the boards are fixed. A diagonal *brace is used to stiffen the structure, holding the door square and stopping the opening edge dropping or sagging. Often the boards are interconnecting where matchboarding is used. In matchboarding the parallel boards, which make up the door, have a rebate cut into one edge and a tongue or male insert formed on the opposite edge. The tongue of one board is inserted into the rebate of the adjacent board. As each board is inserted into the next there are no gaps between boards.

battening Narrow strips of timber fixed to a wall or roof to provide a level surface. Used for fixing dry-lining or tiles.

batten roll A traditional method of jointing used over battens in metal roofing.

batter The inclination from vertical of slope sides or of a ditch.

batterboard (US) A *profile.

batter peg A peg that is knocked into the ground to indicate the edge of a slope.

battery 1. A device that uses a chemical reaction to produce electricity. **2.** A heating or cooling coil in an HVAC system.

battledeck Steel plates stiffened by a grillage of welded sections beneath.

baulk A large square piece of wood functioning as a beam.

Bauschinger effect A type of deformation in metals which results in an increase of the tensile yield strength and a decrease of the compressive yield strength. *See also* COLD ROLLING.

bauxite The ore from which aluminium is extracted, consisting of moderately pure hydrated alumina, the chemical formula being $Al_2O_{3x}\ 2H_2O$.

bay 1. A section or division of a building, e.g. section between columns or beams. **2.** Part of the building used for loading and unloading goods, e.g. loading bay. **3.** Strip of concrete, where concrete is poured in sections. **4.** Section of material plaster, concrete, or brickwork laid during a set period of time.

bayonet fitting A tubular fixing unit that secures two parts together. A tube with studs on opposing faces that slides inside a slightly wider cylinder. The tube is located and locked in the cylindrical socket by twisting the studs into rebates.

bayonet holder Part of the fitting that receives the male part of the locking device, *see* BAYONET FITTING. Light bulbs not secured by a screw fitting are held in place by a bayonet fitting housed into a bayonet holder.

bay window A window that projects outwards from the external wall of a building and usually comprises three or more individual windows. Various types exist depending upon the shape of the projection, e.g. **bow window** (semicircular) and **square bay** (square). If the bay only projects from an upper floor, it is known as an **oriel window**.

bead 1. Semicircular timber moulding, used as a feature or to mask joints. **2.** Defect in paint or varnish caused by excessive accumulation of liquid leading to uncontrolled flow during application.

beading *See* BEAD.

beading router (beader) Machine for making beading and cutting grooves into timber.

beam A horizontal *structural member that resists *bending loadings. Typically formed from wood, steel, or concrete and used to support the storey or roof above.

bearer A timber *beam that carries the floor joist.

bearing 1. Supporting load, at a particular location, over a given area. **2.** A device to allow constrained motions between two parts.

bearing capacity/resistance Bearing capacity, q_{ult}, is the pressure applied to the soil by the foundation to produce general shear failure, whereas bearing resistance, R, is the force applied to the foundation of area A to produce the same failure criterion, and is referred to as the *ultimate limit state. Bearing resistance is now directly applicable to the terminology used in Eurocode 7.

bearing pile A pile that transmits the load from the structure to the ground. An **end-bearing pile** transmits the load to the base of the pile, whereas a **friction-bearing pile** develops skin resistance around the pile to transmit the load. A combination of both end- and frictional-bearing resistance will occur in the majority of piles.

bearing wall A wall that supports a load.

Beaufort scale An international scale for measuring wind speed. It ranges from force 0 (<1 kph) for calm to force 12 (>120 kph) for hurricane.

becquerel *SI unit of radioactivity, corresponding to the decay of one nucleus of radioactive matter per second.

bedding planes The arrangement of layers of sedimentary rock.

bedhead panel Electrical unit or socket, often used in hospitals and positioned above beds, providing an outlet and connection for communication services, nurse call, electric, and other medical or patient equipment.

beetle maul A mallet with a wooden head.

Beggs deformeter A model to determine the reactions of a statically indeterminate structure, developed by **G. E. Beggs** in 1992. From its deflected shape, under the application of a load, the influence of the load along the structure (e.g. shear forces and bending moments) is deduced by using reciprocal theorem.

beginning event The first task, fixture, or item of a project, also called a start event in a *network or *Gantt chart.

behind schedule *See* SCHEDULE.

Belanger's critical velocity A condition of fluid flow in open channels in which the *velocity head equals half the mean depth.

Belbin's roles Aspects of interpersonal behaviour that individual members associate with during teamwork. **Dr Meredith Belbin's** research identified nine clusters of behaviour that can be used to provide a profile of an individual's behaviour within a group. The names given to these roles include plant, resource investigator, coordinator, monitor, evaluator, shaper, implementer, completer, finisher, and specialist. Very few people display the characteristic behaviour of just one role but have strong association with two or three roles. The preferences towards certain behaviour are not fixed but provide a snapshot of team role at the time of completing the self-perception form.

Belfast sink A large traditional rectangular-shaped ceramic sink with deep sides. Usually installed below the top of the work surface.

Belgian truss *See* FINK TRUSS.

bell and spigot A connection joint between two sections of pipe. The *spigot end of one pipe is inserted into a flared-out section (bell) of the adjoining pipe, with a rubber compressible ring or a caulking compound that seals the joint.

bell-and-spigot joint (US) *See* SPIGOT-AND-SOCKET JOINT.

bellcast eaves A roof that is in the shape of a bell, i.e. curving with a more gradual outwards slope at the bottom.

bell curve A curve in the shape (outline) of a bell, e.g. a normal distribution curve.

belled-out bored in-situ concrete pile *See* UNDER-REAMING.

bellmouth overflow A water overflow structure situated in a reservoir, with the opening in the shape of an inverted bell. Used as an alternative to a side *spillway.

below ground level (US below grade) An elevation lower than ground level.

belt conveyor An endless belt driven by rollers used as a *conveyor.

belt sander A *sander, with the *sandpaper in the form of an endless loop/belt.

benched foundation A foundation that is stepped.

benchmark 1. A *datum point used in surveying such as *ordnance datum. **2.** A standard that is aimed at or used for comparison.

bend 1. Something that is curved, e.g. a short length of pipe used for turning corners. **2.** To *yield.

bender A tool used for bending bars and pipes.

bender element A test to determine soil stiffness. Normally undertaken in a *triaxial cell using shear wave velocity.

bending Caused by rotating either end of the length of a material in the opposite direction about its axis. **Bending moment** occurs in a structural element (beam) when a moment (force multiplied by a distance) is applied to the element. A **Bending moment diagram** shows the bending moment along the element.

bending schedule (bar schedule) A list of the shapes and sizes of reinforcement bars required for a section of reinforced concrete.

bending spring Coiled tool used by plumbers to exert an internal force on tubes to maintain the circular shape of copper and plastic tubes whilst the tube is bent. The spring is inserted into the tube to the point where the bend is desired. The tube is then bent and the spring is extracted by means of a draw chord, which is fitted prior to the spring being inserted. The resistance from the spring prevents the pipe folding; sand can be inserted into a pipe and used instead of a spring.

bending stress (flexure stress) The stress along an element that results from bending it. As a result of the load, the beam will experience compressive stress along the top and tensile stress along the bottom. A neutral axis will exist in the middle of the beam where the stress level is zero. Thus, the bending stress will vary linearly with distance from the neutral axis (either in tension or compression). The equation to calculate bending stress is:

$$\sigma_b = \frac{My}{I}$$

where M = bending moment, y = vertical distance from neutral axis, and I = moment of inertia.

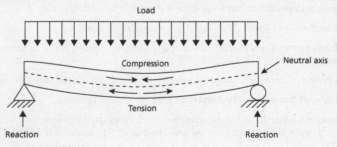

Bending stresses in a loaded beam

bending tool A hand-operated tool used to bend pipes.

bend test A test to determine the ductility of a metal without facture. The sample is bent through a specified arc, and if cracking is considered acceptable the metal is considered ductile.

beneficial occupation An argument sometimes made by contractors that if an employer takes possession of part of a project before practical completion they are prevented from recovering liquidated damages; however, this seems to have no legal merit. Contrary to this argument case, law exists that suggests liquidated damages may be recovered even if the employer has taken possession (BFI Group of Companies *vs* DCB Integration Systems Ltd 1987 CILL 348).

Benioff zone An area at the continental edge, a subduction zone where deep-seated earthquakes originate.

benthic deposits Decaying organic matter found in bottom sediments, e.g. the deposits of animals that live near water.

bentonite A natural clay (aluminium phyllosilicate), which when mixed with water expands to form a thixotropic gel (*see* THIXOTROPY). It can be used to support trenches, for drilling mud, in cements, and adhesives.

BEP *See* BIM EXECUTION PLAN.

berm A long level section such as a shelf (narrow path) or raised barrier (low embankment) separating two areas, which normally contain water.

Bernoulli's theorem An expression for a steady flow of incompressible inviscid fluid, that, in its simplest form, states when the speed of a fluid increases the pressure decreases:

$$\frac{p}{\rho} + \frac{V^2}{2} + gz = \text{constant}$$

where p = pressure, ρ = fluid density, V = velocity of flow, g = acceleration due to gravity, and z = elevation.

berth A *jetty; a docking area for ships.

bespoke Individual, purpose made. A range of individual purpose-made components, elements, or buildings would be termed a **bespoke system**.

Bessel functions (cylindrical functions) Canonical solutions y(x) of Bessel's differential equation:

$$x^2 \frac{d^2y}{dx^2} + x \frac{dy}{dx} + (x^2 - a^2)y = 0$$

Used in wave propagation and boundary value problems.

best endeavours Duty of the contractor to use resources at its disposal to prevent delay. The carrying out of this duty is a precondition to awards of an extension of time under the *JCT and IFC forms of contract.

beta (β) The second letter of the Greek alphabet, used to define many terms in science and engineering, e.g. electrons produced by radioactivity.

betz coefficient Theoretically the maximum efficiency a wind turbine can operate without slowing the wind down too much.

bevel 1. A rounded or angled edge that is not at a right angle. **2.** The point where two surfaces meet at an angle other than 90°. If the angle is at 45°, it is called a **chamfer**.

bevelled closer A brick that is cut at an angle such that it is a half brick's width at one end and a full brick's width at the other end.

bias External influences affect decisions or actions where consideration should be based on logic and merit alone. Under most standard forms of contract, the project administrator is under a duty to act fairly, reasonably, and independently when making decisions that affect matters between the client and contractor.

biaxial Having or about two axes.

bib (bibcock) A tap where the nozzle is bent down towards the ground.

bibliometrics A statistical analysis of published material, typically used to assess the impact of research outputs.

bib tap A tap where the water inlet is horizontal rather than vertical. Examples include a garden tap and traditional Belfast sink taps.

bid (US tender) Price submitted to carry out work described.

bidet A sanitary fitting that is used to wash the genital area.

bidirectional Arranged in two directions, normally perpendicular to each other.

bifold door A door where the opening leaf comprises two panels that fold in two.

bifurcation The division or forking of something into two parts, e.g. a stream splitting into two smaller streams.

big data Enormously large data sets that are typically analysed by computer software to reveal patterns and trends which are particularly associated with human behaviour and communications.

billing Writing or recording the description and quantities in bills of quantities.

bill of sale A document used to transfer property to another party without transferring possession. A bill of sale is a title of possession.

bill of variations Final account and computation of adjustments to the contract sum.

bill of quantities (BQ, BoQ) A list of all the work, labour, and materials required to build a structure or complete the works. The purpose of the bill is to allow rates to be fixed to each item of work in a transparent and comparable method. If variations to the works occur, comparisons can be made between that described and the actual work.

BIM (Building Information Modelling (Model)) A collaborative process of creating multidimensional computer models to plan, design, construct, and maintain the built *asset (e.g. building or infrastructure) throughout a project's life cycle. BIM represents the process of development and use of a computer-generated model to simulate the planning, design, construction, and operation of a building facility. The resulting model is a data-rich, object-oriented, intelligent, and parametric digital representation of that building facility, from which views and data appropriate to various users' needs can be extracted and analysed to generate information that can be used to make decisions and improve the process of delivering the project.

BIM dimensions (levels, BIM maturity level) The maturity level within the supply chain of information digitally. Essentially, the levels are defined as:

- **Level 0 (0D):** No formal collaboration between different disciplines, unmanaged 2D computer-aided design (CAD). Output distributed via paper or electronic means without common standards or processes.
- **Level 1 (1D):** No collaboration between different disciplines, 2D and 3D information used for drafting statutory approval documentations and the development of concept models. Electronic sharing of data is undertaken via a *common data environment (CDE), typically managed by the contractor.
- **Level 2 (2D):** The introduction of standardized structures and formats. However, information at this stage is not automatically shared between project partners.
- **Level 3 (3D):** The process of creating shared information models, graphically and non-graphically, in a common data environment (CDE).
- **Level 4 (4D):** The stage at which scheduling data is added to show how construction will develop sequentially.
- **Level 5 (5D):** The stage at which accurate costing can be extracted from BIM, e.g. purchasing and installing a component, together with its associated running and maintenance costs.
- **Level 6 (6D):** Facilities management: to account for the project whole-life cost.

bimetallic strip (bimetal strip) A metallic material, which is usually brass on one side and steel on the other, with the brass having the higher thermal expansion coefficient. When heat is applied, the strip curves toward the steel side. Conversely, when there is a decrease in temperature the material curves the other way. Bimetallic strips are utilized in devices for detecting and measuring temperature changes.

BIM Execution Plan (BEP) As defined by the PAS 1192-2: '*A plan prepared by the suppliers to explain how the information modelling aspects of a project will be carried out.*' See also SEVEN (7) PILLARS OF BIM.

BIM protocol A document containing information about the BIM models that are required by the project team. It also includes information about specific obligations, liabilities, and associated limitations of these models.

BIM roadmap A document to describe the activities of BSI B/555 (Construction design, modelling, and data exchange), in support of the 2011 BIM Strategy. It refers to the close past, current, and future provision by offering guidance to the UK industry that provides and operates built *assets. In addition, it maintains the vision and mission statement in respect of the reduction of whole-life cost, risk, carbon reduction, and the delivery of buildings and infrastructure.

binder A component of paint which produces the paint form on solidification. Binders are usually linseed oil or alkyd resins.

binder bar (binding) A secondary bar that holds the main bars together in a reinforced cage.

binder course Previously termed **basecourse**; one of the layers in a *flexible pavement construction which is used to provide an even surface for the *surface (wearing) course and contributes strength.

binding wire (tying wire, tie wire) Wire used for fixing (tying) reinforcement bars together to the correct shape.

biocementation A soil-strengthening technique that utilizes the molecules of excreted precipitates formed during metabolic reactions within bacteria. *See also* MICROBIOLOGICALLY INDUCED CALCITE PRECIPITATION.

biodegradation A natural process of decay (rotting), utilizing microorganisms (such as bacteria) to breakdown chemical (or biological) components.

bioenergy Energy created from a *biofuel.

bioengineering The use of engineering principles incorporating biological processes.

biofilm A layer of active microorganisms attached to a surface.

biofiltration A technique using microorganisms to remove and oxidize organic gases. Contaminated air is passed through a **biofilter** (a bed of compost containing indigenous microorganisms) to convert organic compounds to carbon dioxide and water.

biofuel A renewable fuel source originating from biological material.

biogas Methane, carbon dioxide, and other gases given off during the *anaerobic digestion of organic material.

biological Relating to living organisms.

biological catalyst A protein that speeds up the biological process/reaction. *See also* ENZYME.

biological shield A large mass of material (such as concrete) placed around a nuclear reactor to reduce the release of radiation.

biological treatment A process that utilizes bacteria to break down waste into less polluting compounds, e.g. *trickling filters in sewage treatment.

biomass 1. The number of organisms in a given space. **2.** Plant and animal waste that is used as fuel.

biomaterial A material that is formed from a biological process, e.g. extracts from plants or animals.

biophilic design The use of the natural world to inform and contribute directly to design, engineering, landscapes, communities, and construction; learning from the structure in the natural world to bring improvements in performance, the principle being that species have evolved and survived and can therefore offer benefits if they are considered. There is an emphasis on the benefits that can be gained in health and well-being, although other benefits have been realized by examining nature. Where the advanced fitness and survival properties of nature can be harnessed in design and engineering practice, benefits could be achieved. Biophilic design can be accomplished by bringing the forms and patterns of nature into a building.

biopolymers *Molecules joined together in a long chain that are produced by living organisms, e.g. a *protein.

bioreactor A device that supports a biologically active environment. It can be in the form of a tank to cultivate cells and tissues or a vessel to undertake (enhance) a biological process.

bioremediation The addition of *nutrients and/or *cultures of *microorganisms to the subsurface environment to enhance natural *biodegradation processes. Major process types include gaseous nutrient injection, hydrogen peroxide circulation, nitrate enhancement, and bio-augmentation. Typically, injection wells are used to add nutrients in the form of air/oxygen, and additional microbial cultures to increase the bioremediation process. Gaseous by-products from the breakdown of organic material may be extracted using a soil vapour extraction system. Subsurface conditions are more difficult to control than ex-situ processes. Leachate and gaseous by-products are difficult to contain and may contaminate the environment further. This process is not well suited to low permeable soils, e.g. *clays, as *permeability is required to allow nutrients and cultures to reach contaminants.

bioscrubber A *scrubber that uses microorganisms to convert organic compounds from contaminated air or gas to carbon dioxide and water.

biosolids Treated sludge from wastewater treatment plants.

biosphere The part of the Earth and its atmosphere that can support living organisms.

biostabilizer A machine that converts solid waste into compost by grinding and *aeration.

biosynthesis The synthesis of substances by a biological activity.

biotechnology The industrial application of biological processes.

bi-parting door A door where the opening leafs comprise two or more panels that move away from the centre of the door when opened.

BIPV *See* BUILDING-INTEGRATED PV.

bird screen A wire mesh installed over an opening to prevent the entry of birds.

birdsmouth A V-shaped notch at the bottom of a rafter that allows the rafter to rest on the *wall plate.

bi-steel construction A composite construction material, consisting of two steel face plates connected by an array of transverse bar connectors, which can be in-filled with concrete, installing material, or left as a void. Applications include crane beams, lift shafts, and building cores.

bit 1. A binary digit, 0 or 1, used to represent one of two outcomes, e.g. on or off. **2.** A cutting end of a rotary tool.

bittiness Paint finish where dust or other similar small-grained objects cause defects in the surface.

b

bitumen (bituminous) A highly viscous liquid used primarily for roads and pavements. Bitumen originates from crude oil.

bitumen dpc A *damp-proof course formed using a viscous mixture of hydrocarbons (obtained naturally or residue from petroleum distillation). Bitumen dpc can be applied in paint form or preformed roll. The course of impervious bitumen prevents moisture penetration (rising up) through the wall.

bitumen dpm A viscous mixture of hydrocarbon, bitumen, trowelled, painted, or poured over a large area of a floor preventing moisture penetration from the ground.

bitumen felt roof Sheet material comprising glass fibre impregnated with bitumen that is fitted, built up, and bonded to the roof to create an impervious layer. Felt is normally produced in roll form, allowing the material to be transported, stored, and laid.

Blackpool tables A series of tables that relate the amount of decompression time needed, to the time and pressure a person has spent breathing compressed air.

blade A long thin flat part of a tool or machine, e.g. a blade of a propeller or wind turbine.

blank 1. End of a service run left ready for future connection. **2.** Cover for electrical duct or conduit ready to receive a socket, if required.

blank door (blank window) 1. False or painted door or window to give the impression of the real thing. **2.** An opening previously used for a window or door that has been filled with masonry; walled up.

blank wall A wall without any door or window openings.

blasting A process of extracting or mining rock from a quarry.

bleach A corrosive liquid based on hydrogen peroxide utilized mainly for domestic cleaning applications.

bleed valve A valve that enables air or fluids to be released from a pressurized system. Commonly found on radiators in a central heating system.

blended water A mixture of hot and cold water, e.g. from a mixer tap.

blind fixing A concealed fixing system.

blinding layer A base layer of material (sand, fine aggregate, or concrete) which acts to protect and seal the *strata or material above or below the blinding; the material acts as a separating interface which is clean and level. Sand blinding is used on top of *hardcore to help provide a smooth, level surface on top of which a damp-proof course (dpc—typically a visqueen product), rigid insulation, or concrete can be laid. As well as being level, the smooth, fine material reduces the chance of the dpc being punctured or insulation deformed. Concrete blinding is used to seal the hardcore on which it is placed as well as providing a smooth, clean, and level platform on which reinforcement can be accurately set out, positioned, and tied. When used on hardcore, the concrete blinding prevents the water required for hydration and fine materials in the main concrete leaching or seeping between the gaps in the hardcore. The blinding can also reduce the chance of impurities in the ground interacting with the main structural concrete base. Concrete blinding can also be used and placed directly on exposed strata to immediately seal the ground, preventing water entering the strata and reducing swelling and heave.

blind mortise A mortise that only passes part-way through a piece of timber.

blind nailing A method of nailing where the head of the nail is not visible. Also known as **secret nailing**.

blinds Window coverings that are used to control daylight, solar gain, and privacy. Usually comprise horizontal or vertical slats that can be raised, lowered, or rotated to achieve the desired level of control.

blind wall A wall that contains no openings.

bloated clay A light porous material, formed from clay that has expanded during firing.

blob A soft mass of material.

block copolymer A combination or mixture of different polymers.

blocklayer 1. Layer of road pavings, sets, or stones. 2. A semi-skilled tradesperson who specializes in the construction of concrete blocks walls, but may not be capable of constructing a wall of facing bricks. Where a person refers to themselves as a *bricklayer they are skilled and competent at laying *bricks and blocks.

blockmaker (blockmaking machine) Machine used for the manufacture of concrete or clay blocks.

blockout A pocket gap, space, or hole left in a wall or service run ready to receive equipment, structural insertion, or service.

block plan A small-scale drawing that shows the position of buildings and proposed building, general layout of roads and services. Used for planning and as a working drawing, which shows the general outline of each building.

block saw A hand saw for cutting lightweight concrete blocks.

blower door Equipment used to undertake a *fan pressurization test. Comprises a portable variable speed fan, an adjustable doorframe and panel, a fan speed controller, and a pressure and flow gauge.

blower door technique A technique that utilizes a *blower door to measure the *air leakage or *airtightness of a building. *See also* FAN PRESSURIZATION TEST.

blower door testing *See* FAN PRESSURIZATION TEST.

blow gun (lance) A device in the form a long thin pipe that is connected to a compressed-air supply, to undertake *blowing out of formwork.

blowing out A process using a *blow gun to clear rubbish out of formwork by blowing it out by compressed air.

blowlamp (blowtorch) A portable gas burner used for heating materials, e.g. soldering copper pipes.

blueprint A reproduction of a technical drawing or design plan, originally using a contact print process, characterized by white lines on a blue background. Nowadays, used as a more general term for a technical drawing or design plan to provide guidance.

blue-skies research Provides the freedom to carry out flexible, curiosity-led research that can yield outcomes not imagined at the outset. This type of research frequently tests accepted thinking and presents novel fields of study.

BM *See* BENCHMARK.

board A flat piece of wood or other material.

board foot The quantity of timber measuring one square foot with a thickness of one inch.

boarding joists *Joists for flooring.

board metre The quantity of timber measuring one metre square with a thickness of 25 mm.

boaster A chisel used for dressing stone.

boat scaffolding *See* FLYING SCAFFOLD.

bob *See* PLUMB BOB.

body Main (central) part of something.

body-centred cube (BCC) The packing of atoms within a unit cell, where there are atoms at the corners of the cube and one in the centre. Examples of BCC metals are sodium, vanadium, chromium, iron, molybdenum, and tungsten. *See also* FACE-CENTRED CUBIC.

body force The force that is distributed throughout the body, e.g. self-weight.

body of deed The operative part of a contractual document which provides and sets out the terms of agreement between the parties to the contract.

body wave A *seismic wave that travels through the interior of the Earth.

bog An area of *wetlands.

bogie A small wheeled truck that runs on rails, used for carrying heavy loads from tunnel and mining works.

boil 1. To reach *boiling point. **2.** When soil flows into the bottom of an excavation, normally as a result of an upward seepage force. *See also* PIPING.

boiler (US furnace) An appliance used to heat water. The water is then used for space or water heating. Despite the name, the water should not be allowed to boil.

boiler failure A form of degradation associated with boilers, where the metal (usually a ferrous alloy) fails mainly due to the *creep mechanism.

boiling point The temperature at which water vaporizes. The exact temperature depends on atmospheric pressure; at sea level water boils at 100°C.

bollard A strong post that is used to restrict access. Typically used in pedestrian zones to keep cars out; usually constructed from concrete or steel.

Bologna process A series of ministerial meetings and agreements between European countries designed to create comparable standards, hence quality of higher education qualifications throughout the European Union. It was founded in 1999 by 29 European countries at the University of Bologna, hence its name. *See also* DUBLIN ACCORD, SYDNEY ACCORD, WASHINGTON ACCORD.

bolster 1. A steel chisel with an enlarged head used for cutting masonry. Used to cut and trim bricks and blocks (**brick** or **masonry bolster**), or to chase walls (**electricians' bolster**). 2. A *crown plate. 3. A short horizontal piece of timber on top of a column to support and decrease the span of a beam.

bolt A bar with a head at one end and a thread section at the other to receive a nut.

bolt box (bolt sleeve) Formwork in the form of an open-ended box that is used to provide protection for the alignment of holding down bolts in a pad foundation during a concrete pour. The gaps around the holding down bolts are then grouted up.

bolt croppers A device, similar to a pair of shears, used for cutting bolts.

bombproof construction A structure that withstands the effects of a bomb exploding.

bond 1. The way that courses of brick or blockwork are laid. *Stretcher bond, *English bond, *Flemish bond, *English garden wall, and *Flemish garden wall are all examples of brick bonds. 2. A written agreement between two parties in which one party agrees to pay money to the other party. 3. Contract under a seal where a financial house such as a bank or insurance company takes on an obligation to pay a party (normally the client) should a specific event arise. Bonds generally act as a guarantee of payment should a party fail to perform a specific contractual duty. 4. A mechanism holding atoms together in molecules and crystals, involving the sharing and transferring of electrons. Common types include covalent, ionic, metallic, and van der Waals. 5. The adhesion that holds two items together, e.g. the forces that hold concrete and reinforcement or bricks and mortar together.

bond course The course in a brickwork or blockwork wall that bonds the construction together.

bonder A *header stretching across two half bricks, a *bonding brick

bonding brick A brick laid across the width of the wall tying the stretcher courses together. In *English bond or *Flemish bond the bonding bricks are depicted by the *headers seen on the face of the wall. Special cut bricks such as *Queen closers and *King closers are used to ensure the corners of brick joints are properly tied and the corner is fully constructed.

bonding capacity (US) A document provided by a bank or finance house that gives information on the financial stability of a contractor. A rating is provided to give an indication of the amount of work that the contractor is financially capable of handling.

b

bonding energy The amount of energy required to separate two atoms that are chemically bonded to each other. It may be expressed on a per atom basis, or per mole of atoms.

boning 1. Checking alignment with the naked eye. **2.** Using T-shaped rods for *setting out.

bonnet roof The roof over a bay window.

bonnet tile (bonnet hip) A roof tile that is used at the *ridge of a *hipped roof. *See also* HIP TILES.

bonus Extra money awarded to workforce for good work, extra duties, or to help ensure retention of the workforce.

bonus clause A provision within a contract that acts as an incentive encouraging the contractor to complete the works before the contractual completion date; additional money is offered for early completion. The amount of money offered is usually provided as a set amount per day or week that the contractor completes before the end date. Being quite different from a *penalty clause, the amount of money specified does not need to bear any relationship to that which would be saved by the contractor for early completion. If the employer defaults, delays the contract, and prevents the contractor finishing earlier, the employer may have to pay the contractor the amount that they would have been entitled to had they not been delayed.

bonus system Incentive scheme supported by an explicit payment structure that establishes targets and goals, which if met, result in payment of extra money. Money could be awarded by achieving a set level of performance, producing the required quantity of work within a timeframe or before a deadline, for working a period of time without taking sick leave or unauthorized vacation. The extra money awarded is given on top of the normal salary or daily payment.

bookmatching The slicing or splitting of timber so that the two faces of plywood are cut from adjacent timber resulting in the grain on each face of the plywood matching the other face.

boom 1. A *jib of a crane. **2.** The *chord of a truss. **3.** A floating barrier on a water surface, e.g. to prevent an oil slick for spreading.

boom cutter (boom loader) A rock-tunnelling machine with a cutter device that projects out from the machine.

booster A device to increase the power or performance of something.

boot 1. A computer start-up procedure. **2.** The bottom of a bucket excavator. **3.** A step around the perimeter of a concrete floor slab to carry the outer skin of a cavity wall.

boot lintel A *lintel with a boot-shaped projection.

border A strip around the edge of something.

bore 1. Internal diameter of a pipe. **2.** To make or drill a hole. **3.** A tidal wave in a river.

bored pile (cast-in-situ pile, cast-in-place pile) Formed by excavating or boring a hole in the ground. The hole is then filled with concrete; it can be reinforced or *precast.

borehole A hole drilled in the ground, used to investigate the underlying strata conditions of a site during a site/ground investigation. Also, a borehole can be used to extract water and other liquids such as oil and methane from the ground.

bore lock (key-in-knob set, tubular mortise) A door lock that has a T-shaped tubular case that slides into a mortise hole cut with a hole saw or special drill bit.

borescope An optical instrument that is inserted into a small hole in the envelope of a building to inspect the building fabric. It comprises a rigid or flexible tube attached to an eyepiece (analogue device) or a screen (digital device). Used for inspection of voids and cavities, typically used in buildings to explore cavities in walls, floors, and roofs.

boring The process of making holes in wood or the ground by animals or tools.

borosilicate glass A type of glass based on boron oxide and silica. This type of glass is used mainly because of its coefficient of thermal expansion which is about 60% less than ordinary glass; as a result borosilicate glass is more resistant to breaking and imparts superior thermal shock properties. In construction, borosilicate glass is popular for glazing applications.

borrow Material excavated from the ground and used as fill in another place.

borrowed light A window in an internal wall that enables daylight to pass from a room that contains an external window to one that does not.

bossed connection (bossed end) The end of a plastic pipe to which an *O ring is fitted.

bottled (bottled edge) A rounded edge. Can also be an internal angle that has been rounded and looks like it has been created by smoothing a bottle over the surface.

bottled gas *Propane, *butane, or a combination of both, compressed and stored under pressure in a steel container. Used as a portable gas supply.

bottle trap A section on a waste pipe from a sink or basin, having a bottle-shaped cap, designed to be removed to enable rubbish to be cleaned out of the system.

bottom chord The bottom-most truss member.

bottom-hung window A window with a bottom-hinged *sash.

bottom rail The horizontal lower rail of a door or a *sash.

bottom-up construction Standard method of construction, starting with the *substructure (foundations, basements, and services), and finishing with the *superstructure (walls, floors, and roof). This is the opposite of *top-down construction, where excavation for basements takes place at the same time as the superstructure.

boulder A large rounded rock.

boulder clay (till) Material from a glacier deposit, containing clay, boulders, gravels, sands, and silts.

boundary (boundaries) The official line that divides areas of land. The **boundary conditions** relate to the condition at the edge of something.

Boundary Property Federation (BPF) system (BPF system) An organization that represents the interest of property owners, which has drawn up a manual describing how contract management should be organized. The concept divides the process into five stages: concept; preparation of brief; development of design; tender documentation; and construction. The duties of the parties are also specified in the manual. Care should be taken when referring to any such documentation as new legislation may well make some clauses invalid.

bound construction Construction methods that use frame construction where the bricks, stonework, tiles, or blocks are bonded to the walling units.

bound water Water that has become adsorbed in the surface of a material.

Bourdon gauge A mechanical pressure gauge.

Boussinesq equation Used to calculate the stress beneath a point load (but can be extended to accommodate different shapes of foundation) on a semi-infinite, homogeneous, isotropic, elastic medium. The base equation is:

$$\sigma_z = \frac{3P^3}{2\pi R^5} = \frac{3P}{2\pi z^2 \left(1 + \left(\frac{r}{z}\right)^2\right)^{5/2}}$$

Boussinesq equation

bow Bend or sag under the application of load.

Bowditch's method A method used for adjusting a closure error in a closed loop *traverse. Provided that the misclosure is within acceptable limits, the error is proportionally distributed to the distances around the traverse from the start.

bowl The water-containing part of a sanitary appliance.

bowlers Large rounded stones or pebbles used for paving and decoration to walls and floors.

bowl urinal (pod urinal) A wall-hung, bowl-shaped urinal that can be installed at different heights to suit the users.

bow saw A saw with a thin blade held in a U-shaped or H-shaped frame with a narrow handle at each end. Used for cutting curves.

bowser A tank, either fixed on a lorry or towed behind, used for refuelling or carrying water.

Bow's notation A graphical method of representing forces and members in a truss. Letters of the alphabet are used to label internal and external spaces between loads, reactions, and member forces. Two letters define the force that separates those two spaces.

bowstring truss A truss where the top chord arches (like a bow) and meets the bottom straight chord at either side (like the string to the bow).

bow window (bowed window) A *bay window that has a semicircular-shaped projection from the wall.

box beam (box girder) A beam in which the cross-section resembles a rectangular box made from either *steel, *prestressed concrete, or *reinforced concrete. Used in a box-girder bridge typically for *highway bridges.

box frame The frame of a traditional sash window that contains the weights.

box gutter (trough gutter) A U-shaped gutter.

BPE *See* BUILDING PERFORMANCE EVALUATION.

brace A member within a structure to stiffen it.

bracket A support to hold something.

bracketed stairs A *string, diagonal-edge stair support that is cut to produce a more open string with ornamental brackets to support each *tread and *nosing.

bracketing 1. Brackets used to support a plaster cornice or feature that is hung from the wall of ceiling. **2.** Brackets or angles fixed to columns or walls to carry beams, floors, or stairs.

branch Any cable or pipework in an electrical or drainage system that is connected to the mains. A division off the main part of something, e.g. a secondary pipe in the water supply, or a branch sewer that feeds into the main sewer. The point where the branch joins the main in a drainage system is the **branch connection**.

branched polymers (graft polymers) A type of polymer where the internal polymer chains are arranged like branches in a tree. Branched polymers usually exhibit higher stiffness and tensile strength.

branch manhole An inspection chamber over a branch connection.

branch pipe *See* DISCHARGE PIPE.

branch vent A vent on a *branch pipe that connects to a *vent pipe in a drainage system.

brass A non-ferrous copper (Cu) rich copper—zinc alloy. The metal is copper based with zinc (Zn) as the alloying element. Brass is popular in bathroom applications.

brazing A technique used whereby a molten metal is used to join other metal components. Brazing metals usually have a melting temperature greater than about 425°C.

breach of contract Failure to carry out contractual obligations; refusal to perform contractual obligations is known as repudiation. Failure to perform a contractual duty can have varying consequences depending on whether the term broken is a condition or a warranty. Typically, the party wronged can sue for damages, however, if the condition affected goes to the root of the contract and is fundamental, the contract can be treated as a repudiatory breach and the contract discharged.

breast 1. The protruding part of wall above a fireplace that contains the flue. **2.** The portion of wall under a window that extends from the floor to the sill.

breather membrane (building paper) A vapour-permeable *building paper.

BREDEM (Building Research Establishment's Domestic Energy Model) A simplified physically based steady-state model used to estimate the energy use and CO_2 emissions attributable to various end uses within dwellings. Forms the basis of the *Standard Assessment Procedure.

BRE Digests Technical journals on specific topics produced by the *Building Research Establishment.

BREEAM (Building Research Establishment's Environmental Assessment Method) An environmental assessment method undertaken by independent impartial assessors to assess and certify the sustainability of communities, infrastructure, and buildings (new, existing, and refurbished). In total, ten separate categories are assessed, namely energy, health and well-being, innovation, land use, materials, management, pollution, transport, waste, and water. The output is a performance rating and a number of stars. The performance rating ranges from acceptable (in-use schemes only) to pass, good, very good, excellent, and outstanding, and these ratings translate into a one- to six-star rating (one star equating to acceptable and six stars equating to outstanding).

breech fitting A Y-shaped fitting connecting the flow from two pipes into one.

bressummer (breastsummer) A large timber *lintel; however, has largely been superseded by reinforced concrete or steel.

BREHOMES (Building Research Establishment's Housing Model for Energy Studies) A physical steady-state model based upon *BREDEM used to estimate the existing and future energy use and CO_2 emissions attributable to the UK housing stock.

Brewster's fringes An oily-looking rainbow effect (not a defect) in double-glazed insulating units, when both panes of glass are of equal thickness. When the incident light meets the reflected light from the other surface in such a way they are out of phase and thus cancel each other producing this oily-rainbow effect.

bribery and corruption To offer money, reward, gifts, or secret commission to influence a party to give preferential treatment where contractual or legal dealings preclude such influence. Secret dealings of this nature can lead to discharge of contract, damages, and criminal prosecution.

brick A ceramic material commonly used in construction of dwellings and municipal buildings, and produced in a regular block form for building walls. Bricks are usually made from clay and calcium silicate, although other types are also used, e.g. concrete. Bricks are porous materials with up to 25% free volume with compressive strengths up to 100 MPa (N/mm^2) with common UK dimensions of $215 \times 102.5 \times 65$ mm. A common house brick is normally within a range of 20–40 MPa.

brick-and-half wall *See* ONE-AND-A-HALF-BRICK WALL.

brick arches Semicircular rows of bricks used to provide a curved masonry support over an opening, transferring the loads of the wall or structure above to the *jambs of brickwork or the supporting walls or columns. Brick and stone arches provide a traditional method of bridging openings in walls. A blind arch is an arch where the opening has been filled with brickwork, although the arch could still fulfil the purpose of spanning the opening; for blind arches the blind arch feature provides a decorative function. A bullseye is one which forms a full circular opening in the wall. The brickwork arch is completed with a segment below the window opening fully closing the circle. Elliptical arches have brick arches with two centres of changing radii. A flat arch is constructed with a keystone that has sloping sides at the same angle as the skewback. *Gauged arches are formed with tapered voussoirs and thin mortar joints; they are considered to be more stable than arches formed with standard bricks with tapered joints. Gothic arches are more pointed, formed with two arch segments meeting at the keystone. Gothic arches have higher rises and relatively narrower spans than other curved arches (a large rise-to-span ratio). A horseshoe arch has intrados with large radii going beyond the width of the supporting jambs before returning back to the supporting wall, and is also called an Arabic or a Moorish arch. A jack arch is an arch with limited or zero rise. Multicentre arches consist of several arches of circles. A triangular arch has two sloping sides formed with straight edges.

brick axe (bricklayer's hammer) A small hammer with a double axe-type blade used for cutting and shaping bricks.

brick bat 1. A brick that has been shortened along its length to complete a bond. For example, a brick that had been cut in half along its length would be a **half-brick bat**. **2.** Specially designed bricks that are built into a wall to attract bats. Also known as a **bat brick**.

brick building 1. The process of constructing load-bearing brick walls. Also known as **brick construction**. **2.** A space that has been enclosed using load-bearing brick walls.

brick elevator A device used for transporting construction materials, traditionally bricks, vertically up scaffolding.

brickie Slang term used to describe a *bricklayer.

bricklayer Skilled craftsperson capable of laying bricks and blocks to line and level. A craftsperson skilled at laying bricks will be able to work to various brick bonds, e.g. *stretcher, *English, and *Flemish bond, and would be able to do both *common and facing brickwork.

bricklayer's labourer Site operative who works alongside a gang of bricklayers supplying them with bricks and mixed mortar. Mortar needs to be mixed consistently and bricks delivered to the place where the individual bricklayers are working. The labourer's job is demanding and requires a high level of strength and fitness to ensure that the team of bricklayers are constantly supplied with the materials they need. Occasionally, the labourer works alongside the bricklayer observing practice as part of their training for bricklaying – although this is not that common since the labourer's job is demanding, leaving little time to watch over the bricklayer.

bricklayer's line (builder's line) A line of material placed between two points that is used as a guide during bricklaying or building work.

bricklayer's scaffold Tubular supporting structure used by bricklayers to access walls at height. Traditionally *putlog scaffolds were used by bricklayers. Putlogs have a flat end to a steel or aluminium tube, which is built into a wall as the wall is built, the other end of the tube is fixed to and carried by a *ledger, which is supported on a standard.

bricklaying The process of laying bricks in courses to construct a wall.

brick-on-edge A course of *headers laid on their edges.

brick skin (brick veneer) The non-load-bearing brick cladding on an external cavity wall.

brick slip A thin (e.g. 15 mm) non-load-bearing slice of brick used as a cladding material. Designed to give the appearance of traditional brickwork.

brick tie *See* WALL TIE.

brick trowel (bricklayer's trowel, laying trowel) A trowel with a flat triangular-shaped steel blade, one side of which is slightly rounded. Used in bricklaying to pick up and spread mortar.

brick truck A small trolley used for transporting bricks horizontally around a site.

brick veneer *See* BRICK SKIN.

brickwork chaser (keyway miller) *See* WALL CHASER.

bridge 1. A structure that spans a gap. It is used to carry highways, railways, and people on foot over depressions in the ground. **2.** The point where an element crosses a gap in the construction. For example, a bridge in a cavity wall can occur if mortar droppings accumulate on the wall ties and cross the gap between the inner and outer leaf of the cavity. *See also* THERMAL BRIDGE.

bridging Timber or metal members that are placed perpendicular or diagonally to the *bridging joists to brace the joists and spread the loads from above.

bridging joist A *joist that spans across an opening from one support to another. Also known as a **common joist**.

bridle (Scotland bridling) *See* TRIMMING JOIST.

bridle joint A mortise-and-tenon joint where the mortise and the tenon are the full width of the pieces of timber that are being joined. Also known as a **slot mortise-and-tenon**.

brine A water-based liquid saturated with salt (NaCl) used for preservation and transport of heat.

Brinell hardness test A mechanical test utilized to ascertain the hardness of a metal. The hardness is determined by measuring the indentation by a hard steel or carbide ball on the surface of the material.

brise-soleil Shade used above windows with vertical and horizontal slats that reduce glare and solar gain.

British Board of Agrément A body that issues quality certificates for products and installers, which state that the item or organization has achieved compliance with set standards and has been independently assessed.

British Standard (BS) These are documents defining a standardized method or procedure such that specifications are met during testing or manufacture. The documents are produced by the **British Standards Institution (BSI)**, the UK's national standards organization.

(⊕) SEE WEB LINKS
• The official website, where standards can be purchased.

British Thermal Unit (bTU, bthU) A unit of energy, defined as the amount of heat required to raise one pound of water through one degree of Fahrenheit. Replaced by the *joule in the *metric system.

brittle A material that can break or shatter easily.

brittle crack propagation Failure mechanism in brittle materials where the crack expands at an accelerated rate characterized by low energy.

brittle fracture A form of failure exhibited by brittle materials. Brittle material fracture is mainly characterized by rapid crack propagation, low energies, and little or no plastic deformation.

brittleness Measure of how brittle a material is, characterized by little or no plastic deformation.

brittle to ductile transition temperature (for metals) The temperature at which a ductile metal transforms to a brittle material. In metals, this transition occurs at 0.1 to 0.2 of the absolute melting temperature T_m, whilst in ceramics the transition occurs at about 0.5 to 0.7 T_m.

broadband The breadth of the band of frequencies that is available to transmit information as part of a telecommunications system, referring specifically to high-capacity, high-frequency, and high-speed data transmission enabling a large number of messages or volume of data to be communicated simultaneously. The term is often used in reference to communications systems and links that make it possible for large amounts of data to be transferred between devices quickly.

broad-crested weir (wide-crested weir) A weir in an open channel, with a crest length in the direction of flow to ensure that a critical depth of flow occurs somewhere along the crest.

broken bond (irregular bond) Brickwork where the bond has been interrupted; for example, by the insertion of a *brick bat.

bromooxynil A herbicide used to kill unwanted plants. Levels are monitored in drinking water supplies.

bronze A non-ferrous alloy based on copper (Cu) with tin (Sn) as the main alloying element.

brook A stream, small river, or creek.

brown clause Contractual provision that sets out conditions for liquidated damages associated with prolongation costs.

brownfield site A previously developed site, which has been abandoned or has stood idle, or an underused industrial site, where redevelopment is complicated by environmental contamination issues.

Brownian motion The random movement of particles suspended in a liquid or gas.

Brundtland Report A report published in 1987 by the World Commission on Environment and Development, entitled *Our common future*. The report is known for its definition of *sustainable development and address sustainable development issues. The report represents a seminal point in addressing climate change and was presented at a press conference in London, England, on 27 April 1987. The report examines the critical issues of environment, ecosystem, and development. The Commission, in the report, proposed action necessary to limit climate change, proposing far-reaching changes at national and international levels to mitigate the *anthropogenic part of climate change.

(((⊕))) SEE WEB LINKS
• The website contains the full version of the Brundtland report.

brush hand A painter who has not completed his or her apprenticeship or is not properly trained.

brushing out Initial application of paint distributed evenly over the surface in order to achieve a smooth finish.

brush seal A narrow strip of brush-like bristles that are fixed to the opening portion of a door or a letterbox as a weatherstrip.

BS *See* BRITISH STANDARDS.

BSI (British Standards Institution) *See* BRITISH STANDARDS.

bucket-handle joint (concave joint, bar joint) A concave *pointing finish in brickwork formed using a buckle handle, a length of hosepipe, or a trowel handle. The pointing tool concaves and compacts the mortar, improving weather resistance.

buckling A sudden mode of failure, when a structural member experiences high compressive stresses and the structure moves out of the line of the action of the load.

budget price Estimate of the cost of work, and although not precise, is considered to cover the cost of all the contractors' work

buffer 1. A shock absorber, e.g. a railway buffer. **2.** A solution that resists changes in pH when a small quantity of acid or alkali is added to the solution. **3.** A temporary data storage area on a computer, where data is held before being transmitted to an external device such as a printer.

buggy (US) *See* MOTORIZED BARROW.

buildability The ease with which a design can be built. Some ideas and designs are not considered practical to build and would rate very low on a 'buildability' scale.

builder 1. Contractor who undertakes building work *see* BUILDING CONTRACTOR. **2.** Craftsperson who engages in any building activity. Generally, the term is interchangeable referring to any company or operative that engages in the site-based activities.

builder's rubbish (building rubbish) Waste material on-site originating from the process of building.

builder's work Work that needs to be carried out by a trade or skilled operative, organized by the contractor (builder), in order to allow a subcontractor to perform their part of the contract. An example of such work would include the construction of peers for beams or fittings to rest on or the breaking out of holes in walls and floors for the service contractor to feed its services through. The term is used in contractual documentation to describe the activities and work that the contractor must organize and complete before subcontractors or other parties can complete their work.

builder's work drawing Illustration or detailed plan that clearly shows or highlights the *builder's work. The drawing is used as a coordination tool between builder and subcontractor to ensure all necessary work is considered.

builder's work in connection Work that needs to be done to allow other trades or subcontractors to do their work. The builder's work must be complete to allow other connecting or associated works to commence.

building A structure used to provide shelter to humans. The **building envelope** is the skeleton of the building, e.g. columns, beams, floors, and roof, without any internal fittings.

Building Act 1984 Legislation that governs and provides the vehicle for the implementation of the Building Regulations.

(⊕) SEE WEB LINKS
• Communities and local government portal that provides the Building Act and Building Regulations.

Building Centre An organization set up to provide information to the construction industry.

building code Set of guides or legislation that act as a minimum standard for construction or safety – in the UK these would be the *Building Regulations enforced by *building control.

building contractor Organization that undertakes the management of the site-based works and building construction. Traditionally, the organization would enter into a contract to construct the building only; it is now quite common for the *main contractor to enter into a *contract for both the design and construction of the building. Although the main contractor can enter into a contract to design the project, often much of the works are subcontracted out to a design specialist, e.g. architects, structural engineers, and mechanical and electrical designers. Nowadays, main contractors have a limited pool of skilled and unskilled labour, and tend to subcontract a large portion of the actual building works. It is common for the main contractor to package ground works, steel frame fabrication and erection, wall cladding, roof covering, masonry, floors, and mechanical and electrical works into subcontracts.

building control Enforced legislative procedures and processes that govern the minimum standards of building. Applications are made and designs submitted to the building control office to demonstrate and check compliance with legislation. Works are checked by the *building control officer at significant stages in the building process to ensure that the building is being constructed and is consistent with the approved drawings and plans.

building control officer Professionals employed by the local authority to inspect and check that buildings are compliant with the relevant legislation and the drawings that were submitted and approved by *building control. The officer ensures compliance with fire, safety, structural, thermal, acoustic, and environmental legislation. The term originates from the London Local Authority where principal building control officers were employed to inspect all buildings constructed in the area.

building craft A broad description of any one of or all of the skilled work or trade professionals associated with a construction project or industry.

building industry All of the *contractors, *subcontractors, fabricators, materials suppliers, *plant and equipment companies, *labourers, and professionals associated with *construction and *civil engineering works. The business and work directly associated with construction represents around 8–9% of GDP (Gross Domestic Product) in the UK.

building information modelling *See* BIM.

building inspector Professionals who inspect the construction to ensure that works are properly constructed; can be an employee of the *local authority, bank, building society, or insurance company who checks that the building is compliant with legislation and regulations. Inspectors for finance organizations may inspect to assess the value of the property. Local authority inspectors may be called *building control officers although the origins of both terms are slightly different.

building-integrated PV (BIPV) A type of *photovoltaic system where the photovoltaic modules are substituted in place of various elements of the building fabric. Commonly installed on the façades and roofs of buildings.

building management system (bms) (building automation system, intelligent building) Automated central control unit that checks, monitors, adjusts the building's performance. Used to coordinate and operate the lighting, air conditioning, heating, lifts, security, and other building appliances and facilities.

building notice Notification to the local authority that building work of a particular nature is about to commence, e.g. excavation, or drainage, or foundations. This allows the *building control department to inspect and check the works for compliance with approved drawings and the *building regulations.

building operative Site-based person who undertakes the skilled or unskilled work involved in the actual construction of the building; does not include construction professionals.

building owner A *client who takes ownership of the building once it is completed.

building paper 1. Reinforced fabric, which is laid under concrete and used to prevent chemical attack from the ground and assist curing. 2. A breathable fabric membrane, which is fixed over timber walling panels and below roof tiles to allow water vapour to escape from the structure but prevent rain entering from the external environment. The paper helps to prevent condensation. 3. A waterproof sheathing paper, the degree of waterproofing varies depending on purpose.

building performance evaluation (BPE) The collective name given to various techniques and methods that can be used to measure, monitor, assess, and evaluate the performance of a building both during and after construction.

building performance reports Energy performance evaluations and comparisons providing an indication of the relative energy efficiency score or metric of the performance of buildings. Popular rating systems include *BREEAM (UK): Building Research Establishment Environmental Assessment Method; iiSBE GBTool (Canada): International Initiative for a Sustainable Built Environment; LEED (US GBC): Leadership in Energy & Environmental Design. *See also* *ENERGY PERFORMANCE CERTIFICATES.

(⊕) SEE WEB LINKS
- BREEAM official website
- iiSBE official website
- LEED official website

building permit Authorization given by the local authority to allow works to commence.

building regulations The approved documents and guides implemented under the *Building Act 1984 that enforce minimum standards of construction, safety, and performance of buildings. The legislation is enforced by *building control.

Building Research Establishment (BRE) An independent consultancy that specializes in undertaking work in the built environment.

(⊕) SEE WEB LINKS
- The official website of the Building Research Establishment.

Building Research Establishment's Domestic Energy Model *See* BREDEM.

Building Research Establishment's Environmental Assessment Method *See* BREEAM.

Building Research Establishment's Housing Model for Energy Studies *See* BREHOMES.

building society Financial house, with very similar business activities to high-street banks.

building surveyor Construction professional who provides advice and reports on the condition of houses or commercial buildings; the surveyor may also comment on the value of the house. The surveying professional will tend to specialize but can give advice on a range of issues including building condition, defects, restoration, renovation, fire protection, structural integrity, thermal performance, etc. It is normal for a survey to be requested by clients and financial institutions prior to the exchange of contracts for a building's sale. Surveys are particularly important in old or previously owned properties that are not covered by a warranty.

building system The method of producing buildings. Buildings that are produced using modern methods of construction (MMC), with significant portions of the buildings being *prefabricated, may be classed as industrialized buildings.

building team Group of professionals associated with the design and manufacture of the whole process. As a minimum, the building team normally includes the client's representative, *architect, *contractor, *specialist contractors, and designers.

building trades All skilled operatives and practices associated with building, including *joiners, *bricklayers, ground workers, *steel fixers, etc.

building-up of prices The gathering together of projected costs to produce an estimate for the total project.

building work Work carried out during building.

Buildmark A 10-year warranty and insurance cover provided by the NHBC for newly built houses that have been inspected by their approved officers.

 SEE WEB LINKS
• The National House-Building Council.

built environment Man-made structures and facilities used to accommodate societies' activities. Any enclosures, spaces, structures, and infrastructure formed to convert the natural environment into a habitable and useable area for the purpose of living, working, and playing.

built-in (building-in) The process of inserting components into the construction as it progresses.

built-up To make something that consists of different components.

built-up roofing A flat-roof covering comprising multiple layers of sheet materials.

bulb A rounded spherical shape, such as a light bulb.

bulk A large mass of something, e.g. a **bulk excavation** consists of removing a large mass of soil from the ground.

bulk density The weight of a material per unit volume, which includes water, air voids, and solids.

bulkhead A dividing partition or barrier, e.g. a wall within a ship or aircraft; a seawall; a sheet pile.

bulking An increase in volume of a material (e.g. soil) when it is excavated from the ground.

bulk modulus (K) An elastic constant, defining the relationship between an increase in pressure with an associated decrease in volume.

bulldog grip A rope clamp consisting of a U-shaped bar threaded at both ends.

bulldog plate connector *See* TOOTHPLATE CONNECTOR.

bulldozer (dozer) A tracked earth-moving piece of equipment with an adjustable *mouldboard attached to the front.

bullet catch *See* BALL CATCH.

bullet-head nail (US finish nail) A nail with a small rounded head.

bullhead connector A T-shaped pipe fitting, where the branch is larger than the other two run openings.

bullying (harassment) Offensive, intimidating, malicious, or insulting behaviour, an abuse or misuse of power through means intended to undermine, humiliate, denigrate, or injure the recipient; unwanted conduct affecting someone's dignity in the workplace, and/or behaviour that makes someone feel distressed, intimated, or offended. It may be persistent, or an isolated incident (ACAS, 2018).

( ) SEE WEB LINKS
• ACAS advice leaflet on bullying and harassment at work

bunched wires (bundled wires) Wires that have been manufactured for *trunking.

bund *See* EMBANKMENT.

bungalow A single-storey structure in which people live.

bunker 1. A container used for storing materials, such as cement, sand, and aggregate. **2.** An underground shelter.

buoyancy The ability of an object to float. *See also* ARCHIMEDES' PRINCIPLE.

bur A drill bit used for widening drilled holes rather than forming them.

burden of a contract A party's obligation to perform, execute, and complete the contract. A party cannot assign (give its contractual obligation to another party) without the consent of the other contracted party.

Burgers vector (*b*) A value or vector indicating the direction and magnitude of the disruption to the crystal structure caused by the presence of dislocations inside a metal.

burglar alarm A mechanical device comprising sensors and a siren designed to warn off intruders who enter a protected area.

burglar-resistant lock *See* SECURITY LOCK.

buried services (underground services) Gas, water, sewers, electric cables, etc., which are located (buried) *below ground level.

burner A place where fuel is ignited and burns, e.g. a gas ring on a hob.

burning off To remove a surface covering with heat, e.g. an old coat of paint can be burnt off with a blowlamp.

burnt clay A *ceramic material.

burnt lime *See* QUICKLIME.

burring reamer A tool to remove jagged edges (caused by cutting) in pipes.

burst strength 1. The hydraulic pressure needed to burst a container of given thickness. **2.** The pressure needed to break a fabric, by either inflating a diaphragm or forcing a smooth, spherical object against a clamped piece of fabric.

busbar A conductive bar, usually copper or aluminium, that enables a connection to be made between two or more electrical circuits. Used in consumer units, distribution boards, switchboards, and substations. Busbars are connected together using a **bus coupler** and are usually enclosed within a **bus duct** or **busway**.

bush hammer A hand-held device that is either connected to compressed air or an electrical supply. Used to remove the surface layer of concrete to give an exposed aggregate finish.

bushing 1. A pipe fitting that contains both an internal and an external threaded connection. Used to join different diameters of pipe together. **2.** An insulated sleeve used at openings to pass conductors through.

business process re-engineering The redesign of work behaviour, processes, systems, and structure so that greater efficiency and value can be achieved with the resources available.

butane Flammable gaseous alkane (C_4H_{10}) usually obtained from petroleum or natural gas and used as fuel.

butt The process of placing components together without overlapping. The joint that is formed is known as a **butt joint**.

butter coat A soft coat of render used to achieve a **pebble dash** finish. The render is applied to the wall over the *scratch coat and decorative aggregate is thrown and then pressed into the render.

butterfly gate Similar to a *sluice gate, however, the gate opens and closes on a central shaft located in the pipe, to save on the need for headroom to open the gate. A **butterfly valve** turns on the central shaft within the pipe to regulate flow.

butterfly wall tie A galvanized steel wire *wall tie in the shape of a figure eight.

buttering (buttering up) The process of spreading mortar on a material as one would spread butter on bread.

butt gauge (butt gage) A tool used for marking the position of hinges on doors and door jambs.

butt hinge A hinge comprising two leafs that are mortised into the door and the door jamb, and fold together when the door is closed. It is the most common type of hinge used for doors.

button *See* PUSH BUTTON.

button-headed screw (half-round screw) A screw with a semicircular head.

buttress A stone or brick structure built against a wall to support it, i.e. to resist active thrust.

butyl rubber A polymeric material also known as polyisobutylene and PIB $(C_4H_8)_n$. Butyl rubber is a synthetic polymer used as a sealant in roofs, particularly for roof repair work.

buzzer An electrical device that emits a buzzing sound when activated.

buzz saw A circular saw.

byatt A timber beam that carries a walkway in excavation.

bye channel A *spillway.

bypass To go around something, e.g. a bypass road with a route around a town rather than through it; a **bypass channel** diverts water.

by-product A secondary product that is formed as a result of making something else.

cabin Temporary site accommodation for offices, storage, and welfare facilities. The standard of accommodation can vary from something that is secure and provides cover from the rain, to offices that are fully fitted out with furniture, heated, air-conditioned, and insulated. Due to variation in levels that can be experienced on sites, most units are fitted with adjustable legs. Smaller units may be fitted with wheels enabling transportation without the need of a crane.

cable 1. A conductor of electricity usually made from copper and sheathed with insulation. **2.** Steel rope used in winches and pulleys to lift and carry equipment and materials.

cable clip A plastic or metal fixing bracket that holds and secures electrical cables. The fixing is held in the wall or ceiling by a nail or screw.

cable duct (duct pipe, pipe duct, ducting) A tube or conduit which is mounted on surfaces or sunk into the ground, floors, ceilings, or walls and is used to carry service cables. The conduit is usually made from plastic or steel providing a robust containment to carry the services. The duct is laid as construction takes place and a draw cord is placed in the pipes as they are positioned. When needed, electrical cables can be pulled into the ducting with the draw cord. The duct serves a similar purpose to a *conduit.

cable gland A bulb of material used to seal the end of a cable or seal where the cable enters a casing, and prevents water, dust, and dirt penetrating into the cable or duct.

cable percussion (shell and auger) A drilling rig which comprises an A-frame with a top pully wheel and motorized winch unit. The winch is set to enable the cable to be pulled in by means of a motor but then to fall free when the clutch is disengaged.

cable sleeve A protective coating around the wire. The coating may be armoured to provide mechanical protection, or have a polymer-based coating to protect from the environmental conditions.

cable tail 1. Additional cable left at the end of a run to ensure that there is sufficient material to make the connection to the fitting or equipment. **2.** A length of electrical cable attached to a device to enable it to be wired in.

cable tray A long open container within which multiple services can be neatly positioned. Cable trays are often used below raised floors and above suspended ceilings to keep electrical and telecommunication cables together. The trays help to distribute the wires throughout the building in an organized manner.

caisson A watertight structure (chamber) that is sunk through the ground or water to enable dry excavation and placing of the foundation. An **open-caisson** is open at the top and bottom. A **box-caisson** is only open at the top. A **compressed-air** or **pneumatic caisson** contains a chamber where air pressure is maintained above atmospheric pressure to prevent entry of water or mud.

calcination Heating a substance below its fusing and melting point, which leads to a loss of water or oxygen, resulting in a simpler form of the substance. Calcination is used in the extraction of metals from mineral ores.

calcite A carbonate mineral and a form of calcium carbonate with chemical formula $CaCO_3$.

calcium A silvery, moderately hard metallic element, symbol Ca. Calcium is useful as an alloying element; the addition of calcium to steels result in improved mechanical properties. Calcium compounds are used in the production of Portland cement, plaster, and quicklime.

calcium silicate bricks Building blocks (bricks) generally made from sand and lime, also known as sand lime bricks. The bricks are manufactured using a mixture of lime, sand, and other crushed siliceous material such as crushed flint or graded fly ash, with water which is then moulded under high pressure and autoclaved to form a structural and aesthetically pleasing brick. The bricks are often used for decorative purposes, as they can be coloured and have a smooth, clean finish.

calculus The mathematical study of continuous change by calculating the derivatives and integrals of functions, namely by differential or integral calculus.

Caledonides A mountain belt formed during the **Caledonian orogeny**; a mountain-building era roughly 390–490 million years ago as a result of continental drift. The belt lies in the northern parts of the British Isles, western Scandinavia, Svalbard, eastern Greenland, and parts of north-central Europe.

calibre 1. The ability or quality of someone. **2.** The internal diameter of a pipe or cylinder, particularly with reference to the barrel of a firearm.

California bearing ratio test (CBR test) An empirical test, devised by the California Division of Highways in 1929, and used in pavement design to determine the strength of the lower layers in roads (*subgrade and *binder course) and airfields. The ratio of force and penetration is monitored when a cylindrical plunger (49.6 mm diameter) is pushed into soil at a constant rate (1 mm per minute). At given values of penetration the ratio of applied force to standard force is expressed as a percentage. The higher of these two percentages is taken as the CBR value.

calliper An instrument consisting of two hinged legs, which either turn out or in, to measure the internal or external dimensions of an object, e.g. the bore or diameter of a pipe.

calorific value (CV) The amount of energy produced by a fuel that has undergone complete combustion, usually measured in J/kg for a solids or MJ/m^3 for gas.

calorifier A vessel used for heating low-pressure hot water. The water within the vessel is heated indirectly by a coil immersed within the water. A domestic *indirect hot-water cylinder is an example of a small storage calorifier.

calyx boring (short drilling) A *boring method that uses steel shot as the cutting means.

camber A slight upward curvature of a surface, such as a cross-section of a highway, or in structural members (e.g. **camber beams**) to resist dead and imposed loads, such that the beam sits flat and prevents *sag in service.

canal An artificial waterway constructed for navigation, water power, or irrigation. Nowadays, canals are often used for recreational purposes.

candela (cd) The international standard (SI) unit of luminous intensity.

canopy 1. A roof over an open structure (one that is not enclosed by walls). **2.** A hood above a cooker or fire used to direct steam or smoke to a chimney or flue. Sometimes hoods can be decorative and have no function other than aesthetics.

cant (canted) 1. Any flat surface that has a slight incline or tilt that is not at right angles to the main plane. **2.** External surface of brickwork that has a change of direction that is not a right angle. Bay windows often have brickwork that angles out to support the window frame. *See also* CANT BAY.

cant bay The angled profile of brickwork used to form a bay window. *See also* CANT.

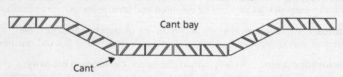

Cant bay

Cant

Cant bay: plan of cant bay brickwork

cant brick A non-standard (special) brick with a slope or angle cut across the head of the brick. Such bricks are sometimes used to change the direction of the brickwork.

cantilever A projection, such as a beam, which is only supported or fixed at one end, the other end is free.

cantledge *See* KENTLEDGE.

cant stop brick A specially trimmed brick used to neatly finish off the end of a wall where a *cant brick has been used. The brick is cut so that the end profile matches that of the cant brick to which it abuts.

cap 1. The uppermost point, head or top component of a member. **2.** Fixing that is pushed or screwed onto a container or vessel to close or seal the container. **3.** A cover placed over the top of an object.

capillarity (capillary action) The action of fluid (usually water in buildings) being drawn up or down pores or closely positioned surfaces due to the surface tension of a fluid. A **capillary break** is a gap between two surfaces that is large enough to prevent capillary action. **Capillary groove** is a rebate cut into a material to produce a surface that prevents capillary action. A **capillary joint** is a pipe joint formed by placing steel or copper pipe into a slightly larger diameter fitting that encourages solder or fixing solvent to be drawn between the tubes by capillary action. The contact surfaces of the pipe and fitting are cleaned with a flux ready to be soldered. The gap between the pipe and fitting is sufficiently close for capillary action to take place when solder is applied. When heated and molten solder is applied, the surface tension of the solder draws the fluid into the space, evenly distributing the solder and filling the joint. Once the solder solidifies it provides a watertight joint. Other adhesives are also available that work on capillary action. Plastic pipes can be joined using solvent-based adhesives, which also flow under capillary action.

capital In classical architecture, the uppermost member of a column.

capping layer Low-cost granular material used to provide a working platform when the CBR (*see* CALIFORNIA BEARING RATIO TEST) of the *subgrade is weak (<5%). Capping is usually required to achieve a CBR of 15%.

capstan A mechanical vertical rotating device, used to move heavy objects by employing ropes, cables, or chains, for example, a rotating drum located on a ship or shipyard to haul in ropes.

carbonation A chemical process occurring when carbon dioxide is dissolved in water or an aqueous solution. This can be harmful for many materials used in construction as carbonation usually causes a decrease in strength.

carbon capture and storage (CCS) A process of capturing *carbon dioxide (CO_2) from industrial processes and storing it safely a few kilometres below the Earth's crust between carefully selected rock formations. Up to 90% of CO_2 emissions can be captured from this process. Carbon sequestration and carbon capture and storage processes assist emitters of CO_2, such as fossil fuel-burning power stations, by capturing and storing the CO_2 they create, preventing it being released into the atmosphere. Instead, it is stored by being injected into underground or undersea geological formations.

carbon credit A tradable certificate that permits *carbon dioxide (CO_2) to be released into the atmosphere to achieve a *carbon-neutral status by investing in environmental products which will remove CO_2 from the atmosphere. (*See also* CARBON SINK.) One carbon credit permits one tonne of CO_2 emissions.

carbon dioxide A substance in gaseous form at ambient temperatures, formula CO_2. It is produced during the combustion process of fossil fuels.

carbon footprint The amount of *carbon dioxide (CO_2) released into the atmosphere from human activities (such as from fossil fuels) expressed in equivalent tonnes of CO_2. A measure of the total amount of carbon dioxide (CO_2) and methane (CH_4) emissions, often linked to a defined population, system, or activity, considering all relevant sources, sinks, and storage within the spatial and temporal boundary of the population, system, or activity of interest. Calculated as carbon dioxide equivalent using the relevant 100-year global warming potential (GWP100).

carbon monoxide A highly toxic, colourless, and odourless gas, chemical formula CO.

carbon-neutral (carbon neutrality) Having a net zero release of *carbon dioxide (CO_2) into the atmosphere, by carbon offsetting.

carbon offsetting (carbon offset) A process for counteracting carbon dioxide (CO_2) being released into the atmosphere by buying *carbon credits.

carbon sink The ability to absorb more *carbon dioxide (CO_2) than is released. For example, oceans, forests, and soils all have the ability to remove and store CO_2 from the atmosphere for an indefinite period of time.

carbon steel A type of steel whose mechanical properties are dependent on the carbon content. The majority of steels contain between 0.1–1.0 weight % carbon and the steels are termed low (0.1–0.25 weight % carbon), medium (0.25–0.55), and high carbon (greater than 0.55) steels in accordance to the carbon content. Carbon has a very marked effect on the properties of steel; an addition of only a fraction of 1.0%, has a very great effect. Increasing the carbon content increases the *tensile strength, *yield strength and *hardness of the steel; however, increasing carbon content also results in decreased ductility. As ductility is an essential property for structural steel (for protection against earthquakes, explosions, and strong winds), most constructional steels contain between 0.15% and 0.4% of carbon.

carborundum A trademark used for an abrasive material of silicon carbide crystals.

carcass The load-bearing structure, skeleton, or framework of a framed building. Used to describe the external load-bearing elements of timber frame construction.

carcassing Rough framed timbers and structural work usually used to describe wall studs, panels, roof rafters, and floor joists.

carpenter Skilled craftsperson who works with wood, usually assembling the larger rough structural wooden components of a building, such as walls, floors, and roofs. The craftsperson may also be trained as a *joiner.

carpenter's hammer (claw hammer) A hammer with a double head, one end for hitting and embedding nails, the other with two splayed tongues for extracting nails.

carpenter's helper A labourer who helps the carpenter, may be an apprentice carpenter.

carpentry The craft of cutting and shaping timber to make and assemble structural frames.

carriage (carriage piece, rough string) A beam that runs centrally underneath wide *staircases to carry the *treads and loads placed on them. Where staircases are narrow the treads are carried by the *strings; however, where the staircase is wide, excessive deflection would occur. The timber, which runs diagonally from the lower floor to the upper floor, may be toothed or stepped out to carry the treads or may have *brackets fixed on either side of a straight string which extend to the underside of the *tread.

carriage bolt A fixing bolt with a rounded head and square section shank that is tightly fitted into the timber. The square shank embeds into the timber and prevents the bolt rotating when the nut is tightened.

carriageway The part of the road that carries vehicles. *See also* DUAL CARRIAGEWAY.

carrier wave (carrier) An electromagnetic waveform that is modulated in terms of frequency, amplitude, or phase, to carry a radio or television transmission signal.

carrying capacity The maximum current that an electrical cable can safely handle without overheating.

carrying out The undertaking of specified work.

carry up The laying of brickwork up to a given height.

Cartesian coordinates A reference system used to define a point on a plane, in relation to an origin and two perpendicular axes (the x-axis and y-axis). An additional axis (the z-axis) can be introduced to define a point in space.

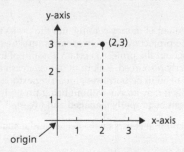

Cartesian coordinates

cartographer A person who makes maps and charts.

cartridge filter Easily locatable and disposable filter with its own frame and carrying unit.

cartridge fuse Metal *fuse in a non-combustible tube or container that is easily fitted and removed from an electrical circuit. The fuse is made from a metal strip or wire that melts or breaks when excessive current passes through it.

cartridge tool (fixing gun, nail gun, stud gun) A handheld pistol-shaped tool for firing nails into wood and steel.

Casagrande, Arthur (1902–81) A pioneer in fundamental developments in soil mechanics, in particular those relating to seepage and liquefaction. He was born in Austria, but relocated to the US in 1926, where he worked for the US Bureau of Public Roads as a research assistant, under Karl Terzaghi at MIT, improving soil-testing apparatus and techniques. He is specifically known for the development of the liquid limit apparatus, also known as the Casagrande cup. In 1932, he moved to Harvard University and became chair (professor) of soil mechanics and foundation engineering in 1946. In 1961 he was given the honour by the Institution of Civil Engineers (UK) to present the Rankine Lecture entitled 'Control of seepage through foundations and abutments of dams', for his outstanding contribution to his profession.

case hardening A process used to harden the outer surface of steel by exposing metal to a specific atmosphere. The hardening process improves the material's resistance to wear and fatigue.

casement A vertically hinged window *sash that can be opened.

casement door (French door, French window) A hinged pair of glazed double doors.

casement fastener *See* SASH FASTENER.

casement stay A bar containing a series of holes that is used to hold a *casement window open in various positions.

casement window A window containing one or more hinged *sashes that can be opened.

cash flow The movement of money coming into the project from clients and the money going out of the project by payment to subcontractors and suppliers. To ensure liquidity throughout the project a certain amount of funds are required to keep the project properly resourced. Failure to manage cash flow will mean additional money will need to be borrowed, incurring extra interest. If income is generated unexpectedly it may accrue without being properly invested. Good cash flow ensures income can be properly managed and invested.

cash flow chart Graph that shows the projected and actual expenditure and income over time.

cash flow monitoring Checking of income and expenditure against that projected. As part of earned value analysis, the amount of money earned, including profit, may also be monitored against each activity.

casing 1. The timber or plastic enclosure used to carry a unit or provide an aesthetic finish. For example, the enclosed unit of a *staircase or *casement windows. **2.** The framing or cover that hides the rough work such as structural beams, service pipes, and cables.

cassiterite A light yellow, reddish-brown, or black mineral, with chemical formula SnO_2, that is an important tin ore. Casserite is sometimes referred to as tinstone.

castellated beam Steel beam, where the web is made from two sections of steel cut in a castellated pattern, a shape resembling the up-and-down profile of a castle turret wall. The two sections are joined with the uppermost parts welded together, leaving holes in the centre of the beam. As the neutral axis of the beam runs down

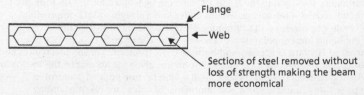

Castellated beam

the centre of the beam, the holes do not result in large strength losses. Where the greatest stresses occur the steel remains.

casting A process used to shape metals that are difficult to form by other means. Can be used to cast complicated structural components for utilization in bridges and other structures.

casting lintels Structural beams for use in walls above doorways, *windows, and other openings. Generally the lintels are precast, formed offsite, and delivered as a whole unit ready to be placed within a wall. *Lintels can be formed with a mould of formwork in situ (on site and in the place where they are to be used). They are constructed from *concrete and *reinforcement bars or cages to ensure that they have the tensile properties to transfer the loads placed along the lintel to the supports (jambs of the wall).

cast iron A ferrous alloy with the carbon content typically between 3.0 and 4.5 weight % C. The chemical composition of cast iron is the same as steel, however, cast iron contains up to ten times more carbon. Carbon gives it excellent casting properties, making it possible to produce many complex items and shapes.

cast iron pipes Pipes made from an alloy of iron which has been cast in a mould to form a pipe. Traditionally these pipes have been used to carry gas, water, oil, and sewage. The pipes can be push-fit or screw-fix often using a gasket to ensure an effective seal between pipes. The cast iron material typically used in these pipes falls into a group of alloys with a carbon content greater than 2%, with varying amounts of silicon and manganese. Generally, the metal is hard and relatively brittle.

catalyst Something or someone that aids change without being altered itself. For example, a chemical that accelerates a chemical reaction without itself undergoing any change.

catch A fastener used on doors and windows to keep them closed.

catch basin *See* CATCH PIT.

catch bolt *See* RETURN LATCH.

catch drain A channel to reroute streams or rivers to avoid flooding.

catch feeder A *ditch used for *irrigation.

catchment area An area of land that drains rainfall into a river or provides water to a *reservoir. The water from this area is referred to as **catchment water**.

catch pit (catch basin, gravel trap, sump) An area at the entrance of a sewer where grit and other obstructive material that would hinder the flow in the sewer, is collected. Periodically, the grit in the catch basin is pumped or dug out.

catenary The U-shaped curve formed by a heavy inextensible cable, chain, or rope of uniform density suspended from its endpoints. The curve is a hyperbolic cosine.

Caterpillar A brand name for a producer of large earth-moving equipment. When used on-site, the more generic term refers to a large vehicle that is propelled using *crawler tracks. The tracked vehicle is often used on construction sites to move earth.

caterpillar gate A large steel gate used for regulating the flow over a *spillway.

cathode 1. The terminal from which a current leaves an electronic cell, voltaic cell, or battery. **2.** The positive terminal of an electronic device, cell, battery, or voltaic cell; opposite to anode.

cathode protection The protection of a metal structure from corrosion by making the metal act as an electrochemical cell. The metal being protected is placed in connection with another metal that is more easily corroded. The more easily corroded metal acts as the anode in the electrochemical cell. The anodic metal corrodes in preference to the other metal; eventually, the more easily corroded metal must be replaced.

cation A positively charged *ion, formed when an atom loses one or more negatively charged electrons in a reaction as in *anions. In *electrolysis, cations move towards the negative electrode. **Cation exchange** is used to soften water. Hard water is passed over *zeolite and cations of like charge are exchanged.

cat ladder Steps used to climb on and over roofs. The ladder has cleats or angled legs fixed to it so that they can be hooked over the roof's eaves safely, holding the steps on the slope of the roof.

catwalk A narrow walkway around the top of tall buildings, which provides access to the roof and eaves so that maintenance work can be undertaken.

caulk To make watertight or airtight by filling or sealing, e.g. caulk a pipe joint; caulking the cracks.

cause and effect analysis Methods used to identify the possible causes of problems. The diagrams, maps, and charts encourage thought about all of the possible causes of a problem rather than focusing on just one aspect of the problem. Diagrams commonly used to pull together causes include Ishikawa and fishbone diagrams.

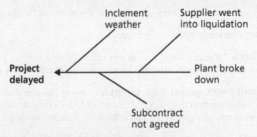

Cause and effect analysis: Ishikawa / fishbone diagram

causeway Raised road, track, or path through a marshland or low-lying area that is often waterlogged.

cavity The space between two leaves of a wall.

cavity barrier A *fire barrier that is inserted horizontally and/or vertically in cavity walls. Usually consists of polythene sleeved mineral wool.

cavity batten A timber batten that is temporarily inserted into a cavity wall to collect mortar droppings during wall construction.

cavity closer A material used at openings in cavity walls to close the cavity.

cavity fill A thermally insulating material used to fully fill the cavity within a cavity wall to reduce heat loss. Usually consists of blown fibre or glass fibre.

cavity flashing *See* CAVITY TRAY.

cavity inspection The process of inspecting a cavity wall for defects, such as *thermal bridging or dampness, either during or after construction.

cavity insulation Thermal insulation material that is placed within the cavity of a cavity wall. The insulation can be inserted to fully or partially fill the cavity.

cavity tie Plastic or galvanized strip used to add stability to the two skins of masonry either side of the *cavity wall. The tie provides a link that restrains the external leaf of masonry and holds it into the load-bearing internal partition. The ties are preformed of varying lengths to suite the width of the cavity. *See also* WALL TIE.

cavity tray A tray-shaped damp-proof course built into a cavity wall above openings to prevent damp penetration. Can be preformed or made on-site using a flat roll material.

cavity wall A wall constructed from two separate leaves separated by a *cavity and tied together using *wall ties. *Thermal insulation can be placed within the cavity to reduce heat loss.

cavity wall insulation Insulation that is placed within the void between the two structural skins of masonry or wall construction. Commonly, the supporting materials in traditional cavity wall construction are masonry (brickwork, stonework, and blockwork). The masonry could be two skins of brickwork, two skins of blockwork, or, more commonly, an outer-facing brickwork leaf and internal blockwork leaf. Within the void, insulation can be fixed to the internal structural leaf using insulation batts or rolls of insulation positioned over the wall ties and held in place using clips. Insulation of this type can be partial-fill (only filling part of the void) or full-fill (filling the total void). The insulation layer can be made of expanded polystyrene EXP, PIR, PUR, glasswool, mineral wool, sheep wool, and other insulating materials. Insulation can also be inserted into walls once the wall is constructed. Insulation is injected or blown into the wall through holes drilled into the bed or perp joints. The thermal conductivity of generic insulations are shown in Table 1. Due to the insulation being used within voids or between parts of the structure, the fire-resisting properties are also important. Table 1 also gives comparison between materials that have undergone comparable fire tests.

Table 1. Generic table describing insulation type, properties, and reaction to fire (Stec and Hull, 2011; Lambda values TIMSA, 2018) TIMSA

Insulation	Density range Kg m^{-3}	Thermal Conductivity range Wm^{-1}K^{-1}(lambda values based on commonly used products)	Reaction to fire Euroclass range
Glass Mineral Wool (GW)	10–100	0.031–0.044	A1–A2
Stone / Rock Mineral Wool	22–180	0.034–0.042	A1–A2
Extruded Polystyrene (XPS)	20–80	0.029–0.039	E–F
Expanded Polystyrene (EPS)	30–40	0.030–0.038	E–F
Phenolic (PhF) (PF)	30–40	0.021–0.024	B–C
Polyurethane (PUR)	30–80	0.022–0.028	D–E
Polyisocyanurate (PIR)	30–80	0.022–0.028	C–D
Cellular glass		0.038–0.050	
Aluminium Foil-Faced Polyethylene			
Expanded Cork			
Expanded Perlite			
Wood Wool			
Wood Fibre (WF)			

(⊕) SEE WEB LINKS

- Different λ-values, based on manufacturer's claimed values can be found on the Greenspec website
- Typical Lambda table

ceiling The uppermost horizontal lining in a room.

ceiling binder A tie that spans between the ceiling joists or trussed rafters.

ceiling fan A fan installed on the ceiling to circulate air within a room.

ceiling finishes Aesthetic features, textures, or coloured materials that are used to enhance the underside surface of the ceiling. Typically, the fixtures are made from plaster, paint, renders, tiles, wood, metal, and dry-lining systems. The materials used may also have functional properties, such as enhancing acoustics, improving thermal performance, and fire resistance.

ceiling hanger A hanger attached to the underside of the floor above to support a suspended ceiling.

ceiling height (headroom) The distance between the finished floor level and the ceiling.

ceiling insulation The insulation placed between and above the ceiling joists. *See also* LOFT INSULATION.

ceiling joist A joist that supports the weight of the ceiling below.

ceiling outlet A surface-mounted electrical outlet in a ceiling used to wire in a light.

ceiling strap *See* CEILING HANGER.

ceiling switch A switch operated by a pull-cord that hangs down from the ceiling.

cell ceiling A lightweight open-cell square grid suspended ceiling.

cellular Materials or products that are cellular are produced with small internal voids or hollow internal structures. Cellular materials contain tiny voids (air pockets) that make them lightweight. The still air within the structure improves their thermal insulation performance, but reduces strength and resistance to impact damage. Cellular concrete (*aircrete) contains tiny voids, which makes it lightweight and improves its thermal insulation. Cellular core doors are hollow in structure, with the hollow structure being formed from honeycombed cardboard or strips of timber that offer some stiffness and rigidity to the door. Cellular glass (often called foam glass) is an insulating material made from glass powder and an air entraining agent that results in a material full of equal-sized closed-cell voids (tiny pockets of air).

CE mark A self-certification mark placed on products by manufacturers to state that the product meets relevant European Directives. The CE mark is not an independent certification that the product has met the safety standards, but is a statement by the manufacturer that the product does meet all the safety requirements imposed by the European Directives. CE is a mandatory mark, under many Directives, that needs to be placed on many products sold on the single market of European Economic Area (EEA). CE is not an official abbreviation, but it is linked to the European Community or *Conformité Européene*.

(((⊕))) SEE WEB LINKS

• The British Standards Institution provides guidance on the conformity assessment procedure and gaining CE.

cement One of the most widely used materials or substances in the construction industry. Cement is essentially a binder used in concrete and mortar. Most construction cements are based on *Portland cement (or ordinary Portland cement). The manufacture of cement is a highly energy intensive process which accounts for 7–10% of global carbon emissions (2008).

cementite (iron carbide or Fe_3C) A compound containing iron and carbon, commonly found in ferrous metals (steels, cast irons, etc.).

cement joggle Mortar or grout positioned or injected into cuts and V-shaped rebates in adjacent stones, bricks, or blocks. Once the mortar or grout sets, it forms a rigid connection between the masonry, helping the stonework, bricks, and blocks to locate and stay aligned, locking the masonry in position, and avoiding uneven settlement. Joggles can also be tongue and groove, table or bed joggle joints, and also formed with slate or metal, or dowels, used for resisting movement and holding the masonry in position.

cement mortar A mixture of sand, cement, and water, with water being used to hydrate the cement. The hydration of cement results in a number of chemical reactions leading to the development of bonds between aggregates. Four common materials that are present in cement in the anhydrous state (when the substance contains no water) include alite, belite, aluminate (C_3A), and a ferrite phase (C_4AF). When water is added, a number of exothermic reactions (heat-emitting reactions) occur.

Three main reactions occur: **1.** The clinker sulphates and gypsum dissolve in the water, leaving an alkaline, sulphate-rich solution. **2.** On mixing, the aluminate (C_3A) reacts with the water to form an aluminate-rich gel. The gel reacts with sulphate in solution to form fine rod crystals of ettringite. **3.** Following the initial exothermic reaction, alite and belite within the cement react to form calcium silicate hydrate and calcium hydroxide, increasing the bonds and strength of the mortar. Individual grains react from the surface of the mortar inwards, and the anhydrous particles become smaller. Hydration of aluminate (C_3A) continues, as crystals access free water.

 SEE WEB LINKS

• An introduction to cement hydration.

cement plaster An internal mix of materials used to provide a smooth, level surface based on a mix of plaster, sand, Portland cement, and water. Cement-based plasters can be mixed with lightweight aggregate or vermiculite to provide an adhesive fireproofing product, which can be trowel- or spray-applied. Cement plasters in the USA were also known by the generic term adamant plaster. Properties such as cement plaster's strength, hardness, durability, quick-setting time, and ability to perform in damp conditions provide advantages over other mixes. It is normally applied in thicknesses of 12–20mm, depending on the conditions, protection required for underlying materials, and levelness of the underlying strata.

cement screeds A mixture of cement, sand (fine aggregate 0–4mm) and water which is laid and levelled on floors to provide a smooth, flat surface. Generally the mixtures are semi-dry (moist to the touch) and able to hold their form with sufficient workability to be trowelled and levelled. Semi-dry mixes are normally applied in thicknesses of 30–60mm and can be used to cover uneven surfaces and services such as underfloor heating. Fluid (more liquid) self-levelling screeds have become common. The self-levelling screeds are poured over the surface, positioned where needed, and allowed to level under gravity.

CEN (Comité Européen de Normalisation) The European committee for the development of standardization across the European Union. By specifying common standards, the committee helps to ensure the flow of materials and working practices across Europe.

SEE WEB LINKS

• European Committee for Standardization

centering *See* CENTRING.

centesimal angular measure A system for angular measurement, where a right angle is divided into 100 centesimal degrees (known grades or GRADS when used on calculators), each grade divides into 100 minutes and each minute divides into 100 seconds. Used mainly in France.

centralized wastewater treatment The large-scale wastewater treatment system which is provided to the combined localities/utilities and municipalities (local authorities). The system requires efficient management and skilled labour for its operation and maintenance and may incur high capital and operational cost.

central meridian An imaginary line that joins the rotational poles of a planet.

central reservation (US median strip) A strip of land that divides traffic going in opposite directions on *dual carriageways or motorways.

centre of gravity A point in a body through which gravitational forces act. **Centre of mass** is the point in a body at which the total mass and sum of external forces (e.g. gravitational forces) act. **Centre of pressure** is the position where the total forces from a fluid can be considered to act.

centres (US centers, centre-to-centre) The spacing between the centre point (**centreline**) of one object to the centre point of the next object.

centrifugal compressor A *compressor in which compression is achieved using a centrifugal force. Suitable for applications that require high volume flow rates, such as large chilled water plants.

centrifugal fan (centrifugal blower) A fan that utilizes a *centrifugal force to increase the pressure of the gas moving through it. Used in air-conditioning systems, warm air-heating systems, and industrial processes.

centrifugal force An apparent force that appears to drag an object away from a centre point around which it is rotating. Roads and railways are designed with a super elevation so that cars and trains resist this force around bends.

centrifugal pump A pump that uses a centrifugal force to move a fluid.

centring (US centering) Temporary support, formwork, or falsework upon which bricks, stones, or blocks are laid. The centring provides the temporary arch support until the mortar is sufficiently set to hold the masonry in place. Some arches may be dry, without mortar, relying on gravity and the shape of the arch to hold the masonry in position. Once all stones and bricks are in position, the centring can be removed.

ceramic A material comprised of metallic and non-metallic constituents. Typically ceramics have high strength and *hardness but low *ductility.

ceramic disc valve A valve containing two polished ceramic plates, which move across each other allowing or preventing the flow of water. They are considered highly durable, drip- and maintenance-free.

ceramic matrix composite (CMC) A composite material based on a ceramic matrix.

certificate Official document that records an achievement, test results, or position.

certificate of practical completion The formal statement that works have been finished to a satisfactory condition. The certificate is issued by a resident engineer, a client's representative, or the project administrator. The issuing of the certificate normally results in payments and triggers the start of the periods for withholding retention monies.

cesspit (cesspool) A pit or underground tank for collecting and temporarily storing wastewater, particularly sewage.

chain-link fence A lightweight metal barrier formed from strands of wire threaded together so that they interlock to provide a fence. The fencing material is delivered in rolls.

chain pump A water pump powered by a series of discs attached to a chain. The chain is drawn through a tube, part of which is open and exposed to water. As the chain is drawn along the tube, the discs are pulled through the water; the water becomes trapped and is drawn into the tube and then discharged.

chainsaw A motorized saw where a chain with individual blades acts as the cutting edge. Each segment of the chain has small blades. The chain, similar to a bike chain, runs around a guide and is powered by a motor. As the chain rotates it is used to make rough cuts in timber and trees.

chain survey A linear survey using a controlled framework of chain lines. The distance between survey points is measured by a chain. The **chain** consists of a connection of 100 metal links of equal length. A **Gunter's chain** or **surveyor's chain** is 20 m (66 ft) and an **engineer's chain** is 30.5 m (100 ft) The *offset distance of objects (tree, buildings, etc.) are measured perpendicular to the chain. The distance along the chain from one *station to the offset is known as the **chainage**.

chair bolt (fang) A bolt fixing with a fishtail end that protrudes from the wall. The fishtail or bracket that is left exposed once the bolt is inserted and secured to the structure allows brickwork and other structural materials to be tied to the fixing.

chair rail See DADO RAIL.

chalk A soft material based on *calcite, formula $CaCO_3$.

chalk line A piece of string covered in chalk and used for marking straight lines on solid flat surfaces. A thin piece of string is pulled from a chalk dust container, then the chalk-covered string can be stretched between two points over a flat surface. Once pulled taut the string is lifted slightly and allowed to snap back onto the surface, resulting in a straight chalk line being transferred to the surface. A chalk line is often used to mark out concrete surfaces for brickwork, blockwork, or for cuts and rebates.

chamber (lock bay) 1. A bedroom. **2.** An enclosed space.

chambered-level tube A *level-tube that has a chamber at one end to adjust the amount of air in the cylinder (i.e. bubble length) for temperature correction.

chamfer See BEVEL.

change An alteration to that which was planned, contracted, priced, or proposed. Standard terms of contract such as the *JCT and *ICE suite of contracts have specific definitions for activities and events that constitute a change or *variation. Changes to work are normally authorized by the client representative or project administrator using a change order.

change agent 1. Something introduced into a substance to induce change. **2.** Personnel brought into a project or organization to manage change.

change face A procedure in surveying to minimize errors by rotating the telescope on a *theodolite through 180° vertically and horizontally so that readings are taken from the opposite side of the circle.

change management The planned process used to make alterations to business or activities. Procedures and guidance are often used to specify the course of action to be taken. New personnel with experience of managing change may be brought in to oversee and lead activities during this period.

change point The position of the levelling staff when a surveying instrument (level or theodolite) is repositioned. *See also* FORESIGHT and BACKSIGHT.

channel An object having a U-shaped cross-section. An **open channel** can be used to convey water for irrigation. A **channel bend** is an open channel which is curved on plan to guide the flow in an *inspection chamber from a *branch sewer. A **channel section** is a rolled *steel section in the shape of a channel. **Channel iron** is a bar with a U-shaped cross-section.

characteristic action (F_k) The principal *representative value of an *action, used throughout the *Eurocodes, i.e. an *action that has not yet had a *partial factor applied to it.

characteristic strength (characteristic stress) A value defined by a certain percentage of material specimens (typically steel or concrete) that exhibit values within a certain probability. Not more than 5% of test specimens should fall below the characteristic strength/stress and as such, is used by engineers as the material design strength.

characteristic value (X_k or R_k) Material (X) properties and resistances (R) are included in the *Eurocodes as characteristic values ($_k$) with a prescribed probability of not being exceeded in a hypothetically unlimited test series. A characteristic material property is converted into a *design value (X_d) by dividing by an appropriate *partial factor (γ_M):

$$X_d = \frac{X_k}{\gamma_M}$$

chargehand (leading hand) On-site gang leader or supervisor.

Charpy test (izod test, notched-bar impact test) A test utilized to measure the impact energy or notch toughness of a standard notched metal specimen. Many metals are ductile at room temperature, however, they become brittle at lower temperatures (e.g. −200°C). The charpy /izod/notched bar impact tests determine whether metals are brittle or ductile (or both) between −100 to −200°C. A small metal specimen (100 mm in length) is supported as a beam in a horizontal position and loaded behind the notch by the impact of a heavy swinging pendulum. Impact pressure is applied to the specimen, which is forced to bend and fracture at a very high strain rate. The principal measurement from the impact test is the energy absorbed in fracturing the specimen, the higher the energy, the more ductile the material. Once the test bar is broken, the pendulum rebounds to a height that decreases as the energy absorbed in fracture increases. The principal advantage of the Charpy V-notch impact test is that it is a relatively simple test that utilizes a relatively cheap, small test specimen. *See also* IZOD TEST.

Chartered Engineer A professional engineer, registered with the Engineering Council UK and who has gained the post-nominal letters 'CEng'. According to the Engineering Council UK 'Chartered Engineers are characterized by their ability to develop appropriate solutions to engineering problems, using new or existing technologies, through innovations, creativity, and change. They might develop and apply new technologies, promote advanced designs and design methods, introduce new and more efficient production techniques, marketing and construction concepts, pioneer new engineering services and management methods.' Typically, a Chartered Engineer would have gained an accredited academic qualification up to Masters level and have demonstrated professional competence through training and experience.

(⊕) SEE WEB LINKS

• The Engineering Council website details the different professional qualifications that can be obtained and is the source of the quotation above.

Chartered Institute of Arbitrators (CIArb) Body that accredits professional arbitrators. Provides governance and guidance in arbitration and allows access to its members who can advise, assist, or conduct arbitration.

(⊕) SEE WEB LINKS

• The official website of the Chartered Institute of Arbitrators

Chartered Institute of Building (CIOB) Professional body that represents construction managers and other related professions. The organization was founded in 1834 as the Builder's Society. A professional builder who has membership (MCIOB) or fellowship (FCIOB) of the CIOB is known as a Chartered Builder.

(⊕) SEE WEB LINKS

• The official website of the Chartered Institute of Building.

Chartered Institute of Building Services Engineers (CIBSE) A UK professional body, which gained its Royal Charter in 1976, to support building services engineers. The CIBSE publishes guidance and codes on controls, equipment, and internal environments within buildings.

(⊕) SEE WEB LINKS

• The CIBSE official website.

chase (chased) A rebate or channel cut into walls to accommodate services. Specialist plant is available to rebate, brickwork, blockwork, and concrete. Ducting, wires, and pipework are clipped into the rebate. When the wall is plastered there is no evidence of the wire or ducts beneath the finish.

check 1. Examining work to ensure that it has been done correctly. **2.** A *rebate (Scotland). **3.** A crack or break in the surface of a material.

checked back Recessed.

checking *See* CRAZING.

check out The process of forming a rebate.

check throat A recessed *throat on a *sill.

check valve (clack valve, non-return valve, reflux valve) A *valve that only allows flow in one direction.

chemical degradation of timber A degradation mechanism for timber material. The higher the density and the greater the impermeability of the wood, the greater its resistance to chemical degradation. The acid resistance of impermeable timbers is greater than that of most common metals; iron begins to corrode below pH 5, whereas attack on wood commences below pH 2, and even at lower pH values, proceeds at a slow rate. In alkaline conditions wood has good resistance up to pH 11.

chemical gauging (chemi-hydrometry) To determine the flow rate and path taken by flowing water by measuring the dilution of a chemical solution introduced into a watercourse upstream.

chequer-board construction (alternate-bay, chequer-board, hit-and-miss, chequer-board slabbing) Concrete floor slab cast in alternative bays giving an appearance of a chess board.

chequer plate (chequerplate, checker plate, checkerplate) An anti-slip flooring material.

cherry picker Mobile elevated working platform (MEWP) with an articulated pneumatic arm that allows personnel to work safely at heights.

chert A type of silica material found in fine-grained sedimentary rock, it may exist as nodules or thin beds. It is a hard material and has similar properties to flint.

chevron drain A V-shaped drain laid in a *herringbone fashion. Typically filled with free-draining granular material and used to collect surface water.

chiller *See* REFRIGERATION UNIT.

chimney A structure containing a vertical *flue.

chimney block *See* FLUE BLOCK.

chimney breast The portion of a chimney that projects from a wall.

chimney shaft A free-standing industrial chimney that usually only contains one flue.

chimney stack The portion of a chimney that projects from a roof. One or more flues can be contained within the stack.

Chinaman chute (loading ramp chute) A colloquial term for a timber-loading ramp used when mining ore.

chipping To break or cut small pieces from the surface of a material.

chipping chisel 1. A chisel bit for use in a pneumatic **chipping hammer. 2.** The chisel end of a chipping hammer.

chipping hammer 1. A portable pneumatic hammer. **2.** A small portable rock hammer used by geologists to extract minerals, fossils, and fragments of rock.

chippings Small pieces of a material, usually stone.

chlorination A process of adding chlorine to water as a method of water purification. Chlorination is also used to sterilize the water in swimming pools and also in sewage works. The term can also apply to the addition of chlorine to other elements, such as the addition of chloride to metals.

choke *See* BALLAST.

chord 1. The horizontal part of a roof truss designed to resist tensile forces. **2.** A straight line which connects two points on an arc or circle.

CHP *See* COMBINED HEAT AND POWER.

chuck An adjustable device that is used to hold tools in the centre of a lathe **(lathe chuck)** or a drill **(drill chuck)**. In a drill, the chuck is adjusted using a **chuck key**.

chute An angled pipe or semicircular channel that uses gravity to transport materials downwards.

CIC BIM protocol A supplementary legal agreement with the prime purpose of enabling the production of *BIM at the design stage. It creates additional rights and obligations to facilitate collaborative working, while safeguarding *intellectual property and liability between all parties. It also provides for the *information manager to be appointed.

cill *See* SILL.

Cipolletti weir A trapezoidal-shaped weir with 1:4 side slopes. Used for measuring the flow in open channels such as streams and rivers.

circuit 1. A circular path of conductors that enables an electric current to flow from the power source to an electrical appliance or component (an **electrical circuit**). **2.** The electrical wiring attached to the consumer unit that supplies electricity to electrical sockets, fixed appliances, and lights. **3.** A series of pipework through which a fluid flows, i.e. a **hot-water circuit**.

circuit breaker A safety device that automatically stops the flow of electricity through an electrical circuit if it becomes overloaded or if a fault develops. Provides greater control than a *fuse and can be manually reset once the overload has been removed. A series of **miniature circuit breakers** (MCBs) are used in a *consumer unit to protect various electrical circuits, such as the central heating system, lighting, cooker, electric shower, electric sockets, and smoke alarms. The number of MCBs contained within the consumer unit will depend upon the size of the building and the equipment installed. In dwellings, modern MCBs are rated at 6, 10, 20, 32, 40, 45 and 50 amps.

circulation The free movement of a fluid through a system.

circulation pipe A pipe used to circulate gas or a liquid, for instance, a flow or return pipe in a central heating system.

circulation space (circulation area, circulating area) The space within a building that enables people to move freely from one part to another.

circumpolar star A star that is always visible above a given latitude, i.e. it never sets.

CIRIA The Construction Industries Research and Information Association established to deliver services, information, and research across the sector.

(((●))) SEE WEB LINKS
• The official website of the Construction Industries Research and Information Association.

CI/SfB Classification system, used in libraries and offices to divide building information and literature into related sections. The system was developed by *RIBA and is used across the industry.

cistern A tank used to store cold water at atmospheric pressure. Usually fitted with a *ball valve and an *overflow pipe. Used to store rainwater, store the water used for flushing a WC, or as a feed and expansion tank in an *open-vented system.

cistern-fed water supply The cistern (receptacle for holding and storing water) is placed higher than the rest of the heating system; in this way water is fed via gravity into the feed tank and boiler and to the rest of the heating system as necessary.

City and Guilds of London Institute (C&G) Awarding body of many vocational qualifications in the UK.

(((●))) SEE WEB LINKS
• The official website of the City and Guilds of London Institute.

civil engineer A person who practises the profession of *civil engineering. He or she may deal with the design, construction, and maintenance of bridges, tunnels, roads, canals, railways, airports, tall structures, dams, docks, and clean/wastewater systems.

civil engineering This is a professional engineering discipline that deals with creating, improving, and protecting the environment. It provides the facilities for the built environment and includes environmental, geotechnical, materials, municipal, structural, surveying, transportation, and water engineering.

Civil Engineering Standard Method of Measurement (CESMM) A document containing set procedures to prepare and price the *bill of quantities, together with how the quantities of work should be expressed and measured for *civil engineering projects.

cladding The non-load-bearing external envelope or skin of a building that provides shelter from the elements. It is designed to carry its own weight plus the loads imposed on it by snow, wind, and during maintenance. It is most commonly used in conjunction with a structural framework.

cladding rail A secondary beam that spans horizontally across the structural frame of the building to support the cladding.

claim Allegation made against another person stating that something has or has not been performed or delivered. The party making the allegation normally takes the action in order to recover money or order a specific performance. The action is normally taken because a party has failed to perform as agreed under the contract.

claims-conscious contracting Unofficial method of contracting where a *contractor enters into a contract with the intention of pursuing *claims for any extra work. Instead of identifying items not mentioned in the tender documents, but deemed necessary to complete the project, the contractor undertakes the contract and makes a claim for the *variations once work starts. Clients may enter into the contract believing everything is covered only to discover the cost of the contract increases as claims are served and granted in order to complete the works. Such procedures only work where the contract would allow for variations. Good practice would encourage contractors to identify the items missing from the documents at the tender stage, allowing the client to assess the extra costs that would be incurred.

clamping screw A screw used to attach a vernier to levels and theodolites.

clamshell (clamshell grab) A toothless grab bucket that is suspended from a crane jib for handling loose material.

clapotis A standing wave that does not break before it impacts on a near-vertical surface, e.g. a seawall.

clariflocculator A treatment unit that combines flocculation and clarification by using a two-chambered clarifier, where the inner column carries out flocculation and outer column serves as a clarifier. The two columns are concentric.

clash detection An integral part of the *BIM process to check that there is no conflict when different models combine when prepared by different disciplines.

clash rendition As defined by PAS 1192-2, '*The rendition of the native format model file to be used specifically for spatial coordination processes.*' It is used to ensure clash avoidance or for *clash detection.

classification test *See* ATTERBERG LIMITS.

clause A description setting down the agreements that form the component parts of a contract. Contracts are divided into individual items covering the various aspects of the agreement; these are the terms that make up the contracts.

clay Fine-grained strata (rock or soil), hydrous aluminium phyllosilicates, with iron, alkali metals, magnesium, alkaline earths, and carbon, in various quantities, with traces of quartz, metal oxides, and organic matter which, due to the fine-grained structure, changes its properties and characteristic behaviour with water and when dry. The material will shrink, become hard and brittle when dry, will expand, swell, become malleable, and slip when pressure is applied when water is present.

clay block A small building module formed of clay that can be dried, fired (burned), or oven-baked to provide a structural walling material. Clay blocks can be pressed into a mould, extruded through preformed cutters, or wire-cut to create a lighter-weight hollow block. The blocks of clay take many different forms, normally have perforated centres to assist drying and baking. The perforations ensure a more even distribution of heat across the block, preventing unwanted drying stress during baking and also making them light. Some blocks are termed low-fired, when they are hollow and require less heat energy to cure them; they may be considered more sustainable as they contain less embodied energy than clay blocks that are fully baked.

clay block floor Structural floor blocks made from dry clay or oven-baked clay blocks positioned on beams to create a suspended lightweight ceramic floor. A surface coating of concrete or screed is usually applied to create a smooth surface and help distribute the loads to the blocks and beams.

clay bricks Small common building wall modules formed of clay. Bricks tend to be formed out of clay or concrete, with clay being the more common. In warm climates clay may simply be cut into regular shapes and exposed to the sun to dry out. Greater structural properties are achieved through selective sourcing of clays and the use of large ovens to fire and bake the bricks. Perforations, recesses, frogs, and grooves can be cut into the clay to improve firing, ensuring more even baking of the clay and to improve the bedding and bonding when used with mortars to form walls.

clay floor tiles Flat modules formed of clay and used to provide floor covering. The tiles can be regular, commonly square or rectangle, or any shape, laid on a cement of resign base and positioned adjacent to each other to provide a decorative, functional, and level surface. Tiles can be glazed, painted, left in their natural state, waxed, oiled, or finished with other covering. Where used to carry pedestrians and other traffic, they should be slip-resistant.

clay pipes Tubular pipes, formed using vitrified clay creating an inert pipe resistant to chemical degradation. The pipes, when connected in a drainage system, are used to carry surface water, sewage, and effluent. They are joined together using flexible coupling pieces, male and female joints, spigot. The joints are often sealed with rubber-type flexible ring gaskets, clay pugging, and other jointing products.

clay puddle *See* PUDDLE.

clearance The space between two objects, e.g. the space between a door and a doorframe to ensure that the door can open freely.

clear span The unobstructed distance between two supports, e.g. the distance under a beam support between two columns.

cleat A bracket fixed to a structural component used to secure another component.

cleaving Splitting a material along its cleavage.

clench nailing (clenching, clinching) A type of nailing where the point of the nail is driven through and out of the material, and then the projecting point is bent over and flattened against the material.

clerestory (clerestorey, clearstorey) The upper part of a wall that contains windows to produce extra light and sometimes ventilation. Used in churches and tall building where the centre of the structure rises above the rooflines of perimeter sections of the structure.

clerk of works (COW) An inspector used to check the project as it progresses. As a client's representative the COW may have to agree *day works items, check setting out, inspect excavations, and check levels. The COW takes no active part in running the project.

client (employer) The person or organization that requires and pays for the building or works.

climate change The shifting of global and regional weather patterns as a result of the amount of *carbon dioxide (CO_2) in the atmosphere due to the burning of *fossil fuels, typically resulting in more extreme and erratic weather conditions being observed.

Climate Change Act 2008 The UK approach to address *climate change. It stipulates that carbon dioxide (CO_2) and other greenhouse emissions should be reduced, and the potential risks from *climate change prepared for.

climate measurement Metrics based on temperature (maximum, minimum, lower-ground, and soil), wind, relative humidity, rainfall, atmospheric pressure, irradiation, or sunshine hours. The climate of a particular region is determined through analysis of climatological variables in the region which is usually undertaken over a period of thirty years or more.

clinker 1. Large round lumps of cement-like substance, formed during the cement production and often ground down to become cement. It is produced during the cement kiln stage from sintering limestone and alumino silicate. **2.** Lumps of vitrified brick or burnt clay fused together.

clinograph An instrument that records the derivation from vertical, used in boreholes, shafts, and wells.

clinometer A handled instrument for measuring the dip of a bedding plane or the angle of a slope, used in geology and surveying.

clip A type of fixing used to hold materials or components in position, such as cladding panels, electrical cables, pipes, or roof tiles.

clip anchor *See* ANGLE CLEAT.

clo *See* CLOTHING INSULATION UNIT.

clogging 1. Blocking up of a system with debris. **2.** Blocking of wastewater treatment through the saturation of the filter media or blocking the channel in water treatment. The blocking (clogging) is a result of the accumulation of debris or settling particles over time. Clogging physically impedes further movement of water and may lead to overflow or reduction in effluent of a treatment unit.

close boarding Timber boards laid abutting one another such that there is no gap in-between. Usually used for fencing and trench support.

close board trench support Excavation support where the earth-retaining face is formed of *close boarding, where the boards are positioned abutting each other so that no material of any significant size can pass through.

close-coupled cistern A *water closet where the *cistern is connected directly to the *pan.

close couple roof A roof structure where the diagonal timbers (rafters) are tied together at the base of the rafter feet by a horizontal length of timber. The horizontal timber normally also acts at the ceiling joists, supporting the plasterboard and finishes which make up the ceiling.

closed loop convective bypass A type of *convective loop bypass where the air located within the building element constantly recirculates within the element in a loop, due to temperature differences, but remains largely unchanged.

closer (closing piece) A brick that is reduced in width by cutting it along its length. Used on a *course to complete the bond. Examples include a *king closer and a *queen closer.

closing device A *fastener used to keep doors and windows closed.

closing error (error of closure, misclosure) The discrepancy between two readings in a survey, which should actually be the same. For example, in a closed loop *traverse the initial readings should be the same as the final reading. Any difference between these two readings (the closing error) is proportionally distributed over the traverse (*see* BOWDITCH'S METHOD).

closing face The part of a door leaf that is closed against the doorframe.

clothing insulation unit (clo) The thermal resistance provided by any layer of clothing that includes the effect of any air trapped between the clothing and the skin. Used as a metric to measure the amount of thermal insulation provided by a person's clothing. Measured in m^2K/W.

clothoid A spiral whose curvature changes in a linear manner with its length. The curvature begins with a zero radius and increases linearly with its length.

cloud computing A computing resource for storing and accessing data together with use of software programs, accessed via the Internet rather than being stored locally on a hard drive, a USB memory stick, or a local network.

clough 1. The sloping sides of a deep narrow valley in a hill. **2.** A *sluice for controlling the flow of water.

clout nail A short galvanized nail with a large flat round head. Used for fixing plasterboard and sheet metal roofing.

coagulant An agent that causes a liquid or solution to semi-solidify.

coal A solid fossil fuel that is burnt providing heat and energy. Coal is usually a black combustible mineral formed about 300 million years ago.

Coanda effect Discovered in 1930 by aircraft engineer Henri Coanda, this effect is the tendency of a moving fluid (air or liquid) to adhere to a curved object as it passes over it. As a fluid passes over a surface it will be subjected to skin friction. This friction causes a decrease in flow rate and pulls the fluid towards the surface. If the surface is curved as the fluid leaves the surface it will have a tendency to continue the curved shape. Amongst other aerodynamic applications, the Coanda effect is employed in certain ceiling diffusers as it is known to be a very effective method for circulating air; these diffusers are also known to leave a dirty stain from the air on the surrounding ceiling.

cob A mixture of clay and straw used as a building material. Cob is a very popular construction material in the developing world as it is cheap and easy to manufacture.

cock *See* STOPCOCK.

Code of Practice (CP) A description of work or procedure that establishes a minimum standard to be achieved or followed. *British Standards and *Eurocodes have now largely replaced CPs.

code of professional conduct A set of principles laid down by a professional body, such as the *Institution of Civil Engineers, to promote competency and good behaviour practices. For example, in respect to health and safety, sustainable management of natural resources, acquisition of professional knowledge, and ethics.

co-digestion of sludge The biological process used for the degradation of organics present in sludge and added materials into stable products. It commonly involves the addition of chicken/poultry waste, food waste, or cattle manure to sludge waste and yields the high volume of byproduct, i.e. biogas.

coefficient 1. A number or constant multiplied by a variable or unknown in an algebraic equation, e.g. 4 is the coefficient in the equation 4xy. **2.** A numerical measure of a physical or chemical property that is constant under specific conditions, various examples are presented below. **Coefficient of active earth pressure (K_a)** the ratio of minimum horizontal effective stress to vertical effective stress that would occur in a soil mass at some depth, e.g. behind a retaining wall where the surface of the wall is considered to move away from the soil it is retaining. It is defined by the equation

$$K_a = (1 - \sin \phi)/(1 + \sin \phi)$$

where ϕ is equal to the angle of internal friction of the soil. *See also* ACTIVE EARTH PRESSURE. **Coefficient of compressibility (m_v)**, the volume change per increase in effective stress, i.e. the inverse of pressure. **Coefficient of consolidation (c_v)**, the ratio of permeability to volume compressibility of a soil. It is used to estimate the rate at which settlement takes place. There are two methods to determine c_v, i.e. root time or log time. **Coefficient of discharge (C_d)** ratio of actual discharge obtained through an orifice, weir, or pipe compared to the theoretical value. **Coefficient of dynamic viscosity (μ)**, a constant linking fluid flow to resistance. **Coefficient of earth pressure at rest (K_o)**, the ratio between horizontal effective stress and vertical effective stress in a soil mass at some depth where the horizontal strain is zero, e.g. behind a retaining wall where the surface is not considered to move. **Coefficient of expansion or coefficient of thermal expansion (α)**, the unit increase in length per degree increase in temperature. **Coefficient of friction (μ)**, the ratio between the force causing a body to slide along a plane and the force normal to the plane. **Coefficient of passive earth pressure (K_p)**, the ratio of maximum horizontal effective stress to vertical effective stress that would occur in a soil mass at some depth, e.g. behind a retaining wall where the surface of the wall is considered to move towards the soil it is retaining. It is defined by the equation

$$K_p = (1 + \sin \phi)/(1 - \sin \phi)$$

where ϕ is equal to the angle of internal friction of the soil. *See also* PASSIVE EARTH PRESSURE. **Coefficient of permeability (k)**, the mean discharge velocity of water flow under the action of a unit *hydraulic gradient. **Coefficient of uniformity (C_u)**, the slope of a soil's grading curve, defined as

$$C_u = d_{60}/d_{10}.$$

coefficient of performance (COP) The ratio of total heat output to total electrical energy input (both in watts). Used to measure the efficiency of heat pumps. The higher the COP, the greater the efficiency of the heat pump.

cofferdam A temporary structure constructed from sheet piles. Used to support the surrounding ground and minimize pumping while construction within the ground takes place.

coheating test *See* ELECTRIC COHEATING TEST.

cohesion The molecular attractive forces that bind soil minerals (such as clays) together, sometimes referred to as the 'stickiness' of the material particles. It is a component of shear strength in *Coulomb's equation.

cohesionless soil (non-cohesive soil) *See* GRANULAR.

cohesive soil A soil that exhibits *cohesion, i.e. granular soil.

coil 1. A length of material wound into a spiral. **2.** A curved pipe used within air-conditioning systems or ductwork to heat **(heating coil)** or cool **(cooling coil)** the air passing across it. **3.** An electrical device comprising a spiral of insulated electrical wire.

coincidence effect A reduction of the sound insulation of a sheet of material (such as partition wall) as a result of the critical frequency of a sound wave being the same as the flexural bending wavelength of the sheet of material. The resultant is an increase in the transmission of sound through the material.

coir A fibrous material obtained from the outer husk of coconut. Used to manufacture rope, doormats, and floor coverings.

coke A material derived from coal comprised mainly of carbon.

cold bridge *See* THERMAL BRIDGE.

cold roof Where the insulation at roof level is positioned within the horizontal ceiling rafters, creating a cold roof space in the *attic (Figure b), being different from the warm roof, where the insulation is positioned in the diagonal sloping aspect of the roof, creating a warm roof area which can then be used as a habitable space for a compression wave room in the roof (Figure a). In cold climates the traditional pitched roof is known as cold roof construction, where the insulation is placed at the horizontal ceiling level, normally between and over the top of the ceiling joists or the bottom of the ceiling truss tie (see Figure b). At the ceiling, a vapour control layer should be used to prevent warm moist air moving from the habitable area in the roof space.

Breathable vapour control under counter-
battens, battens treated and seasoned,
supporting single lap interlocking tiles.

Counter battens, running up the roof
50mm ventilation

Vapour control layer
behind plasterboard

Seal gap between floor
and ceiling

WARM ROOF SPACE

Ensure ventilation at the eaves and
ridge.

Continuous or flexible insulation laid over the
eaves, preventing thermal bridge at the
junction

Figure a: Warm (cold climate): Traditional roof structure: thermal
insulation reduces the heat energy entering the roof

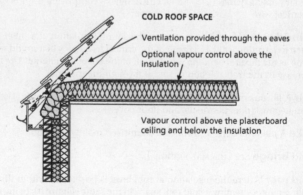

COLD ROOF SPACE

Ventilation provided through the eaves

Optional vapour control above the
insulation

Vapour control above the plasterboard
ceiling and below the insulation

Figure b: Cold roof (cold climate): Traditional roof structure: thermal
insulation reduces the heat energy entering the roof

Cold roof

cold joint A visible joint in concrete that occurs when concrete has been laid at different times.

cold rolling (cold forming, cold working) A process that deforms steel or metal at room temperature (i.e. below its re-crystallization temperature) by squeezing it between a set of rollers. It causes defects to be formed in the crystalline structure, which in turn increases the yield strength and hardness of the metal.

cold-water supply system The main system of water coming into a property.

cold weather working A period of working when the condition (temperature or wind) dictates a substantially different working environment; this can mean that the works become substantially different, meaning that different tools, hours of work, rates of pay, and welfare come into action.

collapse The excessive and irregular shrinkage of hardwoods when dried in a kiln too quickly or for too long.

collapse grading The stability of a fire door during a rated period of resistance under specified fire conditions.

collapsible form *Formwork that is designed to fold up or telescope inwards to take up less space or allow partial stripping of the formwork to take place.

collar A ring-shaped part of an object that guides, seats, or restricts another object.

collar beam A horizontal tie, used in a **collar-beam roof**, to join similar *rafters on opposite sides of the roof to prevent the roof from spreading.

collar boss A pipe fitting where sections can be drilled out to allow for future connections.

collar roof A roof structure using a horizontal tie to link the diagonal members; the *collar beam or tie (horizontal piece) can be positioned relatively high in the structure or tying the diagonals at the eaves level.

collector well A vertical well with horizontal collection pipes radiating outwards at the base of the well.

collimation line The line of sight through the cross-hairs of a surveying instrument (e.g. on a theodolite or level). **Collimation error** is the inaccuracy caused by the line of sight not being horizontal or out-of-alignment, e.g. error in corresponding *face left and face right readings. **Collimation method** is a technique used in levelling where the staff readings are subtracted from the collimation line to determine the reduced levels. A **collimation test** can be undertaken to determine the accuracy of the instrument. This test involves setting the instrument in the middle of two stations, with the stations being approximately 60 m apart on level ground. A staff reading is taken from each station and a true level is obtained, and any collimation errors will cancel each other out. The instrument is then placed about 3 m beyond one of the stations, still in line with the stations. A staff reading is taken from the near station and using the true level the reading that should be obtained from the far station is calculated. The reading is then taken at the far station and if the two correspond, the instrument is considered accurate, if not adjustment to the instrument is required.

collision avoidance system (post-crash intervention system, forward collision warning) A safety system that is automatically applied to prevent or reduce the severity of a potential collision. Such systems include, for example, automatic braking systems (automatic emergency braking, AEB), and forward automatic braking (FAB).

collision load The load cause to a structure when a vehicle crashes into it.

colloid A mixture where one element is mixed evenly with or throughout another. Examples of colloids include paint and pigmented ink.

colonnade A row of evenly spaced columns. In reference to classical architecture these columns are joined by their entablature.

column A vertical structural component that acts as a strut or support.

columnar structure (metallurgical) The appearance of a particular type of grain on a metal surface when viewed under a microscope. When metals and alloy surfaces are magnified and viewed under microscopic examination it is possible to see the chemical structure (or phase) of the material. Each phase has a distinct appearance; a common appearance is the presence of columnar grains. The morphology of columnar structure is elongated grains similar to that of columns. This appearance results when a metal is heated to a high temperature (when it is in liquid form) and very rapidly cooled to room temperature.

column clamp Braces and yolks used to secure and tighten column formwork to ensure that no concrete leaks from the formwork.

column form (column formwork) *Formwork used to provide the mould or case for *in-situ concrete columns.

column head Top of a vertical strut or support. The top of a column that supports a floor may be wider at the top to help distribute the loads and prevent the column puncturing through the floor.

column splice The joining of two columns end-on-end. Gusset supports and bolts or rivets are used to brace and hold the two columns together.

column starter (kicker) A concrete upstand cast on a concrete base used to secure column formwork. The raised piece of concrete is cast to the same size and profile as the main column. The formwork can then be fixed to the kicker and held firmly in position while the concrete column is being poured.

column yoke Brackets or braces that are clamped together to hold column formwork in position.

combed joint (tooth joint) A prefabricated joint between two comb- or saw-tooth pieces of material.

comb hammer (scutch) A single- or doubled-headed hammer where small trimming combs can be inserted into the head. Used to cut and trim blocks to the required shape and size.

combination factor (ψ) A value which is taken as less than or equal to one for each variable *action (not applied to permanent actions), by considering different lead variables to obtain the most onerous case.

combined heat and power (CHP) A device that simultaneously produces heat and electrical power. As the electrical power is being generated, the useful waste heat from the generation is captured and used to provide space heating or hot-water heating.

combined loop convective bypass A combination of a *closed loop convective bypass and an *open loop convective bypass.

combined system A below ground drainage system where the surface water and the foul water are discharged through a combined surface and foul water drain and sewer. Results in relatively low installation costs (less pipework) but increased sewerage costs. *See also* SEPARATE SYSTEM and PARTIALLY SEPARATE SYSTEM.

combplate The tooth-shaped plate at the entrance and exit to an escalator.

comfort conditions and metrics Six primary metrics include the environmental—air temperature, humidity, radiant temperature, air velocity—and the personal—clothing insulation and metabolic heat.

(⊕) SEE WEB LINKS
• The HSE website details the six basic factors which affect an individual's thermal comfort

comfort index A single index that attempts to reflect human perceptions of thermal comfort. A number of different comfort indices are available including predicted mean vote (PMV), globe temperature, *environmental temperature, *effective temperature, corrected effective temperature, *equivalent temperature, and *operative temperature. However, it is difficult to find one single index where everyone is thermally comfortable under all possible conditions.

comfort zone A range of air temperatures, mean radiant temperatures, humidities, and air velocities where the majority of people achieve thermal comfort.

commencement of work Start of a construction project. For contractual purposes, this is often the date when the contractor has possession of the site.

commissioning The process of running the building services systems installed within a building for the first time and performing various tests and checks to ensure that they operate correctly.

common arrangement of works sections (CAWS) A common arrangement for specifications and *bills of quantities for building projects, developed by the *Construction Information Committee (CIC).

common bond Brick bonding, where bricks are laid end to end with the *course above and below staggered so that brick joints fall at the halfway point of the brick above and below. As the *stretcher face of the brick is exposed, this bond is also known as stretcher bond.

common brick A general-purpose brick that is not considered decorative. Although it may be used in exposed positions it is not considered to be aesthetically pleasing.

common brickwork Brickwork that does not need to be properly finished or faced as it will not be exposed. Brickwork to be covered in plaster and cladding is usually common. Bricks that are laid quickly, cut roughly, and not pointed fall under this category. Brickwork that is exposed, pointed, and neatly laid is called facing brickwork.

common data environment (CDE) A single online storage system used to collect, manage, and disseminate information (data, drawings, and models) for the whole *BIM project team.

common joist (common rafter, common stud) One of a number of joists, rafters, or studs of uniform size placed at regular intervals within a construction.

commons *See* common brick.

common trench A trench that contains multiple services. Mains electricity, gas, street lighting, and telecommunications are usually laid at different levels and are covered by sand and warning tape so that future excavations can avoid or locate the services safely.

communication pipe The cold-water service pipe that runs from the water main to a stop valve that is usually located next to the property boundary.

compaction A rolling, ramming, or vibration process to make particles form a density configuration. Used in soil and concrete to expel air voids; the result is a decrease in volume but an increase in density.

compaction factor test A test to establish the workability of fresh concrete. It is considered more accurate than the *slump test, but the apparatus is quite cumbersome and is therefore used more frequently in a laboratory than on site. The apparatus consists of two hoppers above a cylinder: the top hopper is gently filled with freshly mixed concrete so that no compaction occurs, the door is released at the bottom of the top hopper, and the concrete falls freely into the lower smaller hopper. The door is then released on the lower hopper and the concrete falls into the cylinder. Excess concrete is wiped clear and the density of the concrete in the cylinder is then calculated. This value is then divided by the density of the concrete of a fully compacted cylinder. This value is termed the **compaction factor**. The density of the fully compacted cylinder is obtained by compacting concrete in the cylinder in four equal layers.

compactness The ratio of the volume of a building to the external envelope surface area. Measured in m^3/m^2.

comparator An instrument for comparing the properties of an object, system, or finish with standard properties.

compartment 1. A contained unit. **2.** A fire-protected room or floor used to conceal and limit the spread of fire in the building. The compartment room has walls, floors, ceilings, and self-closing doors built to a specified fire resistance. The Building Regulations limit the size of compartment rooms and the travel distances to the nearest fire escape. Depending on the type of building, size, and the number of people using the building, the fire resistance will vary. The fire resistance of walls, floors, and doors are specified in terms of the time a component will resist the passage of fire. For compartment rooms a minimum size of door openings and escape routes are specified depending on the number of people who use the building. Approved Document B of the *Building Regulations provides guidance on the compartmentation.

(⊕) SEE WEB LINKS

- Online government planning portal: the Building Regulations can be downloaded from this site.

compass (magnetic compass) A device that has a magnetized pointer, which swings to always point to Magnetic North. Used as a navigational aid or in surveys such as a **compass** *traverse, where horizontal angles are determined by *magnetic bearings.

compatibility The suitability of a substance to be used in connection with another substance. For example, the compatibility of paint or plaster finishes to the base coat, so that it adheres correctly without forming defects such as pinholing (*see* PINHOLE) or *crawling.

compensating errors Mistakes that would cancel each other out, with the end result being accurate.

compensation Payment or recompense for something by one party that has an effect on the other. Contracts often state what type of events will entitle a party to payment. **Compensable delays** are delays that result from specific actions or defaults specified in the contract. Where contracts specify the type of delays that will result in additional payment or extension of time, any other action would not result in payment under the contract. Also where contracts state events that are compensable, only those **compensable events** will result in additional payment. The *Engineering and Construction Contract (previously the New Engineering Contract) lists events such as variations, tests, and other similar items as compensation events.

competitive tendering (competitive bidding) Multiple parties bid for a project with the contract being awarded to the lowest price or the company that offers the service that is considered the best value.

completion 1. The point when all the tasks in a project have been successfully accomplished and the work is finished. **2.** For contractual and payment purposes, the time when the project is sufficiently performed and is considered to have reached practical completion. When a project is considered to have reached practical completion, other events within the contract are triggered, for example, practical completion will result in either the release of *retention monies or the time period to hold retention monies commences. All standard forms of construction contracts have a provision for stating the date when all the work is expected to be finished; this is called the **completion date**.

component A small self-contained part of a building, constructed from a number of smaller parts, and designed to perform a particular function, for example, a window or a pump.

component air leakage *Air leakage that occurs through various components installed in the building envelope, such as the loft hatch, permanent vents, windows, doors, and window and door surrounds.

composite beam A beam constructed from two or more materials in order to improve its load-bearing properties, for example, a timber beam clad in steel.

composite board (sandwich panel) A cladding panel, usually faced in metal, that contains a core of insulation material.

composite construction (sandwich construction, mixed construction) A component constructed from two or more materials in order to improve its properties.

composite decking 1. Decking used to construct a *composite floor. **2.** Decking manufactured from two or more materials.

composite floor A floor that consists of a profiled metal deck spanning across the structure that is covered with concrete. Commonly used in multistorey dwellings, particularly when the speed of construction is important.

composite material A mixture of different types of materials. One of the most commonly used composite materials used globally and also in the construction industry is *reinforced concrete (made from steel and concrete). Most composites are made up of just two materials, however, more are possible. It is also possible for a single generic material to be a composite material, e.g. *dual phase steel that combines a ductile and lower strength phase with a hard high strength brittle phase.

compound beam (compound plated beam, flitch beam, flitched beam) A *composite timber beam comprising a steel plate sandwiched between two pieces of timber that are joined together at regular intervals.

compound girder (built girder, sandwich girder) The main supporting *compound beam.

compressed air A specific form of air under greater than atmospheric pressure. Compressed air is usually utilized to power a mechanical device or to provide a portable supply of oxygen.

compression The effect of squeezing, i.e. particles within a material being pushed closer together. A **compression** *boom, *chord, or *flange is one that is subjected to compression, as opposed to *tension. **Compression failure** occurs in something (e.g. a column) that has failed to take the amount of compressive forces imposed on it.

compression index (Cc) The slope of the linear portion of a curve of void ratio *vs* logarithmic effective pressure, using data obtained from a *consolidation test.

compression settling A stage of settling, in wastewater treatment, where the weight of particles settled in upper layers is exerted on layers below, resulting in further release of water in interstices and leading to compression of settled particles.

compression wave (compressional wave) A mechanical longitudinal wave that propagates along or parallel to the direction of travel, e.g. a sound wave in air.

compressive strength The ability of a material to withstand *compression, defined as a value quoted in terms of N/m^2.

compressor A mechanical device used to increase the pressure of a gas. Used in air-conditioning systems, heat pumps, and to produce compressed air. Various types of compressor are available, including axial flow, centrifugal, lobe, reciprocating, and rotary vane compressors.

compriband A generic name given to a wide range of joint-sealing materials (tapes, strips, and profiles) designed to accommodate movement. Used in a wide range of internal and external applications such as sealing windows and doors, sealing air-conditioning and ventilation ducts, and sealing movement joints in brick walls.

computer-aided project management Use of software capable of producing fully resourced networks, resource plans, *Gantt charts, and resources histograms. The software uses algorithms to analyse the tasks, resources available, and fixed time activities to calculate the shortest and most cost-effective plan of work.

concentrated load (point load) A *load that acts or is applied over a very small area.

concentration decay method A type of *tracer gas decay method where a predetermined volume of a tracer gas is injected into a space and mixed with the air inside the space. The concentration of the tracer gas within the space is then measured as it decays over time.

concentrator collector system A type of *solar collector where a series of mirrors and lenses are designed to concentrate the solar radiation onto the collector surface. By concentrating the solar radiation, higher collector temperatures can be achieved, higher conversion rates per surface area can be achieved, and the heat losses per surface area are reduced. It can incorporate tracking systems that track the path of the sun throughout the day.

conceptual design An early stage of the design process, which details an outline of form and function.

conceptual site model A site investigation model that conceptualizes the association between contaminant sources and receptors by considering the potential or actual migration and exposure pathways.

conchoidal facture A facture surface shaped like a shell, which is either concave or convex. Used to describe the irregular broken surface of minerals and rocks.

conciliation A dispute resolution process where an independent party acts as an intermediary between two parties attempting to find an arrangement that will overcome a dispute.

concrete A commonly utilized construction material composed of cement and aggregates. Concrete is used in many civil engineering applications such as pavements, structures, foundations, motorways, roads, and bridges due to its excellent compressive strength (up to 100 N/mm^2). The raw ingredients include *sand, *cement, and *aggregates.

concrete blocks Building blocks made out of concrete.

concurrency Activities running at the same time. Where delays resulting from both client and contractor are running at the same time, the contract is considered to be in a period of **concurrent delay**. Where the design and manufacturing overlap and run at the same time this is called **concurrent engineering**.

condensation The process in which a gas cools down and changes into a liquid. In the case of moisture-laden air, if the temperature of the air is reduced below the *dew point temperature, the air can no longer retain the original quantity of water vapour, so water will be deposited as condensation. *See also* SURFACE CONDENSATION and INTERSTITIAL CONDENSATION.

condition-responsive maintenance Unplanned maintenance, which is based on the exceedance of threshold values of identified performance indicators with respect to the condition of an asset, and would be undertaken to restore an asset to a safe, serviceable operating condition. An example of this process would be maintenance actions resulting from structural health monitoring.

conditions of contract (general conditions of contract, conditions of engagement) The legally binding terms and conditions that parties agree upon and which govern the expected actions of each party. Default of the terms by one party will usually entitle the other party to compensation or repudiation. The *clauses and terms that make up the conditions can be found in the main body of the contract, usually placed between the recitals and the appendix.

conduction The process by which heat flows along or through a material or from one material to another by being in direct contact. The rate at which conduction occurs varies considerably according to the substance and its state. Metals (e.g. copper) are good conductors. Gases are poor conductors. It is one of the three main heat-transfer mechanisms. *See also* CONVECTION and RADIATION.

conductivity The ability of a material to transmit electricity. *See also* THERMAL CONDUCTIVITY.

conductor A material that allows heat, electricity, light, or sound to pass along or through it. The **conductance** of a material is the degree at which it conducts heat, electricity, light, or sound.

conduit A small covered channel, usually metal or plastic, used to distribute pipework and cables around buildings.

cone penetrometer The apparatus used in a **cone penetration test** to determine the *liquid limit of soil. The apparatus consists of a polished stainless steel cone, with a mass of 80 g having an angle of 30° and a dial gauge calibrated to 0.1 mm. In the cone penetration test the cone is allowed to penetrate a soil sample, at a range of different moisture contents, confined within a tin 55 mm in diameter and 40 mm high. Penetration values are plotted against moisture content values and a best-fit line is drawn. The liquid limit of the sample is taken as the moisture content corresponding to 20 mm penetration.

configuration management A process which identifies how each part of a project fits together and makes clear links with those responsible for each activity. By having clear links between tasks and those responsible, it is easier to manage change, those parties that need to make changes, and those affected by the change.

confined aquifer (artesian aquifer) An *aquifer whose water table lies above its upper boundary, which could either be above ground level or above an overlying strata of low permeability. *See also* ARTESIAN WELL.

confined compression test *See* TRIAXIAL TEST.

conflict resolution Informal, formal, and legally binding processes used to resolve disputes. *See also* ADJUDICATION, mediation, CONCILIATION, and ARBITRATION

conformity Compliance with the relevant design procedure or code.

conformity mark A manufacturer's stamp, tag, identifier, or trademark that signifies a standard or certification status.

congestion Excessive volumes of traffic or people making movement difficult.

connector 1. A device that enables one item to be joined to another. Often enables the items to be joined together and taken apart when required. **2.** (US) A highway that connects one highway to another.

conservation The efficient and careful use of existing natural and human resources such as energy, the environment, and buildings, so that they are preserved for future generations.

conservation area An area, usually of architectural, historical, or natural significance, that is designated by the local authority in order to ensure that it is protected or enhanced. Building and development work within the area is usually tightly controlled.

conservatory 1. A room attached to a building that comprises mainly glazed walls and a transparent or translucent roof. Also known as a **sunroom. 2.** A glass constructed building used to grow plants.

consideration The part of a contract that has something of worth. For a contract there needs to be an offer, acceptance, and something performed or something of worth, e.g. agreeing to undertake a task for payment or buy something is the consideration.

consistency (consistency index, consistency limits) *See* ATTERBERG LIMITS.

consolidated-drained test A *triaxial test where free drainage of the sample is allowed during both the consolidation and axial loading stage. Loading is carried out at a slow rate so that no increase in pore pressure occurs. *Effective stress parameters (c'_d and ϕ'_d) are obtained from such a test and typically used for long-term stability analysis.

consolidated-undrained test A *triaxial test where free drainage of the sample is allowed (usually for 24 hours) under cell pressure, to allow the sample to consolidate or to become saturated. Drainage is prevented and pore pressure readings (u) are taken during axial loading. *Total stress parameters (c_{cu} and ϕ_{cu}) and *effective stress parameters ('c' and 'ϕ') are obtained. Typically, such parameters are used to analyse problems regarding sudden change in load after an initial stable period, e.g. rapid drawdown of water behind a dam, or where effective stress analysis is required, e.g. slope stability.

consolidation Decrease in volume of voids (expulsion of air or water) in a soil that is subjected to *compression. In a fully saturated soil the resulting increase in pressure is first taken by the pore water and is then gradually transferred to the soil particles over a period of time. The rate at which this transfer of load occurs depends upon the *permeability of the soil. In a high permeable soil (e.g. silty clay) the transfer of pressure from the pore water to the soil skeleton will occur very quickly, thus consolidation will be rapid. In a low permeable soil (e.g. clay) the transfer of pressure from the pore water to the soil skeleton will be slow, thus consolidation would occur over many years.

consortium A group of people or organizations that come together for a specific task, project, or programme.

constant concentration decay method A type of *tracer gas decay method where the volume of tracer gas injected into the space is varied to maintain a pre-determined concentration of tracer gas within the space. The concentration of the tracer gas within the space is then measured as it decays over time.

constant-head permeability test A test to determine the *coefficient of permeability (k) of highly permeable soils such as sands and gravels (i.e. $k > 10^{-4}$ m/s), fall-head permeability test. Water from a constant-head supply flows through a soil sample until steady flow conditions prevail. The head loss (Δh) across the samples is then obtained by determining the difference between two manometers: one connected to the top of the sample and the other at the bottom. The head loss is divided by the length (L) of the sample between the manometers (device for measuring pressure) to obtain the *hydraulic gradient (i). The rate of flow (q) is obtained by measuring the quantity of water (Q) collected in a measuring cylinder over a measured time (t). Given that the cross-sectional area (A) of the sample is known, the coefficient of permeability can then be calculated from Darcy's equation:

$$k = q/Ai$$

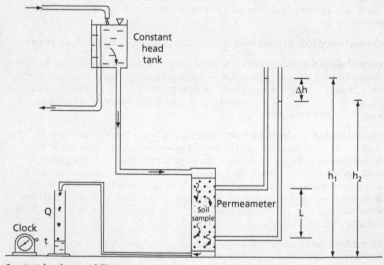

Constant-head permeability test

constant injection decay method A type of *tracer gas decay method where a predetermined concentration of a tracer gas is constantly injected into the space. The concentration of the tracer gas within the space is then measured as it decays over time.

constant velocity grit channel A long narrow channel, parabolic in cross-section, that is designed to give constant *velocity with varying depth of flow. They are used in sewage treatment plants to allow grit particles to settle out. The maximum depth of the channel is calculated on the basis of a design particle which will settle at 0.02 m/s when the velocity is slowed to 0.3 m/s. This particle must reach the bottom within the length of the channel.

construction The process of creating or altering a building, structure, or object.

Construction 2025 A joint, long-term strategy between the UK Government and industry to transform the construction sector, focusing on smart technologies, *green construction, and overseas trade.

construction contracts Standard forms of construction contract created specifically for building and civil engineering projects. The term is also used in the Housing Grants Construction and Regeneration Act 1996 meaning an agreement for carrying out construction operations.

Construction Design and Management Regulations (CDM regs) A health and safety regulation that specifies health and safety responsibilities and management duties that must be performed by the client, designers, and contractors.

Construction Industry Council (CIC) A UK forum that collectively represents and supports professionals within the construction industry.

((⊕)) SEE WEB LINKS
• The CIC official website

Construction Industry Development Board (CIDB) An entity responsible for the facilitation of construction industry performance improvement, including the development of contractors.

((⊕)) SEE WEB LINKS
• The CIDB official website

Construction Industry Training Board (CITB) Organization that organizes and subsidizes training and skills development for the construction industry. Construction companies pay training levies to the CITB, now largely replaced by Construction Skills. The companies can then draw from the organization subsidies for training employees.

construction joint A deliberately made joint in concrete that is formed when one area of concrete is laid at a different time to another. The joint should provide a good bond between the areas of concrete and should not allow movement.

construction load The loads imposed on an element during construction.

construction management (construction management contracts) The management of a building project where one organization takes a fixed fee or percentage of the project value to organize and manage the designers, contractors, and subcontractors.

construction manager The person who organizes and leads the on-site building process. Different companies and organizations may have different meanings for this, but the term is also interchangeable with *construction project manager, site manager, and *site agent.

Construction Operations Building Information Exchange (COBie) A non-proprietary data format for the publication of a subset of *BIM, which is directed towards providing asset data rather than geometric information.

construction process A physical item of work that is undertaken during *construction.

Construction Project Information Committee (CPIC) A committee to provide best practice and guidance on the content, form, and preparation of construction production information (CPI).

construction project manager A professional who oversees the completion of upstream and downstream construction tasks, including the coordination of consultations on a project.

constructive dismissal The process where an employee is wrongfully forced out of their job or position or is placed in a position that makes their job untenable. The illegal process means that an employee can bring an action against the employer for wrongful dismissal and should be entitled to compensation.

consultant A professional person providing expert advice in a given subject area. For example, a **consultant engineer** is a chartered engineer who advises the client on certain or all aspects of a project such as the design, suitability, and/or construction. They can be self-employed or work for a consulting firm.

consumer *See* CLIENT.

consumer unit A box located inside a dwelling containing a *mains isolation switch and a series of *fuses or miniature *circuit breakers. It should be located as close as possible to the *cable tail intake on the internal wall.

consumptive use The amount of water lost through evaporation, taken up by vegetation, or other natural processes.

contact bed *See* TRICKLING FILTER.

contact pressure The distribution of the load beneath a foundation slab. In sand the actual contact pressure decreases from the centre to the edge, while in clays the contact pressure is higher at the edge than the centre.

contaminated land Section 78a (2) of Part IIA of the Environment Protection Act 1990 presents a legal definition of contaminated land as: 'any such land which appears to the local authority in whose area it is situated to be in such a condition, by reason of substances in, on or under the land that—

(a) significant harm is being caused or there is a significant possibility of such harm being caused; or

(b) pollution of controlled waters is being, or is likely to be caused'.

SEE WEB LINKS

- The GOV.UK website details the above Act
- Department for Environment Food and Rural Affairs (DEFRA) circular regarding contaminated land

contemporaneous documents All of the paper-based documents that can act as evidence of events and transactions. In delay and disruption claims all such evidence shows the events and consequences relating to relevant events and the resulting delay and disruption.

contiguous bored piles *See* PILE.

contingency Something that is set aside in case of events that have a likelihood of occurring but are not planned into the contract in the usual way, or an allowance made for unexpected events. A **contingency plan** identifies a specific course of action should the need arise; this may be for an event that may occur or an unexpected event. A **contingency sum** is an allowance of money for certain events that may occur

Continuing Professional Development (CPD) The process of updating one's knowledge and skills during a professional career. It is a requirement of all professional bodies that the members keep their knowledge and skills up to date and maintain records of their attendance on training courses.

continuous flight auger (CFA) *See* AUGER.

continuous grading *See* GAP-GRADED AGGREGATE.

continuously reinforced concrete pavement A *pavement made from Portland cement, which does not have any transverse expansion or contraction joints, but has continuous longitudinal steel reinforcement. Transverse cracking occurs, but these cracks are held tightly together by the reinforcement.

contour line A line on a *map or plan joining points of equal height above datum, usually *ordnance datum but sometimes a local datum.

contract The agreement between two or more parties comprising an offer, acceptance, and consideration. The contract may be a standard form of agreement between the *contractor and *client, such as a *construction contract or maybe a simple contract formed between the client and contractor.

contract administrator The person who oversees the construction project on behalf of the client, ensuring that the process, terms, and agreement are followed in accordance with the agreement.

contract bills The contractual documents that identify the rates, prices and quantities, and materials that should be delivered.

contract documents All of the drawings, details, bills, terms, and conditions that make up the information submitted for tender and used to agree the contract.

contract drawings The illustrations, working details, and other graphics used to support the tender and contract documents at the point when the contract was signed and agreed. These drawings emanate from the architectural *drawings and are subsequently used during the contract period as a basis for the drawings agreed.

contracted weir A *weir that does not extend the full width of the channel and thus has side and end contractions. Used for measuring the flow in open channels such as streams and rivers. *See also* SUPPRESSED WEIR.

contract hire An agreement to hire plant and equipment from a supplier for a fixed term.

contraction joint (shrinkage joint) A joint within a concrete structure or slab allowing shrinkage to occur on drying.

contractor A person or organization that agrees to undertake the works.

contractor's obligations The terms, conditions, and warranties that the builder agrees to undertake for the client. The builder can assign benefits but cannot assign any contractual burdens (things the contractor agrees to do for the client) without express (written) agreement.

contract programme 1. The plan of work contained within the package submitted with the tender documents and agreed upon. The start, finish, and critical milestones are normally the only fixed parameters of the plan of works. Unless specified within the contract documents it is reasonable to expect other items or work to change as the project unfolds, weather conditions vary, and change orders are issued. Normally illustrated in *Gantt chart format. **2.** The live plan of work that is referred to during the contract. The plan of work will vary and develop slightly, making use of float, accommodating changes, and adjusting tasks to ensure full utilization of project resources.

contract sum (contract price) The monetary figure fixed and agreed upon in the contract. The contract value can also have the same meaning or might have a different connotation depending on whether it is the expected profit, loss, or earned value that is being discussed.

contract time The period of the project proposed, fixed, and agreed in the bid and accepted contract. The period is measured from when the contractor is expected to enter the land to establish site set-up, to the date when the contractor hands over the completed project.

contra proferentem A rule used to decide the meaning of a term where it could have two or more different meanings. Where an ambiguous term with two or more meanings arises, the courts can construe the meaning such that the party against whom the term is used is allowed the more favourable definition. The party putting forward the claim would not get the benefit of the interpretation that they were implying and relying on.

contributory negligence Where the actions of an individual have increased their own risk of injury and may have added to it. In such situations, the injured party is still able to claim against the party that was negligent; however, the claim may be proportionally reduced to take into account the injured party's actions that negligently exposed themselves to and increased the risk of injury. Contributory negligence is governed by the Law Reform (Contributory Negligence) Act 1945.

control gear The electrical and electronic equipment used to control discharge lighting, such as fluorescent, sodium, mercury, and metal halide lamps. Usually includes a starter and *ballast, but may include other items such as a power factor correction capacitor and dimming equipment.

controlled water Any watercourse such as a canal, river, stream, including underground water. It is an offence to cause pollution to any of these.

controlled waste In the UK, all waste (commercial, household, or industrial) that requires a licence for its disposal or treatment.

controlled waste regulations Regulations governing controlled waste in the UK.

(((●))) SEE WEB LINKS

• Archive of the original Controlled Waste Regulation.
• The Controlled Waste (England and Wales) Regulations 2012

Control of Major Accident Hazards (COMAH) Regulations which implement the *Seveso III Directive to prevent and mitigate the effects of major accidents concerning dangerous substances which would cause serious damage or harm to people and the environment.

Control of Substances Hazardous to Health Regulations (COSHH) The regulations that ensure all materials are described so that the risks associated with handling and using them are clear to those who are likely to come into contact with these hazards.

controls Electrical or mechanical devices capable of operating items of equipment according to predefined conditions.

control valve *See* DISCHARGE VALVE.

convection The process by which heat is transferred in a gas or liquid. As the gas or liquid is heated, it expands and becomes less dense. It then rises and is replaced with colder, more dense gas or liquid which in turn is then heated and rises. This results in a continuous flow or convection current from one area to another. It is one of the three main heat-transfer mechanisms. *See also* CONDUCTION and RADIATION.

convective loop The flow of air that occurs due to *convection. As the air is heated and expands, it becomes less dense and rises. Once the hot air reaches the top, it cools down, increases in density, and then sinks to the floor. *See also* CONVECTION.

convective loop bypass Convective heat loss that occurs within a building element and bypasses the benefit of the thermal insulation.

convector (convector heater) A space heating device that heats the space using *convection.

conversion The process of cutting up tree trunks into sections prior to *seasoning. The aim of conversion is to maximize output in financial rather than volume terms.

conveyor A mechanical device used to transport materials from one place to another. Usually comprises a continuous wide rubber belt supported on rollers.

cooling tower A device in which circulating air is used to cool warm water by
*evaporation. Used to dispose of waste heat from the cooling coils of air-conditioning
chillers and remove heat from coolant water in electricity generating plants.

coordinates A set of numbers that details the exact position of something with
reference to a set of axes. *See also* CARTESIAN COORDINATES.

coordination The act of aligning tasks, activities, and resources so that they are
managed effectively. **Coordination meetings** are used to bring together all of the
contractors, subcontractors, and designers necessary to ensure that tasks and
activities can be organized so that they are properly integrated together and
managed effectively.

COP24 The informal name and acronym for the 24th Conference of the Parties to
the United Nations Framework Convention on Climate Change held in Poland.

• The COP24 official website

coping A protective covering placed on top of a wall or parapet to prevent
rainwater penetration. Usually projects slightly from the face of the wall.

copper (copper alloys) A non-ferrous metal, chemical formula Cu. Copper is used
in buildings for plumbing pipes (due to its excellent corrosion resistance) and
fittings, and sometimes in sheet form for roofing. Due to its electrical conductance
and malleability, copper is the predominant material used in electrical wiring in
buildings and dwellings. Copper can be alloyed with zinc to form *brass or with *tin
to form *bronze. These alloys, especially brass, find many applications in plumbing
and electrical fittings. Pure copper has a density of 8,900 kg/m³.

corbel A section of brickwork, masonry, timber, or other material that projects
outwards from the face of a wall to support a load above.

corbelling Layers of brickwork or other materials that successively project out
from the face of a wall. *See also* CORBEL.

cordierite A mineral silicate derived from magnesium, aluminium, and
sometimes iron. Cordierite has a violet or grey colour and is sometimes also called
dichroite.

corduroy road A road formed from logs tied together at each end. Used for
crossing muddy or swampy ground.

core 1. The central part of something. **2.** A cylinder column of material, typically
soil or concrete, obtained during a sampling. A **core barrel** is a steel tube with a
coring bit (i.e. a cutting bit with pieces of tungsten carbide or industrial diamonds
set in a soft metal), which is used to obtain an undisturbed sample of strata during a
site investigation, *see* SOIL SAMPLER. A **core box** is a container with dividers, used to
keep cores from a borehole. A **core catcher** is a steel spring in a soil sampler that is
used to retain a sand sample. A **core cutter** is an attachment that can be fitted to a
core barrel to break and grip the core so that it can be extracted from the ground for
examination. A **core drill** is a water-cooled industrial diamond-bitted power-drill
tool, used to extract rock cores from the ground. A **core test** is a crushing test
undertaken on a concrete core sample.

cornice 1. An internal decorative moulding used at the junction between the wall and the ceiling. **2.** An external decorative moulding used over openings or at the top of an external wall.

corridor An enclosed passageway inside a building used to access rooms that lead off it.

corrosion The most common form of degradation mechanism for materials, especially metals.

corrugated sheeting A sinusoidal-shaped sheeting material used for cladding, and available in a range of materials including aluminium, steel, GRP, and PVC.

corundum An extremely hard mineral, aluminium oxide Al_2O_3, sometimes containing iron, magnesia, or silica, that occurs in different forms and colours and is used mainly in abrasives.

cost breakdown The separation of items, resources, tasks, and projects and their associated costs and values. The ability to separate costs provides greater financial control. Comparisons of estimated against actual cost for individual items ensure that those items that result in loss and those that make a profit can be identified.

costing The calculation of the monetary worth of work by measuring the materials and labour used and multiplying the quantities by the rate agreed. The actual cost can be compared with the amount fixed in the contract documents.

cost-plus contracts Projects where the price of the work is covered and a rate is agreed to cover the service or profit.

cost-plus-fixed-fee contract The cost of the work performed by the contractor is covered by the client and a set amount of profit and contractor's overheads are agreed.

cost-plus-percentage contract The cost of the work is covered and a percentage rate for overheads and profit is calculated against the total amount of work completed. Such remuneration schemes offer little incentive for work to be undertaken quickly and efficiently, the more resources used the greater the cost and profit. Work of this nature should be performed with close supervision

cost-reimbursement contract Contracts that are completed and the price of the work is paid as the work is completed, profit and overheads may be covered by a fixed figure or percentage of the value of the work paid. Such contracts are normally only performed where there is insufficient time to measure work and prepare bills.

Coulomb's earth pressure theory A concept (developed in 1776) used in retaining wall analysis for cohesionless soils (*see* GRANULAR) in which a failure wedge develops behind a retaining wall; the failure plane rises at an angle from the base of the wall, *see* RANKINE THEORY.

Coulomb's equation An equation developed by Charles Augustin Coulomb (1736–1806) that defines the shear strength (τ_f) of soil to the normal stress (σ_n) on a failure plane as illustrated below. Where c is *cohesion and ϕ is *angle of shearing resistance, the equation can be represented in terms of *total and *effective stress parameters.

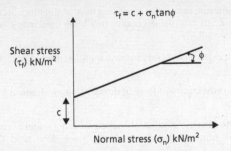

$$\tau_f = c + \sigma_n \tan\phi$$

Shear stress (τ_f) kN/m^2

ϕ

c

Normal stress (σ_n) kN/m^2

Coulomb's equation: failure envelope

couple roof The simplest form of roof construction, in which the diagonal timbers (rafters) are simply leant and fixed to each other, with the rafters fixed at the wall plate and at the ridge.

Council of Engineering Institutions (CEI) An association founded in 1965 for the engineering institutions, but was replaced in 1983 by the *Engineering Council.

counter ceiling *See* FALSE CEILING.

counterclaim The defending party to a claim serves a claim for damages against the claimant.

counterjib The rear part of a crane's *jib that carries the counterweight.

counter-offer Following an initial offer by one party, an alternative or different offer is made by the other party. For a contract to be made there must be consideration and an offer and acceptance. If an offer to do something is made and the other party adds or alters the offer, this is not a contract, but it is a counter-offer and the party undertaking the work must agree to the alterations for the contract to be complete.

countersink 1. A shallow enlarged hole drilled in a material to enable the screw head to sit flush with the surrounding surface. **2.** A tool used to produce a countersink.

couple Two horizontal forces that are equal but opposite in magnitude, which causes a turning effect.

coupler (coupling) 1. A pipe fitting that enables two pipes to be joined together. **2.** A device that enables scaffolding poles to be connected together.

course A horizontal layer of masonry, one unit in height, including any bedding material. Also used to describe horizontal layers of materials such as tiles or damp-proofing. Materials that are laid in courses are said to be **coursed**, for instance, **coursed brickwork**, **coursed blockwork**, etc.

cove A concave-shaped moulding placed at the junction between a ceiling and a wall.

cover 1. A lid or piece of material that can be placed over a surface, either to protect or to hide it. **2.** The thickness of concrete that is required to be placed over steel reinforcement to protect it.

covering Materials used on walls, floors, and roofs to hide the structure.

covering-up The process of covering over previous work.

cover meter A non-destructive test instrument, used to check the position and direction of reinforcement in concrete.

cowboy A slang term used to describe an unskilled worker.

cowl A device, usually terracotta or aluminium, that is placed at the top of a chimney to prevent down-draughts, assist ventilation, and reduce rain penetration and bird entry. Available in a wide range of shapes and sizes and can be static or revolving (wind-driven).

crab A small portable winch.

craft A skilled occupation or trade, for instance, bricklaying, carpentry, joinery, painting, plastering, or plumbing. The person who undertakes a particular occupation or trade is known as a **craftsperson** or **craft operative**. A **craft foreman** supervises the work undertaken by the craft operatives.

cramp 1. A hand-operated tool comprising two jaws and an adjustable screw that is used to hold pieces of wood together while they are being glued. **2.** A metal strap attached to a door or window-frame that is used to fix the frame into an opening.

crawler track The tracks used on construction and earth-moving equipment.

crawling A defect in paintwork where the paint tends to pull away (crawl) and bead on the surface. Caused by a lack of adhesion to the substrate

crawl space A shallow space under the ground floor of a house or in the loft just under the roof that is not large enough to allow a person to stand up. Used to gain access to services, such as ductwork, electrical cables, and pipework.

crawlway A duct containing services that is large enough for a person to crawl through.

crazing A defect in paint, plaster, or concrete where a random series of tiny hairline cracks appear on the finished surface.

creasing One or more courses of tiles that project out from the face of a wall. Used as a coping and at window sills to project water away from the wall.

creep Permanent deformation of materials that takes place usually at elevated temperatures, especially in metals. Creep in concrete occurs at ambient temperature.

creosote An oily liquid containing phenols and creosols. Creosote is usually obtained from wood tar, especially from the wood of beech. It is usually colourless or yellow.

crest The top of something, e.g. the **crest of a wave** is the top of a wave or the **crest sight distance** is used in the design of carriageways where the rate of change of the summit is designed to accommodate adequate sight for overtaking.

crib Steel or timber members laid at right angle to each other, used to spread a foundation load. *See also* GRILLAGE. A **crib dam** is a retaining structure formed from stacked stones or *precast concrete units. A **crib wall** is a series of pens made from prefabricated timber, precast concrete, or steel filled with granular material. It acts like a gravity wall but can withstand large differential settlements. **Cribwork** is a series of timber cells filled with concrete and are sunk to form bridge foundations.

Crimp and Bruges' formula Used to calculate flow in sewers:

$$v = 56m^{0.67}\sqrt{i}$$

where v = velocity (m/sec), m = hydraulic mean depth (m) and i = the slope of the sewer.

It is now considered that the equation yields conservative estimates (of the order of 20%) for the value of v when compared to experimental data. *See also* BARNES FORMULA.

crippling load (buckling load) The load at which a column starts *buckling.

critical Something that has extreme importance at a certain state. **Critical angle** is when complete reflection of a ray of light from a surface occurs. **Critical circle** is a failure slip circle in a slope stability analysis with the lowest factor of safety. **Critical height** is the height to which a *cohesive soil will stand without support. **Critical slope** is the maximum angle a slope can stand without deformation. **Critical velocity** is when the *Froude number equals one for flow in an open channel or when laminar flow changes to turbulent flow in a pressure pipe. *See also* BELANGER'S CRITICAL VELOCITY and KENNEDY'S CRITICAL VELOCITY.

critical path The longest path of events that has no float between them, which dictates the total duration of the project. Through critical path analysis logical links are established between task and a network of events. Tasks that do not fall on the critical path have float and potential for deviation without affecting the total duration of the project; these tasks are not critical.

critical path method (CPM) (critical path analysis (CPA), critical path scheduling) Logical links between tasks are established and the start, finish, duration, and float of each task is determined. The links and activities are analysed to determine which tasks are critical and have no float and those where some changes in start times can take place. Float is the ability of an event to slip backwards without affecting other events. Events that can change, be brought forward, or moved back, are non-critical. Slippage in non-critical events does not affect the overall completion time, unless excessive slippage makes them critical to the delivery. The critical path analysis determines the earliest start time, latest start time, earliest finish, and latest finish of each task. Where the earliest finish and latest finish are the same, the events are critical. Those events that have different earliest and latest start and finish times have a certain amount of float and are not critical.

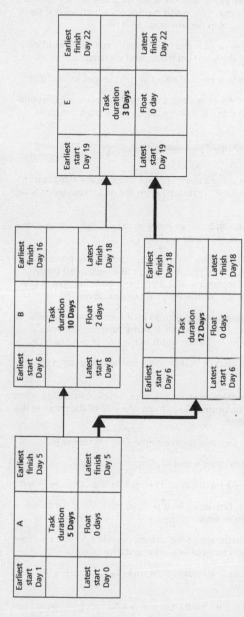

Critical path method: Critical path through precedence network

critical state (critical state theory) A condition at which phases of substances are identical or in equilibrium. For example, in soil mechanics when soil is sheared without incurring any further changes in density or stress level.

cross-bracing Diagonal bracing or tie rods that resist shear forces from wind loads etc. to provide internal stability/stiffness in roof truces and rectangular frames.

cross-cut A cut made with a saw in a piece of wood at right angles to the *grain.

crossetted arch An arch built of stones cut into a wedge shape allowing the joints between the stones to form a semicircular arch, radiating out from a common centre.

crossfall A slight gradient on a footpath or a carriageway to prevent ponding of surface water.

cross-hairs Fine lines that cross at right angles (in the shape of a + sign) in a telescope or theodolite, which are used to precisely align the horizontal and vertical lines of sight.

crossing Applying coats of paint with brush or roller strokes at right angles to the direction of the previous coat to ensure a smooth finish.

cross-laminated timber (CLT) Large sections of structural timber produced from dried, quick-growing spruce and fir (larch or pine are alternatives) boards, stacked at right angles, and glued together under a pressure-bonding system over their entire surface in generally 3, 5, 7, or more board or panel layers.

cross-machine direction (x-machine direction) The direction perpendicular to the *machine direction that determines the width of something, for example, a geotextile. Typically, this is determined by the width of the bed of the machine.

cross-section A view of a component or building as if the item was cut in order to provide an internal perspective.

cross-ventilation A *natural ventilation strategy where *ventilation openings are provided on both sides of a space. Air then flows from one side of the space to the other in response to the *wind effect.

cross-wall A structural internal wall that intersects the external wall at 90°.

cross-welt A type of joint used in sheet-metal roofing.

crown 1. The upper-most part of a building. **2.** The middle of a cambered road.

crown course The top *course of masonry on a wall. In brick walls, the crown course often comprises a brick on its edge.

crown plate A purlin that sits on the top of the *crown posts in a *crown post roof to support the *collars. Also known as a **collar purlin** or **collar plate**.

crown-post The vertical member in a *crown-post roof that spans from the *tie beam to the *crown plate.

crown-post roof A traditional timber *roof truss where a *crown-post spans from the centre of the tie beam to the *crown plate, which in turn supports the *collars.

crown rafter The central *common joist in a *hipped end roof.

crown-strut A vertical timber roof truss member that spans from the *tie beam to the *collar.

cryogeology A term used with general reference to frozen/freezing the ground. The prefix cryo- is derived from the Greek word *krous*, meaning 'icy cold'.

crystalline Any material that has an ordered internal atomic structure.

crystallinity A measure of how *crystalline a material is.

crystallization A process whereby the internal structure of a material becomes more ordered. Crystallization is an act of forming crystals or bodies during solidification, the crystals are normally symmetrically arranged.

crystal structure How atoms or ions are arranged or spaced inside a material. For example, most metals have either a *body-centred (bcc) or a *face-centred cubic (fcc) structure.

cubic A three-dimensional shape in the form of a cube. A **cubic parabola** is a *polynomial curve of the third degree, e.g. $y = x^3$. A **cubic spiral** is a *polynomial curve. *See also* CLOTHOID.

cubicle A small enclosure, i.e. a changing cubicle or a shower cubicle.

Culmann line A graphically constructed curved line, which represents values of earth pressure for randomly chosen failure planes behind a retaining wall.

culpable delay Part of the delay for which the client is responsible. A delay may be a result of changes and delays by the contractor, client, or designer. Where a delay is a result of actions taken by more than one party, the amount that results from the client's actions needs to be determined.

culvert A covered *channel used to convey a watercourse below ground, mainly under roads and railways.

curling The uplifting of the edges of a laminate or wooden floor due to the upper surface having a greater rate of shrinkage than the lower surface.

curtailment The bending of reinforcement bars at the end of the bar to ensure greater anchorage within the concrete.

curtaining A defect in painting that occurs on vertical surfaces where there has been excessive movement in the paint film, resulting in it sagging into thick lines, caused by the uneven application of the paint.

curtain wall A non-load-bearing wall used to clad a building. Usually comprises a secondary framework that supports cladding panels or sheets of glass. Frequently used to clad non-domestic buildings.

curtilage The enclosed area of land occupied by a dwelling, its outbuildings, and garden.

curvature 1. The amount to which something curves, e.g. the curvature of the Earth. **2.** The rate of change of a curve, e.g. in a circle, curvature is the reciprocal of radius.

cusec The rate of flow given in one cubic foot per second.

cushioned flooring A durable and tough floor covering comprising a vinyl or rubber flooring mat topped with an acrylic coating.

cut *See* CUTTING.

cut and fill A process of excavating earth (cut) and placing it (fill) to form the vertical alignment of road, railway, or other forms of earthwork.

cut and fit *See* SCRIBED JOINT

cut brick *See* ROUGH CUTTING and FAIR CUTTING.

cut corner A 90° corner that has been cut to form an angle of 45°.

cut nail A nail of constant thickness that has been cut out of a steel plate.

cut-off A construction formed below the ground to minimize water seepage into an excavation, such as a **cut-off trench** (or **cut-off wall**), which is filled with (made from) impervious material. The depth below excavation to the cut-off is termed the **cut-off depth.**

cut-out 1. An automatic device that disconnects a circuit, e.g. a *circuit breaker. **2.** A section that has been removed from a building or wall after the original construction has finished.

cut stone *See* DIMENSION STONE.

cut string (open string) At one side of the staircase the *string is cut out so that the *treads and *risers profile can be viewed from the side, i.e. where a side of staircase is open rather than adjacent to a wall.

cutter block A steel block housing two or more knives that is fitted to woodcarving machines. It rotates at high speeds to shape or smooth timber.

cutting A permanent excavation with given side slopes to allow the formation of a railway or road.

cutting-in A process of painting a clean edge at the perimeter of a painted area, for example where ceilings meet walls or where walls intersect.

cutting list A list of steel bars (detailing diameters and length) from which reinforcement is ordered.

cutwater (starling) The pointed projection at the base of a bridge pier, shaped to allow the water to part around the base.

cyanide A highly hazardous compound which is a form of hydrogen cyanide. Cyanide is extremely poisonous (with fatal consequences if consumed by human beings) in the form of compounds potassium cyanide and sodium cyanide.

cyclic loading A loading effect (that represents a sine wave) which is repeated in cycles and used in *fatigue tests.

cylinder A tube-shaped container with straight sides and circular ends of equal size. A **cylinder test** involves casting concrete in a steel or cast-iron cylindrical mould; the cast concrete is then tested to destruction by compression. It can be tested with the top and bottom surfaces of the cylinder against the loading platens to determine compressive strength. Alternatively, the cylinder can be placed with its horizontal axis between the platens to determine its tensile strength, referred to as the **splitting tension test**.

dabs Small amounts of soft adhesive material used for fixing sheet or board materials to walls.

dado 1. The lower part of an internal wall in a room, decorated differently from the rest of the wall. For instance, it may be painted differently or be clad in wood panelling. **2.** A rectangular groove cut across the grain of a piece of timber to allow one piece to connect to another.

dado capping *See* DADO RAIL.

dado rail (dado capping, chair rail, rail chair) A protective moulding that runs around the internal walls of a room at chair-back height. Also used for decorative purposes.

dado trunking Surface-mounted *trunking that runs along the internal walls of a room at *dado rail height.

dairy Farm building for milking cows.

dam A barrier of concrete or earth that is constructed to retain water, typically to form a *reservoir.

damage Physical destruction that reduces the usefulness, value, or function of something. The act results in damage, reducing the value of a service or product. The effect is usually caused by the actions of a person or the effects of the environment.

damages Losses suffered as a result of failure to deliver the contract as specified. For example, where a party has failed to complete their part of the contract on time, the additional costs incurred as a result of the delay could be claimed.

damages for delay (US penalty clause) The loss and expense associated with a delay.

damp (dampness) A state of matter containing moisture, usually higher than the equilibrium amount.

damp course *See* DAMP-PROOF COURSE.

damping A process of reducing the effects of an oscillation or vibration in a system.

damp-proof course (damp course, dpc, DPC) A horizontal strip of impervious material that is built into a wall and is designed to prevent moisture penetration by capillary action. DPCs should be laid a minimum of 150 mm above ground level.

A variety of materials can be used for a DPC including slate, lead, bitumen-polymer, and polyethylene. The most commonly used materials are those that are flexible (**flexible damp-proof course**). Care must be taken when installing a **flexible damp-proof course** to avoid puncturing the material. DPCs can also be retrofitted into existing walls by injecting a suitable chemical, such as silicone, into the wall at regular intervals.

damp-proof course brick A brick used to construct a *damp-proof course.

damp-proofing To prevent damp by, for example, using a *damp-proof course or a *damp-proof membrane.

damp-proof membrane (dpm, DPM) An impervious layer of material, usually 1200 gauge polythene sheet, that is built into a solid ground-bearing concrete floor and designed to prevent moisture penetration by capillary action. Should lap with the damp-proof course in the wall. *See also* DAMP-PROOF COURSE.

darby (darby float) A large twin-handled aluminium float used by plasterers to level plaster.

Darcy's Law The flow properties of water through soil was first investigated in 1856 by Henry Darcy. He showed that under steady flow conditions through beds of sand of varying thickness and under various pressures, the rate of flow was always proportional to the hydraulic gradient (the fall in hydraulic head per unit thickness of sand). This principal is known as Darcy's Law. The equation has been found to be valid for most types of fluid flow through soils.

$$q = Ak\,i \quad \text{where}: \quad q = \text{rate of flow}$$
$$A = \text{area of flow}$$
$$k = \text{coefficient of permeability}$$
$$i = \text{hydraulic gradient}$$

Darcy-Weisbach formula An expression to derive the head loss due to friction in a pipe as a result of its roughness

$$h = f\left(\frac{L}{D}\right)\left(\frac{v^2}{2g}\right) \quad \text{where}: \quad h = \text{head loss}$$
$$f = \text{Darcy frictional factor}$$
$$L = \text{length of pipe}$$
$$D = \text{diameter of pipe}$$
$$v = \text{velocity of flow}$$
$$g = \text{acceleration due to gravity}$$

dart valve A valve, on the bottom of a *bailer, that employs a pin-like mechanism, which closes on the up-stroke and opens on the down-stoke.

dashed finish A type of render finish used on external walls to shed water. Can comprise either a dry finish, such as *pebbledash or a wet finish, such as *roughcast.

dashpot A device to dampen vibration. It consists of a piston enclosed in a fluid-filled cylinder.

data Unprocessed factual information. It is frequently captured in laboratory experiments by a **data logger**, an electronic device that records data from a sensor or instrument over a period of time. **Data processing** then occurs, involving entering, storing, uploading, and analysing the data using a computer. Such data can be grouped to form a **database**, a collection of computer data that can be sorted or manipulated in various ways.

data acquisition (DAQ) A process of measuring signals of electrical or physical conditions (such as voltage, current, temperature, pressure, or sound) and representing them numerically. A DAQ system consists of sensors to monitor the value, and measurement hardware and software to interrogate the values.

data cabling A collection of computer cables or optical fibres that are bound together.

data drop The exchange of information within *BIM, equivalent to the stage reports pre-BIM, including (1) brief, (2) concept, (3) definition, (4) build and commission, (5) handover and closeout, and (6) operation and in use. Each data drop offers the chance for the client to check and validate compliance of the project with the initial brief and employer's information requirements, as well as costs.

data exchange specification An electronic file format which is used for the exchange of digital data for a variety of *BIM software applications.

data logger A compact electronic device equipped with an internal microprocessor, data storage, and one or more sensors. The devices can be deployed in a variety of environments to record measurements at set intervals. They automatically monitor and record data (including environmental parameters) over time, allowing conditions to be measured, documented, analysed, and validated. The data logger contains a sensor to receive the information and a computer chip to store it. Then the information stored in the data logger is transferred to a computer for analysis.

datum A fixed point of reference in surveys, upon which other measurements or calculations are based. *See also* BENCHMARK and ORDNANCE BENCHMARK.

day *See* DAYLIGHT SIZE.

daylight (natural light) The light obtained directly from the sun (sunlight) and reflected off clouds (skylight).

daylight factor The ratio of the interior illuminance at a particular point within a space divided by the simultaneous horizontal unobstructed exterior illuminance. Measured as a percentage (%). It is a combination of three separate components:

- The *sky component
- The *externally reflected component
- The *internally reflected component.

daylight prediction (daylight analysis) A method of estimating the amount of daylight within a space.

daylight protractor A calculation method developed by the *BRE to predict daylight factors.

daylight size (daylight width, sight size, sight width) The total area of a window opening that admits daylight.

days The glazed elements of a window.

daywork (dayworks, US force account) Payment agreement to undertake work for the price of the labour and materials, with a percentage to cover overheads and profit. The method is generally used for unforeseen events or variations where the work was not specified in detail within the contract, or due to the circumstances were not covered. Where work is undertaken using this method, the daywork sheets, describing the work done, material used, and the labour, are signed off at the end of the day, or period of work, by the *resident engineer, *clerk of works, or *client representative.

dB(A) A method of measuring the loudness of a sound which is weighted according to the frequency response of a human's ear. *See also* DECIBEL.

dead 1. A material that is no longer suitable for use. **2.** Disconnected from a distribution system or supply, for instance, the electrical circuit or telephone line is dead. **3.** Does not contain an electrical charge, for instance, a dead battery.

dead bolt (deadbolt) The part of a dead lock that engages into the door jamb when the key is turned.

dead end (US false exit) A street or corridor with only one means of gaining entry and exit.

dead leg A length of pipe within a hot-water system where the water only circulates when it is drawn off.

deadlight *See* FIXED LIGHT.

dead load The weight of a structure and any permanent fixtures and fittings. *See also* LIVE LOAD.

dead lock (deadlock) A type of security lock incorporating a *dead bolt that can only be operated using a key.

dead shore A supporting timber or prop positioned directly underneath *needles to support the weight of a wall. Where a wall needs to be temporarily supported while work is carried out below the wall, beams are inserted into and through the walls, props are then positioned under the beams (needles) so that the weight of the wall can be carried and secured, enabling work below to be undertaken.

deathwatch beetle (*Xestobium rufovillosum*) A wood-boring beetle that burrows deeply into the sapwood, which makes the beetle difficult to kill. The beetle tends to be found in decaying hardwoods, but will bore into adjacent softwoods. The entrance and exit holes are approximately 3 mm in diameter. The beetle is known for the ticking or knocking noise that it makes as it bores.

debris-collection fan (protection fan) A temporary canopy projecting from the face of a building that is designed to protect pedestrians by collecting any falling debris.

decametre (US decameter) A unit of length equal to 10 meters.

decanting 1. A method of pouring fluid from one container to another without disturbing the sediment. **2.** To transfer from one place to another. **3.** A method of decompression.

decarbonization The reduction of the carbon content from, for example, energy sources.

decay 1. The weakening of timber that has or is being attacked by fungus **(fungal decay, rot)**. Both *wet rot and *dry rot are types of decay, both requiring moisture to be present in the timber. Well-seasoned timbers have a natural resistance to decay, preservative treatments are also available to reduce the potential of the timber being attacked. **2.** The natural deterioration of something.

decentralized wastewater treatment A small-scale wastewater treatment system designated to a few localities/utilities based on the composition/quantity of influent wastewater which is produced. The system is generally preferred for small communities and is easily manageable.

dechlorination The removal of chlorine from water; for example, by the use of granular activated carbon filters.

deci- Prefix denoting a tenth.

decibel (dB) A logarithmic scale used to measure the intensity of a *sound or a sound-pressure *level.

deck 1. An outdoor platform, usually constructed above the ground. **2.** The structural component of a floor or roof.

decking 1. The material used to form the finished surface over a deck. **2.** The profiled metal sheeting used to construct in-situ concrete slabs.

declination 1. A downwards slope or bend. **2.** The angular distance to a point on a celestial object, measured in degrees from the celestial equator along the great circle passing through the celestial poles.

decorator Skilled worker who applies finishes such as paint, varnish, wallpaper, and stippled finishes.

deduction 1. The action of removing or subtracting one thing from something else. **2.** Coming to or reaching a conclusion or theory based on the known facts, or reaching a conclusion in respect of how a prior decision or outcome was reached. If used in construction research, it is based on scientific principles, the use of theory to direct theory, and research with the use data to conform. Quantitative data are used to explain and explore a causal relationship.

deed An agreement or *contract.

deeping The process of cutting timber to the required thickness by sawing through the thickest section of the timber. The opposite of **flatting**.

deep learning A form of *machine learning based on learning data representations rather than performing task-specific algorithms.

deep-plan building A building which has a high ratio of internal floor area to external wall area, resulting in large areas of floor space that are greater than 4 m away from the external walls. Such buildings can be difficult to light and ventilate naturally, so are commonly electrically lit and mechanically ventilated or air-conditioned.

deep-seal trap A U-shaped anti-siphon *trap having a seal of approximately 75 mm or more.

default Failure to undertake an agreed action or contractual duty.

defect Something that renders an object less than specification, not as required, and not perfect. Defects include scratches to surfaces, poor workmanship, products that are not within tolerance or specification.

defect action sheet Papers produced by the BRE that list and describe recognized problems, often in houses.

defect correction period The time allowed to make good or correct all defects listed.

defective work Work that does not comply with standards, regulation, or specification. The work may be a result of incorrect design, poor workmanship, error, or damage to the work.

defect structure A reference relating to the irregularities in the structure of a metallic or ceramic compound. These include the presence of vacancies and interstitials at a microscopic level.

definite integral The *integral between two set limits.

deflection The deformation from the horizontal due to loading.

deforestation The removal of trees from a forest.

degradation 1. The breakdown of chemical compounds into simpler compounds. 2. The erosion of the Earth's surface by *weathering. 3. The loss of quality or performance of something.

degree 1. The extent or amount of something. 2. An educational qualification. 3. A unit of angular measurement quoted between 0° and 360°. 4. A unit of temperature measurement as defined on the Celsius or Fahrenheit scale. **Degree Celsius** (previously degree centigrade) is the *SI unit of temperature at which water freezes at 0°C and boils at 100°C, under normal atmospheric conditions. **Degree Fahrenheit** is referred to as the non-metric scale at which water freezes at 32°F and boils at 212°F, under normal atmospheric conditions. 5. The highest power of a mathematical variable.

degree-day The number of degrees in a day that the average temperature is above or below a particular reference value, known as the base temperature. In the UK, the base temperature is 15.5°C. Degree-days are cumulative so can be added together over periods of time, for instance a month or a year. They are used to estimate the amount of heating or cooling that a building is likely to require.

degree of compaction A measure of a soil's compactness. *See also* VOID RATIO.

degree of freedom The number of independent coordinates that determines the orientation of a body or system, e.g. in a two-dimensional system a point will have two degrees of freedom.

degree of polymerization The extent to which a material has been polymerized, usually determined by either chain lengths or repeat units.

degree of saturation The ratio between the volume of water to the total volume of voids in a soil expressed as a percentage.

degree of sensitivity A measure of a clay's sensitivity to remoulding. It is the ratio of undistributed shear strength to the disturbed.

degree of utilization (Δ) A percentage value that represents the design *effect of actions on their corresponding design resistance, which should be $\leq 100\%$ to satisfy the *limit state design.

$$\Delta = \frac{E_d}{R_d} \quad or \quad \Delta = \frac{1}{\Gamma}$$

dehumidifier A mechanical device that reduces the moisture content of air either mechanically using a refrigerant, or chemically using a desiccant. Used in some types of air-conditioning systems.

dehydration A condition whereby an object or material has the concentration of water reduced to a level that is lower than what is considered to be equilibrium.

delamination A type of failure whereby one of the layers separate from the object.

delay The period that the project or events take place after the contractual or predetermined completion date.

delay damages Loss incurred as a result of a *delay.

delta 1. The fourth letter of the Greek alphabet. **2.** Used to represent a change in a variable; symbol Δ or δ. **3.** An area of land, normally triangular in shape, found at the mouth of a river as it branches out into a number of outlets.

delta iron (ferrite) A form of iron stable above 1400°C and with a *body-centred cubic (BCC) lattice structure. Iron/steel exists in different phases depending upon the temperature and carbon content (C). For pure iron and low carbon steel (containing about 0.1% C) the phase changes (upon heating from room temperature) from ferrite phase to austenite (at 900°C) and then to delta iron (ferrite).

deluge sprinkler A type of *sprinkler that always remains in the open position and when activated enables large amounts of water to flow through it to extinguish the fire. Used in areas that contain hazardous materials.

demand factor *See* DIVERSITY FACTOR.

demand matrix A document that identifies the information to be incorporated in the COBie (*See* CONSTRUCTION OPERATIONS BUILDING INFORMATION EXCHANGE) file that forms part of each *data drop.

demineralized The removal or loss of salts.

demolition The act of taking apart, disassembling, breaking down, and removing buildings, structures, and works. The act of taking down buildings has become much more sophisticated, as health and safety regulations require the structure to be safely dismantled and the materials safely disposed. Recycling of waste is now a regulated process demanding the segregation and reuse of materials.

demountable partition A prefabricated unit used to form a wall that is capable of being erected, taken down, and repositioned. Demountable partitions offer greater flexibility for office and building space.

denier A textile unit to denote the weight in grams of 900 metres of yarn, superseded by *tex, both of which are used to indicate the fineness of the yarns.

denitrification The conversion of nitrate to nitrogen gas. Used in wastewater treatment plants.

densification To increase the *density, the number of soil particles per unit volume, of an in-situ soil by the use of any form of ground improvement technique.

density Mass (m) per unit volume (V) for an object measured as kg/m_3.

depth The distance from the top to bottom or from the front to back of something.

derelict A building or structure in a state of disrepair or run-down condition that renders it beyond practical use. Such properties may be demolished, *recycled, or refurbished.

derrick crane A simple crane with controls at the bottom of a vertical mast, used for lifting and moving objects.

description Definitive statement of the product or service referred to in the contract to ensure the desired standard is achieved. It is important that sufficient detail is described to ensure that the required service and standard is met. The description serves two key purposes: it defines the work set that is to be completed; and if the nature, quality, standard, or type of work that has been carried out is questioned, it provides a benchmark statement against which the work can be compared.

design A model, sketch, drawing, outline, description, or specification used to create the vision of that which is to be created—a working item, product, building, or structure. Design can be considered as: **1.** Concept design: information that conveys the general idea or vision. **2.** Working design: information developed by pulling together details and specifications from the various trades and professionals building and integrating the content so that all of the information can function and fit together without gaps. The developing design is often broken down into stages with different stages being completed as the information from different groups or professionals is brought together. **3.** Final design: possesses sufficient detail to ensure that a fully functional product can be created.

design action An *action, namely a *representative action, that has had a *partial factor applied.

design and build contract *See* DESIGN-BUILD CONTRACT.

design approach (DA) 1. The method used the *Eurocodes, e.g. Eurocode 7 (EC7), to accommodate different ways to combine actions (A), materials (M), and resistances (R) to account for both location (where in Europe) and application (e.g. foundations, slope, piles, or retaining walls). There are principally three design approaches for EC7, which are designated as:

- DA 1: Combination 1: A1 + M1 + R1 and Combination 2*: A2 + M2 + R1
  Except for the design of axially loaded piles and anchors, where Combination 2 = A2 + (M1 or M2) + R4
- DA 2: A1 + M1 + R2
- DA 3: A1 or A2 + M2 + R3

The UK has adopted design approach 1 (DA1), which requires two calculations based on different combinations of *partial factors. **2.** Principles-driven design that sees the unique characteristics of each place and project as a source of inspiration and innovation. The foundational principles brought to each project derive from a vision of the future: a delightfully diverse, safe, healthy, and just world—with clean air, soil, water, and power—economically, equitably, ecologically, and elegantly enjoyed.

design–build contract (design and build contract, package deal, turnkey project) A complete contract that includes development, coordination and delivery of design, and all assembly aspects of a project. The contracting party is considered wholly responsible for taking whatever concept, design, and specification information that is given to them and developing it into a fully functional structure or building. As both the design and construction are contained within a single contract, the *contractor is exposed to greater risk than is carried for contracts that separate design and construction.

design code A document that provides rules for design. Such documents may or may not be statutory documents.

Design Engineer A *Chartered Engineer who is qualified to undertake the design of a structure.

design factor (overdesign factor) (Γ) The ratio of *design resistance (R_d) to the corresponding effects of the *design actions (E_d). To satisfy the *limit state design, $\Gamma \geq 1$.

design intent The original intention of a design.

design leader The professional who heads up the *design process ensuring the information is taken from concept to the final design, and ensuring the information brought together is properly integrated and coordinated.

design life The period of time, as expected by its designers, during which a structure or component is expected to perform within the specified parameters.

design resistance The *resistance offered, for example by the ground (such as the soil shear strength parameters), when a *partial factor has been applied.

design value (F$_d$) A value obtained by applying a *partial factor to the *representative value (F$_{rep}$).

desludging The removal of sludge from an item of equipment.

desuperheater 1. A device that reduces the temperature of superheated steam to its saturation temperature. **2.** A type of heat exchanger that extracts the waste heat from the compressor in an air-conditioning system or heat pump, and uses the waste heat to preheat water.

detached A building, dwelling, or structure that is separate from other buildings.

detail Drawing or written specification that conveys more intricate information about the assembly, arrangement, or components to be used. By showing more detail, more information is given, and a greater understanding should be achieved.

detail drawings Drawings that show arrangements of components and provide supporting information that ensures the assemblies can be achieved.

details 1. Drawings that show the *detail. **2.** Intricate information.

determination (termination) The end of a contract or agreement before all of the work has been completed. The end of a contract may be brought into force by either the employer or the contractor, for breach of contract, to the extent that it is considered reasonable to bring the contract to an end and seek damages for the breach. Suspension of works may allow the parties to continue to work together once matters have been agreed, whereas determination is final.

deterioration The reduction in the quality, value, or strength of something.

detritus Fragments of rock that have been produced due to erosion.

deviation 1. A change from an original course. **2.** The difference between a particular value in a set and a fixed number, e.g. the average of all the other values in the set.

deviatoric stress (differential stress) A condition where the stress components are not the same in all directions, e.g. in a *triaxial test on soil.

devil float (devilling float, nail float) A wooden hand *float containing nail points that project outwards from each corner of the float. Used by plasterers to scratch the surface of the plaster to provide a *key for the next coat. This process is known as **devilling**.

dewater To removal water, for example, to lower *groundwater by pumping to allow excavation to be undertake in dry conditions.

dew point temperature The temperature at which air is fully saturated with water vapour (i.e. at 100% *relative humidity).

dew point temperature gradient The change in the magnitude of the *dew point temperature that occurs through a construction.

diagnostic The identification of the cause of something by investigating and examining its symptoms.

diagonal 1. Something that is slanted. **2.** A line that joins two opposite or non-adjacent angles or corners of a *polygon or polyhedron.

diagram A schematic drawing of something.

dial gauge A mechanical instrument used to measure small displacements of a plunger by means of a pointer moving round a circular scale.

diameter A line running from one side of a circle or sphere to the other that passes through the centre.

diamond A material based on carbon with exceptional *hardness.

diamond break stiffening A very slight fold, running from corner to corner, in a panel of sheet metal, which reduces the tendency of the panel to vibrate, preventing a drumming noise.

diamond saw A tool that has industrial diamonds embedded along its cutting edge.

diamond washer A curved washer used with a hook bolt to secure corrugated roof sheet. It can be manufactured from aluminium, galvanized steel, or plastic.

diaphragm A taut thin membrane. It is can be used in a **diaphragm pump**, which is a reciprocal pump that has a flexible membrane. It can also be used in a **diaphragm tank**, which is a closed container having an upper and lower chamber separated by a diaphragm—the upper chamber containing air or nitrogen, the lower water.

diaphragm wall A retaining wall that has been formed within an excavated trench in the ground. As the trench is excavated it is filled with *bentonite slurry to prevent collapse. A reinforcement cage is then inserted into the trench. Concrete is placed via a *tremie at the bottom of the trench. Uses include basement walls, earth-retaining walls, water storage tanks, and cut-off walls.

dicamba A herbicide that is found in *groundwater and monitored in drinking water supplies.

die 1. A metal form used as a permanent mould for die-casting. A die is a tool responsible for cutting the shape of the label out of the material. **2.** A threaded cutting block that is used to cut threads into the outer surface of bars and pipes. *See also* TAP, which is used to cut threads into the inner surfaces of pipes or holes drilled into metal.

dielectric Any material that is electrically insulating.

differs (key changes, variations) The amount of alternative keys that can be made for a given lock. *See also* KEY CHANGES.

diffuse solar radiation Radiation from the sun that is scattered in the atmosphere before it reaches the Earth's surface. *See also* DIRECT SOLAR RADIATION.

diffuser 1. A grille attached to an air supply duct that distributes the air in a particular direction. **2**. A baffle placed in front of a luminaire to distribute the light evenly and minimize glare.

diffusion The rate at which atomic entities pass through a liquid, solid, and gas medium.

digestion The breakdown of substances (e.g. sewage) as a result of heat, water, chemicals, enzymes, or bacteria.

digging The breaking up and loosening of ground by hand, tools, or machine. The excavation using a hand, tools, and machines, such as a *spade or digger may be considered different from the loosening of the ground or spoil.

digital Representation of data as numerical digits.

digital plan of work (dPoW) *See* SEVEN (7) PILLARS OF BIM.

digital procurement The management of purchasing of third-party goods and services through electronic means, e.g. through the Internet.

digital railways The use of digital technology to replace conventional signalling and train control solutions to optimize performance and efficiency.

digital twin A digital representation of a physical asset or process. Digital twins consist of three main elements: a data model, a set of algorithms, and knowledge. Computer simulations are run bridging the physical and digital world. They are used to understand, predict, and optimize performance and reflect change in the environment or processes.

dilatancy Becoming large in volume, particularly in reference to fine sands and inorganic silts. When a pat of such material is placed in an open palm of one hand and the side of the hand tapped, moisture will come to the surface of the pat. When the pat is gently pressed the water will disappear again.

dilatometer An instrument used to measure the expansion of a liquid.

diluting receiver (laboratory receiver) A container on a waste drain from a laboratory sink that helps to reduce the concentration of chemicals entering a sewage system.

dimension 1. The size of something in terms of length, width, or height. **2**. A property defining a physical quantity, e.g. mass or time.

dimensional analysis The compatibility of physical quantities in terms of their units, i.e. mass (M), length (L), time (T), electric charge (I), and temperature in a given equation. For example, velocity is expressed as LT^{-1}.

dimensional coordination The size of various building components to ensure consistency in design and build of modular systems, e.g. a standard door size and opening.

dimensional stability Materials and products with little movement when exposed to different moisture contents and temperatures. The materials are stable in size, and do not suffer movement or *creep in different environments.

dimension stone (cut stone) Natural stone that is cut to shape, normally to form rectangular blocks.

diminished stile Door stile that is narrower at the top allowing greater space for windows.

diminishing courses Stone courses laid with the larger courses at the bottom and smaller courses towards the top of the wall.

diminishing stile *See* DIMINISHED STILE.

DIN (Deutsches Institut für Normung) German standards institute.

dip In most cases the strata will be dipping at some angle between the vertical and horizontal. There are two aspects to the dip of a plane, which are: **direction of dip**, i.e. the direction that water would flow if poured onto the surface, measured using a compass; and **angle of dip**, i.e. from $0°$ for horizontal bedding to $90°$ from vertical bedding, measured using a *clinometer. To record the dip of a plane, both the direction and angle of dip are used. Hence, 140/38 is a plane that dips at $38°$ in the direction $140°$ north.

direct air leakage point A point in the building envelope where *air leakage occurs directly through the *primary air barrier from inside the thermal envelope to outside or vice versa. Common points include around trickle ventilators and through poorly closing trickle ventilators, around and through the loft hatch, around poorly fitting windows and doors, through gaps at bay windows and around sliding mechanism of patio doors, at thresholds, and around services at the point where they penetrate through the primary air barrier. *See also* INDIRECT AIR LEAKAGE POINT.

direct cold-water supply A cold-water system where all of the appliances are supplied with cold water directly from the mains.

direct current (DC) An electric current that flows continuously in one direction. *See also* ALTERNATING CURRENT.

direct cylinder A cylinder used for storing hot water in a direct hot-water system. *See also* INDIRECT HOT-WATER CYLINDER.

direct gain A type of *passive solar heating system where the living space acts as the collector, storage medium, and heat emitter. The solar energy enters the thermal envelope directly and is absorbed by the thermal mass located within the living space. The thermal mass then re-emits the stored energy in to the living space. The simplest system is a well-insulated building with a large expanse of south-facing windows. *See also* INDIRECT GAIN; ISOLATED GAIN.

direct hot-water system A hot-water system where the water that is heated is the same water that is drawn off at the hot-water taps. Not suitable for hot-water central heating and only used in soft water areas. *See also* INDIRECT HOT WATER SYSTEM.

directive An official instruction which national governments have to legislate to enforce.

direct labour Workers who are employed through the company rather than being subcontracted or hired through an agency for short periods of time or for single projects. Labour that is employed directly is considered to be 'on the books', receiving the full benefits of being a full-time or part-time employee who has a contract directly with the employer. Being employed directly may result in additional benefits with regard to bonus pay, welfare, and holiday entitlement.

direct reading tacheometer A tacheometer (*see* TACHEOMETRY) fitted with a scale to enable horizontal distance and difference in level to be obtained without the need to measure vertical angles.

direct solar radiation Solar radiation that reaches the Earth's surface directly from the sun. *See also* DIFFUSE SOLAR RADIATION.

direct stress (normal stress) Tensile or compressive stress without bending or shear.

direct supply *See* DIRECT COLD-WATER SUPPLY.

direct transmission of sound Sound (airborne or impact) that is transmitted directly through a separating wall or floor. *See also* FLANKING TRANSMISSION.

dirt-depreciation factor (US) Lux or light loss factor.

dirty money Extra money awarded for having to work in difficult or messy conditions.

disabled facilities (handicapped facilities) Adaptations and equipment within a building that enable the building to accommodate disabled needs. As part of increasing accessibility, stairs, ramps, toilets, signage, and hearing loops are provided to ensure people with a disability are better able to use the building.

disc grinder Hand-held or fixed machinery with a rotation disk that can be used to smooth, grind, and cut metal and ceramic materials, such as concrete, stone, tiles, etc. Different discs are used, depending on the material that is to be cut.

discharge Something that is emitted from a pipe, duct, or conduit, or something that permeates away from its source.

discharge lamp A light source emitted from an electric discharge passing through a tube containing mercury or sodium vapour or argon, krypton, or neon gas.

discharge pipe A *branch that carries waste from a sanitary appliance to the *soil stack.

discharge valve (control valve, regulating valve) A valve used to control the flow of a fluid through pipework.

disconnecting trap (disconnector, intercepting trap) A water trap on a drainage system that connects a rainwater pipe to a sewer—used to prevent foul smells being released into the environment from the sewer.

discontinuity An interruption in a continuous process, e.g. any interruption in the deposition of rock stratum or a boundary between rock types.

discontinuous construction (isolation) Structure or assemblies with breaks that separate construction to reduce the transmission of vibration, sound, and thermal energy.

discrete element method (DEM) A numerical technique that models the movement and interaction of circular particles, individually or clumped together to form arbitrary shapes. Computer programs employing this technique include, e.g. PFC (Particle Flow Code), which can be used to simulate the behaviour of particle flow, rock fall, etc.

discrimination Unlawful behaviour, such as:
- direct discrimination—treating someone with a protected characteristic less favourably than others
- indirect discrimination—putting rules or arrangements in place that apply to everyone, but that put someone with a protected characteristic at an unfair disadvantage
- harassment—unwanted behaviour linked to a protected characteristic that violates someone's dignity or creates an offensive environment for them
- victimization—treating someone unfairly because they have complained about discrimination or harassment.

The law protects people against discrimination at work, including dismissal, employment terms and conditions, pay and benefits, promotion and transfer opportunities, training, recruitment, redundancy, trade union membership (and non-membership), and being a fixed-term or part-time worker. The protected characteristics vary by country. In the UK they include age, gender reassignment, being married or in a civil partnership, being pregnant or on maternity leave, disability, race, including colour, nationality, ethnic or national origin, religion or belief, sex, and sexual orientation.

((())) SEE WEB LINKS
- The UK Government's advice page on discrimination

disc sander Hand-held machine with a rotating shaft onto which sanding discs can be attached. The sanding discs are made to different grades for different materials.

disc tumbler lock A type of cylinder lock that contains rotating disc retainers.

disease A specific disorder affecting the structure or function of an organism, such as a human, animal, or plant.

disinfection A tertiary stage in a *water treatment process, where chlorination or ultraviolet light is used to remove pathogens (disease-causing bacteria, viruses, and protozoa) so that it is fit for human consumption.

disinfection by-products (DBPs) The undesired chemical, organic, or inorganic compounds formed as a result of the reaction of disinfectant with naturally existing organic content in water. Some DBPs can be carcinogenic in nature, therefore making it compulsory to limit/control their generation. For instance, chlorine/bromine/iodine react with precursors (organic compounds), resulting in formation of carcinogenic trihalomethanes (THMs).

dislocation line A line created inside a metal at an atomic level by the presence of *dislocations.

dislocations 1. A large fracture (*fault) in the Earth's surface. **2.** An imperfection in the lattice structure of a crystalline material, such as a metal where, for example, an atom is misaligned. Dislocation can be in the form of *edge, screw, or a mixture of both.

displacement 1. An amount of movement in a particular direction. **2.** The fluid displaced by a floating body, e.g. a ship. **3.** The amount of movement that has occurred in a geological fault, with reference to two corresponding points on either side of the fault. **4.** The volume displaced by a piston or a ram in an engine every stroke.

displacement pile (driven pile) A *pile that has been pushed into the ground, displacing the soil rather than excavating the soil.

displacement pump A pump that draws fluid in then pushes it out, such as a *lift-and-force pump, reciprocating pump, or diaphragm pump.

dispute Difference of opinion on a point of technicality or contract causing the contracting parties to disagree. The dispute threatens the relationship between the parties such that if a compromise, settlement, or agreement does not occur a legal dispute may occur.

disruption Events that occur that cause events to be delayed and/or to run out of sequence.

dissolve 1. To be dispersed in a liquid. **2.** To break down and disappear.

distance piece A wedge fitted into a glazing rebate to prevent displacement.

distillation A process of separating, concentrating, or purifying a liquid by condensing the boiling vapour.

distilled A liquid that has been subjected to the distillation process.

distinct-element modelling *See* DISCRETE ELEMENT METHOD.

distribution board (distribution panel, distribution fuse board, sub-board) An electrical board that subdivides the feed from the mains into the various electrical circuits contained within the building. Contains a number of fuses or circuit breakers that protect each electrical circuit.

distribution pipe A pipe that is used to move water or steam from one place to another.

district heating A system that utilizes a central heat source to heat a number of buildings within a town, city, or district. The heat from the central heat source (usually in the form of hot water) is distributed to the buildings via a series of underground pipes.

district surveyor Building control officer.

ditch A long narrow channel that has been formed in the ground, typically for drainage and irrigation purposes.

diuron A herbicide that can contaminate *groundwater.

diversity factor (demand factor, use factor) The sum of the individual maximum demands in an electrical distribution system divided by the total maximum demand of the system, usually always greater than one.

diverter A three-way valve to direct/split the flow.

divided responsibility Duty or obligation shared between two or more parties. Duties need to be clearly stated to manage work and to avoid disputes.

division wall 1. A load-bearing wall that is used to subdivide a space into rooms. **2.** A boundary wall between two properties, which is not part of either property. **3.** A *firewall.

DNA (deoxyribonucleic acid) A double helix molecule formed from nucleic acid that carries genetic information.

do-and-charge contract Cost reimbursement contract.

dock An area where ships and boats moor, load, and unload, which can be cut off from the rise and fall of the tide by dock gates.

docking saw A type of circular saw mounted on a stand that cuts the timber by lowering the saw blade onto it.

documents Contract documents, papers, drawings, and bills referred to in the contract.

dog A U-shaped metal fastener used to fix large timbers together.

dog bolt See HINGE BOLT.

dogleg stair (dog-legged stair) Stair with two flights between a floor where the stair returns on itself.

dolly See DRIVING CAP.

dolphin 1. A pile or group of piles used for mooring ships or as a *fender to protect a *dock.

dome 1. A structure that has a hemispherical roof. **2.** A curved layer of rock strata, formed by an upward *fold.

dome-cover screw (domed top screw) A screw where the head has a small threaded hole into which a domed cover is inserted.

dome-head screw A screw with a dome-shaped head.

domelight A dome-shaped roof light.

domestic hot water (DHW) Hot water supplied to a dwelling for washing, bathing, etc.

domestic subcontractor A subcontractor who is selected by the main contractor, being different from a *nominated subcontract who is specified in the contract documents.

domical grating Wire mesh or grating shaped in a dome, which acts as a strainer over the end of a rainwater outlet.

domus A Roman dwelling.

dooking The process of plugging a hole in a wall with a suitable material to hold a fixing. The material used to plug the hole is known as a **dook**.

door A solid barrier to an opening or an enclosed space that can be opened to allow access or closed to deny access.

door assembly *See* DOOR SET.

door blank (blank door) A door-sized recess in a wall either for aesthetic reasons or to enable the simpler insertion of a door at a later date.

door buck *See* DOOR CASING.

door buffer A device attached to a door to allow it to close softly and prevent banging.

door casing The *architrave around a door.

door closer (door check) A spring-operated mechanical device that closes a door.

door control A system for controlling entry through a door.

door face The visible portion of a *door leaf.

doorframe The frame that surrounds a door and on which the door is hung. It comprises a head and two jambs. Mainly used to hang external doors or doors in solid partitions. The frame itself has sufficient strength to support the weight of the door.

door furniture (door hardware) The collective name given to all the items of ironmongery that are fixed to the door. For instance, door handle, *escutcheon, locks, etc.

door handle A device located on the inside and/or outside of a door used to open it.

door hardware *See* DOOR FURNITURE.

door head (head member) The horizontal top member of a *doorframe or *door lining.

door holder A device that is used to hold a door in the open position.

door jamb The vertical side members of a *doorframe or *door lining.

doorkit All of the individual components used to make a *door set.

doorknob A knob-shaped *door handle.

door latch 1. A spring-operated door lock, operated either by key or by turning the door handle or knob. **2.** A door fastener comprising a metal bar that is either slid across or lowered to prevent the door from opening.

door leaf The opening portion of a door.

door lining (US door buck) The surround that covers the *reveal around an internal door and on which the door is hung, usually of thinner cross-section than a *doorframe. The lining does not normally have enough strength to support the weight of the door on its own, so relies on additional support from the surrounding wall or partition.

door lock A *lock for a door.

doormat A mat placed in front of or just inside an external door that is used to wipe dirt from shoes.

doormat frame A frame sunk in the floor just inside an external door to hold a doormat.

door opener An electronic device for automatically opening a door.

doorpost A door jamb.

door pull A handle used to pull open a door.

door rail The horizontal member that subdivides a door.

door schedule A list, usually in tabular form, containing information about all the doors that will be used on a project. It usually contains details of their size, specifications, location, etc.

door set (doorset, door unit) A pre-assembled door and frame that can be installed on-site immediately. Comprises a door hung on its frame, linings, architraves, and door furniture.

door sill *See* THRESHOLD.

door stile The vertical side members of a door.

door stop 1. Strips of wood attached to the head and/or jambs of a doorframe or lining. Used to prevent the door leaf moving through the frame and past the closed position. **2.** A device that is attached to the floor, wall, or skirting board to prevent the door opening past a particular position. Used to protect the skirting board and wall from damage caused by the door or the door handle.

door switch A switch used to open or close a door.

door threshold *See* THRESHOLD.

door unit *See* DOOR SET.

doorway width The distance between the jambs of a *doorframe.

doping 1. The modification of semiconducting materials to attain a certain property/application. **2.** The intentional alloying of semiconducting materials with controlled concentrations of different impurities. Electron doping is a procedure that makes semiconductors such as silicon and germanium ready for diodes and transistors. Semiconductors in their undoped form are electrical insulators that do not insulate very well.

Doppler effect The change in apparent frequency of a wave (such as a light wave or sound wave) due to distance between the source and observer.

dormer (dormer window) A vertical window that projects out from a sloping roof.

dormer cheek The vertical side of a dormer window.

dormer ventilator (dormer vent) A dormer-shaped roof ventilator designed to be installed on a sloping roof.

dose A measured quantity of something that is administered.

dot and dab The process of attaching plasterboard to a wall using dots of plasterboard adhesive.

dote (doat) Dots, spots, and speckles on timber that indicate early signs of decay.

dots and screeds Mounds, spots, and strips of screed that are levelled to the correct height and placed at regular intervals so that infill screed can be laid and levelled between.

double-acting hinge A hinge that enables a door to open in both directions.

double-action door A door that can open in both directions.

double connector A short pipe connector containing a thread at each end.

double curvature Curved in two directions.

double door A pair of doors.

double-eaves course (doubling course) A double row of tiles at the eaves of a roof, the lower tiles are half-length to tie in with the longer tiles that overlap.

double-faced door A door where each face of the door is decorated differently.

double glazing Glazing comprising two panes of glass that are separated by a cavity. The cavity can be filled with air or an inert gas such as argon or krypton.

double-handed lock *See* REVERSIBLE LOCK.

double-hung sash window A sash window where the sash slides vertically.

double-inlet fan (double-width fan, double suction blower) A *centrifugal fan where air can enter from both sides.

double insulation A method of insulating an electrical appliance that does not require it to have an earth connection.

double-leaf separating wall A cavity wall where the two separate leaves of the wall are not connected together in any way. Used to provide sound insulation between spaces. *See also* DOUBLE PARTITION.

double lining Two layers of lining paper hung on a wall, one horizontally and the other vertically, to provide a smoother wall surface.

double-lock seam (double-lock welt, double welt, (Scotland) **cling)** A *single-lock seam with an additional fold. Provides greater strength and weather resistance than a single-lock seam. Used to join sheets in sheet-metal roofing.

double margin door A single door that is divided into two halves to give the appearance of a double door.

double partition A partition wall comprising two separate leaves that are not connected together in any way. Used to provide sound insulation between spaces. *See also* DOUBLE-LEAF SEPARATING WALL.

double-rebated doorframe A *doorframe that contains a rebate on both sides.

double Roman tile Single lap clay tile 420 x 360 x 16 mm, laps onto two half tiles making two waterways or channels. The edges of the tile are normally lapped 75 mm.

double-suction blower *See* DOUBLE-INLET FAN.

double swing door A door that swings/opens in both directions, useful in kitchens or other rooms where occupants may have their hands full when moving through doorways.

double time Paid at twice the normal rate for work outside normal working hours or work carried out in difficult conditions.

double triangle tie Galvanized or stainless steel wall tie with two triangles formed in the wire at each end to provide a bond in the mortar course. The tie is used to link two separate skins of *cavity wall together.

double window A window that contains two glazed sashes with an air space in between.

dovetail A fan-shaped joint used to join the corners of boxes and drawers, the tenons are cut in a fan shape and the same shape is cut out of the end of the timber to be joined. The thicker end of the dovetail is positioned at the far end of the tenon, so that when the two parts are joined together, the shape of the dovetail prevents pullout.

dovetail cramp A cramp used for stonework shaped in a dovetail.

dowel A round wooden peg inserted into holes to join to pieces of timber together. The holes are drilled so that the tight fit causes friction enabling the two pieces of timber to be joined together. The joint may also be glued.

dowel (dowel pin) A wooden or plastic headless peg, sometimes with ridges down the side, that fits into aligned holes on two separate pieces of wood or metal to hold them together.

dowel bar A smooth reinforcement rod that is placed across a contraction joint in large ground slabs to transfer the load from one slab to the next while accommodating axis expansion.

dowel laminated timber Large structural sections of timber fabricated from smaller sections of softwood timber lamellae that are stacked in one plane and connected with hardwood timber dowels. This relatively simple manufacturing method has the potential to utilize lower-grade timber to form load-bearing solid timber wall, floor, and roof panels.

doweller A machine for drilling *dowel holes.

dowelling The process of drilling dowel holes by using a *doweller and inserting *dowels into the holes.

downcomer A pipe or passage in which fluids flow downwards.

down conductor The cable or rod that runs vertically down the outside of a building to transfer any lightning discharge safely to the ground. Usually made of copper.

downlighter A *luminaire designed to direct light downwards.

downstand An edge that has been folded down.

downstand beam A beam that extends beneath a floor slab.

downthrow The downward displacement of the rock strata adjacent to a fault line. *See also* NORMAL FAULT.

downtime A period when it is not possible to work due to adverse conditions or equipment failure.

downwards construction *See* TOP-DOWN CONSTRUCTION.

dowsing The use of a forked stick, a divining rod, to search for underground water.

dozer *See* BULLDOZER.

dpc (DPC) *See* DAMP-PROOF COURSE.

dpm (DPM) *See* DAMP-PROOF MEMBRANE.

draft (US) *See* DRAUGHT.

drafted margin A uniform border chiselled around the edges of the face of a stone.

drag 1. Part of a mould used in metal castings. **2.** The bottom half of a horizontally parted mould used in metal castings. **3.** The resistance a body experiences when it is moved through a fluid.

dragline A tracked or wheel excavation machine, where the bucket is operated by ropes. The bucket is cast out from the boom; excavations can be undertaken below the level of the tracks and under water. The excavation and dumping can be widely separated.

dragon beam (dragon piece, dragon tie) 1. A diagonal timber support to a *jetty or to the corner of the foot of a hip-rafter.

drain A channel or pipe (**drainpipe**) that carries surface and foul water.

drainage A process to convey fluids or gases to an egress point.

drainage easement A condition assigned to the deeds of a property in respect to drainage. For example, it might give rights to a third party to maintain a drain or culvert under a stretch of land or mean that runoff cannot be restricted from an area of land.

drainboard A drainer, used for allowing cutlery, plates, cups, and dinner services to dry after being washed.

drain channel A water course, recess, groove, duct, conduit, or culvert to divert surface water or sewage along a dedicated path.

drain chute A specially designed drainpipe that makes *rodding easy.

drain clearing *See* RODDING.

drain cock (drain valve, drain plug, drain stopper) A device for controlling (e.g. stopping or draining) the flow of a liquid in a *drain.

drained test *See* TRIAXIAL TEST.

drainer (drainboard, draining board) A sloping surface adjacent to a kitchen sink that enables water to drain into the sink.

drain layer (pipelayer) A tradesperson who lays drains.

drainline A series of pipes that are adjoined to form a *drain.

drainpipe *See* DRAIN.

drain rods A series of flexible rods that are screwed together and pushed down a *drain for cleaning purposes or to remove a blockage. *See also* RODDING.

drain shoe A drainage fitting that has an access cover and is connected to a *down drain.

drain test A *hydraulic test on a newly laid drain to test the water tightness of each drainage run before backfilling.

drape Rock strata that has the appearance of a fold.

draught (US draft) The flow of cold air into a space that causes discomfort.

draught bead A bead placed on the sill of a sash window to prevent draughts at the sill.

draught stop (US draft stop) *See* FIRE STOP.

draught strip (draught stripping) A strip of compressible material fitted between the opening leaf of a door and the frame, or a window casement and frame, to prevent draughts.

draw bolt *See* BARREL BOLT.

drawbore pinning (drawboring) When the hole in the *tenon is slightly offset from the hole in the mortise, causing the pin to be hammered into place through the holes so that it pulls the pieces tightly together.

drawbridge A bridge that is hinged at one or both ends (in the latter case the deck will be split in the middle), such that the deck can be raised and lowered to allow, for example, ships to pass beneath.

draw cable (draw wire, (US) fish tape) A thin wire that is used to pull other wires or cables through a cable duct or conduit. It can be inserted during construction or blown through using compressed air.

drawdown The lowering of the groundwater due to pumping from a well or surrounding wells.

draw-in box (pull box) An access point into ducting, cable tuns, or pipework that allows services to be pulled through the service tun; forms part of the draw-in system.

drawing An illustration of something, e.g. a cross-section of a bridge.

drawings Normally refers to the working details used to construct the building or structure.

drawings and details Images and supporting information, such as specifications and descriptions, that are used to inform the assembly of the building, structure, or works.

drawing symbols Recognized shapes, used in drawings to depict components.

draw knife Steel blade used to smooth and shape the surface of wood.

draw-off pipe A pipe that enables a fluid to be removed from a system.

draw-off tap A tap that controls the amount of fluid that is removed from a system. A **water tap** is a type of draw-off tap.

draw-off temperature The temperature of the hot water that is drawn off from a hot-water storage cylinder.

dredge (dredger) A floating vessel that allows underwater excavation to be made.

drencher sprinkler *See* DELUGE SPRINKLER.

dress 1. To plane, smooth, and finish timber. **2.** To smooth or cut stone to its finished dimensions. **3.** To cut and fold lead or other malleable roofing materials, so that they are embedded into the mortar joints of walls, and are overlapped and folded to prevent water penetration. The folded lead often provides a decorative finish to the roof.

dressed dimension (neat size) The size of stone, masonry, or timber once it has been cut or shaped to the desired dimensions.

dressed stone *Stone that has been cut or shaped to the desired dimensions.

dressed timber *Timber that has been cut or shaped to the desired dimensions.

dressing Shaping and cutting materials to their finished dimensions.

dressing compound 1. Hot or cold bituminous liquid poured or levelled over roofing felt to provide an adhesive surface onto which limestone chippings can be scattered. The chippings provide some protection against degradation caused by the sun. **2.** Levelling compound.

drift The distribution of engineering soil (e.g. alluvium, peat, terrace gravels, marine/estuarine deposits, and glacial deposits such as boulder clay) at the surface of the ground.

drift bolt A steel pin that is driven into a timber hole with a smaller diameter than the shaft diameter of the pin.

drift map A map that shows the distribution of rocks and engineering soil at the surface of the ground. These maps are particularly valuable as they offer information on soils likely to be encountered in shallow excavations. They are often used to assist in the design of site investigations. *See also* SOLID MAPS.

drill A tool for boring **(drilling)** holes. The **drill bit**, a long pointed piece of metal, is the boring part of the drill.

D-ring flexible 1. D-shaped sealing ring with one flat surface that prevents the ring rolling when it is slid over the male end of a pipe (*spigot). Once the ring is in place over the spigot, the pipe is pushed into the larger female pipe (socket). The rounded part of the ring makes a water- and air-tight seal by connecting with the socket. **2.** Elasticated ring with a D-shape in cross section allowing the rounded edge of the ring to run and connect with rounded grooves in pulleys and wheels.

drinking fountain A device that provides water for drinking.

drinking water (potable water) Water that is suitable for human consumption.

drip 1. Part of a product or component that is shaped down towards the ground to encourage water to drip from it. **2.** The front-lipped edge of the windowsill. **3.** The 'V' bent or formed in the middle of a stainless steel, galvanized steel, or plastic wall tie that prevents water passing along the tie and crossing the cavity.

drip edge The lipped edge of a windowsill or doorsill that encourages the water to run and drip away from the building.

drip-free paint *See* NON-DRIP PAINT.

dripping eaves The edge or eaves of a roof that has no gutter to collect rainwater.

drip tray Tray used to catch drips from condensing water from combustion, also known as a condensate pan.

driven pile A precast *pile or pile casing (for a cast-in-situ pile) that has been pushed into the ground and displaces the soil around it; *see also* BORED PILE.

driverless car *See* AUTONOMOUS CAR.

drive screw (screw nail) A square section nail that has a twisted shank, giving the nail a screw effect. The nail, which is 2 mm or greater in cross section, is driven into wood in the same way as a conventional nail, but is more difficult to extract.

driving cap A cover that is placed over the top of a pile to protect it from impact damage when it is being driven (hammered) into the ground.

driving rain (wind-driven rain) Rain that is carried along by the wind and blown onto the envelope of a building.

driving rain index An index that measures the amount of driving rain on a vertical surface. Calculated by multiplying the average annual wind speed by the average annual rainfall on the horizontal. Used to determine the exposure of a site.

drone An aircraft without an onboard human pilot. A drone can either be controlled remotely or can fly autonomously via software-controlled flight plans together with its onboard sensors and GPS system.

drop apron (drip edge) A strip of metal fixed at the edge, eaves, or gutter to act as a drip at the edge of metal-sheet roofing.

drop manhole An *inspection chamber where the inlet pipe is at a considerably higher height than the outlet pipe; a drop pipe is used to take the inlet down to the *invert level of the *channel in the manhole. The manhole may use a backdrop pipe or a drop channel to transfer the inlet to the channel and outlet.

droppings Materials such as mortar, snots, and other debris that have fallen down the inside face of a cavity. To ensure optimum efficiency of the cavity and ensure insulation can be placed so that it is in direct contact with the masonry, cavities must be free from debris and clean. Debris within the cavity causes cold bridges and prevents insulation butting up to surfaces. The droppings within the cavity reduce thermal efficiency, increase sound transmittance, and can allow water to cross the cavity.

drop system A hot-water heating system that is fed from a sub-tank at the top of the system.

drove Scottish term for a *boaster

drunken saw (wobble saw) A circular saw where the saw blades contain slanting flanges resulting in the blade not being set perpendicular to the drive shaft. Used for cutting a groove wider than the thickness of the saw blade.

dry 1. Paint after it has dried, either touch dry or dust dry. **2. Dry construction**, a prefabricated building constructed on-site with no wet trades.

dry area An area not expected to be exposed to water or moisture, being the opposite of a *wet area.

dry bulb temperature *See* AIR TEMPERATURE.

dry concrete A type of concrete having a low proportion of water so that the plastic mixture is relatively stiff; this type of concrete is suitable for use in dry locations, and is especially advantageous where large masses are poured and compacted on sloping surfaces.

dry deposition The process where acid gases and particles in the Earth's atmosphere are either absorbed by plants, trees, and soils or deposited on land, surfaces, and water as a dust. *See also* WET DEPOSITION.

dry hip tile A hip tile that does not require bedding in mortar.

dry hydrate A finely ground hydrated lime, made from calcium or from dolomitic limestone.

dry ice A solid comprised of carbon dioxide.

drying (drying-out, drying of screeds, drying of mortar) An operation in which a liquid, usually water, is removed from a wet solid. In construction this usually refers to the process of water evaporation (water loss) from a screed or mortar.

drying shrinkage The contraction of plaster, cement paste, mortar, or concrete caused by loss of moisture.

dry joint A joint in masonry that does not use mortar.

dry lining (US drywall) An internal wall finish comprising plasterboard sheets that are fixed to *dabs of plasterboard adhesive or timber battens.

dry masonry Masonry without mortar. In other words, blocks are held together without the presence of mortar.

dry mix A mixture of mortar or concrete that contains little water in relation to its other components.

dry partition Partition made from prefabricated components, so wet trades, such as wet plaster, in-situ concrete, or masonry are avoided, keeping the process fast and dry. Traditionally, dry partitions are constructed from a timber frame and plasterboard, but can be constructed with steel frames with cellular cardboard cores, and finished with various prefabricated panels.

dry resultant temperature (operative temperature) A simple metric used to measure *thermal comfort, based upon a combination of the *air temperature, *mean radiant temperature, and wind speed.

dry riser (US dry standpipe) The main dry vertical pipe in a fire protection system that is used by the fire brigade to distribute water to the various floors within a building if a fire develops.

dry rot (*Pula lacrymans*) A fungal attack on timber that occurs in wood with high moisture content, normally around 22% moisture content.

dry sprinkler A *sprinkler where the water is supplied from the *dry riser.

drystone walling Walls constructed of stone without the use of mortar. This was the main technique of construction during the prehistoric era and still prevails in some regions today. Extremely high-quality and solid walls can be built with the careful selection and bedding of the stones.

dry verge tile A verge tile fixed with clips rather than mortar or adhesive.

drywall (drystone wall) *See* DRY STONE WALLING.

DSC (differential scanning calorimetry) An item of apparatus used to determine the *glass transition temperature (Tg) and melting point for polymers.

dual carriageway A road with a least two lanes in each direction, separated by a central reservation barrier or grass verge.

dual-duct system A full air-conditioning system where two separate supplies of air are provided to all the *zones within the building. The temperature of each zone is altered by mixing appropriate amounts of hot and cold air in a thermostatically controlled mixing box, which is usually located in a false ceiling. In a basic dual-duct system, the temperature and humidity of the air are controlled by the central plant, so there may be variations in humidity between zones. In its constant volume form, this system will often mix air that has been heated with air that has been cooled. Although it can provide multi-zone control of temperature in a building, it has high space requirements, and high capital and running costs, so is no longer installed.

dual-flush cistern A water-efficient cistern that contains two buttons for flushing. One button gives a full flush and the other button gives a half-flush.

dual phase steel A type of steel consisting of two distinct phases, allowing a compromise between strength and ductility.

dual system *See* TWO-PIPE SYSTEM.

dubbing-out The filling up of holes and deformations in walls before skimming with plaster.

Dublin Accord An international agreement for the recognition of Engineering Technician qualifications. It was founded in May 2002 by four signatories, namely Canada, United Kingdom, the Republic of Ireland, and South Africa. *See also* SYDNEY ACCORD; WASHINGTON ACCORD.

duckbill nail A nail with a chisel point.

duckboard The use of wooden boards to provide a temporary solution, e.g. a walkway over muddy ground or a floor over joists.

duckfoot bend (rest bend) A 90° bend (vertical to horizontal) in a pipe that is supported by a flanged base.

duct A channel, tube, or pipe through which something flows or passes, e.g. air for ventilation or electrical cables for services.

Duct Blaster A trademark for equipment used to test the *airtightness of ductwork and dwellings with low levels of *air leakage.

duct cover A steel plate, or mesh, or a concrete slab that fits flush over the end of a *duct.

ducted flue (appliance ventilation duct) A flue that supplies fresh air to a *room-sealed appliance for combustion.

ductile fracture A type of failure/fracture in ductile materials involving extensive plastic deformation.

ductile iron A type of cast iron with increased ductility, sometimes known as nodular iron.

ductile to brittle transition The temperature at which a material undergoes a change from brittle to ductile behaviour.

ductility The ability of a material to undergo *plastic deformation.

ductwork The pipes used to move air around a building. Usually metal or plastic and can be flexible **(flexible ductwork)** or rigid **(rigid ductwork)**.

due date Date when the building is expected to be complete.

due time Expected time required and contractually bound to complete the project.

dummy activity An activity built into a network that is used to show a start, finish, or key point in the programme; it need have no properties, and time and resources do not need to be associated with the activity but can be. Start and finish activities are normally dummy activities.

dummy frame A temporary structure that is used to provide the correct spacing for a door set when a brick wall is being constructed.

dump truck A heavy, sometimes articulated, lorry that is used to haul material such as earth, aggregate, and rock over long distances. It has an open back bed that can tilt up to dump its load.

dumper A four-wheeled vehicle that is used on construction sites to haul material and tools over short distances. The driver sits at the back of the machine overlooking a front-tipping hopper.

dumpy level An optical surveying instrument with a telescope that only moves in the vertical direction, used in levelling.

dune A mound of sand or other sediment formed by wind or water action; found in coastal and desert regions.

dungeon Medieval cells used to imprison and torture people.

dunnage Packaging material used to protect a ship's cargo.

duodecimal system System where 12 divisions or small units make one whole unit, such as an inch, in the imperial system.

duo-pitched roof A roof with two slopes that meet at a central ridge.

duplex apartment A modern term for a maisonette.

duplex control Lift control system that sends one car to a floor when requested, even though there are two or more lifts.

duplication of plant The doubling-up of plant so that one piece of plant is on standby in the event of the other piece of equipment failing or requiring maintenance.

durability The ability of a material to resist wear, and loss of material through continual use. Durability of a material tells us how long a material will last in service. This is one of the most important properties an engineer must consider before selecting a material for any application.

duration The time taken to complete an activity or project.

Dutch bond A brickwork bond in which the *stretchers (long side facing outwards) are in line in every other course.

Dutch door (US) *See* STABLE DOOR.

Dutchman A piece of timber used to cover up a defect or an error.

duty of care The degree or standard of consideration that should be given to third parties, members of the public, and contracting parties. There is a duty of care to ensure other parties are not exposed to risks to their health or risks of injury when carrying out contracts. Employers have a duty of care to ensure that the health and welfare of their workers is considered when undertaking their work.

dwang (Scotland) A horizontal member that is fitted between wall studs, floor joists, or rafters to provide rigidity. Usually timber, but can be aluminium or steel. *See also* NOGGING; STRUTTING.

dwarf wall 1. A low external wall used as a border for gardens. **2.** A low internal wall that does not extend all the way up to the ceiling. If the wall is a partition wall, it is known as a **dwarf partition**. **3.** A low wall used in suspended timber floor construction to support the floor joists. Also known as a **dwarf supporting wall** or **sleeper wall**.

dwelling Building or property where people reside.

dye A substance used to colour a liquid or solid element, e.g. paints.

dyke 1. A watercourse cut into land. **2.** A minor intrusion of igneous rock formed in the shape of vertical wall masses within the Earth's crust from slowly cooling magma.

dynamic Relating to a physical change over time. **Dynamic consolidation** is the compaction of soil by repeatedly dropping a heavy weight over an area of land. **Dynamic factor** is an impact factor taking into account the effects of a sudden load to that of a static load. **Dynamic load** is an on–off loading regime, such as from the effects of wind. **Dynamic penetration test** is a standard penetration test where a rod is hammered into the ground. **Dynamic response** relates to continuous time-dependent movement, such as vibrations. **Dynamic viscosity** is the resistance to flow when a layer of fluid moves over an adjacent layer at a given speed.

dynamics The study of motion—forces that produce change.

dynamite A powerful explosive used for blasting rock.

dyne An old unit of force equivalent to 10^{-5} *newton.

e (Euler number) A mathematical constant, named after Leonhard Euler, that has a transcendental number (i.e. impossible to express as an integer—a whole number) of 2.718281828 . . .

E *See* YOUNG'S MODULUS.

earliest date (earliest event time) Earliest time that an activity or event can start. The activities and links in a logical network affect when subsequent tasks can start. Some tasks have *float. A task with two days' float could start as soon as the preceding task has finished, or could start two days after the previous task has finished and would not delay the remaining tasks.

early warning meeting Some standard construction contracts, such as the *Engineering and Construction Contract (previously the New Engineering Contract), require each party to call for an early warning meeting as soon as a problem is recognized. The meetings have been used to help overcome potential problems, reducing the potential for disputes. Such meetings are recognized as a proactive method of dispute management.

ear protectors A pair of ear covers or muffs, normally connected to an adjustable headband, that are used to protect the ears from loud noise.

earth (US ground) 1. The ground or topsoil. **2.** An electrical connection between an electrical device and the earth. The earth is assumed to have a voltage of zero.

earth bar A copper or brass bar that provides a common earthing point for electrical installations.

earth bond (earth bonding) The connection of all metal objects within a building, such as metallic water pipes, to earth, in order to provide protection against shock from the electrical system.

earth building (earth wall construction) External walls constructed from rammed earth (*see* RAMMED-EARTH CONSTRUCTION), *adobe, or *cob.

earth conductor (earthing conductor) An insulated conductor that connects an item of electrical equipment, such as a *consumer unit, to the *earth electrode.

earth dam A water-retaining structure (such as to retain water in a reservoir), formed from compacted earth material. Typically, it has an impervious clay core surrounded by sand, gravel, or rock, with the rock being placed on the upstream side (termed rip-rap) to prevent erosion from wave movement.

earthed concentric wiring A mineral insulated cable where the conductor is totally surrounded by an earthed metallic copper sheath.

earth electrode (earth termination) A conductor that provides an electrical connection directly to earth. Usually takes the form of a rod, stake, or plate.

earthenware *See* CERAMIC.

earthing The process of connecting electrical equipment to earth to provide protection against shock from the electrical equipment.

earthing lead The conductor that connects a device to the *earth electrode.

earth-leakage circuit-breaker (ELCB) A *circuit breaker that switches off the power when a current is detected leaking to earth.

earth pressure The pressure exerted by the mass of the soil under gravity. The **lateral earth pressure** is the sideways pressure from the soil, which will act on a retaining structure (wall). The *active earth pressure and *passive earth pressure are the minimum and maximum values of lateral earth pressures respectively.

earthquake A sudden violent movement of the earth's crust due to tectonic activity, e.g. a slip on a *fault plane or *volcanic activity. The *Richter scale is used to measure the intensity of the earthquake.

earth-sheltered construction Buildings that are designed with earth covering or adjacent to part or most of the building. The practice of using earth to cover and shelter buildings is provides protection from the elements, increases external thermal mass, reduces heat loss and the impact of internal solar gains, and provides a natural aesthetic and a natural habitat for plants and animals (e.g. bees). Many argue that the shelters are more resilient with respect to the environment and offer environmentally friendly passive structures.

earth termination *See* EARTH ELECTRODE.

earthworks 1. Engineering work associated with the movement and processing of quantities of soil, e.g. excavation, backfilling, compaction, and grading. **2.** The process of moving large quantities of earth and soil. **3.** A structure made from soil, for instance an embankment or a berm.

earthworks support The supports used to retain soil during excavations or earthworks.

ease To loosen, free.

eased arris An *arris that has been rounded off.

easements A right held by a person to use land that belongs to another person for a specific or restricted purpose. The power or agreement is often used in civil engineering works, such as right to access services, to use land adjacent to a carriageway to undertake necessary works, the right to discharge water onto neighbouring property, right to services, etc. The specific acts are usually covered under an Act of Parliament, express grant under deed, express reservation when land is sold, or by prescription.

easing To lessen the pressure, slacken, allow movement, or reduce tension, to make room for works or access such as *easement of land.

eaves (eave) The part of a pitched or flat roof that projects over the external wall. Comprises a *fascia and a *soffit. Protects the external wall from rain.

eaves board *See* TILTING FILLET.

eaves course (US starting course) The first *course of tiles at the *eaves of a roof.

eaves drip (roof drip) The end of the *eaves of a roof where water drips off.

eaves flashing (drip edge flashing) A *flashing installed at the *eaves of a roof to prevent water penetration.

eaves gutter A gutter installed at the *eaves of a roof.

eaves lining *See* SOFFIT BOARD.

eaves overhang The portion of a roof that overhangs the external wall at the *eaves.

eaves plate A *wall plate.

eaves soffit The underside of an *eaves overhang. It may be lined with a *soffit board or left open.

eaves tile A tile used to form the *eaves course, it is usually shorter than the other roof tiles.

eaves vent (eaves ventilator) A ventilator installed at the *eaves to ventilate a roof.

ebb Receding from the land or shore, referring particularly to the sea and tidal movements.

EC *See* EUROCODE.

eccentric Away from the centre. An **eccentric load** will create a bending moment.

eccentric circles Not of the same centre point. Circles that deviate or do not all project from the same point, they may have different centre points and radii.

echo The repetition of a sound that is produced when the sound wave is reflected from a solid object.

ecoduct A wildlife crossing, such as a channel under a highway for badgers, small mammals, amphibians, insects, spiders, etc. to pass through safely.

E. coli (*Escherichia coli*) A normally harmless gut commensal found in the large intestine of mammals; however, it causes infection and even death when found in other parts of the body. The detection of *E. coli* in food chains or water supply typically indicates faecal contamination.

ecology The study of the interaction of living organisms and their natural environment.

economic thickness of insulation The maximum thickness of insulation on a component that can be justified on the grounds that the costs of buying and installing the insulation are offset by the amount of energy that the insulation will save over its lifetime.

ECR glass A type of glass with excellent high temperature properties and resistance to corrosion.

eddy When the flow of a liquid or gas doubles back on itself to form a whirl. Eddies can cause problems around tall buildings when the wind whirls back on itself.

edge 1. The perimeter of a surface or object. **2.** A sharp line created by the intersection of two surfaces or objects.

edge beam *See* RING BEAM.

edge bedding *See* FACE BEDDING.

edge dislocation A common type of *dislocation present at a microscopic level in metals and alloys.

edge fixity The structural stability given to a suspended floor slab by an end edge beam or wall.

edge form *Formwork used at the edge of a concrete slab.

edge grain (comb grain, vertical grain) The vertical section of grain which can be seen in wood that is *quarter-sawn.

edge joint 1. A joint along the edges of two materials. **2.** A joint between two materials that has been made in the direction of the grain.

edge nailing *See* SECRET NAILING.

edge tool A tool used for cutting, made of specially hardened steel. Saws, chisels, knives, etc. can all be described as edge tools.

edging strip A thin strip of material, usually timber, used to cover the edge of the facing of a *flush door.

edging tool (edger) 1. A flat concreting trowel with one long edge turned down to make a radius that runs along the length of the tool. The turn-down radius is used to finish the edge of concrete, forming a bevelled edge on external corners. A bevel is often formed on the edge of concrete stairs to take off the hard 90° corner that would otherwise be formed. The bevel also reduces potential chipping that would occur at the corner of sharp concrete forms. **2.** A spade-like tool with a semicircular head used for trimming the edge of grass lawn and turf.

Edison screw cap A threaded fitting used to attach a bulb to a light fitting.

EDM *See* ELECTRONIC DISTANCE MEASUREMENT.

education A process of transferring knowledge, skills, and values. Typically involves the process of teaching, learning, and tutoring to obtain some form of qualification in a specific subject area.

effective angle of internal friction (ϕ') The shear strength value of internal friction in terms of *effective stress.

effective leakage area (ELA) The equivalent cross-sectional area of a theoretical hole that would have the same leakage flow rate as the whole building if it were subjected to an artificially induced pressure difference of 4Pa.

effective span A reduced span, e.g. to take into account end constraints.

effective storage The actual amount of storage within a space.

effective stress (σ') The amount of stress taken purely by the soil structure when the influence of pore pressure has been eliminated. Soil has a skeletal structure containing solid particles with interconnecting pore spaces between them. The pores can be fully or partially filled with water. On the application of load, the total stress is carried initially by the pore water and the soil skeleton. The pore water will become under pressure acting equally in all directions. The effective stress value can be calculated by subtracting the pore pressure (u) from the total stress (σ), i.e. $\sigma' = \sigma - u$. Over time the pore pressure will dissipate and the soil will consolidate and effective stress conditions prevail.

effective temperature (US) A *comfort index that takes into account air temperature, air movement, and humidity.

effect of actions (E) The resulting internal forces, moments, stresses, and strains in the structure or ground from an *action.

efflorescence A white mark on walls (bricks) caused by the evaporation of water.

effluent Liquid waste discharged from a sewage system or an industrial plant.

effusion The flow of gas through a very small aperture.

Egan report The report, Rethinking Construction, was written by Sir John Egan's Task Force in 1998. The report advocated a movement for change within the industry through a focus on the customer, applications of best practice, greater collaboration between client and contractor, a more integrated approach to contracting, and emphasis on radical change through innovation and learning. In 2002 Egan headed up the Strategic Forum for Construction and published the Accelerating Change report. The report identified important drivers for the industry as client leadership, the need for more integrated teams, and the need to address the health and welfare of those involved in construction. Construction Excellence was formed in 2003 to bring together all of the government initiatives.

(⊕) SEE WEB LINKS
• Construction Excellence is an organization charged with driving change in construction.

egg-crate ceiling A *cell ceiling.

eggshell 1. Paint with a finish that gives the very low sheen similar to that of an egg, it's slight shine or gloss makes it suitable for walls, furniture, and other internal finishes. **2.** Shell structures are sometimes referred to as eggshell construction. The curved shell-like structures provide multiple paths for the stresses to be distributed. Although the structures are often thin, they are very strong and resilient, much the same as an eggshell.

E-glass (electrical glass, low-E glass, low-emissivity glass, energy-efficient glass) Glass with a microscopic coating, thinner than a strand of hair, on one or more surfaces that reflects long-wave heat energy and in some cases short-wave solar energy. E-glass is commonly used in buildings to control and reflect heat.

egress 1. In property law, this is the right of a person to leave a property. **2.** The movement of people or substances from an internal to an external environment. In the event of a fire, people within a property should be able to evacuate to a place of safety. A safe egress route must be provided to support evacuation procedures.

EIA *See* ENVIRONMENTAL IMPACT ASSESSMENT.

Eichleay formula A calculation for overheads claimed as a part or percentage of a contractor's delay claim. The formula is often used in Federal cases in the US. Apart from this, it sees occasional use to justify head-office overheads.

ejusdem generis rule A statutory interpretation that restricts the meaning of a general word to the meaning of the preceding groups of words. Originates from the Latin *ejusdem generis*, meaning 'of the same kind' and is now commonly used in contacts to give clarification and/or restriction of meaning. For example, if a contract refers to automobiles, dumper trucks, excavators, forklifts, and other motor-powered vehicles, the rule could be applied to say that aircraft and ships are not included because the list only details land-based vehicles.

ELA *See* EFFECTIVE LEAKAGE AREA.

elastic The capability of a material to regain its original shape after the load has been removed. **Elastic deformation** is a non-permanent deformation of a material that is recovered on release of the applied load. Deformation where both stress and strain are proportional occurs within the **elastic range** below the **elastic limit** (*yield stress). After this limit, deformations are permanent, even on the removal of load.

elasticity The ability of a material to be *elastic.

elastic modulus *See* YOUNG'S MODULUS.

elastomeric sealants A type of *sealant made from a specific group of polymers called *elastomers. Sealants are used to seal gaps or joints between materials or building elements, usually to prevent the ingress of water. Elastomeric sealants are considerably more expensive than other types of sealants, however the extra cost is offset by the superior durability. These sealants have the advantage of being much more resilient than mastics, together with the ability to withstand much greater joint movements (in the region of 20%). The anticipated life is about 25 years in the case of polysulphide sealants. Elastomeric sealants bond well to metals, brick, glass, and to many plastics. The maximum depth of sealant is usually about 15 mm; elastomeric sealants do not change in volume when stressed.

elastomers A classification of polymers which are completely elastic and cannot be plastically deformed. The molecular structure of elastomers consists of long, but helically wound carbon chains, thus imparting their remarkable ability to extend elastically by over 300%, and then to spring back to their original length. Apart from their excellent elasticity (hence their name), elastomers have very low modulus (they can be stretched elastically with a very low load). They have no capacity for even minimal plastic (permanent) deformation. In the construction industry, elastomers can be moulded to form sealing rings for waste pipe and rainwater systems, and sheet materials and gaskets for other seals.

elbow (elbow joint, knee) An angled pipe fitting, usually at 90°.

elbow board (elbow lining) *See* WINDOW BOARD.

electrical distribution The final stage in delivering electrical power, carrying electricity from transmission to consumers.

electrical resistance strain gauge A device used to measure the amount of *strain of an object. A *Wheatstone bridge is used to measure the change in electrical resistance of a flat coil of very fine wire that is glued to the surface of the object.

electrical riser A vertical shaft or duct that contains electrical cables and services. Used in multistorey buildings.

electrical services (electrical engineering, electrical installation) A term that encompasses all of the electrical services found within a building, such as lighting, small power, alarm systems, etc.

electric boiler A boiler that is powered by electricity. Two main types of domestic electric boiler are available; dry-core and wet storage. Dry-core boilers have a core of high-capacity thermal bricks, that are heated electrically, and the heat emitted from the bricks is used to heat pipes that are filled with water. Wet-storage boilers comprise a series of immersion heaters that are used to heat a large volume of stored water.

electric charge A charge, either positive or negative, caused by an excess or deficiency of electrons within a body.

electric coheating test A quasi-steady-state test method designed to determine the aggregate heat-transfer coefficient of an unoccupied building in the field. It involves electrically heating the building to a constant mean elevated and homogeneous internal temperature for a period of between 7 to 21 days using a series of strategically positioned electric resistance point heaters. To minimize temperature stratification within the building, a series of electric air circulation fans are also installed to mix the internal air. In the UK, tests are undertaken during the heating season (October/November to March/April) to ensure that there is a sufficient temperature difference between the inside and the outside of the dwelling (10K) and the mean internal temperature is set to 25°C. During the test, various internal and external parameters, such as internal and external temperatures, solar radiation, and the total electrical power input to the building are measured. Thus, the total daily heat input to the building in watts that is required to obtain a particular ΔT in K can be established.

electric current The rate at which an *electric charge flows through a conductor. Measured in *amperes (A).

electric drill A hand-held electrically powered tool used to drill holes in materials. Can be mains or battery powered.

electric heating (electric resistance heating) The process of providing space or water heating by passing electricity through some form of resistance, such as a coil, panel, or wire.

electrician A qualified tradesperson employed to install, repair, and maintain electrical circuits, plant, and equipment. In the US an electrician may be referred to as an inside wireman or outside lineman. The inside wireman works on building and plant with the outside lineman installing utility cables for the national distribution of electricity.

electricity An electric charge caused by the behaviour of photons and electrons.

electric motor A motor that is driven by electricity which it converts to mechanical work.

electric panel heater A type of space heater that uses an electrical resistance to heat a panel.

electric power The rate at which electrical energy is produced. It is calculated using the following equation:

$$P = IV$$

where P is the power in * watts, I is the current in * amperes and V is the *voltage in volts

electric screwdriver (screwgun) A hand-held electrically powered tool used to tighten or remove screws. Usually battery powered.

electric shock An electric current passing through the body. It causes nerve stimulation and muscle contraction. Currents around 100 mA are fatal.

electric-storage floor heating Electric *underfloor heating.

electric striking plate A *striking plate where the locking mechanism can be opened remotely by applying an electric current.

electric tools Portable tools that are electrically powered from the mains or a battery.

electric water heater A device that heats water using electricity. Can either be a hot-water cylinder or an instantaneous water heater.

electrochemical reaction A reaction between chemicals which produces electrons.

electrode An electrical contact in a circuit; can be either anode or cathode.

electrodeposition Deposits on a plate due to an electric current. The following result from electrodeposition: electroplating, electroforming, electrorefining, and electrotwinning.

electrolier (US) *See* PENDANT.

electrolysis A process used to separate elements by the usage of an electrical current.

electrolyte A liquid allowing ionic conduction.

electromagnet A type of magnet that consists of a soft iron inner core surrounded by a coil. It becomes temporarily magnetized when an electric current is passed through the coil—the magnetic field fades away when the current ceases.

electromagnetic cover meter *See* COVER METER.

electromotive force (emf) 1. The force that causes electricity to flow.
2. Chemical, mechanical, or thermal energy that has been converted into electricity, e.g. energy from a battery. Measured in *volts.

electron A negatively charged particle that orbits the nucleus of an atom and is responsible for carrying the electrical charge in a conductor.

electronegative Terminology used for non-metallic elements.

electronic distance measurement (EDM) A surveying instrument that uses an infrared or laser beam to measure the distance. The instrument transmits the beam to a reflector, which reflects the beam back to the instrument. The difference between transmitted and received signal is converted into a distance.

electro-osmosis A ground improvement process that involves passing an electric current through the ground so that groundwater migrates to the *cathode. Used in silty soils to lower the groundwater level or to divert the groundwater flow away from excavations.

electropositive A terminology used for non-metallic elements.

electrostatic Electricity that is stationary.

electrostatic filter A filter that uses an electric charge to attract particles.

element A major part of a building that has its own functional requirements, such as walls, floors, roofs, windows, doors, stairs, and services. Constructed from various components and materials.

elemental bills of quantities Descriptions of the building works that are organized by dividing the quantities into building elements. The labour and material quantities are rolled up into the main building elements with less detail than standard *bills of quantities, saving time on the detail of estimation.

elevated gravity tank A water storage tank located above ground level, usually on a roof, where the water is distributed via gravity.

elevated road (elevated railway) Located above ground level on an embankment or supported by columns. Reasons for elevation could be to prevent flooding or to allow passage of vehicles/pedestrians beneath.

elevation 1. The vertical distance between a particular point and sea level. **2.** A two-dimensional drawing of the side of a three-dimensional object as seen from the front, back, left, or right. Commonly used to show the exterior of a building.

elevator (US) A *lift.

elevonics (US) The electronic controls for a lift.

ell An extension to a building that is normally at right angles to the main building, producing a L-shape.

ellipse 1. An oval shape. **2.** The shape obtained when a cone is cut on an oblique plane that does not intersect its base.

elm A hardwood, which is dull brown in colour, coming from deciduous and semi-deciduous elm trees. Deciduous trees shed their leaves seasonally and semi-deciduous trees shed their leaves for short periods as new foliage is growing. Elm suffers from warping if not seasoned properly; it should only be used in consistent environments and not ones which experience changes in moisture content. The wood has a twisted grain that makes it harder to split than oak, although it has other characteristics that make it weaker than oak. It is often used as a *veneer and for woodblock flooring.

El Niño An irregular climate pattern that occurs over a period of three to seven years, which influences ocean and atmospheric currents across the tropical zone of the Pacific Ocean.

elongation 1. The lengthening of something. **2.** The angle between the sun and a celestial body.

embankment A compacted earth structure, which is typically trapezium in cross-section. Used to retain water, i.e. acting as a dam or preventing flooding, or can be used to carry an *elevated road or railway.

embedded wall A retaining wall (such as a sheet pile) that is driven into the ground to produce *passive earth pressure on the front of the wall.

embodied energy The amount of energy consumed in the production of a building. It includes all stages from the mining and processing of the natural material to manufacture, transportation, maintenance, and disposal.

embossed A raised image or design on a material.

embossed carpet Decorative pattern formed on carpet by cutting the strands of fabric at different lengths.

embrasure An opening in a wall where the sides of the opening are angled so that the external edge of the opening is larger than the internal opening. This type of opening was originally used by the military so that one could shoot from a small opening while still achieving good vision and increasing the angle of fire.

Internal face of the wall

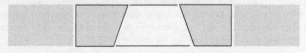

Outside opening wider than the internal opening

Embrasure: Plan section of an embrasure opening

Emden formula Used to calculate damages due to a delay in a construction contract. It applies the average Head Office Overheads and Profit (HOP) that would be achieved elsewhere on a job as a whole to the reimbursable period of delay:

$$HOP - 100 \times Contract\ sum - Contract\ period\ (weeks) \times Delay\ period\ (weeks)$$

emergency exit indicator lighting *See* SELF-ILLUMINATING EXIT SIGN.

emergency powers Government powers that may be used in case of an emergency, national danger, or other exceptional circumstances. The power is derived from an Act of Parliament—the Emergency Powers Act 1920 and 1964. The power could be used exceptionally during peacetime in the case of a major strike which could potentially affect the stability or health and well-being of the country.

emergency shutdown A button, usually red in colour, that when activated in an emergency, cuts the power or stops items of machinery. *See also* MUSHROOM-HEADED PUSH BUTTON.

emissivity A measure of a surface's ability to emit radiation. It is calculated by comparing the amount of energy radiated by a surface to the amount a black body would radiate at the same temperature.

empirical Based on observation and experimental testing rather than theory.

employee A person working for an organization according to the terms set within a contract of employment who will carry out the work personally (*see also* WORKER). A contract exists when terms such as pay, annual leave, and working hours are agreed (not always in writing). Employees are entitled to a wide range of employment rights, including:

- a written statement of employment
- an itemized pay slip
- the National Minimum Wage
- holiday pay, maternity and paternity pay, etc.
- the right to request flexible working hours
- the right not to be treated less favourably if they work part-time
- the right not to be discriminated against (ACAS, 2018).

(🌐) SEE WEB LINKS
- ACAS employment status advice

employee engagement An approach to managing people that results in the right conditions for all members of an organization to give of their best each day, committed to their organization's goals and values, motivated to contribute to organizational success, with an enhanced sense of their own well-being. It is based on trust, integrity, two-way commitment, and communication between an organization and its members. It is an approach that increases the chances of business success, contributing to organizational and individual performance, productivity, and well-being. (Engage for Success, 2018)

(🌐) SEE WEB LINKS
- Information on employee engagement

employer (client) 1. A company that engages and employs the workers. **2.** The person or organization that pays for the project and for whom the building or structure is being provided. The employer is not necessarily the body that will use the building once constructed, although it will often own the land on which the structure is being built.

employer's agent (employer's representative) The representative who acts on behalf of the client, administering the contract and from whom the construction professionals take instruction and report to. Under the standard forms of contract different terms are used. *JCT WCD (With Contractors Design) refer to the employer's agent, and JCT 98 refers to the employer's representative.

employer's requirements Includes performance requirements, service agreements, contract brief, functional requirements, and express instructions provided under contract.

emulsion (emulsion paints) A liquid of usually two immiscible components, e.g. paints.

EN *See* EUROPEAN STANDARD.

enamel paints A special type of paint based on *polyurethane or *alkyd resins to give very durable, impact-resistant, easily cleaned, hard gloss surfaces. Colours are usually bright. Such paints are suitable for machinery and plant in interior and exterior locations.

encapsulation Enclosing or encasing something, e.g. fibre in resin in a composite.

encasement The process of covering or enclosing an element with a fire-resistant material in order to increase the level of *fire protection.

encasing 1. Encasement. **2.** Enclosing elements or components, such as columns or pipework.

encastre support A supporting member (beam, column, etc.) that is fixed, and prevents all movement.

enclosed stage (closing in) The completion of the building envelope sufficient to allow finishing trades and internal works to begin. *Fast-track building methods often mean that internal building works start well before the external envelope is fully watertight. Traditionally, to eliminate any risk of damage to the *finishes from the elements, the external envelope would have been complete before the finishing trades commenced.

enclosed stair *See* PROTECTED STAIR.

enclosure 1. An enclosed space. **2.** The external weatherproof layer of a building. **3.** A box, cabinet, or cupboard that houses electrical equipment.

encroachment Beyond specified limits, e.g. where property enters the land owned by another.

encrustation (incrustation) A surface coating or covering.

end-bearing pile A pile that distributes its load to the strata at the base of the pile; *see also* FRICTION PILE.

end event The final point or task on a project network. As it is the final milestone, the last thing to complete, it will form part of the *critical path.

end grain The exposed grain on a piece of timber that has been cut perpendicular to the length of the log.

end joint The joint that is formed when the ends of two components have been joined together.

endlap (end lap) The amount of overlap at an *endlap joint.

endlap joint (end lap joint) A joint where the ends of two materials are halved in thickness and then lapped.

endoscope An optical instrument that can be inserted into pipes and orifices to inspect the internal conditions.

endothermic A chemical reaction that absorbs heat; *see also* EXOTHERMIC.

endrin A toxic pesticide, which is harmful to human health when found in domestic water supplies.

energy The capacity to do work. *See also* KINETIC ENERGY; POTENTIAL ENERGY; STRAIN ENERGY.

energy flexibility The process of designing demand and supply systems to generate, deliver, and draw down energy variably in order to level out demand and optimize energy delivery across different green and carbon energy providers.

energy from waste (EfW) Waste that can be converted into energy in the following ways, through anaerobic digestion—microorganisms break down biodegradable waste into biogas; incineration—waste is burned to generate electricity; gasification—waste is super-heated to produce syngas (synthetic gas), a renewable fuel; pyrolysis—process of thermal decomposition to produce fuel (including charcoal); or thermal depolymerization—mixed waste is super-heated in water to produce crude oil.

Energy Performance of Buildings Directive (EPBD) Legislation approved by the European Union to encourage cost-effective improvements to the energy performance of buildings.

EnerPHit Standard An energy efficiency refurbishment standard devised by the Passive House Institute (PHI) for existing dwellings which recognizes that it is often more challenging to refurbish any existing dwelling to the *Passivhaus Standard.

engaged column (applied column, semi-detached column) A load-bearing column that is partially set within a wall.

engine A machine that converts energy into power or motion. An engine can be driven electrically, hydraulically, by compressed air, steam, or internal combustion.

engineer A professional with applied mathematical knowledge or technical skills in either *civil, mechanical, structural, electrical, or marine engineering that enables the person to design or work with the tools of the profession. Civil engineers work on roads, sewers, and infrastructure; structural engineers design buildings and structures; electrical and mechanical engineers work with electrics, plant, and equipment; and marine engineers work with structures and plant based in the sea or that which is part of other watercourses.

engineered brick (US) *See* BRICK.

engineering The use of science to design things.

Engineering and Construction Contract (New Engineering Contract (NEC)) Standard form of construction and civil engineering contract published by the Institute of Civil Engineers. The contract has attempted to remove traditional barriers. There are a number of terms and *conditions which encourage greater collaboration especially in the areas of dispute or problem resolution. Identifying and attempting to resolve problems through proactive planning and early warning meetings are two ways that encourage the team to own the problem and attempt to mitigate the effects.

engineering brick (UK, engineered brick, US) *See* BRICK.

Engineering Council A regulatory body that holds a national register of Chartered Engineers (CEng), Incorporated Engineers (IEng), Engineering Technicians (EngTech), and Information and Communications Technology Technicians (ICTTech). It sets and maintains internationally recognized standards to ensure professional competence and ethics.

engineering geology A branch of geology related to *engineering, e.g. the design of a foundation related to the underlying strata (geological) conditions.

engineering installations Collective name given to the electrical and mechanical services within a building.

engineering wood A mixture of different timber products to produce a generic timber material, e.g. plywood.

engineer's chain *See* CHAIN SURVEY.

English bond (Old English bond, Dutch bond) A type of brickwork bond comprising alternating courses of *headers and *stretchers.

English garden wall bond (American bond) A type of brickwork bond comprising stretchers, where every sixth course is a header.

enhanced greenhouse effect A man-made phenomenon in which various *anthropogenic activities have resulted in an increase in the concentration of the Earth's *greenhouse gases. As a consequence, more longwave radiation that is emitted from the Earth's surface is trapped in the Earth's atmosphere, resulting in increased warming. This is also referred to as *global warming. *See also* NATURAL GREENHOUSE EFFECT.

en-suite bathroom A bathroom that is attached to a bedroom.

ensure To do to one's best ability.

entire completion The satisfactory conclusion of all works.

entire contract The contract should be considered in its entirety, a term cannot be read on its own when it forms part of a contract; it must be considered against other terms within the contract which could mean that the term itself has a different meaning.

entrapped air Irregular and undesirable air voids in concrete. Can be caused by poor mix design, segregation during placing, and inadequate vibration; *see also* AIR-ENTRAINED CONCRETE.

entry 1. An entrance to a building. **2.** A point where a component, such as a gas pipe or an electrical cable, enters a building.

entryphone A phone that is used to gain entry to a particular space. Usually used in conjunction with an *electric striking plate so that remote entry can be gained.

entry system An electronic system used to gain entry to a particular space. Various systems are available including card-operated systems, code-operated systems, and entryphone systems.

envelope 1. The external enclosure of a building, often the part of the building that resists the weather. A **building envelope** is the exterior surface of a building that separates the inside of the building from the outside. **2.** The cover or surround that encases an element.

environment Our physical surroundings such as people, building, structures, land, water, atmosphere, climate, sound, smell, and taste.

environmental audit An assessment of the environment and the potential impact of intervention. *See also* ENVIRONMENTAL IMPACT ASSESSMENT.

environmental impact assessment (EIA) An investigation to determine the potential effects (both positive and negative) on the environment resulting from an existing or proposed development.

environmental protection The practice of looking after the natural environment (e.g. air, land, and water) by individuals, organizations, and governments. It also includes, where possible, reversing damage which has already been caused to the environment.

environmental temperature A *comfort index that takes into account air temperature and mean radiant temperature.

enzyme A protein produced by living cells that acts as a catalyst to produce a specific biochemical reaction.

EPBD *See* ENERGY PERFORMANCE OF BUILDINGS DIRECTIVE.

EPDM *See* ETHYLENE PROPYLENE DIENE MONOMER RUBBER.

ephemeral stream A stream that only flows after a period of rain, i.e. it does not have a *base flow.

epicentre The point of the earth's surface which is directly above the focus (centre) of an *earthquake.

epilimnion The uppermost temperature-sensitive water layer of a lake.

epistemology Reality viewed and understood through deduction, induction, and through the use of research tools; the reality of the situation is inferred or obtained from evidence.

epoch A division or a period of geological time characterized by a particular rock formation.

epoxy A polymeric material (*see* POLYMER) used in adhesives and pipeline coatings for corrosion protection.

epoxy paints Special types of paint based on *epoxy polymers which are highly resistant to abrasion and spillages of oils, detergents, or dilute aqueous chemicals. Often applied to concrete, stone, metal, or wood in heavily trafficked workshops and factories.

EqLA *See* EQUIVALENT LEAKAGE AREA.

equal The same in size, quantity, value, or standard.

equiaxed A specific type of morphology present in metals or powder particles at high magnifications where the grain shapes are of approximately equal dimensions. Many metals when viewed at a microscopic level have equiaxed-shaped grains, which can influence the mechanical properties.

equilibrium The point where opposing forces are in balance, resulting in a stable situation.

equilibrium moisture content The point at which the moisture content of a material is in *equilibrium at a given temperature and humidity.

equilibrium phase In metallurgy, the state of a system where the phase characteristics remain constant over indefinite time periods. When a material is in equilibrium, it is considered to be stable, i.e. its properties will not change unless it is subjected to a change in external conditions i.e. force, temperature, ambience, etc.

equipment 1. Building services systems. **2.** Any items used to undertake a task.

equipment noise The noise emitted from building services systems.

equipotential Of equal potential at all points in a system. **Equipotential lines** are used, in a *flow net, to represent contours of equal water pressure around a water retaining structure, such as a dam. Equipotential lines follow permeable boundaries and open water surfaces, and are at right-angles to impermeable boundaries and *flow lines.

equity The fairness stance that provides reliefs and remedies for wrongs done; it has managed to secure a place in the legal system even though it falls outside the strict scope of common law, which is based on statutes and case precedence.

equivalent 1. Items or services of the same standard or value. **2.** Legal language or terms that have the same meaning. In different countries or contracts certain terms are considered to have the same meaning or similar meaning 'near equivalents'.

equivalent leakage area (EqLA) The equivalent cross-sectional area of a theoretical hole that would have the same leakage flow rate as the whole building if it were subjected to an artificially induced pressure difference of 10Pa.

equivalent temperature A *comfort index that takes into account air movement, dry-bulb temperature, and mean radiant temperature.

erection The process of raising, positioning, constructing, and fixing building *elements or *components into their final position.

erg Unit used to measure energy or work. It is equal to work done by a force of one *dye moving through 1 cm in the line of action of the force.

erosion A physical or chemical process of wearing away, for example, through water, wind, ice.

erratic 1. Irregular, has no consistency. **2.** Rock that has been carried by a glacier and deposited a distance from its source once the glacier has melted.

error A mistake, something that should not exist.

escalation of contract prices (US) Fixed price contract that allows for an increase in the contract price if certain events or price fluctuations take place.

escalator A set of moving stairs between two floors. The stairs are attached to a continuously circulating belt. Commonly found in shops where they are used to move people quickly from one level to another.

escape chute A fabric tube used as a vertical escape route from upper floors of a building (not used in the UK). The tube is normally angled and twisted by a helper on the ground to prevent anyone entering the chute and falling straight to the ground. Once a person enters the tube, it is slowly untwisted allowing a person to steadily descend to the ground.

escape route The method of egress out of a building in the event of a fire or emergency.

escape stair (fire escape, fire stair) Flight of steps in a building used as the main exit from the building in the event of fire or other emergency.

Escherichia coli See E. COLI.

escrow Account that is set up separately from the main company to distribute money or funds. An escrow fund or court-administered fund may be used as part of a judicial action where the defendant pays the money into an account, which is then administered by the court. An escrow account may also be set up and held separately until a successful transaction occurs.

escutcheon (key plate) A thin metal plate surrounding a keyhole. Used for decoration and to protect the surrounding door.

espagnolette bolt (shutter bolt) A long vertical bolt used to secure casements, doors, or shutters. Operated by turning a handle halfway along the length of the bolt.

essence of the contract Something that is at the heart of the contract or aspect that goes to the root of the contract, a core term, or condition of contract.

estate 1. Property and land, belonging to an organization or individual. **2.** A person's net worth, their entitlements, obligations, and net assets.

esteem value The functions of prestige, appearance and/or other non-quantifiable benefits, such as purchasing something simply for the sake of possession. Architecture and design work often carry high esteem value. Evaluation of esteem value is subjective, and thus very difficult to measure.

estimate Price given for works. If an estimate is not to be considered a legally binding offer of a price for service or works, it is prudent for the party suggesting the probable price to state so. In construction *bills of quantities are used to fix the approximate costs of the works. Changes to labour costs, and/or variations in the standard, quality, and quantity of work may alter the final cost.

estimating The practice of pricing works—fixing the probable cost and expected profit. The *estimate is fixed by taking off the works (*see* TAKE-OFF), measuring the work from the drawings, descriptions, and quantities, and projecting future labour and material costs.

estimator The person who works out the probable costs and fixes the estimate. The estimator is often a quantity surveyor.

estoppel Legal term used to prevent a party making claims under equitable grounds, when this contradicts the original intent of the contract. The doctrine prevents a party being harmed as a result of another party's promises that later undermine that which was originally stated.

estuary Where a river meets the sea causing fresh and salt water to mix.

etching A surface treatment applied to metals to reveal the morphology of grains and microstructure.

ethics The study of moral conduct. **Engineering ethics** is regulated by the *Engineering Council and relates to what is right and wrong with regards to carrying out the requested work, e.g. it is unethical to falsify gathering data.

ethylene propylene diene monomer rubber An elastomeric polymer also called EPDM and EPM. EPDM is widely used in construction applications due to its retention of properties (strength) at elevated temperatures.

ethylene vinyl acetate copolymers (EVA) A polymer with high density and toughness. EVA is primarily used in construction as hot melt and heat seal adhesives.

eukaryotic organism An organism that contains a cell nucleus and a membrane. It includes all living organisms apart from bacteria and viruses.

Euler buckling load Critical buckling load of a pin-ended member, calculated from:

$$F_{cr} = \frac{EI\pi^2}{L^2}$$

where E is the *Young's modulus of the member material, I is the cross-section area moment of inertia, and L is the length of the member.

Euler number *See* E.

Eur. Ing. The professional title for an engineer registered with the *Fédération Européenne d'Associations Nationales d'Ingénieurs* (FEANI). The federation brings together national engineering associations from European countries.

Eurocodes (EC) A set of European standards published and controlled by the *European Committee for Standardization. These codes fix a minimum standard of quality that must be achieved. In many cases Eurocodes have replaced *British Standards, which has helped to ensure greater consistency across Europe, making it easier to specify goods and services in other European countries. A *limit state design philosophy is adopted throughout the Eurocodes, as opposed to the 'lumped' factor of safety approach traditionally used in the UK. There are ten European standards, namely:

- EC: Basis of structural design (EN 1990)
- EC1: Actions on structures (EN 1991)
- EC2: Design of concrete structures (EN 1992)
- EC3: Design of steel structures (EN 1993)
- EC4: Design of composite steel and concrete structures (EN 1994)
- EC5: Design of timber structures (EN 1995)
- EC6: Design of masonry structures (EN 1996)
- EC7: Geotechnical design (EN 1997)
- EC8: Design of structures for earthquake resistance (EN 1998)
- EC9: Design of aluminium structures (EN 1999)

Apart from EN 1990, each code is subdivided into various parts detailing specific elements. There are 58 parts in total. For example, EC7 contains two parts: 'Part 1: General rules' and 'Part 2: Ground investigation and testing', each supplemented by their own *National Annex (NA).

The European standards (EN; harmonized technical rules) specify how structural design should be conducted within the European Union (EU). These were developed by the European Committee for Standardization on behalf of the European Commission.

The purpose of the Eurocode is to provide:

- a means to prove compliance with the requirements for mechanical strength and stability and safety in case of fire established by European Union law
- a basis for construction and engineering contract specifications
- a framework for creating harmonized technical specifications for building products (CE mark)

Euro NCAP safety rating A safety performance assessment programme to provide consumers with information about the safety of new cars, via a five-star rating system. The more stars awarded, the better the car's performance in respect of crash protection, as well as being equipped with crash avoidance technology.

European Committee for Standardization (CEN) *Comité Européen de Normalisation*, the French body that coordinates Eurocodes and European standards. The organization works with other member states to establish and control standards. The common European standard should help with standardization across Europe, allowing free trade, and reducing international boundaries.

European Standard (EuroNorm (EN)) A standard that has national recognition by all *European Committee for Standardization states.

eutectic In metal alloys, upon cooling from above the melting point, a liquid phase transforms into two solid phases. This is observable only at a microscopic level in metals and corresponds to the chemical and physical properties of each alloying element in a metal alloy. The eutectic point is the lowest melting composition in a metal alloy system. This type of reaction is common in most types of steels.

eutectoid In metal alloys, upon cooling, one solid phase transforms into two new solid phases. This is observable only at a microscopic level in metals and corresponds to the chemical and physical properties of each alloying element in a metal alloy. This type of reaction is common in most types of steels.

eutrophication When the nutrient levels (particularly nitrogen and phosphorus) in a freshwater source become too high and cause an imbalance in the quality of the water, for example, an overgrowth of algae and depletion of oxygen.

EVA *See* ETHYLENE VINYL ACETATE COPOLYMERS.

evaporation (evaporate) The process by which a liquid is converted into a gas.

evapo-transpiration The loss of water from soil resulting from both *evaporation and plant transpiration.

event Task or activity on a project network or plan.

evidence The records, artefacts, documents, and statements providing the material that illuminates a situation, which can then be used to support a claim.

evolution The process of gradual development from earlier forms. Typically related to the biological development of all life forms, but also related to the development of more advanced new materials, construction methods, etc.

examination of site The inspection or survey of a site. The examination can take the form of a site reconnaissance, walk-over survey, desk-top study, and *soil investigation.

excavation 1. The process of digging holes in the ground. **2.** A hole dug in the ground.

excavator A machine or person that digs an *excavation.

excepted risks (employer's risks) The risks that are *expressly excluded from the contractor's responsibilities; these are within the risks that the employer has agreed to.

exceptionally adverse weather Weather conditions that impede or restrict construction activities.

excess current *See* OVERCURRENT.

excess excavation An *excavation that is deeper than required.

excess voltage *See* VOLTAGE OVERLOAD.

execute work Direction to commence an activity or task.

execution The undertaking of an event, activity, or task.

exemption or exclusion clause Provision within a contract that seeks to limit the parties' rights within the contract. The effect of exclusion clauses has been limited by The Unfair Contract Terms Act 1977, and by a number of other statutory instruments that attempt to ensure fairer practice within contracts.

exergy The amount of useful work in a system, related to thermodynamics.

exfiltration The process where air exits a building through cracks, gaps, and other unintentional openings in the building envelope. It is driven by the wind and stack effect.

ex gratia Claims and payments are made without giving any legal liability, obligation, responsibility, or position. Payments of this nature are considered to have been paid voluntarily.

exhaust 1. Combustion gases that are ejected from a device, such as a boiler. **2.** The air that is extracted from a room or building.

exhaust shaft A vertical duct used to extract air from a room or building.

exhaust system 1. Various parts of an appliance that discharges combustion gases to the outside. **2.** An *extract system.

exit 1. An opening or passage that enables occupants to leave a building. **2.** A protected passage that is used as a means of escape in the event of a fire.

exit door An outwards opening door that connects the exit to outside.

exit sign An illuminated sign that identifies the location of an exit.

ex officio A person who has gained a right or membership to a body by holding a particular office or position.

exosphere The outermost region of the earth's atmosphere.

exothermic A chemical reaction that gives off heat; *see also* ENDOTHERMIC.

expanded clay (bloated clay, expanded shale, expanded slate) A material made from common brick clays by grinding, screening, and then feeding through a gas burner at about 1482°C, thus changing the ferric oxide to ferrous oxide and causing the formation of bubbles.

expanded polystyrene (EPS, PS foam) *See* POLYSTYRENE.

expanded PVC (PVC foam) *See* POLYVINYL CHLORIDE.

expanding bit (expansion bit) A drill bit that can be adjusted to drill holes of differing diameter.

expanding plug A plug containing a rubber ring that is used to seal a pipe. Used for gas purging and drain testing.

expansion A process which causes an increase in size. *See also* THERMAL EXPANSION.

expansion cistern (expansion tank) *See* FEED CISTERN.

expansion joint (EJ) A joint between two components that allows a degree of movement due to expansion.

expansion pipe (vent pipe) A pipe that runs from the hot-water cylinder to the feed and expansion cistern in a vented system. Used to discharge any boiling hot water or steam that may be produced if the cylinder thermostat fails.

expansion sleeve *See* SLEEVE.

expansion strip (expansion tape, edge isolating strip) A strip of material used to form an *expansion joint.

expansion tank *See* FEED CISTERN.

expansion vessel A small tank that allows the expansion of hot water in a sealed central heating system.

expense The economic cost incurred to undertake an operation, service, item of work, or task. In some contracts, the term expense will be used in reference to the contractor's delay and disruption claim.

experimental 1. Relating to scientific testing. **2.** Based on an experiment. **3.** Relating to something novel and not yet tested.

expert Person with a recognized specialism or professional called on because of their authoritative knowledge or skills in a field.

expert witness A person with specific knowledge who is called upon to provide their specialist opinion on a subject. The duty of the *expert is to assist the court by providing the relevant facts and perspective based on the person's requisite skill and experience. The professional's opinion may be relied on as evidence—in a number of cases the degree that a professional's opinion on a subject can be relied on as evidence may be questioned.

explosive fixing *See* SHOTFIRED FIXING.

exponent A number or variable, written in superscript, relating to the power a number or expression is raised to, e.g. 5^3 and x^2.

exponential 1. Relating to the *exponent—used to define a mathematical curve, function equation, or series. **2.** Used to describe the natural logarithm to the base e. **3.** Something that is increasingly rapidly.

exponential distribution A continuous probability distribution that describes events which occur independently at a contract rate. *See also* POISSON DISTRIBUTION.

exponential function The function e^x, where *e is approximately 2.718281828. It is used to model situations where a constant change in an independent variable represents the same relative change in the dependent variable.

export To take material off-site.

exposed aggregate finish Concrete with coarse stones and rock exposed at the surface. Prior to the concrete maturing, the exposed surface of the concrete has a retarding agent applied so that the fine aggregate does not set and can be removed by spraying with water; alternatively the fine material can be brushed away. The removal of the fine aggregate at the surface of the concrete exposes the coarse aggregate.

exposure 1. Exposing a building or materials to the elements of weather, such as wind, rain, snow, and sunlight. **2.** A method used to describe the orientation of a building, e.g. the dwelling has a southern exposure.

expressly The items and terms that are provided by and written within the contract. They are normally written and therefore are explicitly and unambiguously included in the contract.

express terms Those terms that are normally written down within the contract. Terms that are oral need to be explicitly agreed between the two parties to be included within the express terms of the contract.

ex-situ Away from its original location, e.g. ex-situ remediation techniques relates to cleaning the soil away from its *in-situ location.

extended price (extension) The price or cost that has been calculated in a *bill of quantities by multiplying the measured quantities by the unit rate.

extension 1. The addition of extra space to an existing building. **2.** An agreed increase in time or money to undertake work.

extension bolt (monkey-tail bolt) An elongated vertical bolt that is attached to a window casement or door leaf. When operated, the bolt slides into a socket on the head or sill of the window or door.

extension ladder A ladder whose length can be extended. Either comprises two or more sections that slide together for storage and are slid apart and locked in position to increase the length of the ladder, or telescopic sections that can be extended and locked into position.

extension of preliminaries Addition to works generally described in the preliminaries section of the contract. The preliminaries are distributed as they occur, where they are sufficiently described and itemized, or where the descriptions are very general, the sum of money allocated may be divided equally amongst the interim certificates or monthly valuations. If there is a delay or an extension of time, there is no automatic claim to adjust the monthly figure for the preliminaries, the contractor must prove loss by reference to records and evidence. The preliminary items of a contract are for the work that has been described in general terms.

extension of time The agreement to extend the contract duration by a specified time, administered by the client's agent or representative, under the terms of the contract. Under the contract, additional time may be awarded as a result of client variations, exceptionally inclement or adverse weather, or terms that are *expressly agreed within the contract provisions that deal with *variations and extension of time.

external angle A special brick that is angled part-way along its length to enable the brickwork to turn corners of less than 90°.

external glazing Glazing that is located on the external faces of a building.

external insulation Insulation that is applied to the external faces of a building.

external leaf The outer leaf of a cavity wall.

externally reflected component The light received on an internal surface directly by reflection from buildings, the ground, and obstructions outside a room. Used in conjunction with the *sky component and the internally *reflected component to calculate the daylight factor.

external plumbing Plumbing fixtures and fittings that are located outside the building, for instance, rainwater goods.

external rendering *Render that is applied to external areas and surfaces.

external wall A wall that has one face located outside.

external works Construction work that is undertaken outside.

extinguisher Something that ends or removes something else, such as a *fire extinguisher to put out a flame.

extra (extra work) Additional work which was not included within the original contract bills. The additional work is ordered under *variation order by the client's agent or representative.

extract air Air that is removed from a space.

extract fan (extractor fan, exhaust fan) A fan used to extract air from a space.

extractor hood A device located above a cooker that extracts moisture and cooking smells.

extract system (exhaust system) A type of mechanical ventilation system that uses grilles, ductwork, and fans to extract air from a space.

extrados The convex outer curve of an arch.

extra-low voltage Electric current that is less than 50 volts AC.

extra-over Additional sum of money allowed for items of work, normally used where the work has become slightly different from that agreed or described in the contract documents. For example, where excavation has become more difficult due to different ground conditions being experienced, e.g. bedrock where excavation through clay was described. Part of the original term and price for the excavation can be used with the addition of an allowance to provide for the difference.

extrapolate 1. To estimate a value, which lies outside a set of known values that is based on a trend. **2.** To use facts as a starting point and conclude about something that is unknown.

extras (extra work) Additional task or service required by the client that is not described in the *contract documents. The work would be ordered by the architect, engineer, or other client representative and treated as a *variation and priced in accordance with the contract documents. For a contractor to be paid for the work there must be a provision within the contract, and the variation order must be administered by the nominated representative.

extrinsic evidence (parol evidence) Oral and other preliminary evidence that is outside the contract. Once the contract is signed the basic rule of interpretation is that verbal evidence and preliminary negotiations between the contracting parties are outside the scope of the contract and are not used to change, verify, qualify, or interpret the contract. An exception to this is where there has been misrepresentation, where it is claimed that the written contract does not reflect the agreement and needs to be rectified, or where a party claims that a collateral contract was formed. Extrinsic evidence may be used where a term is imprecise and is difficult to define. The evidence may be admitted to help explain the written agreement. To understand a term or terminology used, the court will turn to the English language but, if still uncertain the court will also turn to technical, trade, or customary meanings which the parties intended to bear on the term. Thus, the main situations where such evidence is used are where the terms are not easily defined by the court or where another agreement, such as a collateral contract, was formed.

extruded brick A type of brick produced by the extrusion process. *See also* BRICK.

extruded gasket A preformed gasket that has been formed by *extrusion.

extruded section (extrusion) A component, usually metal or plastic, that has been formed by *extrusion.

extrusion The process of forming something from a semi-soft material by pushing it through something, e.g. the process of forming igneous rock when magma is pushed through cracks in the earth's crust.

exudation The discharge of a liquid via pores or a surface cut, e.g. the resin from a tree.

eye 1. The middle of a building component or element. **2.** A small hole in a metal object.

eye bolt A bolt in which the head of the nut is replaced by a steel loop, such that when it has been screwed into place the loop can be used for lifting.

eyebrow (eyebrow dormer) A curved *dormer window in the shape of an eyebrow that contains no sides.

eye protection *Personal protective equipment in the form of glasses, goggles, and face masks (particularly for welding) essentially used so that vision is not impaired due to the activity being undertaken.

fabric 1. Materials from which a building is constructed. **2.** A textile structure produced by weaving or knitting techniques. *See also* GEOTEXTILE.

fabrication The construction or manufacture of something, which involves a number of consecutive steps or procedures.

fabric heat loss Heat energy emitted or passing through the fabric (envelope) of a building.

fabridam An air- or water-filled tube made from neoprene, laminated rubber, and nylon, anchored to a concrete plinth and used as an inflatable dam. It is most suitable for applications where the width-to-length ratio is high, such as irrigation systems, tidal barriers, etc.

façade The external *face of a building, usually the front.

façade panel A prefabricated cladding panel designed to be installed on the façade of a building.

façade retention The process of retaining an existing façade while the other elements of the building (roof, floors, and internal walls) are removed and replaced.

face 1. The front of a building or wall. **2.** The exposed external surface of a material that has the best appearance. **3.** The working surface of a tool, such as the face of an axe.

face bedding (edge bedding) Stone placed vertically so that its vertical 'grain' or natural bed is also vertical. The rock is placed on its grain so that a clean stone face is shown. Some rocks, such as sedimentary rocks, have natural beds formed by compression. Sedimentary stone is formed by rock and organic particles laid on top of each other and compressed by further deposits and the weight of the seas over thousands of years. Such rocks have natural horizontal beds. If such rocks are positioned vertically, with the grain (or beds) also running vertically, the face of the rock may delaminate as it weathers. Over the years, as the rock goes through seasonal cycles, including rainwater freezing in the pores and expanding, placing internal pressure on the rock, the beds of the rock may delaminate, with the top layer of the rock (the face of the rock) falling away.

face brick *See* FACING BRICK.

face-centred cubic (FCC) At a microscopic level in metals, where the atoms are more closely packed, the atoms are situated in each corner of the crystal and on the face-centred locations. Most metals have this kind of atomic arrangement including most *ferrous metals.

faced wall A wall clad in a *facing.

face edge (face side) The exposed edge of a material that has the best appearance.

face joint A joint in the *face of a material or object.

face left (face right) The position of the telescope on a *theodolite in reference to the vertical scale. Face left is when the scale is on the left when viewed through the eyepiece—the converse for face right.

facelift (face lift) Improving the external appearance or face of a building without undertaking major changes to the existing structure.

face mark A temporary mark placed on a material to identify the face of the material.

face shovel (crowd shovel, forward shovel) An excavation machine that has a rope or hydraulically operated bucket. It removes soil from the base of excavations in a direction away from the machine; *see also* BACKHOE.

facing 1. Material used to cover a less decorative material to protect or decorate a building. **2.** The process of smoothing or finishing the surface of a material. **3.** The visible portion of a *door leaf.

facing brick (US face brick) Bricks with an aesthetically pleasing face, used where the brick is exposed and can be seen. Facing bricks are used to clad buildings to produce a desirable finish.

facing hammer *See* BUSH HAMMER.

factor of safety (FOS) A design margin between maximum or failure loading and working loading conditions. Values below one will result in instant failure—typically values of 1.3 and above are normally requested, as a FOS value of one does not allow any room for errors in the design or material properties. **Factored loads** have become more commonplace in modern design codes (*see* EUROCODES), where, for example, the design load is multiplied by a factor during the design calculation stages, rather than just at the end with the traditional FOS approach.

factory inspectorate Division or sector within the Health and Safety Executive.

faecal sludge management The safe collection, treatment, and disposal of sewage from household waste in the form of a sludge. The sewage matter (faecal sludge) can be collected in septic tanks or using an onsite sanitation system.

failure When maximum load or serviceability is exceeded, which can result in a structure or item suddenly collapsing or being unable to perform its requested task.

fair cutting Cutting fair-faced material.

fair face The neatly built surface of a material that is on show. Common materials used include brickwork **(fair-faced brickwork)**, blockwork **(fair-faced blockwork)**, and concrete **(fair-faced concrete)**.

fall A slope used to prevent water ponding, e.g. in gutters, on flat roofs, and in pipework.

falls Stability failures of steep (nearly vertical-sided) slopes. These occur in coastal regions, e.g. by the wave action undercutting a rock cliff.

fall velocity The rate at which water flows down a pipe to prevent sediments from settling and blocking the pipe.

false body Due to the high viscosity of non-drip paint, it maintains a shape when undisturbed. The shape that is formed by stirring or applying the pressure of a brush. The thixotropic characteristic, holding the shape into which it is moved, means that it does not readily drip.

false ceiling (dropped ceiling) A secondary ceiling that has been installed below the existing ceiling resulting in a reduction in *ceiling height. Generally not a *suspended ceiling.

false exit A dead end, corridor, or passage that does not lead out of the building.

false leaders The supporting framework (legs) of a pile-driving rig.

false tenon A separate tenon. Usually used to replace a tenon that has failed.

falsework A temporary scaffolding structure, used to support *formwork and the construction of arch bridges.

fan An electrically powered mechanical device that uses a series of blades to move air.

fan coil unit An air-conditioning unit comprising two coils that are supplied with hot and cold water from a central *boiler and *refrigeration plant. A fan built into the unit recirculates heated or cooled air to the space. Fresh air from a separate system can be introduced directly to the space, or may be ducted to the inlet of the unit. Air *filtration can be incorporated within the fan coil units. Fan coil units are usually located within a *suspended ceiling or are floor-mounted and are commonly used in multi-zone buildings, such as hotels and cellular offices.

fan convector A type of space heater that uses an electric fan to draw cool air in at the bottom, passes it over a heating element or coil, and then blows the heated air out into the room. *See also* CONVECTOR.

Fanki pile A reinforced concrete cast in-situ pile, with an enlarged base, which has been formed by a bottom-driving displacement and concrete-ramming technique.

fanlight (transom window) A small window located above a door to admit daylight and provide ventilation. Traditionally, semi circular in shape with a number of glazing bars radiating out from the centre of the semi circle resembling an open fan.

fan pressurization test A simple and widely used technique to determine the *airtightness of a building. The technique involves sealing a portable variable speed fan into an external doorway, using an adjustable doorframe and panel. A fan speed controller is then used to pressurize and/or depressurize the building. The airflow rate that is required to maintain a number of particular pressure differences across the building envelope is measured and recorded. The leakier the building, the greater the air flow required to maintain a given pressure differential. In the UK, a pressure differential of 50 Pa is used.

fan truss A timber *roof truss that contains vertical and inclined struts.

fan vault A fan-shaped *vault.

fascia (facia board, eaves facia) Vertical board placed at the *eaves of a *roof, and providing a neat finish to the edge of the roof, covering the edge of the *rafters. The facia board also provides a suitable surface onto which the gutter can be fixed.

fastener Device, fitting, or unit used to secure two or more objects together.

fast to light A colour that does not degrade in light.

fast-track Describes construction projects that are programmed so that they are completed in the shortest possible time; the building works commence before the building's detailed drawings are totally complete. As soon as sufficient design information is produced, the on-site construction starts with further design information being produced so that subsequent elements of the building can be procured and built. The overlapping of design and construction reduces the overall project duration.

fat board Slang term for mortar board used to carry mortar for pointing.

fatigue failure A common failure mechanism in metals due to repeated loading cycles. This type of failure is prevalent in both brittle and ductile metals, and usually occurs without any or with little warning with rapid acceleration of crack length.

faucet A small water tap, the type used in a domestic household sink.

fault A facture in rock strata caused by large displacements from a build-up of stress. *See also* RIFT.

feasibility study Investigation undertaken to determine whether a site or building is economically viable or functionally suitable for a specific development.

feather edge board (feather edge) A tapered board used for cladding and fencing.

feather edge rule (featheredge rule) A tapered straight-edged metal rule used to smooth, straighten, and level plaster and render.

feathering The process of reducing a material from one thickness to another. Used in plastering to smooth out lumps in walls.

federated model The integrating of different discipline models to create one complete model in *BIM.

feed The gradual supply of something to ensure that it remains operational, e.g. water to a boiler, electricity to equipment, etc.

feed cistern A cold-water storage *cistern that supplies cold water to a boiler or hot-water cylinder. In indirect hot-water systems, a vent pipe may be required to feed into the cistern to accommodate any expansion in the hot-water system. Cisterns that incorporate expansion are known as **feed and expansion cisterns.**

feeder (metallurgy) Part of the gating system that forms the reservoir of molten metal necessary to compensate for losses due to shrinkage as the metal solidifies;

sometimes referred to as a riser. The feeder is part of the machinery used for heat treating metals to achieve the desired mechanical properties.

Fellenius circular-arc method A method to determine the ultimate bearing capacity of a homogeneous cohesive soil. **Fellenius method of slices**, a slope stability analysis for rotation failures where interslice forces are ignored and the *factor of safety is calculated from:

$$Factor\ of\ safety = \frac{\sum Resisting\ Forces}{\sum Sliding\ Forces}$$

felt See BITUMEN.

felt and gravel roof Flat roof with weather protection formed with a waterproof bitumen layer covered in gravel to protect it from the sun and to help prevent it lifting off the roof surface during strong winds.

felt nail Fixing nail with a large head to hold roofing felt in place. The large head prevents the felt pulling through or ripping over the nail. See also CLOUT NAIL.

feltwork The process of using bonded layers of *bitumen felt.

fence An enclosure around something, e.g. an area of land, to act as a barrier. **Fence posts** are the main vertical members sunk or concreted into the ground to support the fence.

fender A guard or cushion that provides protection for something against impact. A **fender wall** is a low wall that carries the hearth slab of a fireplace. A **fender pile** is a pile driven into the seabed to provide protection to the berth from impacts of mooring ships.

fenestral A *window.

fenestration A term used in architecture referring to the way in which the windows have been arranged on the outside of a building.

ferrimagnetism (ferrimagnetic) A type of magnetic behaviour found in many ceramic materials. Ferrimagnetism is permanent and the magnitude of the magnetization is usually very high.

ferrite (iron) A phase found in ferrous metals having a *body-centred cubic crystal structure. This type of phase is found in most commercial steels.

ferrite ceramic A ceramic material based on iron with magnetic properties.

ferrocement A thin layer of *Portland cement and sand that contains a steel wire mesh. Used to form curved sheets, such as hulls for boats, canoes, water tanks, and sculptures.

ferromagnetism (ferromagnetic) A ferrous material with magnetic properties under the influence of a magnetic field.

ferrous metal A metal based on iron.

ferrous metal alloy A metallic material for which iron is the dominant constituent. A ferrous metal must be based on iron, not merely contain iron.

Examples of ferrous metals include steel, stainless steel, galvanized steel, cast iron, and wrought iron.

ferrule 1. A metal cap or ring to protect the end of a shaft, e.g. to prevent the end of a wooden pole from splitting. **2.** A metal cylinder used to join pipes.

festooned cable A flexible electrical cable that is supported, at intervals, on an overhead bogie, allowing the cable to concertina back on itself.

fetch A distance the wind or a wave travels without interruption.

fettle 1. The finishing of any trade work. Final adjustments, corrections, and finishing to complete the work. **2.** The hanging and striking with a hammer of a casting; for example, something made out of cast iron is struck with a tone hammer to check that it is free from flaws. Something that is free from flaws will have a solid ring sound, whereas something with flaws will produce a dull short tone.

ffl *See* FINISHED FLOOR LEVEL.

fibre A long slender thread of filament of natural or synthetic origin from which yarns are spun to make a fabric structure, such as a *geotextile.

fibreboard (fibre building board) A timber product comprised of wood fibres, e.g. medium-density fibreboard (MDF).

fibre cement board A high-quality calcium, cement fibreboard with additives.

fibreglass Material made from extremely fine fibres of glass. The fibres are embedded in a matrix, e.g. in polymers, better known as glass reinforced plastic (GRP). The resultant material has improved strength and toughness as the inclusion of these fibres helps accommodate the stress imposed. Furthermore, materials and composites with fibreglass reinforcements have excellent strength : weight ratios.

fibre optic A flexible, transparent fibre formed by a very thin glass (silica) core through which light signals can pass with hardly any loss of strength.

fibre-reinforced concrete (FRC) Concrete with fibres embedded in the cement matrix. Usually polymer fibres are used, but steel fibres are also used. The addition of these fibres results in an enhancement of the mechanical properties, i.e. strength and toughness.

fibre-reinforced polymer (FRP) A composite material consisting of a polymer (an epoxy, vinylester, or polyester thermosetting plastic) reinforced with fibres, such as fibreglass, carbon, or aramid.

fibre saturation point The moisture content at which the strength of timber remains constant—around 30%. Below this value, as the timber dries, it becomes stronger and shrinks.

fibrescope (borescope) A flexible *endoscope that uses fibre-optic technology to view inaccessible places, e.g. to inspect cavity ties in a cavity wall.

Fick's first law For metals at a microscopic level, when a chemical reaction is taking place within a metal, the diffusion of the atoms is proportional to the concentration of new or foreign atoms.

Fick's second law For metals at a microscopic level, when a chemical reaction is taking place within a metal, the diffusion of the atoms is not directly proportional to the concentration of new or foreign atoms. This usually occurs under unstable conditions.

Fidler's gear The lifting mechanics, used on block-setting cranes, which allow stone blocks to be lifted flat and placed at an angle.

field 1. An area of land used for agricultural or playing purposes. **2.** An area rich in natural resources that can be mined. **3.** A subject discipline area—speciality. **4.** An area of force, e.g. magnetic field.

field book A notebook to record measurements and observations while on site.

field drain A series of interconnecting porous or perforated pipes, which have been buried in the ground to remove groundwater so that the surrounding ground is less boggy or waterlogged.

fielded (fielded panel) A panel with a raised central surface where the edge is rebated, recessed, or shaped away from the surface.

field moisture equivalent The moisture content of the soil when water remains on the surface of the ground instead of being immediately absorbed.

field order A site order given by an engineer in the US.

field settling test A test to determine the cleanness of fine aggregate/sand. This type of aggregate should be free from dust, clay, silt, and organic materials, because such impurities reduce the bond between the cement and aggregate, hence reducing the net strength of the hardened concrete or mortar. The test involves placing 50 ml of 1% solution of common salt water in a 250 ml measuring cylinder. Sand is added slowly until the 100 ml mark is reached. Further solution is then added to bring the level to 150 ml. The cylinder is then shaken and allowed to settle for 3 hours. The thickness of the silt layer is then measured and expressed as a percentage of the sand layer. If this value is less than 10%, the sand is classed as acceptable. If it exceeds 10%, it cannot be classed as failed as this test only gives an approximate estimate, thus further laboratory testing would be required.

field splice A structural steel *splice that is bolted on on-site.

field superintendent (US) Engineer in the US who often acts as the on-site project manager. This person may also manage the design process.

figure 1. A geometrical two- or three-dimensional shape. **2.** A *drawing or *diagram. **3.** A symbol representing something, e.g. a number. **4.** The natural markings and colour of timber.

figured dimension Dimensions on a drawing where the drawing or sketch is not to scale.

filament A slender strand of wire or fibre.

fill Material such as sand, gravel, or stone used in *earthworks to raise the level of, for example, an embankment to the desired level.

filler A specific type of additive utilized to increase the bulk density/volume of an object or material.

filler beam floor (filler joist floor, filler concrete slab) An element that fits between the main component, which can also be made of a different material. For example, smaller reinforced concrete beams spanning between main steel beams.

fillet 1. A triangular-shaped piece of material inserted into the angle between two materials to strengthen the joint. **2.** A timber moulding used to cover the joint between two members.

fillet chisel A tool used by a mason for working stone.

fillet saw A small hand-saw.

fillet weld A triangular weld that can join two pieces of metal at right angles to each other.

filling knife A *stopping knife that has a thin, flexible blade, used to apply fillers.

filter A straining device to remove larger particles. A **filter bed** is a layer of sand, gravel, or any other pervious material used to remove sewage or other impurities from wastewater. A **filter drain** is a *French drain. **Filter material** is washed and graded sand, gravel, or any other pervious material.

filtrate Filtered material.

filtration A process to permit the flow of liquids and gases, but to prevent major passage of particles, through a *filter.

final account The total costs of the development based on the work that has been measured and undertaken. Any variations or fluctuations that are allowed under the contract would be included in this total. The work is normally measured and recorded by the contractor's *surveyor and checked and agreed by the client's *quantity surveyor.

final certificate Document issued by the client's representative, traditionally the *architect. The paperwork acts as a contractual trigger stating that the work is complete and entitles the contractor to payment of the *final account and part of the retention monies.

final completion The point when the end of the defects liability period is reached, all defects have been made good, and the final certificate is issued, releasing the second part of the retention monies.

final fixing *See* SECOND FIXING.

final grade (US) *See* FORMATION LEVEL.

final inspection An inspection of a building prior to completion.

financial control Checking the cost of the actual works against that budgeted and priced *bills of quantities; includes the forecasting, reporting, and setting new targets where necessary. Through the examination of unit costs and elements that are costing more than anticipated, plans can be produced to ensure that the future work, labour, materials, and services do not exceed the budget.

fin drain A flexible three-dimensional *geocomposite drain, consisting of a polypropylene drainage core surrounded by a nonwoven *geotextile, which acts as a *filter, and prevents the drain becoming blocked.

fine aggregate Material used in concrete, usually in the form of sand. Fine aggregates are usually small in size, typically smaller than 4 mm in diameter.

finger joint A type of glued timber joint comprising a series of interlocking zig-zag shaped tapered fingers.

finger plate (fingerplate, doorplate, pushplate) A protective plate, usually decorative, fixed to the face of an internal door to prevent finger marks.

finial A decorative element used at the top of a gable, newel post, or fence post.

fining off The process of applying a finish coat to the external *render.

finish The external coat or surface of material. The exposed surface could be the material's natural appearance or be enhanced by applying another material, or polishing the surface.

finished floor level (ffl) The final level of a finished floor.

finisher Tradesperson who trowels or tamps the surface of concrete to give it the final finish.

finish hardware (US) Builder's hardware that is seen.

finishing coat (fining coat, setting coat, skimming coat, white coat) Usually the final layer of paint applied to provide a durable and decorative layer of the required colour. Typically, the gloss, silk, or matt (eggshell) finishes are based on oils and alkyd resins, although water-borne products are increasingly becoming available. Waterborne gloss finishes are more moisture permeable than the traditional solvent-borne hard glosses. Waterborne glosses have the advantage of quick drying without giving off solvent odour, and generally they do not yellow on ageing. The matt and silk finishes are usually vinyl or acrylic emulsions.

finishing schedule List of surface coats and the programme of works necessary to complete the project and ensure that all of the walls, ceilings, floors, doors, and windows have the desired appearance. The paint, varnishes, wallpaper, and other fittings are listed against each room. The order in which the operations are to be completed is also listed.

finishing trades Painters, joiners, decorators, plasterers, carpet and tile fitters, tillers, and other trades necessary to give the building and the rooms the final appearance.

finish nail (US) *See* BULLET-HEAD NAIL.

finite difference method (FDM) A numerical computational method used to solve differential equations. In contrast to the *finite element method, every derivative in the set of governing equations is replaced by an algebraic expression in terms of the required variables (such as stress and displacement) at discrete points in space, without being defined within elements. It is used extensively to model and design geotechnical problems.

finite element method (FEM) A numerical computational method for solving differential equations, from which the finite differences approximate the derivatives. In contrast to the *finite difference method, it has a requirement that field quantities (such as stress and displacement) vary throughout each element in a prescribed fashion, which involves the adjustment of these parameters to minimize error. It is used extensively to model and design structures.

finite resource (non-renewable resource) A resource that does not renew itself at a sufficient or sustainable rate for economic extraction. For example, *fossil fuels (such as coal and oil) are a finite resource, whereas, wind or solar power is from a renewable resource.

Fink truss (Belgian/French truss) A simple roof truss of the configuration shown below:

Fink truss, Belgian

fin wall A wall that projects outwards at a right angle to the structure.

fir An evergreen tree, with flat needle-shaped leaves.

fire 1. The process of combustion. **2.** The process of baking materials in a kiln. **3.** A space-heating device located within a fireplace.

fire alarm A mechanical, electromechanical, or electronic device that is activated by fire or smoke to warn the occupants of a building that an unwanted fire has broken out.

fire alarm indicator A panel that indicates where in a building a fire has broken out.

fireback 1. A wall at the rear of a *fireplace. **2.** A decorative cast-iron plate inserted at the rear of a *fireplace to protect the masonry wall. This also stores heat from the fire, which is radiated back to the space once the fire has died down.

fire barrier A strip of non-combustible material that is inserted into a construction to restrict the movement of smoke and flames.

fire block (US) A *fire barrier used inside wooden walls and floors.

fire booster A pump used to increase the pressure of the water within a *fire riser.

firebox The part of a *fireplace or the space inside a boiler where the fuel is burned.

fire break (firebreak) A gap between buildings, a fire-resisting wall between *compartments, or a strip of land where the vegetation has been cleared. Used to stop or slow down the spread of fire.

fire brick A brick made for use in burning stoves and fireplaces. Although they are normally used for a base, they can be used for surrounds and hearths.

fire calculations Mathematical calculations relating to the science and engineering principles of fire.

fire cell *See* COMPARTMENT.

fire certificate Under the Fire Precautions Act all non-domestic buildings must be inspected by a fire officer to check that the building complies with the Act. The certificate confirms that the building is compliant.

fire check door A fire-resistant door where its period of integrity is less than its period of stability. *See* FIRE-RESISTING GLASS.

fire compartmentation (compartmentation) The sealing of a room, corridor, apartment, or floor with fire-resisting materials to ensure that the occupants within that room or dwelling are protected from smoke and fire for a given period of time. Fire-resisting doors, floors, walls, and ceilings are specified to give fire protection from 30 minutes up to 4 hours.

fire damper A fire-resisting damper installed within the ductwork of an air-conditioning or ventilation system to prevent the spread of fire and smoke.

fire degradation A form of degradation of a material due to the presence of fire, which ultimately leads to destruction of the component or structure.

fire degradation of timber A form of degradation of timber due to the presence of fire. Wood is an organic (hydrocarbon) material and in its dry state will ignite. When the wood is heated to 160–200°C, it will pyrolyze and give off flammable gases. However, wood is also a very poor conductor of heat, and so heat does not easily soak into the wood to raise its temperature enough to cause pyrolysis. Therefore, thick-section timber behaves surprisingly well in a building fire. Thin sections, in contrast, burn through much more easily.

fire detector An electrical device that is designed to detect the presence of a fire and set off an alarm.

fire door (firebreak door) A door, including the frame, which is designed to provide fire resistance for specific periods of time depending on their grading, for instance, a FD_{30} is designed to give 30 minutes of resistance.

fire engineering (fire protection engineering, fire safety engineering) The use of science and engineering principles, such as measurement and calculation, to provide fire safety in buildings.

fire escape (fire exit) An egress point, stairwell, or pathway to allow people to exit a building in the safest, quickest route if there is a fire. *See also* ESCAPE STAIR.

fire extinguisher (portable fire extinguisher) A cylindrical metal container containing foam or other extinguishable liquid that can be sprayed onto a fire. A **fire hydrant** allows a fixed connection to a continuous water supply via a valve such that a **fire-hose** can be attached by fire-fighters. A **fire-hose reel** is a roll of pipe that is often positioned adjacent to a fire hydrant. A **fire-extinguishing system** is the fixed equipment within a building, such as hoses, hydrants, and a sprinklers system, used to extinguish a fire.

fire-fighting shaft A fire-protected stairwell, and sometimes a lift shaft, in tall buildings that allows fire-fighters, together with their equipment, to again access to the floor on which the fire is situated. The lift will have its own back-up power supply.

fire floor (firebreak floor) A fire-resistant floor.

fire grading *See* FIRE RATING.

fire hazard Something that is liable to catch fire.

fire inspection The examination of a non-domestic building by the fire officer to check that the building has all the required measures necessary to comply with the Fire Precautions Act. If compliant, a *fire certificate will be issued.

fire lift A lift installed within a protective shaft that is designed to be used by fire-fighters in the event of a fire.

fire load density The amount of potentially combustible material within a structure. It is used to predict the *fire severity, which is measured in megajoules (MJ) per square metre. Anything up to 1135 MJ/m^2 would be classed as having a low fire load density and would be applicable to domestic dwellings. Moderate and high fire load density values would be 1135–2270 MJ/m^2 and 2270–4540 MJ/m^2, respectively, and could be associated with various types of warehouses.

fire lobby A *lobby that leads to a fire lift or forms part of an escape route from a building in the event of a fire.

fire main A horizontal or vertical fixed pipe within a structure used to supply water to a fire-extinguishing system. It is normally made from galvanized steel and colour-coded red.

fireman's panel A diagram on a wall, near the entrance of a large building, to give information to fire-fighters about the layout of each floor, particularly the location of *fire-fighting equipment and *fire escapes.

fireman's switch A *fire switch.

fire modelling The use of computer programs to determine how a structure will perform in a fire and to assess the potential consequences.

fire officer (fire prevention officer) Senior member of the fire service with duties to inspect buildings, issue *fire certificates, and provide advice on necessary changes. Employed by the fire service, the officer's work is to enforce the fire prevention laws and issue fire certificates for buildings that are compliant.

fire performance (fire behaviour) A general term used to describe how a structure or a material performs when subjected to fire.

fireplace A point within a building that has been designed to contain a fire or heating element for a building. Where open combustion is to take place, as in coal, wood, or gas-fuelled fires, a *chimney and *flue is required to remove the fumes and smoke. The fireplace is normally set on an incombustible *hearth.

fire point The location in a building where fire-extinguishing equipment is kept.

fire precautions Measures taken to avoid the risk and effects of fire, both to people and property.

Fire Precautions Act 1971 (UK) The legislation and Parliamentary Act provided provision for protecting persons within a building from fire. The Act requires all new and existing buildings to be inspected by a fire officer and receive a fire certificate before they can be occupied and used.

(⊕) SEE WEB LINKS
• Information on fire regulations from the Health & Safety Advisory Service.

fire prevention Taking measures to avoid or reduce the risk of a fire occurring.

fireproof *See* FIRE RESISTANCE.

fire propagation index A measure or rating of the heat index from a surface material or wall lining when it is subjected to a fire test or furnace test.

fire protection Steps taken to protect materials and buildings from the effects of fire, heat, and smoke. The measures are taken so that the building is sufficiently resistant to the effects of fire to provide sufficient time to safely evacuate a building in the event of a fire. The materials will also limit the combustibility of the building, offering protection to the building, which can also be advantageous in reducing insurance premiums.

Fire Protection Association (FPA) A body that publishes advice on fire and fire protection.

(⊕) SEE WEB LINKS
• The UK Fire Protection Association home page—provides information on the protection of people and property from the effects of fire. It collects and publishes statistics, and provides guidance and information on regulations and best practice.

fire rating The amount of time various building elements (*see* FIRE DOOR) have the ability to resist fire. It is determined in accordance with BS 476, which assesses absence of collapse, flame penetration, and excessive temperature rise on the cool face to determine a fire rating.

fire reserve Water that is specifically kept (stored in a tank) and used for fire-fighting.

fire resistance (fireproof) The ability not to catch fire or to burn so slowly that it will cause minimum impact, e.g. a *fire door.

fire-resisting glass Glass that holds its form, does not fall away from the frame, and maintains a barrier in the event of a fire. The glass may crack but it will maintain sufficient integrity to ensure that it prevents the passage of smoke and fire for a given time.

fire retardant An element used as a deterrent to combustion. Fire retardants either prevent or delay the commencement of ignition/combustion. Examples include intumescent coatings on steel structures.

fire riser Water pipes travelling up and down a building that deliver large quantities of water for use by the fire service in the event of a fire. Both dry and wet risers can be used. A dry riser provides an empty pipe run which can be connected to the water supply by opening a valve, whereas a wet riser is kept permanently charged with water so that it is available for immediate use in the event of a fire.

fire safety sign Signage required to make users aware of building laws, fire fighting equipment, fire exits, and procedures necessary in the event of a fire.

fire setting An old method of mining where a fire is lit against the rock face, then when the face is red hot it is doused in cold water. This procedure causes thermal cracks to develop in the rock—the cracks make excavating the rock easier.

fire severity This is a measure of the fuel load and the anticipated effect and growth of a fire. By calculating the fuel load density in a compartment, and taking into account the room's shape, size, and ventilation, the fire's potential severity can be calculated. This is used to determine the amount of fire resistance the fabric (walls, floors, and ceilings) of the compartment will need in order to survive full burn out of the contents of the room.

fire shutter Vertical and horizontal roller doors and walls that provide additional fire resistance by acting in a similar way to that of a *fire door. Can be used to seal lift shaft doors, shop fronts, and other openings.

fire stair A set of stairs that can be used as a means of escape in the event of a fire.

fire stop (fire stopping) Non-combustible material used to fill gaps and voids in order to prevent fire passing through the void. A fire stop may be a solid block of non-combustible material used below *raised floors, a blanket or non-combustible quilt used above suspended ceilings, or it can be a strip of material that expands when heated to close gaps around services and doors.

fire stop sleeve (pipe closer) A collar that fits around a pipe or duct at the point where the duct passes through a wall; in the event that a fire burns through the pipe, the intumescent foam within the collar expands and seals the pipe. Where walls are compartment walls offering a specified fire resistance, all openings within the wall should be sealed with a fire stop.

fire-suppression system *See* FIRE EXTINGUISHER.

fire switch (fire lift priority switch, fireman's switch) A switch that enables fire-fighters to operate a *fire lift.

fire testing Calculations, experiments, and tests undertaken to determine the fire resistance and grading of building elements.

fire tower (US) A compartmentalized or protected stairway.

fire tube A long steel tube in a boiler that carries the hot combustion gases.

fire vent A damper located in a roof that is opened for *fire venting.

fire venting The process of enabling the hot gases and smoke from a fire to escape from a building using the *fire vents.

firewall (firebreak wall, division wall) A fire-resistant wall.

firing A process whereby an object is heated to elevated temperatures with the objective of increasing density and/or strength, e.g. ceramics. A process of baking ceramics in a kiln to partly or fully vitrify (harden/glaze) them.

firm clay (firm silt) A material that does not remould easily.

firmer chisel A sturdy general-purpose chisel designed to be used by hand and not to be hit with a hammer or mallet. Used by carpenters, joiners, and wood carvers. Also known as a **forming chisel**.

first angle A method of presenting an orthographic drawing. First angle projection is the UK method of presenting a two-dimensional drawing of a side elevation, front view, and plan view of an object. In the US, third angle projection is generally used.

(⊕) SEE WEB LINKS
- The graphics section on the technology students' website provides detailed drawings of orthographic projects.

first fix (1st fix, first fixing) The installation of various items within a building after the primary elements of the building have been constructed (floors, walls, and roof); these include electrical cables, pattress boxes, plumbing, ductwork, and partitions.

first floor The floor immediately above ground level.

first moment of area A measure of a shape's area in relation to an axis, calculated by the area multiplied by the distance (lever arm).

first storey The space between the first floor and the floor or roof above.

fish ladder (fish pass, fishway) A channel that bypasses a dam or weir so that fish can swim up- and down-stream.

fishplate A *splice bar.

fish screen A barrier to prevent fish swimming past.

fishtail fixing lug (fang, tang) A fishtail-shaped *lug that is either cast into concrete or built into masonry.

fish tape *See* DRAW CABLE.

fissured clay A cohesive soil that contains a network of cracks, which particularly open up when the material is dry.

fissured surface A flat surface that contains shallow irregular cracks, used for decorative processes, or to absorb sound.

fitment A factory-made element that is installed and can be removed from buildings, such as baths, basins, and cupboards. *See also* FITTINGS.

fit-out (fitting-out, US outfitting) The process of installing finishes such as suspended ceilings, floor coverings, and partitioning systems.

fitted bolt A bolt that has a small clearance in a hole, and is used to carry high loads.

fitted carpet A carpet that is fixed to the floor and covers the floor completely.

fitter A person who assembles machines or maintains machinery.

fittings 1. A term used to refer to various small items of pipework, such as a bend, coupling, or tee. **2.** Items within a building that can be removed without causing damage.

fit-up Formwork that can be reused, typically it is used repeatedly to form sections of a concrete wall, column, beam, etc.

fixed beam A beam that is restrained from rotating and moving vertically. A beam with a **fixed end** cannot move or rotate. A **fixed end moment** is set up at the restraint end because the beam cannot rotate.

fixed-form paving train (fixed-form paver) A machine that is used to lay, compact, and finish a concrete road surface.

fixed-in The process of inserting components into the construction after it is complete. *See also* BUILT-IN.

fixed light (dead light) A window that does not open.

fixed-price contract Contracts that help secure the final price of the contract by either agreeing to a *lump sum, or schedule of rates and prices, or by a measure and value process. Clients generally prefer such arrangements, since much of the financial risk is transferred to the contractor. Generally, fixed-price contracts have positive benefits in terms of fast completion times and known price, as the contractor wants to complete the work as quickly as possible in order to reduce overheads and labour. However, without good supervision quality can suffer.

fixed retaining wall A *retaining wall that is held in place at the top and bottom, such as a basement wall.

fixer A tradesperson who fixes elements and components to the building on-site.

fixing 1. A device used to hold something in place. **2.** The process of attaching, adjusting, or repairing something.

fixing brick *See* NOG.

fixing channel A channel that enables the location of a fixing to be adjusted. It can be surface-mounted or embedded into a floor, wall, or ceiling.

fixing device A *fixing.

fixing fillet (fixing slip, pad, pallet, slip) A thin piece of wood inserted into the mortar joint in masonry construction to act as a fixing.

fixing gun *See* CARTRIDGE TOOL.

fixing moment (fixed-end moment) The moment created at the support because the beam has been prevented from rotating.

fixing strap A long narrow piece of material that is placed around an object to fasten it in place.

fixing strip A long narrow piece of a material that an object is attached to.

fixture 1. Any item that is fixed to a building or land, and would cause damage if removed; for instance, plumbing, electrical cables, trees, fences, etc. **2.** A *luminaire.

FLAC (fast Lagrangian analysis of continua) A geo-mechanical explicit finite difference modelling program. The program simulates the behaviour of structures built of materials such as soil and rock that undergo plastic flow when their yield limits are reached. Materials are represented by elements, or zones, which form a grid that is adjusted to fit the shape of the object to be modelled. Each element behaves according to a prescribed linear or nonlinear stress/strain (Newtonian) law in response to the applied forces or boundary conditions. Any local disturbance of equilibrium is propagated at a materially dependent rate consistent with Newton's Laws of motion. It also has the ability to model groundwater flow. This can be coupled or uncoupled to the mechanical analysis of deformations. **FLAC/slope** is a mini version of FLAC, which is specifically designed for FOS calculations.

flag (flagstone) A rectangular piece of material used for paving.

flake ice A type of ice made from a frozen sheet of water that has been fractured into irregular shapes. Used as a substitute for water in large concrete pours to reduce the heat of hydration.

flame cutting An oxygen-cutting process for metals that have been heated to a high temperature by the use of a gas flame from a fuel supply such as acetylene. The oxyacetylene flame firstly heats the metal to the required temperature. At this temperature an oxygen stream is then released through the cutting touch—the oxygen reacts to produce the cut.

flame detector A type of fire detector that uses infra-red and/or ultra-violet sensors to detect the presence of flames.

flame retardant A material or object that does not ignite or burn easily.

flame-retardant paints A special type of paint that emits non-combustible gases when subject to fire, the usual active ingredient being antimony oxide. Normally, combustible substrates such as plywood and particleboard can have their fire resistance increased substantially with the application of an appropriate protective paint.

flamespread (surface spread of flame) A measure of a material's ability to burn and spread flames.

flammable A term used to describe materials that easily ignite and will burn rapidly.

flammable liquids A liquid that easily ignites or burns rapidly.

flange A projecting piece for fixing or strengthening, e.g. a flat plate on the end of a pipe used for fixing, or the top and bottom parts of an *I-section for strengthening.

flank The side of something such as the side of a carriageway, the hard shoulder, or the intrados of an arch. A **flank wall** is a side wall of a building.

flanking transmission (flanking sound) The indirect transmission of sound through a separating wall or floor via elements that are adjacent to the wall or floor.

flanking window (wing) 1. A side window, adjacent to a door or main window, with its sill at the same level as the door or main window.

flap trap A drainage fitting that contains a hinged plate, which only allows flow in one direction to prevent backflow.

flared column A column that widens outwards at the top.

flash The process of protecting a joint from the weather using thin continuous pieces of metal. The pieces of metal are known as a **flashing**.

flash distillation A *distillation process used to desalinate (remove salt from) seawater.

flashing A strip of impervious material to protect roof joints and angles from the ingress of water. It can be made from lead, aluminium, galvanized steel, or bitumen felt.

flashover When a material reaches its *flashpoint and ignites.

flashpoint The temperature at which the vapour of a flammable liquid will ignite.

flat arch (French arch, jack arch, straight arch) An arch above a window or door that is almost completely horizontal.

flat-bottomed rail A rail with a rounded top and flanged bottom.

flat cost Cost of the materials and labour without overhead or profit.

flat glass A reference to rolled glass, plate glass, float glass, and sheet glass.

flat interlocking tiles A flat roof tile that interlocks with an adjacent tile.

flat jack A steel flat capsule that expands, pushing the thrust plates apart when oil, water, or grout is pumped in.

flat joint (flush joint) A mortar joint where the mortar is finished flush with the masonry.

flat jointed brickwork Brickwork constructed using a *flat joint.

flat-pin plug A three-pin *plug that contains rectangular flat pins.

flat plate collector A type of *solar collector where a flat surface is used to collect the solar radiation. Although they tend to be fixed in a static position, they can incorporate systems that track the path of the sun throughout the day. It typically comprises a series of pipes attached to a black heat-absorbing plate, all of which are enclosed in a glass-fronted box.

flat pointing Pointing masonry to achieve a flat joint.

flat roof A roof that has a pitch of less than 10°. Flat roofs should not be completely horizontal but should have a pitch that enables water to drain off the roof. *See also* PITCHED ROOF.

flat slab A concrete floor that contains the beams within the slab, such that the underside of the floor slab is level.

flaunching A cement mortar fillet used at the top of a chimney stack to prevent rain penetration.

Flemish bond A type of brickwork bond formed by laying alternating *headers and *stretchers within the same course.

Flemish garden wall bond A type of brickwork bond based on the *Flemish bond, that is formed by laying one header for every three stretchers on the same course. On even numbered courses, the headers are positioned under the centre of the middle stretcher on the odd numbered course.

fleur-de-lis An iris or lily-shaped ornament often used as a *finial.

fleuron General term for any carved floral decoration.

flex The flexible cable between an electrical appliance or device and the plug that is inserted into the electrical socket.

flexible armoured revetment (FAR) Precast concrete blocks connected by plastic cables underlain by a geotextile filter.

flexible coupling A rubberized connection, such as a V-belt between, for example, a mortar and pump.

flexible damp-proof course *See* DAMP-PROOF COURSE.

flexible ductwork *See* DUCTWORK.

flexible membrane An impervious plastic or rubberized sheet.

flexible metal conduit A spiral-shaped steel or aluminium *conduit that can be bent easily and moved around.

flexible metal roofing *See* SUPPORTED SHEET-METAL ROOFING.

flexible mounting *See* ANTI-VIBRATION MOUNTING.

flexible pavement A *pavement where the surfacing and base materials are bound with bituminous binder, e.g. Dense Bitumen Macadam (DBM); Hot Rolled Asphalt (HRA); Dense Tar Macadam (DTM); and Heavy Duty Macadam (HDM).

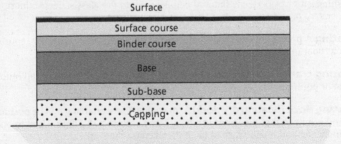

Flexible pavement: Typical cross-section of a flexible pavement

flexible pipe A pipe that allows a degree of movement; used in plumbing applications, allowing a water supply to be connected to the base of a tap with ease.

flexible wall A wall that acts as a cantilever along its vertical axis.

flexible working Deployment of different patterns of work (such as condensed working hours or working from home) or contractual working arrangements (for example part-time or temporary work) most commonly aimed at enhancing employees' *work-life balance and staff retention.

flexure rigidity The resistance to bending.

flight A set of steps between each landing.

flint A very hard greyish-black fined-grained quartz mineral that can be used to make a spark.

flint wall A wall constructed using flint.

flitch A large piece of timber cut lengthways from a tree, which is ready for processing into planks.

float 1. A tool for floating the surface of plaster or a screed. **2.** A shallow sloping drain pipe suspended from a floor soffit or housed within a floor duct, used to carry wastewater from sanitary fittings. **3.** To reset on or move over the surface of a liquid. **4.** A *ball valve. **5.** In scheduling programmes, using the *critical path method, it is the spare time to undertake certain activities. It is calculated from the difference between the earliest start and the latest time finish minus the duration of time assigned to undertake the activity.

float glass process A common process utilized for the production of *flat glass invented in 1959 by Pilkington's Glass. This makes it possible to produce perfectly flat, uniform thickness plates of glass without the need for any expensive grinding and polishing. The float glass process just uses heat and gravity. The ribbon of glass is then drawn across the surface of a molten tin bath. The hot glass, due to its density, floats on the molten tin, and therefore becomes perfectly flat, smooth, and of uniform thickness. As the heat is applied to both faces of the glass—this is fire polishing—it gives perfectly flat and parallel faces to the glass sheet. An inert atmosphere protects the tin from oxidation.

floating A process of compacting the surface of plaster or a screed by scouring it with a *float to leave a rough surface.

floating crane (semi-submersible crane vessel (SSCV)) A crane fixed onto a barge or pontoon.

floating dock (floating dry dock) A *dock that floats, enabling it to be positioned around a ship so that the water can be drained to form a dry dock. Used to enable maintenance and repair work to be undertaken on the ship.

floating floor A concrete screed or floorboards positioned on top of a floor structure with insulating material between them to reduce the transmission of noise.

floating foundation A reinforced concrete raft foundation designed to be buoyant; used to support structures built over silty and muddy ground conditions.

floating pipeline Long lengths of interconnecting pipes that float, with the use of buoyancy aids, on the surface of water. Used to convey water, oil, gas, or other petroleum products.

float switch A device to detect the level of a liquid surface in a tank. It can be used to trigger a pump when the level gets too low or to raise an alarm if the level becomes too high.

float valve *See* BALL VALVE.

floc A woolly mass that forms in a liquid due to the accumulation of suspended particles.

flocculation Occurs in the primary stage of the *water treatment process, when a *coagulant has been added and rapidly mixed to allow dispersion. It is the formation and enlarging of the floc due to gentle stirring, followed by settling. The process in water treatment where particles after addition of coagulant agglomerate/clump together forming flocs that enable them to settle easily/undergo sedimentation.

flood channel An open-air channel that is normally dry, but when excessive rain has occurred, it fills to allow water to be discharged to a safe egress point, thus preventing areas from flooding. **Flood routing** uses a series of dams, flood channels, storage basins, etc. to reduce the impact of flooding on an area.

floor The lower part of a room in a building—the part upon which people walk.

floor area ratio (US) *See* PLOT RATIO.

floor blocks *See* RIB AND BLOCK SUSPENDED FLOOR.

floor boarding (timber flooring) A *floor made of *floorboards.

floorboards A wooden floor covering, normally with tongued and grooved edges, that is fixed directly to the floor joists.

floor centre (telescopic centre) A small extendable steel beam, used to provide *falsework support to the reinforced-concrete floor slab's *formwork.

floor check A *door closer with a water *rebate.

floor chisel A tool used to remove *floorboards.

floor clamp (flooring clamp, floor cramp, flooring cramp) A *clamp used to hold tongue and grooved floorboards together tightly while they are being laid.

floor clip (bulldog clip, US sleeper clip) A U-connector that is inserted into concrete floor slabs before the concrete is set to provide a fixing point for flooring battens.

floor coverings (floor finish, floor finishing) A decorative hard-wearing finish applied to the floor level, such as carpet, laminate wood, etc. Sometimes sound-reducing and moisture-resistant.

floor framing The framework that provides the support to the floor, e.g. *joists and *strutting.

floor grinder (concrete grinder) A machine used to provide a finish to the floor covering once it has been laid, e.g. grinding marble or stone that has been set in mortar to give a polished mosaic effect, known as terrazzo.

floor guide A runner for a sliding door located on the floor.

floor heating *See* UNDERFLOOR HEATING.

flooring The material that the floor is finished with; *see also* FLOOR COVERINGS.

flooring saw (floor saw) A diamond-bladed saw used to cut concrete floors.

flooring tile (floor tile) A thin regularly shaped *floor covering laid side-by-side on an adhesive bed and grouted between. Floor tiles can be manufactured from ceramic, concrete, rubber, asphalt, etc.

floor joist A *joist that provides structural support to a floor.

floor outlet An electrical socket that is flush with the floor.

floor plan A *drawing detailing the layout for a room, noting the position of the walls, wall thicknesses, etc.

floor plate *See* ANCHOR PLATE.

floor scabbler (floor scabber) A machine used to 'scrape' floors, e.g. to remove existing coatings.

floor slab A ground or suspended reinforced concrete floor slab.

floor spring (spring hinge) A pivot with a spring sunk in the floor to act as a *door closer. *See also* FLOOR CHECK.

floor stop A *door stop that has been fixed to the floor.

floor strutting Stiffening between *floor joists.

floor trap *See* GULLY.

floor void 1. A hole made through a floor to allow pipework, cables, etc. to pass. **2.** A hollow space within a floor slab.

floriated Area of carved floral decoration.

flotation 1. The ability to remain buoyant in a liquid. **2.** The ability of a tyre to keep traction on a wet or slippery surface. **3.** A process of separating *floc from treated water in a water treatment plant.

floury soil A fine-grained silty soil containing flocculated clay, which has a smooth dusty texture.

flow chart A diagram that shows the logical steps in a process. The process diagram is depicted by different shaped boxes connected with arrows.

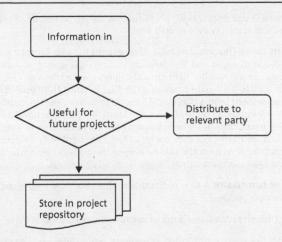

Flow chart

flow curve A graph to show the shear stress of a fluid in respect to time.

flow enhancer (polymer injection) The use of automated equipment to dose a pipeline (such as a sewer) with a chemical lubrication so that it reduces *drag and increases velocity.

flow Index A measure of the ease of flow of a *thermoplastic.

flue A duct or passage that is used to vent the products of combustion from an appliance or fire.

flue block (flue brick, chimney block) A special type of block used to construct a *flue.

flue condensation Acidic condensation that occurs within a *flue if the flue gases are allowed to reduce below their *dew point temperature.

flue isolator An automatic damper located in the flue that closes when the boiler switches off. Used to prevent heat loss up the flue.

flue liner (flue lining) A heat- and fire-resistant material used to line the interior of a chimney and transport the products of combustion. Usually made of stainless steel, fire clay, or terracotta pipe.

flue pipe A pipe that connects a fuel-burning appliance to a flue. Usually made of stainless steel.

flue terminal A type of *cowl that is placed on the end of a *flue to help discharge the combustion products and prevent rain penetration.

fluid A liquid or gas whose molecules flow freely.

fluorescence A lighting (optical) phenomenon usually emanating from cold bodies; thus sometimes known as cold light.

fluorescent lamp (fluorescent tube, fluorescent tubular lamp) A tubular glass envelope filled with argon and low-pressure mercury gas and coated internally in phosphor that produces visible light through *fluorescence when an electric current is passed through it. Two main types are available: **tubular fluorescent lamps** and **compact fluorescent lamps (CFLs)**. CFLs are three to five times more efficient than *incandescent lamps, with up to 10 times the lamp life. Full-size tubular fluorescent lamps are five to six times as efficient and may last up to 20 times as long. However, the energy savings from CFLs may not be as large as claimed, because most CFLs are much brighter seen from the side than from the ends, while a typical incandescent light is almost equally bright from all directions.

fluorescent luminaire A type of *luminaire that produces light using one or more *fluorescent lamps.

fluorspar (fluorite) A mineral form of calcium fluoride, used as flux.

flush 1. To make water flow through something, e.g. a toilet, so as to dispose of its contents. **2.** Something that is completely level with its surrounding, e.g. flush door, flush eaves, flush joint, flush panel, flush pipe, flush soffit, etc.

flushing cistern A *cistern that holds water that is used for flushing a *water closet.

flushing mechanism A device used to release the water contained within a *flushing cistern to flush a *water closet. Usually comprises a button on the top of a cistern or a handle on the side of the cistern.

flushing trough (trough cistern) A long rectangular *cistern that serves several *water closet.

flush valve (flushing valve, US flushometer) A valve located at the bottom of the *cistern that separates the water in the cistern from the bowl and regulates the flow of water when the toilet is flushed.

flushwork External walls finished in flat (flush) ashlar or flint, common in medieval buildings. Alternatively, where stones project from a wall, this is called proudwork.

fluting (flutes) Concave vertical grooves set or cut into columns and pillars.

flux A liquid used to clean oxides and facilitate wetting of metals with solder.

flyer A rectangular tread in a straight flight of steps.

flying buttress A *buttress that supports a wall at the *eaves level, associated with Gothic churches and cathedral architecture.

flying Flemish bond A brickwork bond in which the bricks are laid headers (short side) or ¾ bat then stretchers (long side) such that in surrounding courses the headers are directly central to the stretchers.

flying form (flying formwork) *Formwork that is so large it is moved by a crane.

flying scaffold A *scaffold that is hung from the building to give access to the outside walls of tall buildings.

flying shore (flier, horizontal shore) A temporary horizontal support used above ground between two objects, e.g. if a terraced house is demolished, this can be used to span the gap and support the two neighbouring houses.

fly wire A fine plastic or metal wire used for *insect screens.

foam A *polymer with a *porous structure. Foam polymers have extensive applications; in construction they can be used for insulation of buildings.

foam-backed rubber flooring (sponge-backed rubber flooring) Rubber flooring that has a layer of foam or sponge adhered to the back of it. The adhered layer acts as an underlay, increases the resilience of the flooring, and can help reduce impact noise.

foam inlet A socket that fire-fighting equipment can be inserted into in order to provide a supply of foam.

folded floor A floor that has been laid by springing the boards into place rather than using a *floor clamp.

folding door A door that is opened by folding it back in sections. Generally used internally where there is limited space. Different types of folding door are available, such as a ***bifold door** and a **multifolding door.**

folding partition A *partition that folds back on itself allowing a large room to be divided into two smaller separate rooms.

folding wedges Two pieces of wedge-shaped timber or metal (triangular in section) which are used for bracing and securing building components. When the triangular wedges are pointed in the opposite direction, they can be slid over each other to increase their overall thickness. The wedges are placed into gaps and adjusted increasing their thickness, thus trapping and securing objects.

foliage 1. Leaves or any leafage. **2.** Castings or carvings that resemble a leaf or leaves.

foliated Carved leaf decoration.

follower (long dolly) A length of block used to transmit the load from a hammer to a pile head. It is used when the hammer cannot be fitted directly onto the pile head. *See also* DRIVING CAP.

folly A building or structure often with no other useful function than to attract the attention of people passing.

font The vessel, often at the side or rear of a church, which contains water that is blessed and used for baptism.

foot bolt A *barrel bolt at the foot of a door.

foot cut The horizontal cut in a *birdsmouth.

footing (spread footing) The foundation or base of a structure.

footing fabric (trench mesh) Steel-reinforced fabric used in a *footing.

foot-pound A unit of work, equivalent of lifting a mass of one pound through a vertical distance of one foot.

footprints (pipe tongs, US combination pliers) An adjustable wrench used for gripping pipes.

foot screws (levelling screws, plate screws) The three screws connecting the *tribrach of a *theodolite or level to the tripod plate. Used to level the instrument.

foot-ton A unit of work, equivalent of lifting a mass of one *ton through a vertical distance of one foot.

foot valve A *check valve positioned towards the lower end of a length of pipe.

footway (US sidewalk) A path for people to walk along, typically located at the side of a road.

FOPS (falling object protective structure) An enclosed space for an operator to work, e.g. the cab on a hydraulic excavator is a FOPS as it is designed to provide protection to the driver from any falling objects.

force 1. The physical power or strength of something, e.g. the strength of the wind measured on the *Beaufort scale. **2.** Mass multiplied by acceleration—Newton's second law of motion.

force-action mixer A *mixer used for *lean concrete mixes.

forced circulation (mechanical circulation) The use of mechanical fans to induce a positive pressure to circulate air and negative pressure to extract air from a building.

forced draught (mechanical draught) A *draught in a flue that is fan-assisted.

forced drop shaft An old method of sinking a *shaft, where a cast-iron casing is jacked into the ground and the soil then excavated from inside the casing.

forced drying The process of applying a small amount of heat (up to 65°C) to paint to speed up drying.

force majeure Legal language used in standard contracts to describe events or circumstances which are unforeseen and prevent the contract being fulfilled, such as war, crime, strikes, or 'acts of god'. The term acts to free the parties to the contract from obligation or liability when the unforeseen extraordinary event or circumstances occur. It would not cover aspects described in the contract or mentioned as these are foreseen. *Force majeure* does not cover negligence nor weather that is typical e.g. something that is expected within a five-year cycle.

force pump A *pump that can deliver fluid to a great height above its own datum.

foreman Person in charge of site operatives or a gang of tradespeople. Often it is a person who has progressed through a trade background, and has considerable experience with skills that allows them to organize labour and plant, order materials, and ensure activities on-site are organized and controlled.

- **foreman bricklayer** Person in charge of a bricklaying gang. To organize a specialist trade, the person is often a bricklayer with considerable experience.

Duties would include organizing labour, organizing and checking the setting out of brickwork, and ensuring sufficient materials are available for the bricklayers.

- **foreman carpenter**—Person supervising and in charge of carpenters and *joiners. This person organizes the groups of joiners ensuring there is sufficient labour, materials, and plant to perform operations. To undertake the tasks, the foreman has considerable skill in the specialist trade.
- **foreman plasterer**—Person supervising a gang or group of plasters. This person ensures there is sufficient labour, materials, and plant to perform operations. To undertake the tasks, the foreman has considerable skill in the specialist trade.

forepole (forepoling board, spile) A method used in tunnelling to support the ground.

foresight (fore observation) The staff reading to the next point or station in *levelling—it is the last *observation before the instrument is repositioned; *see also* BACKSIGHT.

forging A process used to shape a metal by heating and hammering.

forked tenon A tenon (recess) cut centrally in the end of a rail or piece of wood to slot over another piece of wood.

Forked tenon

forklift truck A motorized vehicle with protruding forks that move vertically. Used for lifting, stacking, and moving equipment and materials, particularly those stacked on a pallet.

form An item of formwork.

formaldehyde An organic element containing a carbonyl group. Formaldehyde can be modified to form a monomer (*see* MER) for *polymerization, e.g. urea formaldehyde.

format A style or standard layout that can be used repeatedly. A pattern or layout of bricks.

formation level The finished level of the ground before concreting takes place.

former 1. A device used to shape something, e.g. to bend pipes. **2.** An item used to create or form a hole or voids in something, such as the use of polystyrene blocks to form a void in concrete.

form lining The facing of *formwork.

form of contract Contract conditions that are typical in the construction industry. The construction industry has a range of standard forms to suit different clients and projects, for example, the JCT Minor Works, Design and Build, or the ICE New Engineering Contact, Measured Contract, or Design and Construct contracts.

(∰) SEE WEB LINKS

- The Joint Contractual Tribunal's home page provides details of the full range of standard JCT contracts and warranties.
- The Institute of Civil Engineers provides details of their range of standard civil and construction contracts.

form of tender Document that provides the project name, tender price, duration, completion date, and other relevant details, such as the proposed schedule, and is signed by the contractor.

formply *Plywood used as lining for formwork.

formwork (casing, shuttering) A mould normally made of timber used to hold freshly placed and compacted concrete until it hardens. A **formwork foreman** is an experienced tradesperson who oversees the construction of formwork on large projects. A **formworker** or **formwork carpenter** is a tradesperson who erects formwork.

Forth Rail Bridge A cantilever bridge opened in 1890. **The Forth Road Bridge** is a suspension bridge opened in 1964. The bridges are adjacent to each other and span the Firth of Forth, in the east of Scotland.

FORTRAN (FORmula TRANslation) A computer programing language developed by IBM in 1954.

fortress A building constructed to act as a major means of defence such as a castle, ministry of defence building, or other modern-day stronghold.

forum 1. An open space within a building. **2.** An open space surrounded by public buildings. **3.** A group of people gathered together for a specific purpose, to discuss a specific issue, or with a specific agenda.

forward collision warning (FCW) *See* COLLISION AVOIDANCE SYSTEM.

forward pass The logical calculation used to work out how long a string of activities, grouped together in a network, will take. The activities are logically linked together in a network. Each activity has a duration and a logical sequence. Each activity's duration is added to the previous activity's duration. If two activities with different durations are linked into a single activity, and the logical connection means the subsequent activity cannot start until all the previous activities are completed, then the longest duration is used to calculate the duration of events. The longest duration through the activities is also the *critical path.

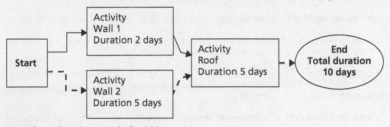

Forward pass through a network of activities

forward shovel *See* FACE SHOVEL.

fossil fuels The remains of organic matter produced by photosynthesis millions of years ago. The main types of fossil fuels include coal, oil, and natural gas, which, when burnt, release heat energy.

foul Waste, dirty, contaminated, unpleasant, etc. A **foul drain** carries wastewater or **foul water** from, for example, a toilet to a **foul sewer**.

foul air (vitiated air) Stale air that is no longer suitable for breathing.

foul air flue A *flue that is used to transport foul air from one place to another.

foul drainage A system of channels or conduits for carrying human waste matter collected from WCs, urinals, showers, sinks, baths, etc. rather than surface water.

foundation The underlying support of a structure. It is normally made from concrete, which is cast below ground level to provide a critical interface between the structure and the ground beneath it. Foundations should safely distribute the load from the structure into the surrounding ground without exceeding *bearing capacity or causing excessive settlement. The three main types of shallow foundations include *strip, *pad, and *raft—a *pile would be classed as a deep foundation.

foundation bolt (holding down bolt) A bolt that has been cast into a concrete foundation, which is to be used to anchor the *base plate of a steel column to the foundation.

foundation inspection A *Building Regulation requirement, requiring the building inspector or *clerk of works to inspect the ground conditions at the bottom excavation on which the foundation is to be placed. Once approved (normally in writing), the work should commence immediately or the ground protected in order to ensure that the ground conditions do not change with, for example, adverse weather conditions.

four-in-one-bucket A multipurpose bucket that can be fitted to hydraulic excavators, which can be used to perform the function of digging, dozing, spreading, and grading.

four-pipe system A heating or air-conditioning system that has separate flow and return pipes.

foxtail wedging Wooden wedges, such as *folding wedges, when used in pairs.

foyer An entrance hall or reception room in a building used as a gathering place for guests who enter or exit the building.

fraction 1. Not a whole number, e.g. ½, ¼, etc. **2.** A small portion of something.

fracture A sudden break or crack.

frame (framework) 1. Rectangular surround used to improve the presentation of an object and make it stand out. Often used in joinery, where straight members are connected together to form a rectangular shape that is then placed or mounted onto a flat surface **2.** A wooden, metal, or plastic surround used to encase an object, such as a window frame. **3.** A load-bearing structure made up of columns, beams, and connections.

framed door *See* BATTEN DOOR.

framed partition (stud partition, stud wall) A walling unit with its own structure or frame.

frame house (US timber-frame construction) *See* TIMBER-FRAME CONSTRUCTION.

frame saw A saw with several blades in one frame, used to cut several pieces of wood in one pass. Such saws can have diamond blades that are capable of cutting stone or slate.

framing gun A large nail gun (*see* CARTRIDGE TOOL), used to insert nails of at least 100 mm long.

Francis turbine *See* REACTION TURBINE.

franking A haunch or step that is cut into a tenon to reduce its width.

frazil ice (slush ice) Small floating ice plates that form in fast-flowing water.

freeboard The level difference between the height of something (e.g. a dam) and normal water level (e.g. in the reservoir).

free cooling The use of cold air from outside the building to act as a chiller to cool equipment and people within the building.

free energy A scientific term describing how the energy levels of a material vary with temperature.

free face An exposed face, a term typically used in rock blasting.

free-falling velocity The speed an object will fall under the influence of gravity.

free float A float that has not been used. Available float is the period between activities within a programme that allow them to be moved or delayed without affecting the subsequent activities. Standard contracts normally allow the float to be used by either the contractor or client.

free haul The distance excavated material can be moved without additional costs; *see* MASS HAUL DIAGRAM.

freestanding (free standing, self-supporting unit) Element of a building, object, or structure that has the necessary strength and form to maintain an upright position without additional support.

freeway (US) A high-speed *highway, such as a motorway.

freezer An electrical appliance designed to store goods below 0°C.

freeze–thaw cycle If water is present in cracks within various materials (e.g. the road surface), it will expand on freezing by 9%. Repeat cycles of freezing will cause the crack to grow, forming a pothole.

freezing point The temperature at which something freezes, e.g. water freezes at 0°C (32°F) under normal atmospheric conditions.

freight elevator (US, trunk lift) *See* SERVICE LIFT.

French casement window (French door) Glazed external double doors that open out onto a garden, patio, or balcony.

French drain A perforated pipe that is laid at a gradient at the bottom of a trench and surrounded by free-draining material, such as gravel. Used, for example, at the side of highways to drain surface water.

French roof *See* MANSARD.

French tile Single lap tiles imported from France. Also used as a generic term for other imported single lap tiles.

Frenkel defect At a microscopic level in materials, this is a description of the vacancies of cations in an ionic solid.

frequency The rate at which something occurs over a period of time.

fresco Decorative painting on fresh plaster or lime mortar. The pigment is absorbed into the wet plaster. Once the plaster has dried, the pigment is fixed into the plaster, and the decorative finish remains.

fresh-air filter A filter designed to remove dust and particles from the fresh outside air as it enters an air-conditioning system.

fresh-air inlet The vent that allows fresh air to enter into an air-conditioning system.

fresh-air rate The percentage of air that is supplied to a space or item of equipment that contains fresh air. Minimum fresh-air rates are contained within CIBSE Guide A.

fresh concrete (wet concrete) Concrete that has just been mixed and has not yet started to set, thus it can be poured and compacted into *formwork, etc.

fret A geometrical decorative pattern of vertical and horizontal lines.

fretsaw A *coping saw, but with a deeper U-shaped frame.

fretting (ravelling) 1. The wearing away of two surfaces, when rubbed together. **2.** The loss of aggregate from a road surface.

friction The amount of resistance generated when one surface moves relative to another.

frictional soils *See* GRANULAR.

friction catch (friction latch) A door spring or ball door fastener.

friction head Energy loss due to friction in a pipe.

friction pile A *pile that transmits the load from the structure to the soil along the length of the pile; *see also* END-BEARING PILE.

frieze 1. A horizontal band between the *architrave and *cornice in Classical architecture. **2.** A long decorative horizontal relief or painting.

frieze panel The panels above the *frieze rail on a panelled door or panelling.

frieze rail The uppermost horizontal intermediate rail on a panelled door or panelling.

frog A recess or indent that is pressed, moulded, or cut into the face of a brick when the brick is being manufactured or hand made. Pressing a brick to form an indent is often used to help compress the clay and fill the mould. The indent also reduces the volume of clay, which helps with the firing process and makes the brick more economical to produce.

frogged A brick or building block which has a *frog (indent or recess) in one or more of the faces.

front The primary face of an object that is meant to be seen or seen first. The side of a building, brick, or wall that is to be the most exposed to public viewing.

frontage The primary face of a building that faces the road. *See* FRONT.

frontage line The line of a building that runs parallel to the road, often the same as the *building line, which is the line imposed by the planning authority, and beyond which no building works can project (there are some exceptions).

front hearth The stone, concrete, or other incombustible platform such as a tiled hearth, that sits at the front of a fireplace and projects into the room. The hearth is often decorative and serves as a feature and functional item.

front lintel A supporting beam which is exposed and sits over a main wall opening at the front of a building.

frost heave The swelling of soil due to water freezing in the pore structure near the surface of the ground—as water freezes it expands by about 9%.

frost-resistant able to be subjected to the *freeze–thaw cycle without incurring damage.

frost-resistant brick A type of brick resistant to freeze–thaw failure.

froststat A type of thermostat, which will start a heating system to prevent the water in the system from freezing.

Froude number A ratio that expresses inertial forces to that of gravity forces.

frustration of contract When an unforeseeable event makes the contracted obligations impossible to complete and the parties are released from their contractual obligations. The events that frustrate the contract must be completely outside the control of the contracting parties, rendering it totally different from that which was contemplated by the parties at the time the contract was formed (signed).

frustum The solid portion of a cone or pyramid that remains when a slice is cut through the top of the pyramid or cone parallel to its base. A **conical frustum** is created when the top of a cone is removed. It is the remaining part of a three-dimensional object when a slice has been removed.

 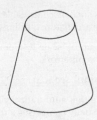

Frustum created from a pyramid and cone

fuel A material liquid or solid that is altered, and often burned, to obtain energy to heat or power plant and equipment.

fuel poverty A condition where the occupants of a dwelling are unable to afford to adequately heat the building.

Fuller's earth A type of clay material that is highly absorbent of oil and grease, used in filtering liquids.

full height The distance from the floor to the ceiling. A full height unit goes from the floor to the ceiling.

full-length bath A standard 1700 mm long by 700 mm wide bath.

full-length pipe The pipe length as manufactured.

full-scale drawing A drawing that has been produced to exactly the same size as the object it represents.

full-way valve A valve that does not reduce the bore of the pipe.

fully fixed Something that has no degree of freedom (rotation), e.g. a beam.

fully quarter-sawn timber A board that has been sawn in such a way to reduce warping.

fully supported metal-sheet roofing Sheet-metal roofing system with a supporting frame.

fume Gas, smoke, or vapour that has an unpleasant smell.

functionalism Design that concentrates mainly on function, ensuring that the building or structure fulfils its functions.

fungi (plural of fungus) A spore-forming organism, such as mildews and moulds, and mushrooms.

funicular railway A mountain railway; a railway that operates by the descending carriages on one line pulling up the ascending carriages on the adjacent line by use of an interconnecting cable.

furnace An enclosed structure where different types of materials (e.g. steel) can be heated to very high temperatures.

furnish and install (US) A contract for the supply and fixing of items.

furniture 1. Items such as chair, table, cupboard, and drawers that furnish a room. **2.** The hardware, fixtures, fittings, and decorative items fixed to a door.

furred up A build-up of *scale in pipework. May result in a restricted flow or a blockage through the pipework.

furring (encrustation) *See* SCALE.

fuse A safety device that automatically prevents electricity flowing through an *electrical circuit. Usually comprises a thin piece of wire **(fuse wire)** or metal strip **(fuse strip)** that is designed to melt when the current flowing through it exceeds its rated capacity. A **rewirable fuse** or **semi-enclosed fuse** is a fuse where the wire or metal strip can be replaced. A **cartridge fuse** is one that is enclosed within a tube. The fuse wire cannot be replaced in a cartridge fuse. *See also* CIRCUIT BREAKER.

fuse board A series of fuses that are mounted together.

fuse box *See* CONSUMER UNIT.

fused switch A switch where a fuse makes the contact.

fusible plug A plug that contains a replaceable fuse.

g *Acceleration of all free-falling objects due to *gravity, which is 9.81 m/s^2.

gabion Wire or plastic mesh crates or boxes filled with large stone blocks linked and stacked on top of each other to form a free-draining retaining wall. As water can naturally filter through the boxes and stone, they do not have to resist any hydrostatic pressure. It acts like a *gravity wall, but can withstand large differential settlements.

gable The triangular portion of the end wall of a building between the sloping sections of a *pitched roof.

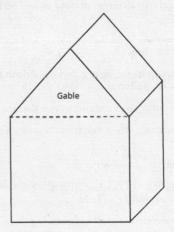

Gable

Gable

gable board Face boarding placed along the edge of tiles that would be exposed at the gable end of the roof. The board covers the tiles providing a neat finish to the roof. *See also* BARGE BOARD.

gable coping Finishing stone that sits on the top of a gable wall. Provides weather protection to the wall, preventing water percolating down the inside and outside faces of the masonry. It is the *coping at the top of a gable wall.

gable end The end wall of a building that has a *pitched roof.

gable post A short post, often decorative, at the apex of a gable. Used for fixing the *barge boards.

gable roof (gabled roof) A roof that contains one or more *gables.

gable shoulder The projecting portion of wall and roof at a *gable wall that is formed by a *gable springer.

gable springer (footstone, skew block, skew corbel) A projecting portion of wall at the base of a *gable. Usually constructed in brick, tiles, or concrete.

gablet The small *gable wall found on a *gambrel roof.

gable wall A wall that has a *gable.

gaboon (gaboon mahogany, okume) Lightweight commercial wood from Africa.

gale A very strong wind, measuring 8 or 9 on the *Beaufort scale, with speeds of 62–88 km/h.

gallet Small pieces of stone or pebbles used to fill the joints in a *rubble wall. The process of inserting the small pieces of stone or pebbles into the joints is known as **galleting**.

galling A condition resulting in localized welding and scuffing due to excessive friction between matching parts.

gallon (gal) An imperial unit relating to volume. A **British gallon** is equal to eight pints or 4.55 litres. One **US gallon** is equivalent to 3.79 litres.

galvanic couple Dissimilar electrical conductors that are in electrical contact.

galvanic current Electrical current that flows between materials in a *galvanic couple.

galvanize To coat a metal surface with *zinc.

gambrel roof 1. A pitched roof that has different slopes on each side of the ridge, and at the *gable. **2.** A *mansard roof (US).

gamification The application of playing games (e.g. scoring points, competing, and achievement) applied to other situations, i.e. a non-gaming environment, such as marketing or the work environment to improve engagement or productivity levels.

gamma (γ or Γ) The third letter of the Greek alphabet. The symbol γ is used to represent unit weight in soil mechanics.

gamma iron (γ iron) An FCC (*face-centred cubic) phase that exists in ferrous metals above 910°C.

gamma radiography A form of *non-destructive testing, where gamma rays are used to photograph specimens, e.g. to inspect the quality of welded joints and voids in concrete.

gamma ray (γ ray) A high-energy, short wavelength, electromagnetic radiation, that is similar to *x-rays but of nuclear origin, emitted by a nuclei of a radioactive atom with a range of wavelength from about 10^{-14} to 10^{-10} m.

gang A group of operatives; skilled workers who perform as a team to complete the work.

ganger The leader of a *gang.

gang form Formwork that is linked together and remains as one unit during stripping and reassembly when on site.

gang rate The unit price for a gang of workers, normally charged by the hour. Rates may vary if additional plant and specialist labour are required.

gang saw A frame saw with several blades.

gangway A temporary walkway or ramp on a building site formed from planks of wood.

gantry A spanning framework structure. Used to span carriageways and rail tracks to display information; carry travelling cranes **(gantry crane)**, *bogies, and *derrick cranes at a raised level in an industrial workplace; carry pipework, or provide a walkway at an elevated level.

Gantt chart A schedule of activities shown on a time-line by a horizontal bar chart. The chart shows the projected start and finish, and can be shaded to show the actual start, finish, and progression. Links from one activity to another, and associated resources may also be shown.

gap An opening between two objects.

gap-filling glue A glue used to join surfaces that cannot be closely fitted together.

gap-graded aggregate An *aggregate with a particle-distribution that is grouped in specific sizes, e.g. having just large and small particles to provide specific properties; *see also* OPEN-GRADED AGGREGATE.

garbage Domestic waste such as unwanted food.

garbage disposal sink (sink grinder) A type of *waste disposal unit that has a grinder fitted directly to the sink outlet.

Garchery sink A type of *waste disposal unit that has its own dedicated waste stack.

garden wall bond Refers to either English garden wall bond, Flemish garden wall bond, or Water bond. **English bond** has three or five courses of stretcher laid on a header course. **Flemish bond** is where three stretchers are laid then one header. **Water bond** has two or three skins of stretcher bond but with staggered joints reducing seepage through what would normally be aligned courses.

gargoyle A *rainwater spout for a roof in the shape of a grotesque person or animal.

garnet A dark red semiprecious stone, consisting of a crystalline silicate mineral, found in metamorphic and a few igneous rocks. It has cubic crystals, specific gravity 3.6–4.3, hardness 6–7, no cleavage, and a subconchoidal fracture.

garnet hinge A T-shaped hinge.

garnet paper A finishing and polishing paper covered with powdered garnet abrasive.

garret An attic.

gas A term commonly used to refer to *natural gas.

gas concrete (aerated concrete) Lightweight concrete produced by developing voids by means of gas generated within the unhardened mix (usually from the action of cement alkalis on aluminium powder—used as an admixture).

gas detector A safety device that can detect the presence of various different gases.

gas fitter Appropriately registered installer of gas fittings and appliances. In the UK fitters need to be approved installers.

(⊕) SEE WEB LINKS
• The Gas Safe Register is the organization that manages and controls gas safety.

gasket A flexible material used to seal the gap between two objects to prevent *air leakage and water penetration.

gas metal arc welding (GMAW) A type of welding process involving heating metals with an arc.

gas pliers Wrench or pliers that are designed to grip tubular objects such as pipes and tubes.

gas tungsten arc welding (GTAW) A type of welding process using a tungsten electrode, also known as tungsten inert gas (TIG) welding.

gate valve (sluice valve) A type of stop valve that uses a sliding disc or gate to stop the flow of fluid in a pipe. It cannot be used to regulate the flow of a fluid in a pipe.

gathering The narrowing section of a chimney located between the *throat and the *flue.

gauge 1. A device used to measure a particular quantity of a material. **2.** The thickness of sheet metal or wire. **3.** The process of mixing specific proportions of materials together; *see* GAUGED BRICKWORK. **4.** The spacing between the tile battens on a pitched roof.

gauge box Container with set dimensions used to either measure or check the volume or quantity of a material.

gauged arch An arch constructed from *gauged brickwork. Various types of gauged arch are available including a **flat gauged arch**, a **segmented gauged arch**, **semi circular gauged arch**, and a **Venetian gauged arch**.

gauged brickwork (gauged mortar) A brick wall, usually decorative, with very narrow mortar joints.

gauged stuff (gauged lime plaster) A type of lime plaster containing plaster of Paris to reduce cracking and shrinkage.

gauge pot Similar to a *gauge box, but a cylindrical container with set dimensions used to either measure or check the volume or quantity of a material.

gauge pressure A method of measuring pressure that uses atmospheric pressure as the zero reference.

gauge rod Pole or length of timber cut to the height of something, for example, a door, floor etc. The rod can be used repeatedly to check height and measurements.

gauging A method of measuring something: volume, length, or weight.

gauging board 1. Flat board or surface used for mixing mortar or plaster. **2.** Board with increments marked on for measuring materials placed on it.

gauging box *See* GAUGE BOX and BATCH BOX.

gauging trowel A round-tipped triangular trowel used for transferring plaster and applying it to a wall; due to the regular size of the trowel it can also be used to measure out quantities of plaster.

gauging water Water of a known quantity and consistency, being free from impurities, that is used for mixing with plaster, mortar, or other substances.

gaul A small depression in the final coat of plaster or render.

Gaussian distribution Normal in shape, *see* NORMAL DISTRIBUTION.

gavel A *gable.

G-cramp Clamp in the shape of a letter 'G' or 'C', fitted with a clamping screw and used to secure two pieces of material together temporarily.

gear A toothed mechanical device that transmits motion, e.g. from an engine to the part that is being driven by it.

gel coat An epoxy-based coating used as a surface finish on composite materials.

general contractor Organization that undertakes a range of construction and building work, the term may be used interchangeably with main contractor.

general foreman Person in charge of a group of tradesmen, which could be made up of mixed trades such as bricklayers, joiners, ground-workers, etc.

general location plans Drawings showing the proposed buildings and roads in relation to boundaries, other property, and other features. The drawings give a general position and orientation.

general operative Labourer who can be used to assist with many trades rather than one specific trade or skill.

genetics A branch of biology relating to heredity and inheriting genetic factors. **Genetic engineering** is the manipulation of an organism's genes.

geo- Relating to the earth. Originates from the Greek word *gē* and is used as a prefix.

geocline A depression in the ocean's crust near a continent.

geocomposite Typically a *geosynthetic consisting of two or more materials such that the properties of each material is exploited in the final product.

geodetic Related to the measurement of the earth or the geometry of curved surfaces.

geographic information system (GIS) A computer-based information system that captures, manages, analyses, and displays data of geographical referenced information.

geogrid A specific type of *geosynthetic, formed from plain sheets of extruded polymer, which are punched and drawn. The drawing process simultaneously extends the punched holes into the required aperture size and induces molecular orientation. These structures are ideal for soil reinforcement applications.

geological map A topographical map overprinted with geological information which illustrates the distribution of rocks and deposits. There are two types of geological maps, these being known as the solid or *drift editions.

geology The study of the earth's crust, with particular reference to its rocks and minerals in relation to their history and origins.

geomagnetism The properties (and the study) of the earth's magnetic field.

geomatics The discipline of collecting, storing, manipulating, presenting, and applying spatial and geographic data in digital form.

geomatrix A three-dimensional grid-like structure used to prevent soil erosion while vegetation becomes established.

geomembrane A thermoplastic *geosynthetic, which is designed to be impermeable. Used to prevent the flow of water or leachate (*see* LEACH) in ground engineered constructions.

geometric stair A curved staircase, normally without a newel post.

geometry In mathematics, the study of the shape, size, and position of figures.

geomorphology The study of the physical features of the earth's movement of soil and rock, for example, rock fall, landslips, mudslides, etc.

geophone A listening device, which is inserted into the surface of the ground to detect the arrival of shock waves during *seismic surveys.

geophysical log A detailed account of the structure and composition of the ground by drilling a borehole.

geophysical survey The remote sensing of the physical property of the ground using instruments on the ground. This can be undertaken by (1) **passive methods,**

such as gravity and magnetic surveys, where the measurements of the Earth's properties are deduced, or (2) **induction methods**, such as seismic, electrical, electromagnetic, and radar surveys, in which a signal is transmitted into the ground and monitored at ground level.

geosphere The solid part of the Earth, e.g. rock and soil.

geostrophic Related to the rotation of the Earth, with reference to wind and sea currents caused by gravitational forces and the Coriolis effect.

geosyncline A large trough along the edge of a continent in which sediment has accumulated.

geosynthetics The universally accepted name for synthetic materials used in intimate association with earth materials. There are four predominant polymer families used as raw materials; polyester, *polyamide, *polypropylene, and *polyethylene. There are also many other names given to geosynthetics outlining more specific uses, e.g. geotextiles, geogrids, geonets, and geomembranes. **Geotextiles** are fabrics, woven, nonwoven, knitted, and knotted grid structures. **Geogrids** are manufactured from extruded, punched, and drawn polymers, and **geomembranes** are manufactured from polymeric sheets. Five fundamental applications exist for geosynthetics: reinforcement, *separation, *filtration, *drainage, and *erosion control.

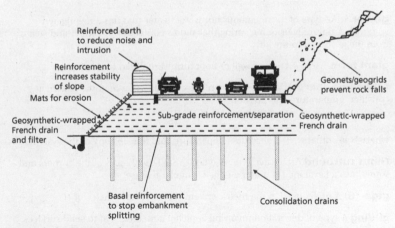

Geosynthetics: Functions and purpose

geotechnics The application of scientific methods and engineering principles related to the ground. It is practised by geotechnical engineering and includes the study of soil and rock mechanics, geology, geophysics, and hydrology.

geotextile A *geosynthetic (or vegetable fibre) product formed by a woven, nonwoven, or knitted process. **Woven geotextiles** are produced by the interlacing of yarns to leave a finished material that has a discernible warp and weft. The main applications include reinforcement, erosion control, drainage, and separation. **Nonwoven geotextiles** are needle-punched or thermal/chemically bonded structures predominantly utilized for filtration, drainage, and erosion control—those for the latter typically manufactured from vegetable fibres, e.g. **Geojute**®. **Knitted geotextiles** involve the looping of the weft or warp yarn to produce a grid structure or to encapsulate inlay yarns for directionally structured fabrics, particularly for reinforcing applications.

geothermal energy Power captured from heat stored within the earth. The heat originates from the radioactive decay of minerals, volcanic activity, and absorbed solar rays. Electricity can be produced by using steam or hot water, extracted from wells sunk a mile or more in the ground, to drive turbines situated at the surface.

German siding (drop siding, novelty siding) A type of *weatherboarding comprised of a concave top edge and a rebate at the bottom edge.

German silver (nickel brass, nickel silver) An alloy containing copper, zinc, and nickel.

geyser 1. An type of instantaneous hot-water heater that has an elongated spout. **2.** A naturally erupting spring that shoots courses of hot water and steam from the ground into the air.

giant form Large section of wall or floor formwork, lifted by cranes.

Gibbs free energy Determines the relative stability of a material system at constant temperature and pressure:

$$G = H - TS$$

where H is enthalpy, T is absolute temperature, and S is entropy of the system.

Gibbs surround Architrave around a door with large blocks at the corners and sometimes a keystone set under a large triangular feature.

giga- (G) A prefix used in the metric system representing a factor of 10^9.

gilding A type of decoration involving applying a metallic leaf to solid surfaces.

gimlet A small hand-tool for drilling holes.

gimlet point A sharp penetrating tip of a screw, used to help penetrate into the surface of a material.

gin block (jinnie wheel, rubbish pulley) Lifting wheel around which a rope or chain can be threaded. The block or frame that secures the wheel can then be hooked or attached to a joist or other secure fixing, and used to hoist objects to a higher level.

gin wheel Pulley used for hoisting small items of plant and material to a higher level.

girder A main supporting beam, usually of steel or concrete, but may also be timber.

girder bracket A bracket used to connect components to a girder.

girder truss A *trussed rafter designed to support heavy loads, such as other trusses.

girt A horizontal structural member that spans between columns or posts in framed construction, and is used to support cladding.

girt board A timber *girt.

girth 1. The circumference of a tree trunk. **2.** A *girt.

GIS *See* GEOGRAPHIC INFORMATION SYSTEM.

glacial 1. Ice-cold or freezing. **2.** Concerning the slow movement of a glacier.

glacial till A mixture of different sediments that have been eroded or entrained by a glacier, which are then deposited farther down the glacier.

gland A watertight seal around a shaft. The seal can be made using an *olive or *packing. *See also* GLAND NUT.

gland nut (packing nut) A nut that is placed around a *gland. As the nut is tightened it compresses the *packing, producing a watertight seal.

glare Any excessive brightness that causes annoyance, a distraction, or reduces vision. It is caused by one part of a room being excessively bright in relation to the rest. Glare can be caused by daylight, by artificial lighting, or a combination of both. There are two different types of glare: disability glare and discomfort glare. Disability glare is where the glare impairs vision, but does not necessarily cause discomfort. Discomfort glare is where the glare causes discomfort, but does not necessarily impair vision. Disability and discomfort glare can occur simultaneously or separately.

Glaser method A steady-state calculation method used to assess the risk of *interstitial condensation within a construction element.

glass A transparent inorganic material based on silica, used primarily in construction for glazing applications.

glass block (glass brick) A brick made from glass. An architectural element made from glass.

glass concrete A type of concrete comprising of glass (usually recycled and crushed).

glass cutter Tool used for cutting flat glass.

glass door 1. A *panelled door where all the panels are made of glass. **2.** A frameless door where the *leaf comprises one sheet of glass.

glassfibre Fibre manufactured as a continuous filament from molten glass, normally used for reinforcement due to its high tensile strength.

glasshouse A *greenhouse.

glasspaper A paper with powdered glass on the surface, used for smoothing other surfaces.

glass size *See* GLAZING SIZE.

glass stop A *glazing bead.

glass transition temperature (Tg) The temperature at which a polymer transforms from rubber-like behaviour to a brittle glass-like solid on cooling; this is a reversible process and only occurs in non-crystalline polymers.

glass wool (glass silk) A material comprised of glass fibres used for insulating purposes.

glaze A thin, smooth, shiny coating.

glazed door A door that incorporates some glass.

glazed tile A type of tile made from clay, usually with a glossy finish for moisture resistance.

glazier Skilled tradesperson who cuts, installs, and fits glass.

glazier's chisel Knife with a squared-off blade for applying putty.

glazier's putty Putty used to secure the glass in timber windows and seal in the unit.

glazing 1. Transparent or translucent sheets of glass or plastic that are installed within a frame. **2.** The process of fixing transparent or translucent materials into a frame.

glazing bar A metal or wooden bar that subdivides and supports *glazing within a frame.

glazing bead A small moulding applied around a window frame to hold the glazing in place.

glazing block Small plastic blocks or packers used to hold glazing centrally within the frame.

glazing fillet A *glazing bead.

glazing putty A type of putty that holds glass panels in place and provides resistance to moisture penetration.

glazing rebate An L-shaped recess in a window- or doorframe for fixing glazing.

glazing size (glass size) The actual dimensions of a pane of glass or other glazing material. The visible dimensions are likely to be smaller than these dimensions.

glazing sprig (glazier's point, brad) A small headless nail used to secure glazing in place while the *putty sets.

global positioning system (GPS) A global navigation satellite system based on a system of satellites (*see* NAVSTAR) that provides a method of establishing the position of points on the earth's surface, whenever there is an unobstructed line of sight to four or more GPS satellites.

global warming A gradual increase in the earth's average temperature due to the release of carbon dioxide, CFCs, and other polluting gases. *See also* GREENHOUSE EFFECT.

globe temperature (black bulb temperature) The temperature measured using a *globe thermometer.

globe thermometer A black-coated hollow 150 mm diameter copper sphere with a thermometer located in the centre of the sphere. Takes into account the effects of air and *mean radiant temperature. *See also* GLOBE TEMPERATURE; MEAN RADIANT TEMPERATURE.

glory hole *See* SPILLWAY.

gloss (gloss paint) A surface shine or lustre applied in paint form that has the ability to reflect light in a specular direction. The ability and optical property of gloss paint to reflect light has led to different definitions based on the percentage of light returned and dispersed. At one end of the continuum is flat paint with low reflection, to gloss with a high reflection; common names used in the continuum include flat (low reflection) sheen, eggshell, semi-gloss, and gloss (high reflection).

glue A material used for adhesion or bonding objects together.

glue block *See* ANGLE BLOCK.

glued-laminated timber (glulam) A timber product comprised of several layers glued together.

glue line The thickness of a joint that has been glued.

gluing clamp Adjustable clamp used for securing two or more objects tightly together while the glue sets.

GMAW *See* GAS METAL ARC WELDING.

goal Ultimate project target.

goat's foot clip A clip in the shape of a goat's foot used on steel reinforcement bars.

goggles (safety goggles) Eye protection used when workers are exposed to the risk of small flying objects or working with hazardous chemicals that could damage eyes. Hazardous tasks when eye protection should be used include: using cutting or grinding tools; working in particularly dusty environments; or working with chemicals.

going A staircase measurement of the horizontal distance between the front of one step (also called the *nosing) to the front of the next step. This measurement can be different from the tread, which measures the total length of horizontal platform that is available for the user to step on. The going can be shorter than the tread as the nosings can overlap the step below.

going rod A measurement stick, cut to the length of a *going used to ensure the length of each going is consistent.

gold leaf A very thin form of gold foil.

good building guides (good practice guides) Publications issued by the *BRE providing technical guidance for designers and builders on the construction of buildings.

good ground (good bearing ground) Ground that has suitable *bearing capacity to support foundations without requiring excessive excavation.

good practice (current trade practice, site practice) The accepted method of building, with the standard and quality being satisfactory to all requirements and regulations.

goods lift *See* SERVICE LIFT.

gooseneck A 180° pipefitting, in the shape of a goose's neck, used to connect a communication pipe to the water main or a gas service pipe to the gas service main. Used to overcome differential settlement.

gorge A steep narrow valley, which has been carved out along a weak plane by fast flowing water.

gouge A long curved bladed chisel used for hollowing out timber, and cutting into rebate when turning wood.

Government Soft Landing (GSL or soft landing) *See* SEVEN (7) PILLARS OF BIM.

GPS *See* GLOBAL POSITIONING SYSTEM.

graben Part of the Earth's crust that has become depressed between two normal faults, forming a broad valley.

grab rail A securely mounted handle or rail fixed to the wall at the side of a toilet or bath to assist the elderly and those with limited mobility.

grab tensile strength Determines the tensile strength of a fabric in a specific width by taking into account the additional strength contributed by the adjacent material.

grade 1. Stated quality of a material, that meets with specified strengths and standards. Concrete has different mixes including high strength and pumping grade. Timber is stress-graded to give an accepted strength and quality relating to the number of natural defects such as knots present in the material. **2.** The flat *formation level of a building from which all building works commence.

grader Typically a six-wheeled earth-moving machine used for grading/levelling the formation level of a site. A wide shallow blade (the mouldboard) is moulded below the vehicle, whose height and angle can be adjusted. The front wheels on the vehicle lean to resist side forces.

gradient A measure of the rate of incline or decline on a slope. Expressed as a ratio, such as 1:5.

grading The levelling off of ground to provide a flat surface.

graffiti Drawings and writing on buildings, walls, and other surfaces that have been undertaken without the owner's consent. Can be minimized by using *anti-graffiti treatments.

graffito A decorative technique where the surface of a material is scratched to produce a contrasting colour underneath.

grain A part of a metal, timber, or ceramic material representing an individual crystal. Usually grains are visible with an optical aid, for example, a microscope.

graining The process of painting the non-wood surface of a material to simulate the grain of wood.

gramme (gram, g) A metric unit of mass equivalent to one thousandth of a kilogram or approximately 0.035 oz.

granary A storehouse for grain.

grand master key The key that unlocks all doors on suited locks. The ground master key opens all locks that belong to the same group, with each lock also having its own unique key.

grange A farm.

granite An igneous rock composed of feldspar, mica, and quartz

grant Money made available to encourage upgrading certain aspects of buildings, e.g. home improvements to improve energy efficiency, and development in deprived areas.

granular (granular fill/soil) A non-plastic material (e.g. sand and gravel) deriving its shear strength from *friction as opposed to *cohesion.

granularity The amount something can be sub-divided, e.g. a metre into centimetres has a coarser granularity than a metre into millimetres.

graph A diagrammatic representation of two or more sets of data.

Grashof number A dimensionless number that relates buoyancy to viscosity in fluid dynamics and heat-transfer calculations. Named after the German engineer Franz Grashof (1826–93).

grate (grating) A frame of criss-crossed interlocking bars that cover an outlet or opening. This allows fluids, gases, and small objects to pass through the outlet or opening.

grated waste A waste pipe that has a grate over the inlet or outlet.

gravel board A horizontal board fixed between fence posts and below the fencing panels at ground level. Used to protect the fencing panels from moisture and retain any soil or gravel at ground level.

gravitation The mutual force of attraction between objects that have mass.

gravity The attractive force that exists between all objects that have mass. Measured in metres per second squared.

gravity circulation (thermosiphon) The circulation of hot water through a system by *convection.

gravity dam (gravity retaining wall) A dam or wall whose self-weight is so great as to resist sliding and overturning. Typically constructed of mass concrete; *see also* EMBEDDED WALL.

gravity flow The movement of a fluid due to the effect of gravity. *See also* GRAVITY PIPE.

gravity pipe A pipe designed for *gravity flow.

green 1. A park or open area. **2.** A term used to describe buildings or products that have low environmental impact.

green belt An area of park, farmland, or undeveloped land around an urban area.

green concrete (green mortar) Concrete that has set, but has not thoroughly hardened. It will not support any significant load without deformation.

green construction A structure and the application of processes that are environmentally responsible and resource-efficient from the point of view of design, construction, operation, maintenance, and disposal.

greenfield site A site that has not been previously developed, is free from contaminates, and is normally in a rural setting; *see also* BROWNFIELD SITE.

greenhouse A building with glass walls and roof that is used to grow plants.

greenhouse effect *See* NATURAL GREENHOUSE EFFECT.

greenhouse gas Any gas present in the Earth's atmosphere that absorbs infrared (longwave) radiation emitted from the Earth's surface. The main greenhouse gases are carbon dioxide (CO_2), methane, nitrous oxide (N_2O), water vapour, and ozone. Other powerful greenhouse gases include chlorofluorocarbons (CFCs), hydrofluorocarbons (HFCs), hydrochlorofluorocarbons (HCFCs), perfluorocarbons (PFCs), related bromide compounds, sulphur hexafluoride (SF_6), and many volatile organic compounds (VOCs).

green timber Timber that has not been seasoned.

grees Steps or a staircase.

grey iron A ferrous metal alloy containing 1.7 to 4.5% carbon and 1 to 3% silicon.

greywater Wastewater from domestic activities such as washing machines, dishwashers, showers, but not from toilets, which is classified as *sewage.

GRG (glassfibre-reinforced gypsum) A composite of high-strength alpha gypsum reinforced with glass fibres; used as an alternative to plaster castings because of its lightweight, superior strength and ease of installation.

grid 1. The establishment of points of reference on-site to mark out a network from which other coordinates and lines can be set. **2.** Distribution network for electricity, gas, and communications across the country. **3.** Horizontal and vertical reference lines marked at regular intervals on drawings.

grid line A line on a drawing that is part of a larger *grid. Each line is given a unique label, e.g. A1, A2, C4, with the letters and numbers relating to different points on the drawing's horizontal and vertical axes. On scale drawings, the lines are spaced at known distances apart, the grid lines can then be transferred to the site and used to determine offsets to the building structure and features.

grid plan A drawing where the grid lines are set so that they cut through the structural frame. The frame is set at the same regular intervals as the grid.

grillage A foundation or support comprising a criss-cross framework of beams, used to spread heavy loads over a larger area.

grille (grill) 1. A ventilation *grate used to supply air to or extract air from a space. **2.** A perforated screen used for decorative or security purposes.

grinder A power tool with an abrasive cutting wheel, which rotates at high speed and is used for shaping and cutting materials. Different graders and discs are available for cutting steel, concrete, stone, and plastic.

grinding 1. A process of rubbing the surface of a material in order to wear away a layer, such as to sharpen an object. **2.** A process of crushing a material to obtain finer particles.

grinding disc An thin abrasive cutting wheel used with a grinder or angle grinder.

grinding wheel An abrasive wheel used with a grinder.

grinning through A defect in painting where the undercoat can be seen through the finishing coat of paint.

gripper strip (carpet gripper) A thin metal or wooden strip with metal teeth used to secure the edges of a fitted carpet.

grit A hard coarse-grained siliceous sandstone.

grit arrestor A device that can pre-clean the supply of air or flue gases by removing grit, sand, and particulates.

grommet A metal, plastic, or rubber ring inserted into a hole that a cable or pipe passes through. Used to strengthen or seal the hole.

groove A long narrow slot in a material.

groover A hand tool for making grooves in concrete, used to help control the location of possible cracks.

gross feature (US characteristics) Visual features of timber.

gross load The total load on the underlying soil, without taking into account the soil removed during excavation; *see also* NET LOAD.

ground 1. The material on top of the earth's surface. **2.** In electricity, an earth connection. **3.** A background surface.

ground breaking The ceremonial start of works; a hole is dug into the ground using a highly polished stainless steel spade, which can later be framed and displayed stating the name of the prominent person who broke the ground and officially commenced the building project.

ground coat The base coat of paint that is applied prior to the *glaze coat or *graining.

ground duct (trench duct) A duct set within the ground that houses cables and pipes.

grounded theory The construction of theories through a systematic and inductive methodological gathering and analysis of data enabling the conceptual development of social patterns, relationships, and behaviours.

ground failure Instability of the ground, e.g. exceeding bearing capacity, formation of landslips, etc.

ground floor The floor of a building that is closest to the external ground level.

ground-floor plan A *plan of the ground floor of a building.

ground freezing A method of freezing groundwater so that excavation can take place. A freezing agent, such as calcium chloride, brine, or liquid nitrogen is pumped through pipes in the ground.

ground investigations Site investigation to determine the characteristics of the soil, nature of the ground, and site features.

ground level (GL) The elevation of the ground above a *datum, usually sea level.

ground-penetrating radar (GPR) A device containing a transmitter and receiver, which is used at the surface of the ground to build up a picture of the subsurface. It transmits pulses of electromagnetic energy at microwave frequencies (50–1000 MHz) into the ground. The receiver measures the amplitude and travel-time of the return signals. Depth of penetration in dry sand is about 20 m; however, in wet clay the depth can be as little as 2 m.

ground plate (sole plate) See SOLE PLATE.

grounds (groundwork) Timber battens fixed to walls to provide a flat surface onto which plaster board can be fixed by nails.

ground shaping *Landscaping of the ground.

ground slab (US slab-on-grade) A concrete slab that is supported by the ground beneath.

ground-source heat pump (GSHP) A type of *heat pump that transfers heat from the ground or groundwater to the inside of a building during the winter. In summer, it can be reversed so excess heat from inside the building can be dumped in to the ground or groundwater. As the ground is at a relatively constant temperature all year round, the *coefficients of performance for ground-source heat pumps are higher than those for air-sourced heat pumps. They can either be open-loop (ground water) systems or closed-loop (ground-coupled systems). Open-loop systems take groundwater or another water source, pass this water through the heat exchanger directly, and then discharge the water in to another groundwater well, a stream, or a lake. In closed-loop systems, a horizontal or vertical loop of pipework is placed in the ground and a water-antifreeze solution is circulated through the pipes to collect the heat from the ground.

ground storey The level of a building that is above the ground and below the first floor.

ground terminal A terminal that connects an electrical device or lightning conductor to the earth.

groundwater Occurs in two distinct zones separated by the *water table (or phreatic surface). The unsaturated zone above the water table is known as **vadose water** and contains slow-moving water percolating downwards to the water table. This zone also contains capillary water held above the water table by surface tension forces, with internal pore pressure less than atmospheric pressure. The saturated zone below the water table is known as **gravitational water** (or **phreatic water**). It is subject to gravitational forces, has an internal pore pressure greater than atmospheric pressure, and has a tendency to flow laterally.

groundwork (earthworks, work in the ground) Any work that takes place in the ground, includes excavation, laying of services, foundations, kerbs, paths, and roads.

grouped columns More than one column located on the same pedestal.

group supervisory control Lift control system used to synchronize lifts so that they arrive at destinations allowing passengers to transfer to other levels, or return to a position that allows the cars to be moved quickly and effectively to the desired destination when called.

grout A liquid cement mortar used to embed rebar in masonry walls, connect sections of pre cast concrete, fill voids, and seal joints.

grouting A process of injecting *grout into ground (soil or rock) or concrete to improve strength, stability, and impermeability characteristics, by filling and binding cracks, fissures, and voids.

growth ring *See* ANNUAL RING.

groyne A piled wall built to prevent erosion from the actions of the sea or a river.

GRP (glass fibre-reinforced plastic) A polymer strengthened by the addition of glass fibres. GRPs are renowned for their exceptional toughness.

grub screw A screw or bolt with no head, driven into the threaded hole using an Allen key. The screw is driven in so that it embeds below the surface, and covers can be inserted so that there is limited evidence of the fixing.

GSHP *See* GROUND-SOURCE HEAT PUMP.

GTAW (gas tungsten arc welding) An *arc welding process using inert tungsten gas to prevent contamination of the metal welding surfaces.

guard A device that is used to provide protection against accidents, such as the guard on a circular saw, or a guard for an open fire.

guard bead (corner bead) 1. An *angle bead. **2.** A vertical moulding used to protect corners.

guard board See TOE BOARD.

guard rail (safety rail) A temporary rail installed during building work to prevent workers from falling.

guestimate A rough estimate based on a few known facts and assumptions.

guide coat A thin coat of paint applied to a surface, particularly one that has just been filled, to highlight any bumps and imperfections.

guide price A price given to indicate the perceived sale price and value.

guide rails T-shaped vertical steel rails located on the inside of a lift shaft. Used to guide the lift car and the counterweight.

guide shoes (runners) Devices used on elevators to guide the lift car and counterweight along the *guide rails.

guillotine Bladed tool used for cutting materials by a shearing action.

gullet The distance between the teeth of a saw or file.

gully (gulley, floor drain) 1. A narrow channel formed by running water. **2.** A type of underground drainage fitting that receives the discharge from rainwater or waste pipes.

gun Hand tool with a hand grip, shaped like a pistol. Various guns are available for firing nails, discharging sealant, or glues.

gun metal An alloy of copper, tin, and zinc used primarily for making guns, with excellent resistance to corrosion.

gunstock stile See DIMINISHED STILE.

gusset piece A small triangular piece of sheet metal used at an internal corner to make the corner watertight. It can be lead welded or soldered.

gutter (guttering) 1. A narrow, usually semi circular channel at the eaves of a roof used to convey rainwater. **2.** A narrow channel at the side of a road used to convey surface water.

gutter bearer A short piece of timber that supports the *layer boards of a *box gutter.

gutter bed Sheet metal inserted behind a gutter at the eaves to prevent any rainwater that may overflow from the gutter entering into the wall.

gutter board See LAYER BOARD.

gutter bolt A metal bolt used to attach an eaves gutter to a *fascia board.

gutter end See STOP END.

gutter plate A beam used to support a sheet metal eaves gutter.

guyot A flat-topped undersea mountain, thought to be an extinct volcano.

gypsum A white mineral (hydrated calcium sulphate) used to produce cements and plasters.

gypsum plaster A type of plaster based on *gypsum.

gypsum plasterboard A structure comprising gypsum plaster sandwiched between sheets of paper, and used commonly in interior walls and ceilings.

gyrocompass A non-magnetic navigational compass fitted with a gyroscope—a self-stabilizing device.

habitable room A room used for living in, excludes kitchens, toilets, bathrooms, and other similar areas.

habitat The natural home and environment in which a particular animal, plant, or other type of organism lives.

HAC (high alumina cement) A type of cement that hardens at a rapid rate.

hachures Short parallel lines that are used on a map to indicate the direction of a slope; the closer or thicker the lines, the steeper the slope.

hacking 1. Creating a keyed surface, by making indentations using a comb hammer. **2.** The laying of a course of bricks where the corner of the brick is set inwards to the brickwork line above and below. **3.** The digging out or removal of existing putty.

hacking knife Blade used to remove old putty from window and doorframes.

hacking out Removing putty from a window or doorframe where the window pane has been broken or damaged. Once the glass and putty have been removed, the window frame can be reglazed.

hacksaw A saw for cutting steel. The cutting blade is stretched and held in a frame. The releasable clamps enable the saw blade to be removed and replaced when the teeth are broken or worn.

hade The angle between a fault plane or vein and the vertical; *see also* DIP.

haematite A ferric oxide (Fe_2O_3) mineral, reddish brown to black in colour; it is the main ore of iron.

haft The handle of a small hand tool.

Hagen-Poiseuille flow A parabola laminar flow profile in a pipe with maximum flow at the centre of the pipe, dropping to zero flow at the wall of the pipe.

ha-ha (sunken fence) A ditch used to contain or prevent people and animals from crossing a boundary; it has the advantage over a fence that it does not obscure views.

hair Animal hair used as natural binder in lime plaster. The animal hair acts as a fine reinforcing agent reducing cracking in the plaster.

hair cracking (hairline) Fine cracks in the surface of plaster, concrete, screed, and other ceramic materials, caused by shrinkage during hydration, thermal movement of different materials below the substrate, or slight structural movement. The most common cause of hairline cracking is drying shrinkage.

halcocline A zone of the sea where the salinity changes rapidly.

half bat (snap header) A brick cut through the middle to create a half brick. The brick is cut halfway along its length, making a half bat different from a *queen closer or *half header which is cut along its length.

half bond (stretcher bond) Brick courses are staggered so that the joint on each course sits directly above the centre of the brick below.

half-brick wall A brick wall made of a single skin of brickwork, where the width of the wall is equal to half the length of a brick, less a 10 mm joint. The width of a standard brick and a half-brick wall are the same and equal to 102.5 mm.

half header A brick cut in half along its length, also known as a *queen closer, used next to a *quoin to complete the bond.

half-hipped roof A *gambrel roof, which is a roof with four pitches sloping from the eaves to the ridge, the two sides slope in the standard way from eaves to ridge, whereas the ends of the roof slope at a lower pitch than the sides and abut a vertical section of roof that completes the roof to the ridge.

half-hour fire door (FD 30) A door designed to resist the passage of fire for a minimum period of 30 minutes. Fire doors are normally covered in timber with a core made up of less combustible material.

half landing A level platform formed between two flights of stairs, it allows users to rest and change direction. The platform is positioned so that the flights of stairs run into the platform adjacent to each other, enabling the total stair length to be reduced. Users travel down or up the stairs, turn 180° and travel down or up the other *flight. The landing should be at least the depth of the width of one flight and the width of the two flights that it joins.

half-lap joint Method of joining timber by recessing two pieces of timber to half their thickness and lapping them together. If joined together end-to-end the surfaces of the timber are flush and when joined at a 90° angle, only half the end grain timber is exposed.

half pitch roof A pitched roof where the rise is half the span.

half round A pipe cut in two lengthways, often used as a channel.

half-round bead (US astagal) A half-round moulding.

half-round channel A pipe cut along its length into two parts. The channels can be bedded in concrete surrounds to form a watercourse in a manhole.

half-round screw A screw with a dome head, also known as a button-headed screw.

half-round veneer Strips of timber cut from the edge of wood that has been rotated and rounded on a lathe. The radius of the veneer can be altered by the depth of the cut and the radius of the timber on the lathe.

half-space landing *See* HALF LANDING.

half-span roof A roof with a *pitch in one direction, also known as a **single pitch** or **mono pitch**. Similarly, a lean-to roof is a half-span roof, although the roof abuts another building at its highest point.

half-timbered A building where the structural element is formed of timber and brickwork; lath and plaster, wattle and daub, or other materials are used to infill the gaps between the structure.

hall 1. A hallway, passage, or corridor within a building, that leads to the other areas or rooms within a building. **2.** A central open area or grand room used to greet and entertain guests.

halogen lamp (tungsten-halogen lamp) A type of high-pressure incandescent lamp that is filled with halogen. It produces a bright white light, has excellent colour rendering properties, is dimmable, and has a high efficacy, but it does operate at a much higher temperature than conventional incandescent lamps.

halving (halved joint) The joining of timber by recessing two pieces of timber to half the thickness of the original piece and joining the matched joints together. If joined together end-to-end, the surfaces of the timber are flush and when joined at a 90° angle only half the end grain timber is exposed.

hammer Tool used for striking items, such as nails which are driven into timber or for hitting resistant objects so that they are moved into position, or are dislodged from their original position.

hammerbeam (hammer-beam) A beam used to tie the rafters of a roof together, but does not extend across the full width of a roof as a tie beam would.

hammerbeam roof (hammer-beam roof) A roof with a hammer beam tie. A hammer beam tie joins the diagonal roof members together, providing triangulation to the roof structure.

hammer-dressed stone (hammer-faced stone) Stone that is shaped and finished using a hammer.

hammer drill A power tool with a tungsten drill bit that vibrates up and down as it rotates. The movement helps to smash through the concrete as the rotating drill also bores into the surface.

hammerhead crane A tower crane with a horizontal *jib and counter jib.

hammer-headed key A hammer-head-shaped tenon joint used to secure members in curved joinery. The hammer-headed tenons slot into recesses with corresponding profiles. The hammer-head prevents the joint being pulled apart. The joints are wedged in together.

hammer-headed tenon A hammer-head-shaped *tenon formed into the end of a jamb. The joint fits into a corresponding shaped mortise and is traditionally used to tie curved joinery together. *See also* HAMMER-HEADED KEY.

hand The side of a door or window to which the handle is fitted. In Britain, the handle is fitted to the side to open the door inwards; this varies in other countries. The latch handle normally rotates anti-clockwise, which is different from the *dead lock, which rotates clockwise.

hand-augered borehole A simple ground investigation method to establish a shallow strata profile log of the ground by using hand-operated drilling devices, such as a **posthole auger, Dutch auger, gravel auger, helical auger, open/closed spiral auger**, and **flat spiral shoe auger**.

hand drill (wheelbrace) A hand-powered drill, in which a lever rotates a cog that drives a drill. The drill is useful for working in confined space, making slow, precise, and careful movements with delicate materials.

handed Pairs of objects that are identical but opposite, with one side of the pair reflecting the other as a mirror image. The objects, therefore, have a left- and a right-handed part of the pair.

hand float A plasterer's flat surfaced tool for laying on and smoothing the finishing coat of plaster. Measuring 300 x 100 mm, it is made from timber or light plastic.

handhole 1. A hole in carcassing for latching onto. **2.** A hole in an object through which a hand can pass.

handmade brick A brick made by hand, either hand-cut, moulded, or pressed into shape.

handover (handing over) At final completion, the building is passed over to the client, when it is inspected by the client's representative and possession is returned to the client. When the building is handed over, the *defects liability period starts; this is the period in which the building owner is expected to report defects, which the contractor is obliged to return and rectify. At the end of the defects liability period, any retention monies withheld by the client should be released if the defects have been rectified. Any defects that occur after the defects liability period should still be put right by the contractor if they are the result of poor or defective workmanship.

handpull A recessed handle for opening and closing sliding windows.

handrail A rail positioned at waist height or slightly higher to provide support when moving up and down stairs, the rail may also provide security and safety at the edge of platforms and landings. There are different standard dimensions and regulations for handrails that are to be used to assist people who have greater difficulty using stairs. Stairs that serve a specific use and do not need to accommodate older people, disabled users, and children can be built to less demanding standards, although handrails and stairs should be designed to accommodate all users.

handrail bolt (joint bolt) A bolt that has threads at both ends for securing two lengths of handrail timber together.

handrail core Strip of flat steel that is used to support timber handrails, the core steel can also provide the main support to plastic handrails.

handrail screw *See* HANDRAIL BOLT.

handrail scroll A scroll or spiral at the end of a handrail.

handsaw A saw that can be operated manually.

hand tool (small tool) Piece of equipment that can be held and operated by hand, and may be manual or power driven.

handwheel (wheelhead) A handle shaped as a wheel that sits on top of *valves and is used to open and close the valves.

hang 1. To fix a door or window into its frame using hinges or mounts. **2.** To fix a painting, picture, clock, or ornament to a wall. **3.** To fix wallpaper, boards, or tile finishes to a wall or ceiling.

hanger A tie, chain, strap, threaded bar, or fixing that is used to secure fittings, fixtures, mechanical units, and suspended ceilings to the structure. The length of the hanger needs to be adjustable.

hanging To *hang or mount objects on walls.

hanging device The fitting or ironmongery needed to *hang a door or window.

hanging gutter A gutter fixed to the ends of rafters on a roof or the eaves of the roof.

hanging jamb (hanging post) The side of a doorframe to which the door or window is fixed, the sides of the *jamb that the hinges are fixed to.

hanging sash A *sash window.

hanging stile The side of the stile to which a window is hung.

hanging valley A U-shaped valley that has formed from a tributary glacier flowing into the main glacier. The top surfaces of both glaciers would be the same, but the bed of the main glacier would have been far deeper. Where they intercept, the shallower valley appears to be 'hanging' above the main valley and is commonly an outlet of waterfalls.

harbour A sheltered coastal water where ships can anchor, load, and unload goods.

hardboard (high-density fibreboard) A type of fibreboard, which is an engineered wood product. Hardboards are very popular in construction and provide a cheaper option in comparison with plywood.

hard-burnt (well burnt) The description of clay bricks, tiles, and other ceramics that have been fired to the point of vitrification. Bricks which are made of good-quality clay and are well fired have high compressive strength and are durable.

hardcore Crushed limestone or sandstone aggregate used to make compacted hard surfaces. Compacted hardcore is used as the base layers in road construction.

hard dry Paint that has dried to a solid hard surface, has lost its rubbery feel, and is not easily marked with a fingernail.

hardenability How easily a ferrous metal can be hardened due to various heat treatment processes.

hardener An element used specifically to increase the *hardness properties of a material. Different hardeners exist for different types of materials.

harden hardware Ironmongery that has been case-hardened.

hardening A process with the specific objective of increasing the *hardness of a material. There are many ways in which this can be achieved, depending upon the chemistry of the material.

hard hat (hardhat, safety helmet) *Personal protective equipment in the form of a helmet, manufactured out of resilient plastic, to prevent injury to the head from falling objects. It is now compulsory for a hard hat to be worn on all construction sites.

hard landscaping The finishing of external areas in brick, block or clay paving, concrete, bitumen, tarmac, stone or chippings, or anything that creates a firm surface suitable for walking on or driving vehicles over.

hardness A material property measuring how easily the surface can be scratched—the measure of a material's resistance to deformation by surface indentation. The hardness of a material is generally directly proportional to strength and brittleness, i.e. the harder the material the lower the ductility. Hardness also measures a material's resistance to localized *plastic (permanent) deformation.

hard plaster (hard finish) Plasters that are resistant to knocks and scuffs, such as rendering, squash court plaster, and renovation plaster.

hard plating An abrasion-resistant metal coating achieved by electroplating the surface of iron or steel with chromium plating.

hard solder A solder that has copper within it, which raises the melting point.

hardstanding A paved, concreted, or stabilized area that is suitable for storing materials or parking vehicles.

hard stopping Paint formed from a powder mixed with water or oil that is used to fill small holes or defects.

hard surface *See* HARD LANDSCAPING.

hardware 1. The components of a computer (e.g. the processor) as opposed to the *software programs that run on it. **2.** Heavy military machinery and weaponry. **3.** Tools and household implements made of metal, e.g. hammers and cutlery. **4.** Includes *ironmongery, steel, non-ferrous metals; originally a term used to describe ironmongery, now applies to any solid fitting

hard water Alkaline water containing dissolved salts that limit the formation of lather with soap.

hardwood Timber or wood that originates from broad-leaved dicotyledonous trees—not based on the hardness of the material.

harl Render with stones cast in the surface to provide a rough finish. The finish is porous enabling the surface to both absorb moisture and allow moisture to escape.

harmonic Periodic oscillation that relates to an integral multiple of a fundamental frequency. **Harmonic motion** is the simplest form of periodic oscillation and can be represented as a sine wave travelling across a screen.

harmonization The provision of equal access to members of the European Countries through harmonization documents, *European standards, and *Eurocodes.

harmonization document 1. Report or procedure produced to bring together regulations and technical documents produced in different countries or from different organizations. The document aims to adjust terms and descriptions so that it is possible to allow one overall method of governance or procedure. **2.** European standard that helps to address differences in the member countries.

harsh mix A concrete mix with a low percentage of fine aggregate.

hasp and staple A slotted latch that drops over a large raised staple to provide a method of locking doors, gates, and box lids. The hinged latch is secured by positioning a peg or lock through the staple once the latch is closed.

hatch An opening in a wall, roof, or ceiling through which objects can be passed. A loft hatch is used to gain access from the habitable rooms to the roof space.

hatching Parallel lines on a drawing to indicate a change in something (e.g. a material type).

haul The movement or transportation of materials by using *haulage plants, e.g. *dump trucks and *bulldozers.

haulage The transportation of goods and materials.

haunch A triangular brace used to add strength to a frame at the junction of roof members or the junction between columns and beams, or columns and roof beams. The thickening helps create stiffness to resist bending at stress points.

haunched tenon A *tenon that has a thicker section at its base than it does at the tip.

hauncheon The thick or wide section of a *haunched tenon.

haunching The build-up of concrete behind or against a drain or kerb used to hold it securely in place. The concrete sits on top or at the side of the *bedding sloping up to the side of the *kerb, edging, or *drain.

hawk (mortar board) A board used for holding wet plaster and mortar, providing a temporary place to hold the material while it is being applied. The board is usually 300 mm square with a handle enabling the operative to hold the board and apply the material at the same time.

HAZAN Hazard analysis is the use of numerical methods to assess how often a hazard will manifest itself, and the consequences of the hazard for people, process, and plant.

hazard An area, object, or situation that presents a risk to those who work in the proximity of the situation or object.

Hazen-Williams formula An empirical expression that relates the velocity of water in a pipe or open channel to the physical properties of the pipe/channel, causing a pressure drop due to friction.

head 1. The horizontal top member of a door or *window frame. **2.** The striking part of a *hammer. **3.** A measure of the pressure exerted by a column of water, usually measured in feet. **4.** The vertical distance between a *header tank or *storage cistern, and the outlet below (measured in metres).

headboard 1. A board that was positioned on a labourer's head and used to carry bricks or other building materials. **2.** A cushioned board at the top of a bed.

head casing (US) The *architrave that is used at the *head of a door.

header A brick or block with its narrowest facer exposed with the length of the brick, the stretcher face, penetrating into the wall. Traditionally, bricks that penetrate into the walls were used to link the two skins of brickwork together forming a solid single brick wall. With the introduction of cavities and insulation, header bricks rarely penetrate across the cavity as this would form a cold bridge (*see* THERMAL BRIDGE).

header bond (heading bond) Courses of brick laid end on with only the header exposed, used for curved walls and foundation courses.

header tank (head tank) A raised tank that contains cold water. Used in central heating systems, hot-water systems, and cold-water supply systems. Also known as a **feed cistern**, a **feed and expansion cistern**, and a **cold-water cistern**.

head flashing The lead work around an opening in a roof that creates a gutter to remove rainwater from the area around the opening, shedding it to the roof surface.

head guard A damp-proof course positioned over the top of the lintel to a door or window opening.

heading course Layer or course of header bricks, this could sit on top of a stretcher bond as in *English bond or sit on top of another course of headers as in *header bond.

heading joint A joint made to bond two pieces of timber together end-to-end, such as a *finger joint, *butt joint, or *half-lap joint.

head joint (cross-joint) A vertical joint also known as a perpendicular joint (perpendicular end) in brickwork and blockwork.

headlap The lap produced when the bottom of one roof or wall tile sits over the top of the other.

head lining 1. A lining positioned around the top of a doorframe.
2. A *damp-proof course positioned over the top of a door or window opening *head guard.

head loss The drop in pressure between two points on a pipe or a duct.

head of water The pressure exerted by water on an object.

headroom (headway) The space allowed between the floor and ceiling or any obstruction that hangs below the ceiling. In stair construction and design, it is the gap between the *nosing line and ceiling, or obstructions below the ceiling.

headworks Fittings used to supply hot and cold water to a bath.

health and safety The discipline concerning safety, health, and welfare of people, particular in the work place.

Health and Safety at Work Act The primary piece of legislation covering health and safety in the UK. The Act is the primary legislation and the secondary legislation comes in the form of Statutory Instruments, which cover such things as asbestos, working at heights, etc.

(⊕) SEE WEB LINKS
- The full Act is available on the website as well as other supporting information, including posters and pocket cards.

Health and Safety Executive (HSE) The government department that ensures that the *Health and Safety at Work Act is delivered and enforced. The office must be informed if building works are to be carried out, which will take longer than six weeks to construct, and must be informed if any injuries occur.

(⊕) SEE WEB LINKS
- Full information on the primary and secondary powers of the HSE can be found on the website.

Health and Safety Inspector Main inspectors work directly through the *Health and Safety Executive enforcing the *Health and Safety at Work Act; inspectors can also be Local Authority Inspectors. The inspectors have the power to visit organizations and sites to assess arrangements in place for assessing and controlling risks.

(⊕) SEE WEB LINKS
- Full information on the powers and procedures used by the Health and Safety Inspector can be found on the HSE website.

Health, Safety, and Welfare Regulations 1992 (HS&W) An approved code of practice and guidance intended to protect the health and safety of individuals, together with ensuring that adequate welfare facilities are provided at a workplace, with the exception of construction sites, those in or on a ship, or those below ground at a mine. Construction sites are governed by the *Construction Design and Management Regulations.

health, safety, and well-being (HS&W) The totality of what is required to prevent harm to life in construction.

hearing protection *Personal protective equipment in the form of ear plugs/muffs. They are designed to reduce the intensity of noise, to below 85 dB, so that damage to hearing is prevented. The **Noise Regulations 2005** requires employers to take action at specified exposure levels; these relate to the average exposure over a working day/week or the maximum noise in a working day, and are currently defined as follows:
- Lower exposure action values: daily or weekly exposure of 80 dB; peak sound pressure of 135 dB.
- Upper exposure action values: daily or weekly exposure of 85 dB; peak sound pressure of 137 dB.
- Exposure limit values: daily or weekly exposure of 87 dB; peak sound pressure of 140 dB.

heart The centre of a tree.

heart bond A type of brickwork bond where two *headers meet in the centre of the wall and the joint between them is covered by another header.

heart centre (pith) The wood at the centre of a tree that mainly consists of parenchyma. *See also* HEARTWOOD.

hearth The base of a fireplace. It may extend out from the fireplace into the room.

heartwood Wood around the centre of a tree or cut log that has finished growing; it is more durable than the newer sap wood. The wood at the very centre, *heart centre, of the tree may not be as strong and durable.

heat 1. Energy caused by the movement of atoms and molecules that can be transferred from one place to another by *conduction, *convection, or *radiation. **2.** The perception of warmth. **3.** The process of increasing the temperature within a space.

heat bridge *See* THERMAL BRIDGE.

heat capacity *See* THERMAL CAPACITY.

heat detector A device that measures the temperature within a space and reacts when a particular temperature is reached. Used as a *fire detector.

heat diode *See* THERMAL DIODE.

heated floor (heated screed) Heating pipes laid within the screed or under the surface of the floor, also called *underfloor heating.

heater A device that provides heat.

heat exchanger A device that can transfer heat from one medium to another without allowing them to come into contact with one another.

heat flux The rate of heat flow through a surface per unit area per unit of time. Measured in W/m^2. Measured using a *heat flux plate.

heat flux plate (heat flux sensor, heat flux transducer, heat flux gauge) A transducer that produces a voltage proportional to the rate of heat flowing through the sensor. Used to measure *heat flux.

heat fusing (fusion welding, polyfusing) *See* THERMOFUSION WELDING.

heat gun (heat welding gun) A hand-held electrical device that emits a jet of hot air from a nozzle. Used in *thermofusion welding and paint stripping.

heating The process of increasing the temperature within a space or an object.

heating and ventilation engineer A person who is trained to design, install, and maintain heating and ventilation systems.

heating battery A type of *heat exchanger that uses hot-water pipes to heat the air within a duct.

heating coil 1. Wire or pipes arranged in a coil that is used for *heating. **2.** A heating battery.

heating element An electrically heated wire coil.

heating medium A heat-transfer fluid, such as a coolant, used in central-heating systems.

heating panel *See* PANEL HEATING.

heating, ventilation, and air conditioning (HVAC) A single integrated system that provides the heating, ventilation, and air conditioning within a building.

heat load The amount of heat that needs to be removed or added to a space to maintain the required internal temperature.

heat loss A measure of the amount of heat that is transferred through the various elements of building fabric, such as windows, doors, floors, roofs, and walls, due to the exfiltration of warm air.

heat meter A device that measures the amount of thermal energy flowing through a system. It operates by measuring the flow rate of the fluid through the system and the temperature difference between the output and return of the system. The flow can be measured using a turbine or ultrasonically. The temperature difference is usually measured using two *thermocouples.

heat of hydration The amount of heat produced from *hydration.

heat pipe A relatively simple and efficient type of heat-transfer device. Consists of a sealed metal rod or pipe that is partially filled with a fluid. Heat is transferred by vapourizing the fluid at one end and then condensing the fluid at the other.

heat pump An electrical device that moves heat from a low temperature source and upgrades it to a higher, more useful temperature, using the *vapour compression cycle. It comprises a compressor (usually driven by an electric motor), a circuit containing a refrigerant, an expansion valve, a condenser coil, and an evaporator coil. Heat pumps can be used to heat and/or cool a building. Refrigerators and air conditioners are common applications of heat pumps that operate only in the cooling mode.

heat recovery The process of transferring heat that would otherwise be wasted from one medium to another using a heat exchanger.

heat-recovery wheel *See* THERMAL WHEEL.

heat-resisting glass A type of glass with excellent resistance to thermal shock.

heat-resisting paint Usually an aluminium paint having a lustrous metallic finish, and resistant to temperatures up to 230–260°C. A dry film thickness of 15 microns (μm) is typical.

heat-transfer coefficient (HTC) The sum of the *transmission and the *ventilation heat-transfer coefficient. Measured in W/K. The total heat-transfer rate from a building resulting from heat transfer through the envelope (UA-value) per °C of indoor to outdoor temperature difference in W/°C. It is calculated under steady-state assumptions. The thermal transmittance of each building component composed of homogeneous layers (U-value) can be calculated as the inverse of the sum of the thermal resistances of all layers that compose the building component. If the building component is heterogeneous, it must be estimated by means of guarded hot box for opaque elements (ISO 12567-1) and for semi-transparent elements (ISO 12567 2002). Multiplying the U-value by the total area of the specific building component in the building envelope, the UA-value of that building component regarding the whole envelope is obtained. If all UA-values and thermal bridges equivalent UA-values present in the building envelope are added, the building envelope (transmission) heat-transfer coefficient is obtained.

heat-transfer fluid (coolant, heating medium) A fluid that is designed to transfer heat from one medium to another.

heat-transmission value *See* THERMAL CONDUCTIVITY.

heat welding gun *See* HEAT GUN.

heave The upward movement (expansion) of soil due to a reduction in effective overburden pressure or an increase in water pressure; *see also* SETTLEMENT.

heavy Of immense weight/mass.

heavy-body impact test A door strength test. A 30 kg sand bag is swung to impact the door, adjacent to the lock, three times, at increasing standard height. The test is repeated on the other side of the door.

heavy industry A business that requires a large amount of space to operate heavy equipment.

heavy metal A toxic metal with a relative density of 5.0 g/cc or greater, examples include lead, mercury, copper, and cadmium.

heavy protection Aggregate, chippings, asphalt, or mortar laid over roofing to prevent uplift and protect the material from the sun.

hecto- A prefix denoting 100 times.

heel The part of a structure, normally a beam or roofing rafter that rests on a support.

heel strap A galvanized steel strap, produced in a U-shape, which is used to hold down roof trusses and tie beams.

height board (height rod) Strip of timber or rod that is cut to a known length, e.g. the height of a door, window, or floor, and is used to produce a consistent measure of the element of construction.

height-of-collimation method A procedure for recording field observations during *levelling. The collimation level is calculated from the *backsight and *foresight (or *intermediate sight) readings to obtain the reduced level of a position.

helical hinge A sprung hinge, with two coiled springs, that allows doors to be opening and self-closing in both directions. Used for lightweight swing doors.

helical reinforcement Steel reinforcing bars that are used as cross-links in a circular column.

helical stair A spiral stair.

helices Spiral ornamentation.

heliograph 1. A communication signalling device (such as a mirror) that reflects the light from the sun and can be used to send Morse code messages in terms of flashes. **2.** An apparatus to take pictures of the sun. **3.** A surveying instrument used at a station to reflect sunlight, such that a point can be sighted up to 100 km away. **4.** An instrument to determine the intensity of the sun's rays.

helium (He) A gas at ambient temperature found abundantly in the universe.

helix A spiral line that follows the path of a cylinder at a constant angle.

helmet A hard protective covering worn on the head to prevent injury. *See also* HARD HAT.

helper A worker's mate or labourer.

helve A handle of a pickaxe or other similar heavy striking tool that is lifted and thrown using two hands.

hematite *See* HAEMATITE.

hemihydrate plaster Gypsum plaster produced by heating gypsum so that it loses 75% of its chemically bound water: $(CaSO_4)_2H_2O$ becomes $CaSO_4 . \frac{1}{2}H_2O$. *See also* HEMIHYDRITE.

hemihydrite Another term for hemihydrated plaster. Gypsum plaster that has been steadily heated to 150–170°C

hemisphere Half a sphere, particularly in relation to the earth, e.g. northern hemisphere, but also relates to other celestial bodies.

hemp (*Cannabis sativa*) An annual plant having a stem diameter of 4–20 mm and a stem length of 4.5–5 m containing a *bast fibre. Cultivated throughout Russia, Italy, China, Yugoslavia, Romania, Hungary, Poland, France, Netherlands, UK, and Australia. Requires mild climates with high humidity and annual rainfall >700 mm. Uses include: ropes, marine cordage, ships' sails, carpets, rugs, paper, livestock bedding, and drugs as a result of the narcotic THC content of certain varieties (30 varieties in all).

henry (H) The *SI unit of inductance, equivalent to one kilogram metre squared per second squared per ampere squared:

$$(kg.m^2.s^{-2}.A^{-2})$$

heptagon A seven-sided shape, with all seven sides and seven angles being equal.

hermetic Sealing an object to ensure airtightness.

hermetic seal An edge seal that is airtight. Also refers to a barrier to prevent water vapour entering a cavity.

hermitage A small dwelling in a secluded location.

herringbone (herringbone pattern) Bricks, pavers, or timber placed in a diagonal pattern with units moving towards a point and placed at 90° to each other. The pattern is supposed to be similar to that produced by fish bones.

herringbone drain A surface drain filled with granular material (stones) that has been laid in a herringbone fashion. *See also* CHEVRON DRAIN.

herringbone strutting Timber struts placed at angles to provide bracing to the floor beams. The strutting runs from the bottom of one beam to the top of the adjacent beam. The struts cross each other making the beams stable. As the struts cross, a *herringbone pattern is produced.

hertz (Hz) SI unit of frequency that is equal to one cycle per second.

hessian An open-weave fabric, made from jute or hemp fibres, used to produce sacks, also sheets for protecting newly laid brickwork from frost attack. The fabric can also be laid within the ground to provide a ground reinforcement, and can also be introduced into plastic products to provide reinforcement.

hew To shape dressing stones or square timbers with an axe or hatchet.

hexagon A six-sided shape.

hexagonal close-packed (HCP) A particular type of packing of atoms in a metal, appearing hexagonal in shape. The hexagonal close-packed metals include cadmium, magnesium, titanium, and zinc.

hexahedron A three-dimensional shape that has six sides, e.g. a cube. Eight nodded **hexahedron elements** are used in finite element analysis as they are more accurate and easier to visualize than tetrahedral elements.

hexastyle A porch or portico that has six columns at the front.

HFP *See* HEAT FLUX PLATE.

HGV (heavy goods vehicle) A large vehicle used for transporting goods.

H-hinge A parliament hinge with two legs that extend away from the knuckle. The heavy H shaped hinge is used for carrying large doors and gates.

hickey A tool used for bending steel.

high alumina cement To develop a concrete with an early strength that is much higher than ordinary Portland cement (OPC), aluminium oxide is used instead of clay. In the longer term, the cement has been found to be unstable so it is now rarely used.

high bay lighting A type of lighting system mainly used in industrial buildings that have a high floor to ceiling height. Usually comprises a series of high intensity discharge lamps or fluorescent fixtures.

high-calcium lime (fat lime, non-hydraulic lime, rich lime, white chalk lime, white lime) A lime that contains mostly calcium oxide or calcium hydroxide, and not over 5% magnesium oxide or hydroxide.

high-carbon steel A steel alloy containing between 0.6 and 2.0% carbon by weight; usually high-carbon steel alloys contain 0.6–1.0% carbon. The advantage of higher carbon is increased ultimate tensile strength (UTS) (*see* TENSILE STRENGTH), however high-carbon steels have lower ductility in comparison with low or medium carbon steels. Higher carbon steels also have increased *hardness (resistance to surface indentation).

high-commitment management (HCM) An approach to the management of people that emphasizes personal responsibility, independence, and empowerment of employees and workers across all levels of an organization. HCM seeks high performance through developing organizational commitment amongst its employees, psychological alignment, and continuous learning.

high-efficiency particulate air filter (HEPA filter) A high-efficiency air filter that removes at least 99.7% of the airborne particles that have a diameter of 0.3 microns (μm) or more.

highlighting Emphasizing parts of decoration by making areas lighter in colour.

high-pressure hot-water system (HPHW) A type of *central heating system that circulates the hot water through small bore pipes at high temperature and pressure, usually above 200°C and 16 bar. Used in industrial and commercial buildings.

high-pressure system A term commonly used to refer to a *high-pressure hot-water system or a *high-velocity system.

high rise (high rise building) Buildings that are greater than eight storeys are often said to be high rise buildings.

high spot A part of a surface or substrate that is higher than the general profile or surface.

high-strength concrete Concrete with strengths in excess of 40 kN/mm^2. Concretes with additives such as microsilica, superplasticisers, and other strengthening materials can be produced that have strengths in excess of 100 kN/mm^2.

high-tensile steel A steel alloy with a high ultimate tensile strength (UTS) (*see* TENSILE STRENGTH), the tensile strength is so high that in comparison, high-tensile steels may have up to ten times the tensile strength of wood, and more than twice that of mild steel. Normally high carbon steels have increased UTS, however, other special treatments can also impart this property. With increased UTS, the material's toughness is also significantly increased as more energy is required to impart failure.

high-velocity system A type of air-conditioning system that delivers conditioned air to the space at high velocity using small diameter ductwork.

high voltage A cable that operates with a voltage that is greater than 600 volts.

highway A main public road, particularly one connecting large towns or cities. **Highway Engineering** is the discipline that deals with the planning, design, construction, and maintenance of highways.

Hiley formula An expression that enables an estimate of the resistance of the ground to be calculated when driving a *pile from measuring driving variables:

$$R_u = \frac{wh\eta}{s + \frac{1}{2}c}$$

where R_u is ultimate driving resistance; w is the weight of the hammer, h is the drop height of the hammer; η is the efficiency, i.e. the ratio between P/w and the coefficient of restitution (e); s is the penetration or the pile per blow; P is the weight of the pile including any driving cap, c is the total temporary compression, i.e. $(c_c + c_p + c_q)$, where c_c is the temporary compression of the dolly (*see* DRIVING CAP); c_p is the temporary compression of the pile, and c_q is the compression of the ground.

hindered settling The settling at reduced speed due to interactions with adjacent particles. It is calculated with respect to the settling velocity of a single particle that is not influenced by interaction with neighbouring particles.

hinge A pinned joint between two elements that enables one of the elements to pivot relative to the other. Used to hang doors to their frames. Many different types of hinge exist, including butt hinges, back-flap hinges, butterfly hinges, concealed hinges, continuous hinges, cleaning hinges, flush hinges, friction hinges, H-hinges, mortise hinges, parliament hinges, strap hinges, and T-hinges.

hinge bolt (dog bolt) A type of *security hinge where a bolt pin is fitted into the hinged side of the door.

hinge-bound door A door in which the recess for the hinge has been cut too deep.

hip (hip ridge) The upper angle created by the intersection of the two sloping roofs. The rainwater flows away from the hip, down the two faces of the roof.

hip capping The material, felt, or flashing that is placed over the *hip.

hip hook (hip iron) A strip of metal fixed to the *hip rafter that extends under the lowest hip tile, providing support for the tile, and preventing it slipping from the roof. The metal is usually visible and may be angled, twisted, or rolled to provide an ornamental feature.

hip knob A decorative point (finial) fixed on top of the ridge where it intersects with the *hip or *gable.

hipped end (hipped-end roof, hip) The pitched triangular end of a hipped roof. A hipped-end roof is different from a gable-ended roof, which would just be left with the two standard pitched roofs meeting the wall of the dwelling at the *ridge and *eaves, forming a triangular section of external wall.

hipped-gable roof (jerkin roof) A hipped- and gable-ended roof. The pitched roof has a hip at one end, that extends part-way down the roof where it is cut short by a gable end.

hipped roof (hip roof) A pitched roof with a *hipped end.

hippodrome A building that was traditionally used for chariot racing and staging equestrian events, and in Victorian times was a name commonly associated with music halls.

hip rafter (angle rafter, angle ridge) The rafter that is used to create the structure of the roof *hip, with *jack rafters resting on the central hip rafter.

hip rib A curved *hip rafter used to create domed roofs.

hip roll A turned length of wood with straightened sides and a V-cut in the under side of the timber to cover the hip joint. The rounded length of timber is referred to as a roll.

hip tiles Curved, angular, *bonnet or V-profile tiles that are used to cover the *hip of the roof. The tiles are nailed, dowelled, or bedded in mortar, the lowest tile is supported and held in position by the hip iron.

hip truss A trussed rafter that is used to create the *hip of a roof.

histogram A statistical bar graph to denote distribution of data. Vertical rectangles of different heights are drawn up the y-axis to represent corresponding frequencies.

hit-and-miss fencing Boarded fence with gaps placed between the panels.

H-method A method used in finite element analysis to increase the accuracy of the solution by increasing the number of elements.

hoarding A secure tall fence, often closely boarded, that is used to prevent the public gaining access to the construction site. Many of the fences have open areas or viewing windows that allow people to see the construction, while ensuring that access is prevented to keep the public safe. The hoarding should be particularly resistant to the passage of children. Safety notices should be displayed on the boundary of the site, and direction should be given to the entrance and site offices. Passage onto the site should be secured and controlled at all times.

hob 1. An electric or gas device built into a kitchen worktop, and used for cooking. It will contain the controls and the burners or electric plates. Also known as a **cooktop. 2.** A shelf at the rear or side of a fireplace used to keep food warm.

hod A V-shaped trough attached to a pole that is used to carry bricks or mortar.

hod carrier (hodman) Bricklayer's labourer who traditionally carried a *hod of bricks and distributed the bricks around scaffolds and adjacent to the bays where the bricks were to be used.

hoe excavator A wheeled tracked excavator with its bucket operated by a two-part hydraulic arm that digs in a downward action. *See also* BACKHOE.

hog Bending upwards, something that is concave, e.g. an arch. A **hogging bending moment** is a bending moment with tension along the top and compression along the bottom; *see also* SAG.

hoggin Fine aggregate of ballast used as infill material.

hogging A cambered surface, a beam that bends upwards towards its centre.

hogsback A curved ridge tile.

hoist Equipment used for lifting building materials to the level needed.

hoisting Lifting or lowering building materials or equipment using a crane or *hoist.

hoistway *Lift shaft.

holderbat A galvanized steel fixing, with a dovetail end that is built into a masonry joint, and a ring on the other end for securing pipes.

holding-down bolt A long steel bolt that is set in concrete foundations to hold fast the *base plate of columns.

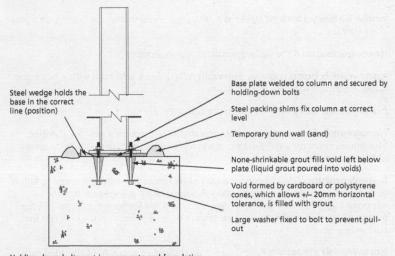

Steel wedge holds the base in the correct line (position)

Base plate welded to column and secured by holding-down bolts

Steel packing shims fix column at correct level

Temporary bund wall (sand)

None-shrinkable grout fills void left below plate (liquid grout poured into voids)

Void formed by cardboard or polystyrene cones, which allows +/- 20mm horizontal tolerance, is filled with grout

Large washer fixed to bolt to prevent pull-out

Holding-down bolts cast in a concrete pad foundation

hole saw (annular bit, tubular saw) A tool used to cut circular holes through wood and plastics.

hollow A sunken or recessed surface.

hollow block A concrete or clay block with a cavity within the block or a cellular core.

hollow brick Cellular brick, normally considered hollow when it has greater than 25% voids within it.

hollow clay block (US structural clay tile) Block with voids or void within it.

hollow clay tile (hollow pot) Tiles used for creating drains, vents, underfloor ducts, or other passages within the tile.

hollow concrete block A concrete block with a cellular core.

hollow-core door A *door that has a hollow core.

hollow floor A hollow pot floor or a suspended floor.

hollow partition A partition with a cellular core.

hollow roll A method of forming a water-resistant joint between two pieces of sheet metal, used on roofing or as a cladding material. The joint is usually used to join the two sheets running down the slope of the roof. The edges of the sheet metal are lifted, positioned together, and rolled to form a cylinder.

hollow wall A *cavity wall construction, traditionally constructed with an air gap within the cavity; it is now becoming more common to totally fill the cavity with insulation.

home battery storage system A *battery located within the home that stores electricity for later use.

homogeneous Of the same consistency throughout.

honeycomb bond (honeycomb wall) A *half-brick wall built with a *stretcher bond, with gaps between the stretchers.

honeycomb door A *door that has a cellular core.

honeycomb fire damper A honeycomb-shaped intumescent *fire damper. The intumescent material expands at a given temperature, preventing the passage of fire and smoke.

honeycombing Gaps in concrete caused by loss of fine aggregate, due to failure to compact concrete properly or over-vibration, causing segregation of the concrete's aggregates. Where concrete is cast against formwork, gaps within the formwork may allow the fine aggregates to escape and leave voids between the coarse aggregate.

honeycomb structure A structure consisting of a collection of six-sided cells, similar to that found in a bee hive.

honeycomb wall *See* HONEYCOMB BOND.

hood (cooker hood) A mechanical device found above a cooker that is used to extract the vapour and smells from cooking.

hood mould (drip cap, hood moulding, label cap) An external moulding that projects from a wall above an opening to throw off rainwater.

hook bolt A bolt with a threaded straight end that is used to hold steel profile sheet roofing material, and with a J-shaped end that hooks around roof *purlins, securing the sheet material to the roof structure.

hook bolt lock A lock built into a mortise in a door stile; the hook locks into the *jamb of the door when closed.

hook curtailment A reinforcement bar with the end bent into a hook shape.

Hooke's law A relationship that states the extension of a spring is directly proportional to the load that is applied to it. This applies to many materials such as steel up to its elastic limit. *See also* YOUNG'S MODULUS.

hook height The maximum height a hook on a crane can be raised; this is also the maximum lifting height of the crane.

hook intake (cable sleeve, turn-down) A duct or pipe through which cable can be added; the duct has a downturn end to prevent rainwater penetrating into the duct.

hook time The time planned for each specialist group or subcontractor to have dedicated access to the crane.

hoop iron Reinforcement occasionally inserted in the bed joints of walls to add strength to the wall.

hopper 1. An open container used for the storage of materials. **2.** The wide square opening to a funnel.

hopper light (hopper window) A window that has a bottom-hung inward-opening casement.

horizon 1. A line where the land or sea meets the sky in the distance. **2.** Circle on a celestial sphere, such as a plane through the centre of the earth. **3.** A distinct layer or band of rock or soil.

horizontal Parallel to the horizon.

horizontal shore A shore that is braced on a horizontal plane; it supports one vertical surface by bracing itself off another vertical surface. *See also* FLYING SHORE.

horn The extended jambs or head of a doorframe that provide additional strength to the mortise and tenon joints during transport. Once the frames are delivered to the site, the horns can be cut off when they are ready to be built in or inserted into the wall.

horsed mould *See* RUN MOULDING.

horsepower (hp) A unit of power equivalent to 550 foot-pounds per second or 746 watts.

horseshoe tunnel A tunnel that has a cross-section in the shape of a horseshoe, typically used when tunnelling through rock.

horst A ridge of the earth's crust caused by two normal faults pulling apart.

hose A flexible tube used to convey liquids and gases.

hose reel A flexible reel of tube used for delivering water to the place needed.

hose union tap (hose cock) A tap with a threaded outlet to which can be fastened the threaded connector of a hose.

hospital door A large flush door used in hospitals.

hospital sink A large sink, often stainless steel, used in hospitals.

hospital window An inward-opening bottom-hung casement window.

hot-air stripping Heated air that is used to soften paintwork, making it easier to remove. Heated paint will soften and often bubble off the surface as it expands and loses its bond.

hot bonding compound A bituminous material that is heated to increase fluidity and used as adhesive between sheets of felt and flat roof substrates.

hot coil *See* HEATING COIL.

hot cupboard A static or portable unit that is used to keep cooked food at a regulated temperature.

hot isostatic pressing A technique for compacting powder at high temperature and constant pressure.

hot pressing A technique used to compact particles at high temperatures and varying pressures to form ceramic materials.

hot-water system A system that supplies hot water to a building.

hot work Work that uses naked flames or glowing hot equipment. The work could be dangerous if chemicals or other flammable materials are in the vicinity. Hot work is normally associated with welding, and bitumen and asphalt work on roofs and roads.

hot working A technique used to alter the properties of a metal by subjecting the material to different heat treatments, thus changing its chemistry.

hot-work permit A permit to undertake *hot work in a specified area at specified times. Where hot work is considered a risk it is necessary to control the operation and activities around the work to ensure the risks are reduced. To ensure the necessary risks have been assessed and effective measures have been taken, a hot-work permit is requested, and the manager or safety officer responsible will issue the permit once appropriate measures have been taken. The hot-work permit helps to avoid fires, explosions, or other dangerous events in areas of high risk, such as chemical or material storage sites.

house A standard description for a small two- or three-storey dwelling, normally used as a home for individuals or families.

house breaking Demolition of houses and other small dwellings.

housed joint (let in) A rebate, recess, or mortise that lets in the end of another piece of timber such as a *tenon or the end of a timber. As the timber is held within another piece of timber, it is said to be housed, or let in.

housed string The side of a staircase that encases the treads and sits against the wall, also known as a close string.

housewrapping Wrapping a house with a breather membrane, may also double up as the air barrier to ensure air tightness. In timber frame buildings the breather membrane is often fitted to individual panels before they arrive on-site, however, the membrane will still need to be lapped and sealed at the panel joins to make an airtight structure.

housing 1. A protective covering. **2.** A number of dwellings. **3.** A *housed joint.

hovel A structure that has an open front and is used to house cattle.

Howe truss A roof truss that has both vertical and diagonal members, distinguished by the diagonal members at each side of the truss sloping up towards the centre so that they are in compression—the reverse of a *Pratt truss.

HPHW *See* HIGH PRESSURE HOT WATER SYSTEM.

HSE *See* HEALTH AND SAFETY EXECUTIVE.

H-section A *universal section that has a cross-section in the shape of an 'H', such as an H-column or an H-pile.

HTC *See* HEAT-TRANSFER COEFFICIENT.

hub (US) The enlarged bell-shaped end of a pipe into which another pipe is joined.

hubless *See* UNSOCKETED PIPE.

human capital management Management through measurement with the aim of establishing a clear line of sight between people, their work and contributions, and organizational success.

human error An unintentional action, mistake, or decision made by humans with consequences that were not intended. There are types of human error: slips, lapses, and mistakes. Human errors occur even with experienced and trained individuals.

(⊕) SEE WEB LINKS
- The HSE website provides information on understanding human error

human failure Stems from errors and violations committed through unsafe acts and conditions in a workplace.

human resource management (HRM) A business-focused approach to effective management of workers intended to help an organization gain competitive advantage; the practice of recruiting, hiring, deploying, and managing workers. Further benefits to the effectiveness of human resource management can be gained by employing strategic human *resource management and *talent management techniques.

human resource planning (HRP) (workforce planning) A process of analysing the current workforce, determining future staffing needs, identifying the gap between the present and the future, and implementing solutions so that an organization can accomplish its mission, goals, and strategic plan. It is about getting the right number of people with the right skills employed in the right place at the right time, at the right cost, and on the right contract to deliver an organization's short- and long-term objectives (CIPD, 2019).

(⊕) SEE WEB LINKS
- CIPD workforce planning factsheet

humidifier A mechanical device that is used to increase the moisture content (humidity) of air.

humidifier fever A flu-like illness that results in a fever, chills, and a headache. Although the exact cause is unknown, it has been linked to the bacteria and fungi found in humidifiers.

humidity A measure of the amount of water vapour in the air.

humidity controller *See* HYGROSTAT.

humus Dark brown organic component of topsoil, which has been produced from decaying vegetable and animal matter.

hung Describes any fixture or fitting that has been attached to a wall, door jamb, ceiling, or roof of a building. The fitting of *doors, opening *windows, curtains, and wallpaper are described as hung, once fitted.

hurdle A difficult obstacle to pass.

hurricane A severe tropical storm with torrential rain and winds exceeding 119 km per hour (i.e. force 12 on the *Beaufort scale).

hush latch A form of *striking plate that allows the latch bolt to close and open under gentle pressure without the need to withdraw the latch bolt.

HVAC *See* HEATING (VENTILATION AND AIR CONDITIONING).

hybrid Containing two different elements. A **hybrid car** is a vehicle that can run on both electricity and petrol.

hydrant A vertical pipe and valve that protrudes up from the ground. Enables water to be taken directly from the water main.

hydrated lime (calcium hydroxide) A caustic substance produced by heating limestone, which is widely utilized in construction. Hydrated lime is one of the primary constituents for *mortars.

hydration When used in connection with concrete, mortar, render, or plaster it refers to the chemical reaction of a substance that combines it with water.

hydraulic Relating to the flow of fluids.

hydraulic breaker A mechanically operated demolition drill with the spring and return action driven by compressed air. Smaller units can be hand-held, although the vibration from the drill and return can lead to conditions such as white finger (poor circulation in the digits on the hand). Larger breakers are often fitted to the *backhoe of excavators.

hydraulic door closer A door closer with a hydraulic damper to prevent the door slamming shut. Equally, rams and pistons can be used to assist opening and closing of doors that are mechanically rather than manually operated.

hydraulic equipment Mechanical equipment that is worked using fluid (hydraulic oil) which is pumped into rams to drive motors or move objects. Hydraulic rams usually have a drive and return valve making sure the units can be opened and closed.

hydraulic excavator An excavator (such as a *backhoe), with the shovel being operated by a hydraulic system.

hydraulic failure In respect to the *Eurocodes, failure caused by pore-water pressure or seepage, namely by:

- **uplift failure (buoyancy):** when the pore-water pressure beneath a structure becomes greater than the overburden pressure.
- **heave failure:** when the upwards seepage force becomes greater than the downward weight of the soil, causing the vertical effective stress to become zero. Soil particles are destabilized, which is known as boiling.
- **internal erosion failure:** migration of soil particles within a soil mass. Initially fine particles at first, being migrated relative to other soil particles or the structure. This then leads to more and more particles being displaced, resulting in the collapse of the structure.
- **piping failure:** a specific type of erosion, typically occurring in dams, where the migration of soil particles starts at the surface and then regresses until a pipe-shaped discharge tunnel is produced.

hydraulic fill Material that has been excavated by high-powered water jets and transported in water through a pipe, then placed and drained to form an embankment.

hydraulic fracture A *fracture in rock produced by fluid pressure. An artificial hydraulic fracture can also be made into rock strata to increase the permeability, e.g. when mining for gas or oil.

hydraulic gradient (i) The ratio of the difference in total head on either side of a soil layer to the thickness of the layer measured in the direction of the flow.

hydraulic hammer A tool that can be placed on the *jib of an *excavator.

hydraulic head (piezometric head) A measurement of liquid above a measurement of liquid pressure above a known reference point or vertical datum. It is the pressure exerted vertically downwards by a length of liquid column above a datum line. If case pumps are not used, hydraulic head in treatment units has to be maintained in order to prevent backflow.

hydraulic jump When the flow in an open channel suddenly changes from high to low velocity, there is a corresponding rise in the liquid surface. The rise causes eddies and turbulence to form with accompanying energy lost, with some of the initial kinetic energy being converted into potential energy.

hydraulic lift A *lift where the car is moved between floors by a hydraulic piston.

hydraulic retention time The average time during which water retains in the treatment system. It is a parameter to design water/wastewater treatment plants. It is generally stated in units of hours or days.

hydraulics The discipline that relates to fluid flow, particularly in rivers, water supply, and drainage networks, as well as irrigation schemes.

hydraulic test A test to ensure water/air tightness—tested to a value just above the design pressure.

hydrocarbon An organic chemical that contains both hydrogen and carbon atoms. Hydrocarbons are generally grouped according to their carbon range. For example, petroleum range organics (PRO) and diesel range organics (DRO) are grouped as follows:

- PROs represent the lighter fraction carbon range C6–C13, such as benzene, toluene, ethylbenzene, and xylene (BTEX), together with phenolic compounds.
- DROs represent groups of heavier hydrocarbon compounds based on the carbon range C10–C36. These include polycyclic aromatic hydrocarbons (PAHs), such as naphthalene and pyrene.

hydrocyclone A piece of equipment used to sort out or separate particles or droplets that are suspended in a liquid. In vortex mixers a hydrocyclone is used to help remove impurities or particles, separating them from the fluid base. Particles are separated based on their difference in size/relative densities, while some particles based on their size and density will be retained in the fluid. The flow pattern in a hydrocyclone is cyclonic, created by tangential injection of liquid into the cylindrical chamber.

hydrodynamic drag The resistance an object will experience when moved through a liquid.

hydrodynamics The study of fluid dynamics that is concerned with water flow, such as over weirs, through pipes, and along channels.

hydroelectric The use of flowing water to turn a turbine, which is attached to a generator to create electricity.

hydrogen A gas at ambient temperature, highly flammable, and constituting 75% of the earth's elemental mass.

hydrogen embrittlement Reduction in ductility of a metal on exposure to a hydrogen-rich atmosphere.

hydrogen sulphide A gas at ambient temperature, toxic, and highly flammable.

hydrogeological map A map that provides information concerning rivers, the groundwater profile (phreatic surface), together with the nature and extent of major aquifers, as well as existing wells.

hydrogeology The study of groundwater flow.

hydrography The study of oceans, lakes, and rivers, with particular reference to recording tidal ranges and river flow.

hydrology The study of the water cycle on the earth, i.e. the circulation of water, namely run-off, infiltration, storage, rivers, lakes, seas, evaporation, and precipitation.

hydrolysis A chemical reaction with water, which causes the decomposition of one compound to form two more other compounds, e.g. starch to glucose.

hydrophilic Having an affinity for water, the opposite is **hydrophobic**, a fear of water.

hydropower The use of flowing water to produce electrical or mechanical power.

hydroseeding The seeding of grass seeds onto an area of land by the use of a jet of water containing nutrients to enhance germination.

hydrostatic Relating to fluid at rest. **Hydrostatic pressure (hydrostatic head)** is the pressure exerted on an object from a column of fluid at atmospheric pressure and at a given height above the object. Also known as **gravitational pressure**

hygrometer An instrument that is used to measure the amount of *relative humidity within a space.

hygroscopic Describes the ability of a material to absorb and desorb moisture from the surrounding environment or materials that readily take up and retain moisture, for example soils. It is also used to refer to moisture taken up and retained under some conditions of humidity and temperature. Wood is referred to as hygroscopic due to its structure.

hygroscopy How a material absorbs and desorbs water vapour in relation to relative humidity. Hygroscopy and capillarity are key components of hygrothermal performance.

hygrostat A device that is used to control *relative humidity.

hygrothermal analysis A systematic study and review of the movement of heat and moisture through buildings. Computer-based modelling offers accurate predictions of hygrothermal performance which can offer data on potential early degradation and reduced service life.

hype cycle A banded graphical representation that the life-cycle technology goes through. After the initial hype peak there is a trough of disillusionment followed by a steady incline defining maturity, adoption, and social acceptability before a level plateau is maintained.

hyperbola A conic section.

hyperbolic paraboloid roof A roof formed from a sweeping curved shape that is made up from straight lines running between sides of a rectangle. The shapes resemble the curves that you get if you bend a piece of paper; in construction such shapes can be achieved by spanning cables, planks of wood (glue-laminated beams), or using reinforced concrete.

hypereutectic alloy An alloy whose concentration of solute is greater than the eutectic composition. If two metals are mixed together, a binary alloy phase system is obtained. At any mix of the two metals, the alloy will not have a definite melting point; instead, it will melt over a temperature range (e.g. over 20°C). Only at one specific mix between the two alloys (assuming alloy A and alloy B), e.g. 50% A and 50% B, will the alloy have one specific melting point. This unique composition whereby a two-metal alloy system has one specific melting point is the **eutectic composition**. Thus assuming the eutectic composition to be 50% A and 50% B, a hypereutectic alloy will contain more than 50% A (or conversely less than 50% B). These types of alloys are very commonly used, especially in structural steels.

hypothesis 1. A theory that has not yet been investigated. **2.** An assumption that is taken as true.

hysteresis The delay or lag in response by an object to a change in the force that acts upon it.

I (intensity) 1. A measure of the average amount of power emitted by sound, solar radiation, or a light source. Measured in watts per square metre (W/m²). *See also* LUMINOUS INTENSITY. **2.** A measure of the brightness of an object. **3.** The symbol used for *electric current.

I-beam *See* I-SECTION.

ice Frozen water; occurs when water cools to below 0°C (32°F) at standard atmospheric pressure. As it freezes, water expands by 9% and can cause damage to roads creating potholes, uplift of shallow foundations, etc.

ICE *See* INSTITUTION OF CIVIL ENGINEERS.

ICE Conditions of Contract An edition of *Conditions of Contract, sponsored jointly by ICE, the Civil Engineering Contractors Association (CECA), and the Association of Consulting Engineers (ACE). The full list of documents includes:

- Measurement Version
- Design and Construct
- Term Version
- Minor Works
- Partnering Addendum
- Tendering for Civil Engineering Contracts
- Agreement for Consultancy Work in Respect of Domestic or Small Works
- Archaeological Investigation
- Target Cost
- Ground Investigation
- Amendments to ICE Conditions of Contract

igneous rock (igneous stone) A rock type that has been formed from solidified *magma through either extrusive or intrusive processes. The character of igneous rock is affected by the chemical constitution of the magma and the rate at which the magma cools. The magma will have a molten temperature between 800°C and 1200°C. When solidified, it crystallizes into a mass of mineral crystals. The main minerals of igneous rocks are quartz, feldspar, muscovite, biotite, and mafics. Extrusive igneous rocks are formed when the magma flows out onto the surface of the land or into the sea, and cools very rapidly. The resulting rock (e.g. obsidian) will be glassy or have very fine crystal grains where minerals have crystallized. Intrusive igneous rocks are formed when the magma is injected into other rocks to form dykes, sills, batholiths, plutons, and stocks. The resulting rock (e.g. granite) is often very coarse-grained as cooling is slower, giving larger mineral crystals time to form.

ignitability The ability of a material to ignite.

ignition The process of lighting a material so that it begins to combust.

ignition temperature The lowest temperature at which a material will ignite.

illuminance (E) The *luminous flux density, or amount of light, incident on a surface. Measured in lux, where 1 lux = 1 lm/m^2. Also known as the **illumination value** or **illumination level**.

illuminated push A push button that illuminates when operated.

illumination Lighting an area or object, usually artificially.

imbrex A curved roof tile laid over the joints between the tegula to provide a waterproof roof covering. Used in ancient Greek and Roman architecture. *See also* TEGULA.

imbricated Overlapped, layered, or woven in a regular pattern.

Imhoff cone A graduated one-litre conical vessel that is used to determine the amount of settled solids in a given time. It gives an indication of the volume of solids that can be removed from wastewater by settling in sedimentation tanks, clarifiers, or ponds.

Imhoff tank A two-storey tank that receives and processes raw sewage. The upper part of the chamber allows sedimentation to take place. Solids slide down the inclined surface of the upper chamber and enter the lower chamber. Anaerobic digestion of the sludge then takes place in the lower chamber. The lower chamber is vented and requires the digested sludge to be emptied periodically.

immersion heater A thermostatically controlled electric element installed within a tank or cylinder that is used to heat a liquid. Commonly found in hot-water cylinders.

immiscible Solutions or materials that do not mix together—unmixable.

impact sound (impact noise) Sound that is generated by something impacting upon the structure of the building. Typical sources include footsteps, slammed doors and windows, noisy pipes, and vibrating machinery.

impact strength The strength of a material or structure to withstand stock loading.

impact test A standardized test that is used to determine the brittleness of a material; *see* CHARPY TEST.

impeller A rotor that transmits motion in a centrifugal or rotary pump, turbine, compressor, or fan.

imperial units Units of measurements that relate to foot, pound, and gallon. Superseded in the UK by the *SI units.

imperishable Not decaying.

impermeable Not allowing a liquid or a gas to pass through. *See also* IMPERVIOUS; PERMEABILITY.

impervious Does not allow a liquid to pass through. *See also* IMPERMEABLE; PERMEABILITY.

impingement filter (viscous impingement filter) A filter that removes particles by forcing them to strike a sticky filter medium.

imposed load *See* LIVE LOAD.

improved nail (ringed shank nail) Fixing nail with raised rings around the shank to increase friction and improve its ability to hold itself within the wood.

improvement line A position from which development will take place in the future, e.g. a point from which buildings will be built and land developed.

improvement notice An order issued by the *Health and Safety Executive under Health and Safety legislation which enforces organizations to take corrective action to avoid serious risk to health. The notice is served when Health and Safety law has been broken and states a period required for improvement.

impulse 1. A force acting briefly to drive something forward. **2.** A measure of momentum by the average force acting over a period of time.

inactive leaf The less frequently used opening leaf on a double door. *See also* ACTIVE LEAF.

incandescent lamp A type of lamp that produces light by passing an electric current through a thin tungsten wire (filament), which heats up. Inexpensive and easily dimmed but has low efficacy, short lifespan, and relatively high running costs.

incentive A factor that is used to provide motivation to increase performance or do something.

incentive scheme A financial reward system that provides extra money for specific and targeted performance.

incineration (soil incineration) High temperatures (up to 1200°C) are used to volatilize and combust organic compounds in contaminated soils, particularly those contaminated with explosives and chlorinated hydrocarbons such as *polychlorinated biphenyls (PCBs). Auxiliary fuels are used to initiate and sustain combustion. Air pollution control systems are employed to remove gases produced. Other combustion residues usually require treatment/disposal. Such gases and combustion residues may be highly toxic, so effective air-pollution control systems are essential. Combustion residues may not be suitable for re-use and may still require disposal to landfill. Contaminated soils may be treated on-site using mobile plant or taken off-site to a static plant.

incise To carve or cut into an object using a sharp implement.

inclement weather (unfavourable weather) Any type of weather, such as rain, snow, high winds, etc. that slows down or prevents work being undertaken on a building site.

inclinator (US) An inclined domestic stairlift.

inclined shore *See* RAKING SHORE.

inclinometer *See* CLINOMETER.

included angle The angle where two lines with a common *vertex meet.

inclusion 1. Something that is embedded or trapped into something else, e.g. a gas or liquid trapped within a rock stratum. **2.** The relationship between two sets such that the second set is a subset of the first.

incombustible (non-combustible) Not able to be burnt.

incompetent strata *Ductile strata, which has a tendency to flow under loading rather than forming a *fault or *fold.

Incorporated Engineer An engineer who is registered with the Engineering Council, UK, and has gained the post-nominal letters 'IEng'. According to the Engineering Council, UK, 'Incorporated Engineers maintain and manage applications of current and developing technology, and may undertake engineering design, development, manufacture, construction, and operation.' In doing so they carry out similar, but less conceptual and detailed work than a *Chartered Engineer.

(⊕) SEE WEB LINKS

• The Engineering Council website details the different professional qualifications that can be obtained and is the source of the above quotation.

increaser A pipe that tapers from one diameter to another allowing pipes of different diameters to be joined together. The pipe is usually fixed with the pipe increasing in size along the direction of the flow.

indefinite integral An *integral which when differentiated equals a given function.

indent A groove or recess that is cut in a material or component.

indented joint A joint that is formed from two matching indents.

indenting 1. The process of cutting an *indent into a material or component. **2.** The grooves, tracks, or marks left in flooring from heavy loads and traffic.

indenture A written contract with two or more parties, typically used for an apprentice to serve a master.

independent float Time before or after an activity that provides some flexibility over when a task can be started or finished, without affecting subsequent tasks. Tasks with float can be started as soon as a preceding task finishes or the task can be started later, within the specified float time, without affecting the start or finish date of any subsequent tasks.

independent scaffolding Tubular support system, with two sets of parallel *standards providing vertical support independent of the building, that allows operatives to work at heights. The support system is made of individual poles and clips that are fastened together to suit the structure that they are tied into. All of the vertical support is provided by the vertical poles in the scaffolding, horizontal support is provided by *braces and ties into the building. The ties are normally fixed into window openings or around columns.

indeterminate structure Where the internal forces or reaction components cannot be calculated by using the equations of equilibrium alone—equilibrium must be combined with compatibility. Two classical approaches to solve such structures include the force method and the slope-deflection method.

index In mathematics, the notation used to state how many times a number is multiplied by itself.

indicating bolt A type of *privacy latch that when engaged indicates whether the door is locked or unlocked or if the room is occupied or vacant.

indicator A device that shows the status of something. It may take the form of a meter, a gauge, or a light **(indicator light)**.

indicator panel (annunciator) A large panel comprising a number of different indicators.

indigenous Originating or belonging to a particular region or country.

indirect Not direct, not in a straight line, or not immediate.

indirect action *See* ACTION.

indirect air leakage point A point in the building envelope where *air leakage occurs indirectly through the *primary air barrier via a series of interconnected voids from inside the thermal envelope to outside or vice versa. Common points include on external and party walls at the ground floor/external wall junction, under kitchen and utility room units, around staircases, into intermediate floor voids and at intermediate floor perimeters, into service voids (e.g. behind bath panels), and at service penetrations where they penetrate the dry-lining and/or internal finish. *See also* DIRECT AIR LEAKAGE POINT.

indirect cold-water system A cold-water system where all of the appliances, except the cold-water drinking outlets, are supplied with cold water indirectly from a cold-water storage tank. All of the cold-water drinking outlets are supplied with water directly from the mains.

indirect gain A type of *passive solar heating system where the solar energy enters the thermal envelope and is absorbed by the thermal mass which is not located within the living space. The thermal mass then re-emits the stored energy to the living space via *conduction. The *Trombe wall, *mass wall, water wall, and *roof pond are all types of indirect-gain system. *See also* DIRECT GAIN; ISOLATED GAIN.

indirect hot-water cylinder A cylinder where the water is heated indirectly by passing hot water from the boiler through a coil within the cylinder. The water from the boiler does not mix with the water within the cylinder. Also known as a **Calorifier**.

indirect hot-water system A hot-water system where the water that is heated by the boiler is not the water that is drawn off at the taps. Instead the water that is heated by the boiler, or other heat source, passes through a coil inserted into the hot-water cylinder, where it heats the cold water that is fed directly into the cylinder. This heated water is then drawn off when the hot-water taps are activated.

indirect lighting A method of lighting a space where the light has been reflected and diffused by a ceiling or wall, rather than falling on the space directly.

indoor adaptive temperature The indoor *air temperature at which the majority of the occupants within a building will feel thermally comfortable. This temperature will vary from day to day depending upon the external environmental conditions.

indoor pool A swimming pool that is located indoors.

induced siphonage The removal of water that forms a seal in a trap due to suction; it is caused by the pressure generated from water flowing through other parts of the drainage system.

inductance (L) The property of an electric circuit or device that relates the electromotive force to the current flowing through it or near it.

induction 1. A progress of introducing information to someone at the start of something, e.g. a structured programme someone attends at the start of a new job, where an employee is provided with a formal period for meeting colleagues, understanding the job, and becoming familiar with the operation and organization system in the company. **2.** Conclusion based on logic. **3.** The production of electric or magnetic forces in a circuit by being in close proximity to (but not touching) an electric or magnetic field, where an object establishes an electrical current due to its proximity to another object with an electric current flowing through it. **4.** Gaining an understanding, through research, of human meanings related to events, understanding of the research context, and understanding that the researcher is part of the process.

induction unit A type of air-conditioning unit where high-velocity air is injected through nozzles to induce the circulation of room air over a coil to which heating or cooling is applied. These tend to have relatively high fan power and may result in a noise nuisance.

industrialized building Systemized construction using prefabricated building modules that are manufactured off-site for quick assembly on-site. Off-site manufacture reduces the time spent on-site constructing the building. When there is a high degree of standardization, the economics and speed of factory building methods are considerably better than traditional bespoke methods. For some years cladding systems, trussed rafters, steel frames, and service modules have been preassembled for construction. There is now a considerable movement to produce full buildings off-site.

industry The activity or employment in construction, trade, and/or manufacture.

inelastic Not stretchy or easily changed—not able to return back to its original shape after deformation; *see also* PLASTIC.

inequality equation In *ultimate limit state design, the effects of *design actions (E_d) should be less than or equal to the corresponding *design resistance (R_d), i.e. $E_d < R_d$. In *serviceability limit state design, the effects of the design actions (E_d) should not exceed the design performance criteria of the structure (C_d), i.e. $E_d < C_d$.

inert A solid, liquid, or gas that is not amenable to a chemical reaction, i.e. is unreactive.

inertia The property of a body such that it remains at rest or continues moving in a straight line unless acted upon by a force.

infectious The ability to spread something, typically related to the spread of diseases caused by bacteria and viruses.

infilling 1. Units or panels of material placed between the structural frame to increase stiffness or provide weather protection. **2.** Material placed within the cavity to improve fire resistance or thermal insulation.

infill wall Non-loadbearing walling units, brickwork, or blockwork positioned between the structural frame.

infiltration 1. The permeation of a liquid through a substance by filtration, such as rainwater entering into a soil's groundwater; *see also* RUN-OFF. **2.** Air entering and leaving a building through openings, cracks, and gaps. Infiltration/ventilation as part of a heat-transfer coefficient is the total heat-transfer rate from a building resulting from infiltration and/or ventilation (Cv-value) per °C of indoor to outdoor temperature difference in W/°C. It can be estimated experimentally by means of the tracer gas ASTM E741-11 Method. When heat recovery devices are present in the ventilation systems, the ventilation heat-transfer rates must be corrected by the heat-recovery exchangers' effectiveness. Both infiltration and ventilation rates can be variable in time and thus the Cv-value has a variable nature. *See also* BACKGROUND VENTILATION.

infinite (∞) Not measurable—having no limits.

inflammable A substance/material that is highly susceptible to ignition, describing the ease of ignition. Inflammable and flammable have the same meaning, inflammable may be misinterpreted as meaning not flammable, as this could be confusing, it is often suggested that the word flammable is used.

inflated structure *See* AIR-INFLATED STRUCTURE.

inflection A change in curvature, e.g. from convex to concave.

inflow The action of flowing in, for example, the point where fresh water flows into a lake.

influent A fluid entering a system, for example, where a stream enters a lake, or where a stream loses its flow by recharging the groundwater, i.e. it does not have a *base flow.

information delivery manual (IDM) A document that identifies the various construction processes together with the information required at each stage.

information exchanges *See* DATA DROP.

information manager Essentially a project manager, appointed by the employer, who is responsible for managing and delivering the *asset using *BIM procedures and protocols.

information paper (IP) A paper published by the *Building Research Establishment (BRE).

⊕ SEE WEB LINKS
• BRE provides publications, research, and consultancy on various aspects of building and construction.

informative In respect to the *Eurocodes, not a requirement, presented for additional information only. *See also* NORMATIVE.

infrared The portion of the invisible electromagnetic spectrum that consists of radiation with wavelengths between about 750 nm and 1 mm.

infrared camera (IR camera, thermal camera, thermographic camera) A device that captures an image using *infrared radiation. The thermographic camera/thermal imaging camera detects infrared radiation and transforms this in to a visible image and spectrum to represent surface temperature, which is then displayed on a monitor.

infrared survey (IR survey) A non-destructive method of measuring and visually mapping surface temperatures using an *infrared camera. The method works on the principle that objects emit electromagnetic radiation of a wavelength which is dependent on the object's temperature. The radiation's frequency is inversely proportional to the temperature, radiation is detected, and coloured metrics and graphs produced with infrared imagers (radiometers). Thermal infrared imaging was commonly used for military purposes before becoming commonly used in buildings. Infrared analysis is a focal plane array-based system that helps identify problems and defects, and can be used as part of a set of survey tools that enable predictive maintenance. Advanced image processing and reporting programs facilitate fast and easy analysis and documentation of survey results.

infrared temperature sensor A monitor that detects temperature change using infrared technology and is capable of capturing and relaying data.

infrastructure The physical public systems, services, and facilities of a country that are necessary for society and economic activity. These include buildings, roads, bridges, and utilities (electricity, gas, water, sewers, and telecommunications).

infrastructure resilience The ability of bridges, roads, buildings, etc. to resist or reduce the impact of disruptive events, such as loading from earthquakes or the damage caused by flooding.

ingenious Clever, imaginative, and original.

inglenook A seated area built into a recess next to a fireplace.

ingo (ingoing Scotland) A window or door *reveal.

ingo plate (Scotland) A *reveal lining.

inherent Something that exists as a permanent or existing feature, for example, *knots are an inherent defect in timber.

initial ground levels The level of the natural ground before any construction operations have taken place.

initial rate of absorption The *absorption rate.

initial set The very early stages of concrete maturity where bonds between the cementious materials have started to form; concrete should not be worked once initial bonds have started to form.

initial surface absorption test (ISAT) A test to determine the *porosity of concrete as defined in BS 1881 part 5. The test employs a plastic cap, which is sealed to the concrete surface. The cap has a water area of 5000 mm² and a head of water of 200 mm. The absorption of water is measured by observing the movement of water in a connecting capillary tube over a fixed time period.

injection A method of introducing a liquid to something under pressure, e.g. spraying fuel into an engine.

injection damp course (chemical injection damp course) A damp-proof course that is formed by injecting a chemical into a wall at regular intervals under pressure. Various chemicals can be used to form the damp course including silicone resins, aluminium stearate, or methyl siliconate. Used where there is no existing damp-proof course (old properties), or if the existing damp-proof course is no longer functioning correctly.

inlaid parquet Parquet *flooring that is set flush within a decorative border.

inlay The process of decorating the surface of a material or component by inserting another material into prepared indentations. The result is a decorative flush finish.

inlet 1. A narrow indentation in a coastline or lake. **2.** A narrow stretch of water between adjacent islands. **3.** An opening through which a liquid or gas passes to enter another device.

innings Land that has be reclaimed from the sea or other waterlogged areas.

innovation To develop something new and originally, rebuild or modify, to create something different from the old that represents a noteworthy development or change to the item or process.

inorganic 1. Formed from minerals rather than matter originating from living things. **2.** Compounds that contain no carbon, however, oxides of carbon, carbon disulphide, cyanides, and associated acids and salts are also considered to be inorganic. *See also* ORGANIC.

input system A type of mechanical ventilation system that supplies fresh air to a building using a fan. In dwellings, if fresh air from the roof space is supplied via a small fan, it is known as **positive input ventilation**.

insect attack (on timber) A common form of degradation mechanism for timber material. Within the UK insect attack is limited to a small number of species of insect and tends to be less serious than fungal attack. Outside the temperate climate regions, including the tropics, termites or white ants cause more damage than all other insect attacks put together. The main damage in the UK comes from beetles, which lay their eggs on timber, and then during the larval stage, bore through the timber, eating the organic material—mainly the sapwood, and cause loss of mechanical strength. Strength is lost because the cellulose cell walls are cut through; this can be very serious if the wood becomes badly infested.

insect screen (fly wire screen) A fine mesh screen used to prevent insects entering a building.

inset The process of placing a material or component within the boundary of another. For example, a kitchen sink will be inset within the kitchen worktop.

inside glazing Term used to describe external *glazing that has been installed from the inside of the building. *See also* OUTSIDE GLAZING.

in-situ In its original position. Refers to components or elements of a building that are formed on-site in their final position, for example in-situ concrete.

inspection The visual checking of work to ensure that operations have been carried out properly and established standards have been satisfied. Various professionals and authorities will be required to check the works as part of their job. Local authorities, clerks of works, architects, and the client's representative may all carry out visual inspections.

inspection certificate Written record that the works have been checked and have achieved the specified standard. The Building Authority will issue such certificates following inspections of drains and other works.

inspection chamber (manhole) A chamber with a removable cover that enables access to be gained to an underground drain for inspection and maintenance.

inspection cover The removable, usually flush-fitting cover over an *inspection chamber.

inspection door A door or panel located in a wall, floor, or ceiling that can be opened or removed to enable access to be gained to an installation or services. Also known as an **access door**.

inspection fitting (inspection eye) *See* ACCESS COVER.

inspection junction A short section of drainpipe with a removable cover that runs at a 45° angle from the main underground drain to the surface. Used to insert a drain rod. Also known as a **rodding point**.

inspection notice Notification given to the local authority to state that an area of works is complete or exposed, and is available for inspection. Certain building stages require the building inspector or nominated authority to inspect and certify that the works are satisfactory.

inspector Person employed to check the quality of the work.

instability The condition of being unstable or not in *equilibrium.

installation 1. The act of installing equipment. 2. A large building or facility, for example, a chemical installation.

instantaneous hot-water heater (single point heater) A device designed to produce hot water instantaneously only when it is required. Usually only supplies hot water to a single point. Can be gas-fired or electric.

instant-start tube (rapid-start tube) A type of fluorescent lamp that can be switched on instantly. It incorporates a ballast that has a continuous input high enough to start an arc through the tube instantly.

institution An important professional or public body, or organization; *see* INSTITUTION OF CIVIL ENGINEERS.

Institution of Civil Engineers (ICE) An international membership organization, founded in 1818, which promotes and advances civil engineering around the world. The purpose is to qualify professionals engaged in civil engineering, exchange knowledge and best practice, and promote their contribution to society. There are around 80,000 members worldwide.

SEE WEB LINKS

• The website details information about the Institution of Civil Engineers and is the reference source for the above text.

Institution of Electrical Engineers (IEE) Now called the **Institution of Engineering and Technology**, this is a leading professional society for engineering and technology, providing a global network to facilitate knowledge exchange and promote science, engineering, and technology.

SEE WEB LINKS

• The knowledge portal for the Institute of Engineering and Technology.

instructions to tenderers (US notice to bidders) Direction and guidance contained within the bills of quantities on how to include prices for materials, labour, and other items of work described

instrument 1. A piece of equipment or tool used to aid performance. **2.** A legal document such as a statutory instrument.

insulate To reduce the rate of heat transfer, sound transmission, or the flow of electric current.

insulation Material or product used to restrict or reduce the passage of heat, fire, or sound. Typical applications of insulation include *acoustic insulation, *thermal insulation, fire insulation, and electrical insulation.

insulator (electric) A non-metallic substance that has very low electrical conductivity at room temperature.

intact clay A clay with no visible fissures; *see also* FISSURED CLAY.

intake An opening or structure through which fluid passes to enter a system, e.g. water intake to a treatment plant.

intake unit (house service cut-out, fuse link) A device that contains the service fuse and connects the incoming electric service cable to the cables that in turn connect to the electricity meter. Usually located within the electricity meter box.

integer Something whole or complete, e.g. a whole number.

integral 1. Important part of something to make it complete. **2.** Whole without missing parts. **3.** Concerning *integers. **4.** Concerning integrals or integration.

integral waterproofing The process of making concrete waterproof by adding the waterproofing component to the cement or the water.

integrity In *fire resistance, the length of time that a component will remain structurally intact in a fire without failing.

intellectual property The product of the human intellect that has the potential to have commercial value, such as a discovery, invention, process, procedure, development/improvement in process/procedure, data, design, formula, model, plan, drawing, documentation, database, and computer software. It is normally protected by a *patent, copyright, or trademark.

intelligent building (smart building) A building in which the control systems for the building, such as heating, lighting, ventilation, security systems, etc. are capable of automatically adapting to changing external conditions.

intelligent fire detector (smart fire detector) A fire detector that monitors a range of parameters in order to respond more quickly to the presence of a fire.

intelligent vehicle highway system (IVHS) A system that allows the interaction of vehicles, highway, and people to be monitored. It can be used to improve safety, reduce wear, and improve transportation times.

intensifier A device that increases the amount of something, e.g. the strength of a signal, the quality of an image, the pressure over the source pressure.

intensity The strength of something, e.g. the intensity of an earthquake (the strength of the quake rather than the magnitude, which defines the energy released).

interaction curves Graphical plots defining limiting conditions for structural members subjected to different forms of loading.

interactive Two-way responsive communication (for example, between user and computer).

interbearing angle The clockwise angle between two bearings.

interbedded Strata that have been laid down in sequence having distinctive beds.

intercept A point where a curve or line crosses an axis.

intercepting trap (interceptor) A *trap that prevents unwanted material entering a drain.

interception The capture of material; in waste management, the phenomenon where freely floating particles are entrapped/entangled in settling flocs and are subsequently removed.

interceptor sewer A large sewer system that is designed to direct dry weather flow to the treatment plant and discharge wet weather flow into a receiving river when the sewer capacity is exceeded.

interchange A major road junction where vehicles pass without stopping, by means of slip roads, bridges, and underpasses.

intercom A device that allows communication between different parts of a building.

interdiffusion Diffusion of atoms of one metal into another metal. This usually occurs at elevated temperatures, for instance during welding of metal and metal alloys.

interface 1. The point at which two or more components of a building meet. **2.** Where building operations come together, and different contractors or professionals have to coordinate and integrate their activities to ensure a joined-up service.

interference An unwanted disturbance that is a hindrance, for example, a signal that interrupts a radio broadcast.

intergranular fracture A failure mechanism in metals whereby the crack propagates along the grain boundaries. The grain boundaries are only visible at a microscopic level, especially in metals and alloys.

intergrown knot A live knot, a natural defect in wood formed where branches of the tree stem from the main trunk.

interim certificate Document that certifies the value of the works, normally issued on a monthly basis.

interim payment Instalments paid to the contractor or partial payment of the contract sum for works performed. Such payments are usually made on a monthly basis with retention money held back to ensure continuation of work, and that any defects can be made good before full payment is made.

interim valuation Calculation of the quantity of work performed and the value of that work so that an *interim payment can be made.

interior adhesive Substance used to stick fabric and building components together. The glue is of reasonable durability but not suitable for external use.

interlock 1. The joining of two components together such that they are interconnected to one another. **2.** A mechanical device or arrangement of controls that prevents a device functioning unless other devices are functioning in a particular way. For example, a boiler interlock prevents the boiler from firing unless other devices (thermostats, programmers, time-switches, and TRVs) indicate that there is a demand for heat.

interlocking joint A joint where a projection in one component connects into a groove on another component.

interlocking polystyrene blocks Insulating hollow building units with a mechanical interface. The blocks are often hollow, allowing concrete and reinforcement fill, creating a structurally insulated wall. The insulated blocks and interface between them create a temporary structure capable of providing a permanent formwork for the structural concrete.

intermediate joist A common *joist that runs from one wall to another.

intermediate rafter A common *rafter that runs at right angles from the wall plate to the ridge. *Jack rafters, which do not run the full slope of the roof, are not included within this classification

intermediate rail A horizontal *rail located between the top and bottom rail of a door.

intermediate rock *Igneous rock that has a chemical composition between *basic rock and acidic rock, for example, andesite.

intermediate sight A staff reading which is taken between a *backsight and a *foresight.

internal angle The internal corner of a room.

internal diameter (ID, bore) The diameter of the inside of a pipe.

internal door A door located inside a building.

internal dormer A vertical *dormer window that does not project above the slope of the roof.

internal glazing Glazing that is located inside a building. *See also* EXTERNAL GLAZING.

internal hazard A fire *hazard that is located inside a building.

internal leaf The internal leaf of a cavity wall. *See also* EXTERNAL LEAF.

internally reflected component (IRC) The light received on an internal surface that is reflected from the surfaces inside a room. *See also* EXTERNALLY REFLECTED COMPONENT and SKY COMPONENT.

internal pipework (internal plumbing) Pipework located inside a building.

International Building Code (IBC) A model building code developed by the International Code Council (ICC) and used throughout the US. The code deals with regulations in regards to design and construction, structural stability, and health and safety.

International Standards Performance requirement established and set by the *International Standards Organization (ISO). The ISO works with the national quality standards organizations, such as the British Standards Institute, the European Committee for Standardization (CEN), etc. to coordinate and set international standards.

International Standards Organization (ISO) The organization that has the world's largest body of published standards. Based in Geneva, it coordinates international standards by working with the national standards organizations, enabling greater consensus between standards and operations.

SEE WEB LINKS
- A non-governmental organization providing a link between public and private sector standardization across many countries.

Internet of things (IoT) A system of interrelated computing devices, machines, objects, etc. that have the ability to transmit data over a network without the need for human intervention.

interpaver A paving brick that interlocks with the other paving bricks and blocks.

interplate The boundary between two *tectonic plates. An **interplate earthquake** occurs at the boundary of two tectonic plates; *see also* INTRAPLATE.

interpolation The calculation of an intermediate point from surrounding information, for example, to obtain the value of a point somewhere between two other known points on a straight line.

intersection 1. The point where two lines meet one another. **2.** The point where two roads meet one another.

interstitial condensation Condensation that occurs within and/or between the individual layers of the building envelope, when the temperature of some part of the building envelope equals or drops below the *dew point temperature. May occur on the surfaces of materials within a structure, particularly on the warm side of relatively vapour-resistance layers, within the material when the dew point and structural temperatures coincide throughout the material, or on more than one surface in a structure. This is because moisture may evaporate from one surface and recondense on a colder one.

interstitial level A level located halfway between two floors of a building that houses mechanical services, such as air conditioning or mechanical ventilation equipment.

intertie A horizontal intermediate member used in framed construction to help strengthen and stiffen the vertical members.

intraplate The interior of a *tectonic plate. An **intraplate earthquake** occurs in the interior of a tectonic plate; *see also* INTERPLATE.

intruder alarm system *See* BURGLAR ALARM.

intrusion Molten *igneous rock that has been injected into older pre-existing rock.

intumescent coatings A special type of protective coating, applied as a paint, offering fire protection to structural steel components. Intumescent coatings are typically 1 or 2 mm in thickness without any noticeable visual effect.

inundate Flooded or overwhelmed.

invar A non-ferrous alloy based on nickel and steel utilized for its low coefficient of thermal expansion.

inverse In mathematics, the opposite or reverse of something. For instance, minus is the inverse of plus and the inverse of multiplication is division.

invert To turn something upside down or reverse its position.

inverted roof (protected membrane roofing, US inverted roof membrane assembly) A type of flat roof where the insulation is applied on top of the weatherproof covering. An earth-sheltered (or turfed) roof is an extreme example of such a roof.

invertor An electrical device that converts *direct current to *alternating current.

investment casting A type of casting process used to mould metal alloys; also known as lost wax and precision casting.

invited bidder (US) *See* SELECTED TENDER.

in vitro Outside a living organism, e.g. a test carried out within a test tube; *see also* IN VIVO.

in vivo Inside a living organism; *see also* IN VITRO.

ion An atom or molecule that has either lost or gained electrons resulting in it being either positively or negatively charged.

ionic bonding A type of chemical bond involving metal atoms/ions, prevalent in metals and ceramics.

ionosphere The earth's upper atmosphere where ionizing radiation from space produces ions and free electrons that are able to reflect radio signals, thus allowing such signals to be transmitted around the globe.

IR camera *See* INFRARED CAMERA.

iron (Fe) A heavy ductile magnetic metallic element, iron is silver white in pure form but readily rusts; it is used extensively in construction, mainly as an alloy in the form of steel. *See also* STEEL.

iron carbide (cementite) A compound found in non-ferrous alloys, i.e. steels and cast iron, chemical formula Fe_3C. The existence of this phase contributes to the hardness and tensile strength of the alloy.

iron fairy A small wheeled crane.

ironmongery A term used to refer to metal window and door fixtures and fittings, such as handles, locks, hinges, catches, etc. Originally, these would have all been made from iron.

irregular coursed rubble Wall built out of rock and stone that differs in thickness, meaning that each subsequent layer of stone has different depths. The bricklayer matches up the rocks on each level so that there is some consistency; however, the subsequent course will be of a different thickness matching the natural thickness of the stone selected.

irregular paving Irregular sizes and shapes of a material that has been laid as *paving.

irreversible Impossible to progress backwards, i.e. to its original state, shape, or place.

irrigation The supply of water via channels and pipes, particularly to enable crops and plants to grow.

IR survey *See* INFRARED SURVEY.

IR temperature sensor A sensor that detects the *infrared radiation emitted from an object and converts it into a temperature measurement.

I-section (I-beam) A rolled steel section with a cross-section in the shape of an 'I'. *See also* UNIVERSAL SECTION.

island 1. A piece of land surrounded by water. **2.** A small area which is surrounded by something else, e.g. a traffic island.

iso- Prefix meaning of equal.

isobar A line on a weather map connecting points of equal atmospheric pressure.

isochromatic Of equal colour.

isochrone A line on a map that joins points of equal time.

isoclinic line A line on a geological map joining points that have the same magnetic dip.

isohel A line on a map that represents the same average number of hours of sunshine in a course of a year.

isohyet A line on a map that represents the same average amount of rainfall in the course of a year.

isolated ceiling A type of false ceiling that is suspended from the floor above on special hangers or clips that are designed to limit the amount of sound transmission. Used to improve sound insulation between floors.

isolated column A column that is located in a different position away from the majority of the other columns.

isolated footing A foundation, such as a *pad, that is not connected to other foundations.

isolated gain (remote gain) A type of *passive solar heating system where the collector and thermal mass are located outside (isolated from) the thermal envelope and living space of the building. The solar energy is absorbed by the thermal mass and is then transmitted to the living space via conduction or convection. An atrium, conservatory, and sunroom are all types of isolated gain system. *See also* DIRECT GAIN; INDIRECT GAIN.

isolating membrane A separating barrier used in flat asphalt roofing to allow asphalt to expand and contract without being affected by the different thermal movement of the roof structure. Breaks and holes in the barrier allow a certain amount of adhesion of the asphalt to the roof. The resulting adhesion and friction ensure that the asphalt does not contract too much in cold weather nor slide off the roof structure.

isolating strip Length of expandable and compressible material that forms a break in the structure, allowing independent movement and preventing cracking.

isolating valve (stop valve) A mechanical device inserted within pipe lengths to enable the water supply to be completely closed off.

isolation Separate from something—remote.

isolator An electrical device that, when activated, ensures that a circuit cannot become live.

isomerism When polymers have the same chemical formula yet different arrangements resulting in different materials. Thus, due to isomerism, it is possible to have the same polymer yet with different mechanical properties.

isometric Equal in dimensions. An **isometric projection** is a three-dimensional drawing that has been produced such that the projected three planes are at equal angles to the three axes.

isopachyte A line on a map connecting points of equal stratum or sediment thickness.

isopleth A line on a map connecting points of equal value, e.g. a contour line.

isoseismal A line connecting points of equal earthquake intensity.

isostasy A state that denotes the equilibrium condition within the earth's crust.

isotach A line on a weather map connecting points of equal wind speed.

isotactic A type of polymer chain configuration where all side groups are positioned on the same side of the chain molecule. This is very important in polymer chemistry as through isotactic polymers it is possible to chemically modify a polymer to yield desirable properties for a specific application.

isotherm A line on a weather map connecting points of equal temperature.

isothermal When chemical reactions take place at a specific or constant temperature, especially in metals (alloys) and polymers.

isotope Two or more forms of a chemical element that have the same atomic number but different numbers of neutrons. Their chemical properties are similar, but their physical properties differ.

isotropic A term referred to materials that have the same mechanical properties irrespective of orientation, i.e. which direction it has been cut. Most metals and polymers, for example, are isotropic, however, timber is not isotropic—it is classed as anisotropic.

isthmus A narrow strip of land, surrounded by water, that connects two larger areas of land.

IStructE (Institution of Structural Engineers) A professional institution that has around 23,000 members worldwide. It promotes professional standards in structural engineering and public safety within the *built environment. Chartered members of this institution gain post-nominal letters 'MIStructE'.

(⊕) SEE WEB LINKS
• The IStructE website details further information about the institution and the different professional qualifications which can be obtained.

IWA (Inland Waterways Association) Organization that campaigns for the use, maintenance, and restoration of British canals.

(⊕) SEE WEB LINKS
• Information on the use and restoration of canals.

IWO (Institution of Water Offices) A professional body that is concerned with the water industry.

Izod test This type of test is commonly utilized in metallurgy to determine the ductile-to-brittle transition temperature for metals and alloys. *See also* CHARPY TEST.

jack A mechanical device for lifting or moving heavy objects by applying a force through a screw thread or hydraulic cylinder.

jack arch (flat arch, Welsh arch) A flat or straight arch formed from bricks or stones that lean outwards away from the *keystone.

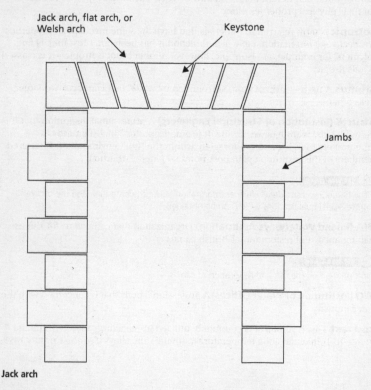

Jack arch, flat arch, or Welsh arch

Keystone

Jambs

Jack arch

jacket The insulated outer cover around a hot-water cylinder.

jacket platform An offshore platform supported by a tubular steel welded framework located on driven pile foundations in the seabed.

jackhammer (pneumatic drill, US rock drill) A hand-held pneumatic drill, used for breaking up and splitting hard density material, such as concrete, rocks, and pavements—typically used in highway works to break up the surface layers of a pavement.

jack-leg cabin A small portable building with adjustable steel legs. Commonly used as accommodation on-site.

jack pile (jacked pile) A short pile used in *underpinning, where the pile is driven into the ground by jacking against the structure.

jack plane A hand-held plane used to shape wood and provide a smooth surface.

jack rafter A short *rafter that spans from the *eaves to the *hip rafter in a hipped roof, or from the *valley rafter to the *ridge in a valley roof.

jack shore (back shore) Part of a raking shore that adds stability to the structure.

Jacobean Architecture or furniture that resembles objects or buildings produced during the reign of James I of England, normally associated with dark oak.

jamb The vertical members of a door or window which are adjacent to the wall; also used to describe the side/s of an opening in a wall.

jamb form Removable form used inside formwork to enable an opening for the window or door to be formed inside a concrete wall. A profile of a timber door or window is inserted on the inside face of the formwork so that when the concrete is poured a space in the concrete wall remains for the door or window.

jamb lining Timber inserted on the face of the *jamb to improve the finish of the door or window.

Japanese saw A hand-held *saw that cuts through material on the pull rather than the push stroke.

jar test A test used to simulate the coagulation process that occurs in a water treatment plant. It is used to determine the optimum dosage of coagulant in relation to pH needed for effective treatment. The test consists of filling four or six 1000 ml beakers with raw water. Different dosages of the coagulant (e.g. *alum) is then added to each beaker—the coagulant is normally added from a stock solution. The beakers are then placed on the jar test equipment, where paddles are inserted into the beakers. The paddles are initially rotated for a short time at high speeds (say, 200 rpm for 60 seconds) to ensure uniform dispersion of the coagulant. After which the speed of the paddles is reduced (say, 20 rpm for 15–30 minutes) to aid the formation of *floc. The paddles are then stopped and the samples are allowed to settle (say, for another 15–30 minutes). The actual timescale is site-specific, but would typically be in the order of periods stated. Too little coagulant and the water will remain cloudy, with hardly any floc forming, together with no settling. Too much coagulant will cause dense fluffy floc to occur, which will not readily settle. Optimum dosage will yield clear water above settled floc. The test is repeated using different dosages until such conditions are obtained—typically the pH of the water is adjusted to aid the coagulation process. The *supernatant can also be filtered through a Whatman No. 1 filter paper to give an indication of the quality of water that would be obtained when filtered through rapid gravity filters in the treatment plant.

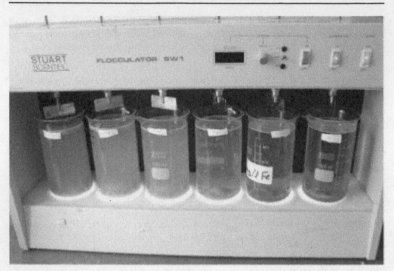

Jar test equipment

jaw breaker A machine that breaks rock.

JBM *See* JOINT BOARD OF MODERATORS.

JCB A proprietary name for a *backhoe excavator.

JCT (Joint Contracts Tribunal) Group responsible for the suite of standard contracts that is commonly used in the UK construction industry.

((())) SEE WEB LINKS
• The Joint Contracts Tribunal website, where the full suite of JCT contracts is described.

jemmy (crowbar, pry bar) A short bar about 400 mm long used to prise or lever two objects apart.

jerkin head roof (hipped gable, shared head) A hipped roof with the top part of the gable end hipped. The gable end is vertical with only the upper part of the roof sloping back towards the ridge.

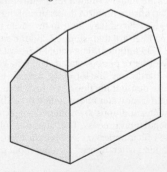

Jerkin head roof

jerry builder A builder who produced poor-quality buildings and cut corners during the First World War were given this name. It is still associated with poor workmanship.

Jersey barrier (Jersey wall) A 3-ft high reinforced concrete *barrier used on highways to separate lanes. It has a distinct shape that tapers outward towards the bottom, and is designed to minimize damage and direct vehicles back into the direction of flow of traffic in the event of a collision.

jet A thin stream of liquid or gas which has been forced at high velocity out of a vessel or nozzle.

jet freezing Method of freezing pipes and ground to temporarily halt the flow of water.

jetty 1. Bank or structure that projects into deep water enabling people to access boats. **2.** A projection from a building beyond the face of the main wall.

jib A crane arm that is used to lift and move objects from one place to another.

jib door (gib door) A door that is concealed to match the surrounding wall.

jig 1. A clamp made to a set profile so that wood can be glued or moulded to a desired shape. **2.** A template made so that material can be repeatedly marked or cut to the same shape.

jigsaw A power saw with a thin protruding blade used for cutting curves in timber.

Jim Crow A hand-operated tool, used for bending rails. The tool is shaped in a 'U' or a 'V' to grip the rails, which is connected to a thread rod to provide leverage. The grip can be mounted on rollers to enable continuous bending to be undertaken.

jobber (handyman, builder's mate) A semi-skilled or experienced operative who can undertake most remedial and repair work on a building project, such as plastering, bricklaying, pointing, joinery, plumbing, and minor electrical works.

job site The building or construction site.

job specification Description of the building project, which might be accompanied by material, labour, and plant required.

joggle Up-and-down shape produced by cutting trapezoidal shapes next to each other.

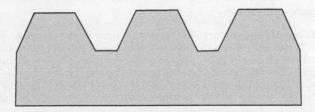

Joggle

joggle joint (concrete stop end, stop end and key) Joint formed at the end of a concrete pour to provide a key for the next pour. A piece of trapezoidal shaped timber is fixed to the *formwork stop end so that a rebate will be formed in the concrete face when the concrete is poured against it.

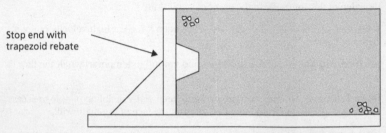

Concrete

Stop end with trapezoid rebate

Joggle joint

Johari's window Management principle that states that there are four ways of viewing human behaviour and knowledge exchange: (a) some aspects are known only to oneself; (b) some aspects are not known by oneself but recognized by others; (c) some aspects are not known by others but known only to the individual; and (d) some aspects are neither recognized by oneself nor others, but do exist.

joiner A skilled tradesperson who works with wood, traditionally at a bench; however, the term is now generally applied to all skilled operatives who work with timber on building projects.

joiner's gauge A tool used for scribing straight lines in timber.

joiner's hammer (Warrington hammer) A hammer with a double head—one with a rounded face and the other with a wedge-shaped head (cross peen head).

joiner's labourer (carpenter's mate) Operative with general knowledge of woodworking equipment and tools such that they can assist with *joinery tasks.

joinery (finish carpentry) Working with timber for furniture, buildings, and structures, and includes assembling, jointing, fitting, and finishing timber.

joint The junction between two components or elements.

joint backing *See* BACKUP STRIP.

Joint Board of Moderators (JBM) Accredits educational programmes to ensure that quality standards are met to develop professional engineers (*see* ENGINEERING COUNCIL) who contribute globally to sustainable economic growth and ethical standards. Formed from four professional bodies, i.e. the *Institution of Civil Engineers (ICE), the Institution of Structural Engineers (*IStructE), the Chartered Institution of Highways and Transportation, and the Institute of Highway Engineers.

joint bolt A bolt used to join two components together.

joint cement (joint filler, joint finish) A plaster-like material that is used in plasterboard dry lining to make joints and cover over nail or screw heads.

Joint Contracts Tribunal *See* JCT.

joint cover Strip of wood, bead, or wood roll used to hide a gap.

jointing Filling and pointing masonry joints in one operation.

jointing compound Malleable putty-like substance that is smoothed around threaded pipework connections to make a water- or gas-tight seal. The seal is formed as one pipe is screwed or located in another, and the substance fills the threads and gaps. Also used to seal other plumbing equipment and appliances. Used in place of *jointing tape. *See* JOINT CEMENT.

jointing fluid Liquid used when making a solvent-welded joint.

jointing mortar Grout used for sealing and finishing tiles.

jointing strip A strip of material used to join two materials.

jointing tape 1. Thin PTFE (polytetrafluoroethylene) tape used around threaded pipework connections to make the joint watertight. Used in place of *jointing compound. **2.** A multi-purpose tape used in plasterboard dry-lining to join boards, reinforce the joint between boards, patch boards, and provide protection to the corners of boards.

jointing tool 1. A tool used to join two components. **2.** A hand-held tool used to produce a *bucket-handle joint in brickwork. **3.** The collective name given to a wide range of hand-held tools used for jointing plasterboard dry-lining and brickwork, such as jointing knives, edging trowels, taping knives, and so on.

jointless flooring A flooring laid as one continuous surface, rather than a number of individual components that are joined together.

joint reinforcement 1. Steel reinforcement used in movement and construction joints. **2.** Steel embedded into horizontal brickwork beds to provide extra tensile strength.

joint ring (rubber ring, sealing ring) Ring of rubber that slides onto the end of clay, plastic, and steel pipes which provides a seal when one pipe is pushed inside another. Push-fit pipes can be sealed with rings of rubber with O profiles ('O' rings), D profiles ('D' rings), and rings that fit on the end of the spigot (lip rings).

joint runner (pouring rope) Asbestos rope or similar fire-resisting rope placed around the outside of a cast iron pipe to act as a guide when forming a lead-caulked joint.

joint tape Adhesive tape, usually 50 mm wide to cover the joints between plaster boards so that they can be easily covered with a light plastered finish.

joist A beam that supports the floor.

joist anchor A tie that fixes the wall to the beam to provide lateral support to the walls, restraining the walls, and preventing bowing.

joist hanger Steel bracket fixed to the internal skin of the wall to fix and secure the floor *joists. The brackets can be fitted with nail holes, which if used, tie the wall to the floor joists, ensuring the walls have greater lateral stability.

joist trimmer (grinder bracket) A special *joist hanger that allows one joist to be fitted at right angles to another joist. The bracket allows one joist to carry another. It is used to form the structural opening for a stairway in timber upper floors.

joule (J) The international unit of energy.

journeyman A craftsperson.

judas A peep hole in a door that allows the occupants to see those who are calling on them.

jumbo brick (US) A brick that has larger dimensions than standard bricks.

jump form Concrete formwork where panels are struck and removed from the bottom, where concrete has previously been cast and lifted to the top, ready for the next pour.

jumping jack A piece of plant used to consolidate ground.

junction 1. The point where two components or elements join together. **2.** The intersection of two or more roads.

junction box A box that contains the connections between various electrical circuits.

just-in-time (JIT) management Producing, making, and delivering only what is needed, when it is needed, and in the amount needed.

jute (*Corchorus capsularis, C. olitorius*) An annual plant with a stem diameter of 20 mm and a stem length of 2.5–3.5 m containing a *bast fibre. Cultivated throughout India, Bangladesh, China, Thailand, Nepal, Indonesia, Burma, Brazil, Vietnam, Taiwan, Africa, Asia, and Central and South America. It requires hot damp climates with annual rainfall >1800 mm with >500 mm occurring during the growing season, high humidity of 70–90%, and temperatures of 20–30°C. Its uses include: ropes, bags, sacks, cloths, and erosion control products such as geo-jute, soil-saver, and anti-wash.

kaolin (China clay) A soft fine white clay with plate-like particles dominated by the hydrated aluminosilicate mineral **kaolinite** $(Al_2Si_2O_5(OH)_4)$. Used for making china, in medicine, and as an inert filler.

Kaplan turbine A water-propelled turbine having adjustable blades that are moved to increase its efficiency; developed in 1913 by Viktor Kaplan.

karst A landscape that has been eroded by the dissolution of soluble bedrock material (such as limestone or dolomite) to produce fissures, sinkholes, and cavities. The surface terrain appears paved and typically lacks vegetation. An example of a **karst topography** is the Burren on the west coast of Ireland.

keel moulding A curved decorative moulding shaped like the keel of a boat.

keep The main tower of a castle, often used as a dungeon.

keeping the gauge (keeping to gauge) A term used in bricklaying where the thickness of the bed joints is maintained over a set area.

keeping the perpends (keeping the perps) A term used in bricklaying where the perpends are maintained in a vertical line.

Kelly (Kelly bar) A hollow bar above a drilling rig used to drive the drilling stem.

Kelvin (K) A unit or scale used to measure temperature. The scale is based on the Celsius scale (°C) but begins at absolute zero (zero Kelvin), which corresponds to −273.15°C. One Kelvin is equivalent to one degree Celsius.

Kennedy's critical velocity The flow of a fluid in an open channel at which it will neither pick up nor deposit silt.

kentledge Heavy material (e.g. large blocks, scrap metal, water tanks, etc.) used to provide weight/stability. Applications include a counterweight on tower cranes, *ballast on a ship, and a reaction point for a jack or sheet pile loading test.

kerb Raised concrete edging between a road and the pavement.

kerf 1. The groove left in a material that is being cut by a saw. **2.** The width of the groove caused by cutting with a saw.

kerfed beam A beam that has been curved by *kerfing.

kerfing The process of curving a material by applying a series of *kerfs.

kerogen A fossilized organic material that occurs in certain sedimentary rocks (e.g. oil shale) and produces a petroleum product when heated.

kerosene A colourless flammable oil, refined from crude oil, used mainly for heating, cooking, and lighting.

Kevlar A trademark for an aromatic polyamide aramid fibre that has good tensile strength and stiffness due to the degree of alignment of modules along the fibre axis during the drawing process, but is weak in compression. Kevlar 49 is currently the most widely used aramid fibre for the reinforcement of plastics.

key 1. A device, usually metal, that is inserted into a lock to open or close it. **2.** A written explanation of the various abbreviations and graphical symbols used on a drawing. **3.** A rough surface that provides a bond for another material, such as paint, plaster, or render.

key changes (key combinations, differs) The total number of keys that can be made to operate a particular type of lock.

key drop See KEYHOLE COVER.

keyed 1. Held in position using a key. **2.** Making a surface rough enough for bonding. *See* KEY.

keyed beam A notched lap jointed beam.

keyed construction joint *See* JOGGLE JOINT.

keyed joint 1. A notched joint between two materials or components. **2.** A joint in brick or blockwork that has been raked out to provide a key for wet plaster.

keyed mortise and tenon *See* TUSK TENON.

key event An important or critical point in a plan. *See also* MILESTONE.

keyhole cover (key drop) A flap, usually round, that is attached to the escutcheon and covers the keyhole on the inside and/or outside of a lock.

keying The process of roughing or scratching the surface of a material so that another material will bond to it.

keying-in The process of joining a new masonry wall to an existing wall.

keying mix A mix applied to a surface to enable another material to bond to it.

key plan A plan showing the location of the development.

key plate *See* ESCUTCHEON.

key saw (compass saw) A hand-held saw, with a small blade that tapers to a point, used for cutting tight curves. The narrow blade can be used to cut out holes in the centre of a piece of timber. A large hole is first drilled into the timber. The saw can then be inserted to cut away from the hole.

key schedule A list of all of the keys and their related rooms, cupboards, and other lockable units. The list is organized into key suites and associated master keys.

keystone The central wedge-shaped stone at the top of a masonry arch.

key suite A set of related keys listed on a key schedule. The suite of keys may be grouped into floors of a building, particular offices, or rooms.

keyway miller A powered hand-held mechanical device for chasing out grooves and rebates in brick and block walls so that service ducts can be set in the walls.

kicker The bottom section of a column that is constructed first, to support the shuttering and/or reinforcement for the remainder of the column.

kicker formwork (kicker frame) The *formwork used to construct the *kicker of a column.

kick plate (kicking plate) A metal plate attached to the bottom of a door to protect it against being kicked.

killed steel Steel that is treated with a strong deoxidizing agent, such as silicon or aluminium, which reduces the oxygen content to a level where no reaction occurs between carbon and oxygen during the solidification process.

kiln An oven or furnace that is used to burn, dry, or fire items such as bricks, ceramics, glass, and timber.

kiln seasoning (kiln drying) A drying process predominantly utilized for reducing the moisture content in timber. The process is carried out in a closed chamber, providing maximum control of air circulation, humidity, and temperature. Drying can be carefully regulated so that shrinkage occurs with the minimum of degradation problems (cracking and splitting). One advantage of kiln seasoning is that lower moisture contents are possible than can be achieved using *air-seasoning. Other advantages include rapidity of turnaround, adaptability, and precision. It is the only way to season timber intended for interior use, where required moisture contents may be as low as 10% or less. It is necessary to regulate kiln drying to suit the particular circumstances, i.e. different species and different sizes of timber require drying at different rates.

kilogramme (kg) The international unit of mass. Equivalent to 1000 grammes.

kilojoule (kJ) The international unit of energy. Equivalent to 1000 joules. *See also* JOULE.

kilometre (km) The international unit of length. Equivalent to 1000 metres. *See also* METRES.

kilonewton (kN) The international unit of force. Equivalent to 1000 newtons. *See also* NEWTONS.

kilowatt hour (kWh) The basic unit of electrical energy, equivalent to 1 joule per second. *See also* JOULE.

kinematics The study of the motion of bodies without considering forces or masses, as opposed to *dynamics.

kinetic energy The amount of energy that an object possesses due to its motion. Equivalent to half of the mass of the object times the square of its speed. Measured in *watts.

king closer A brick closer with a diagonal slope cut from half-way up the *header face to half-way along the *stretcher face. Closer bricks are used to make up the corners, close reveals, and fill in the gaps and joints, often used in and around window and door openings. The brick may also be called a **three-quarter brick** as three-quarters of the brick remains.

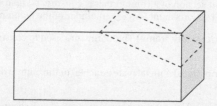

King closer

king post The vertical strut in a traditional *king post truss.

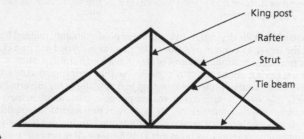

King post

king post truss A traditional timber roof truss consisting of two rafters and a tie beam with a central vertical post (a king post), running from the centre of the tie beam to the apex, and two diagonal struts. Traditional roof construction has now largely been replaced by prefabricated roof trusses.

Kirchoff's law A fundamental law relating to circuits, that states the algebraic sum of all the voltages around a closed path is zero, i.e. the sum of the voltage drops equals the total source voltage.

Kitemark The British Standards Institution trademark, used by manufacturers on their products to show that they comply with the relevant *British Standard.

kite winder A triangular- (kite) shaped stair tread (step) used to turn corners without the need for a landing.

kit home A dwelling constructed on-site from prefabricated parts assembled in a factory.

kit joinery Prefabricated joinery components that are assembled or installed on-site, such as door sets, shelving units, windows, etc.

knapping hammer A masonry hammer for shaping building stones.

knee The convex bend in a handrail found at the top of a flight of stairs where the handrail enters the newel post. *See also* KNEELING.

knee brace (corner brace) A short support that runs diagonally between the post and beam to strengthen the joint.

knee elbow A right-angled pipework bend.

kneeler (skew, skew table) The angled masonry on a *gable wall that carries the *gable coping.

kneeling The concave bend in a handrail found at the bottom of a flight of stairs where the handrail enters the newel post. *See also* KNEE.

knob A round-shaped handle on a door that is used to open or close it.

knob set A pair of internal and external door *knobs.

knocked down Building components that have been cut and shaped ready to be assembled on-site.

knockings A defect in the surface of concrete caused by chips to the exposed faces.

knocking up (retempering) Reworking mortar or concrete that has already started to set. The mortar or concrete may have to be forcefully moved around or have extra water added to make it workable. In practice, concrete or mortar that has started to set should not be reworked then used, as the final strength of the mortar will be significantly reduced.

knockout Parts of a unit that are made so that they can be pushed or knocked out if needed, for example electrical boxes with perforated parts so that they can be pushed out to allow wires and cables in and out.

knoll A small natural rounded hill or mound.

Knoop hardness test An indentation test for ascertaining a material's resistance to surface indentation or scratching. All hardness tests measure how easily diamond can produce an indentation on a material. The Knoop hardness method uses calibrated machines to force a rhombic-based pyramidal diamond indenter, having specified edge angles, into the surface of the test material. By measuring the indentation depth, the hardness value of the material can be determined.

knot The circular dense grain that appears in cuts of timber. The change in grain is the result of branches growing out from the main trunk of the wood. Knots reduce the strength of the wood.

knot area ratio (KAR) The KAR is a part of the visual grading process of timber to identify the number of knots in an area of wood. Timber is graded based on the number of defects it has and is given a grade identifying its potential structural use.

(⊕) SEE WEB LINKS

• The national timber industry website of Australia (Forest & Wood Products Australia).

knot brush A thick paint brush in a round shape for painting walls.

knowledge transfer partnerships (KTP) Launched in 2003, it is a scheme to link industry to academia. It jointly integrates the development of knowledge and implementation of new technology to help industry meet its core strategic needs.

knuckle The cylindrical holes in a hinge through which the pin passes.

knuckle bend A sharp turn in a pipe.

knuckle joint A sharp bend or joint between the two different slopes of a Mansard roof.

K value (K-value, k-value, k value) *See* THERMAL CONDUCTIVITY.

label A drip or mould providing a projection out from the wall over the top of an opening.

laboratory A room equipped with testing equipment, where scientific testing is carried out or taught.

labour The human resource associated with a task or job.

labourer A person employed to undertake and assist with building work, but does not have a skill, trade, or profession.

labour only Description of a worker who is not a direct employee, but is contracted to work on a subcontract basis. The term 'labour' only implies that the person is contracted to complete the work, however, the materials and equipment are to be supplied by the employer or main contractor.

labyrinth A complex interconnected series of maze-like paths or tunnels.

laccolith A sheet formation of intrusive igneous rock which has been injected between two layers of sedimentary rock. The pressure of the magma causes the overlying strata to be pushed upwards forming a dome or mushroom-like ceiling with a normally planar base.

laced valley A *valley created using slates or tiles where the courses run from one roof slope to the next. They do not contain a *valley gutter. Also known as a **swept valley**.

lacer A reinforcing bar that is tied at right angles to the direction of the main reinforcement to create a mesh.

lacing 1. The interweaving of roof slates at an intersection to form a *laced valley. **2.** Tying together the horizontal members in formwork to improve their stability.

lack of cover Inadequate or no concrete around steel reinforcement. For reinforced concrete to act as a matrix structure there must be an adequate level of concrete surrounding the reinforcement. Where the concrete and steel are properly bonded together and positioned, the tensile forces can be transferred to the steel and the compressive forces to the concrete. A layer of concrete over the top of reinforcement also provides protection against corrosion and the effects of fire. Exposed reinforcement is not afforded any protection and does not form a proper bond with the concrete.

ladder A piece of equipment with horizontal rungs encased by two vertical styles that is used to facilitate vertical movement. Ladders can be movable and used for jobs such as cleaning windows and accessing heights for maintenance purposes and inspections, or can be permanently secured to a structure to provide safe, easy access whenever required.

SEE WEB LINKS

• Guidance on the safe use of ladders can be found on the Health and Safety Executive website.

lag bolt (coach bolt) A heavy-duty fixing with a large square head. As the bolt is screwed into the wood, the square head imbeds itself into the surface of the timber, preventing it unscrewing.

lagging 1. Thermal insulation used to wrap around hot surfaces, such as pipes, boilers, and hot-water tanks to prevent heat loss. Also used to prevent heat gain and condensation on cold pipes and surfaces. **2.** A timber frame used to support the sides when constructing an arch.

lagoon 1. A small lake. **2.** An enclosed area of shallow sea water separated from the sea by land. **3.** A shallow pool containing contaminated water or effluent.

lahar A mudflow of volcanic ash, typically occurring after heavy rains.

lake 1. A large area of freshwater. **2.** A pigment that is produced using aluminium hydroxide or another similar insoluble base.

lamella A type of morphology present in metal surfaces when viewed at a microscopic level; usually present in steel alloys.

laminar flow A slow smooth flow in a liquid or gas where adjacent layers move at different velocities without mixing.

laminate (decorative laminate, laminated plastic) An object consisting of a number of layers (laminae) or plies, bonded together. Laminates may be symmetrical or asymmetrical. A symmetrical laminate is one having a mirror plane lying in the plane of the laminate at its mid thickness.

laminated glass A special type of glass that is a combination of layers, and provides a generic material with improved toughness utilized in structures such as balconies, shop fronts, roofs, and the like.

lamination The process of creating *laminated glass.

lamp 1. An electrical device used to provide artificial light. **2.** A lightbulb.

lamp base (lamp cap, lamp holder) The portion of a lightbulb that provides support for the bulb and is used to connect it into place. A variety of bases are available, such as metal, plastic, or glass that incorporate screw, bayonet, pin, or wedge-shaped fittings.

lancet Architectural style that makes use of shapes that resemble a lance.

land drain *See* FIELD DRAIN.

land drain systems Systems to allow excess water within the land to be drained away, rendering the land more usable or fit for a defined purpose, e.g. a playing field, football pitch.

landfill The disposal of waste material by burial.

landing 1. Flat platform at the top and bottom of a flight of stairs, escalator, or a ramp. Allows people to stop, pass, rest, and change direction. A landing between two flights of stairs is called an **intermediate landing**. **2.** A place where a lift comes to rest, and allows passengers to safely enter or depart from the carriage.

landing button The switch (button) used to summon a lift. In multistorey buildings where there is opportunity to ascend or descend, there will be two buttons to indicate the desired direction of travel.

landing valve A *valve attached to the rising main that enables connections to be made to the water supply for fire-fighting purposes. Usually located in the fire cabinet on each floor of a tall building.

landscaped office An open office where the furniture, screens, equipment, plant, and decorations can be moved around to the desired layout.

landscaping The reshaping or restructuring of the external built and natural environment, includes the alteration of existing land, plants, footpaths, roads etc. to provide improved aesthetics.

landslide (landslip) A downwards mass movement of soil or rock, which occurs when the disturbing forces (weight of soil and effects of gravity) exceed the resisting forces (shear strength of the soil) along a potential slip plane. This can be as a result of unsuitable geometry, a change in the groundwater regime, presence of weak planes, bands, or layers, progressive deformation, increase in effective slope height, or additional imposed loading. The four main forms of landslips are falls, transitional slides, rotational slips, and flows.

land survey A topographical survey of the land to document various points by distances and angles. It is undertaken by a **land surveyor**, who is a member of *RICS (Royal Institute of Chartered Surveyors).

lantern 1. A small square, octagonal, or dome-shaped structure found at the top of a roof or dome. Used to admit daylight, provide ventilation, and as a lookout. **2.** A portable lighting device comprising a light source encased within a transparent or translucent cover.

lap (overlap, lapping) 1. The distance that one material overlaps another. **2.** A weather-tight joint used in profiled sheeting and roof tiling. **3.** The thickened film of paint that results when one coat of paint overlaps another.

lap joint (lapped joint) A joint formed between two components, where one component overlaps and joins with the other.

Laplace's equation A second-order partial differential equation:

$$\nabla^2 \phi = 0$$

where ∇^2 = the Laplace operator and ϕ = a scalar function.

Laplace transform An integral transform, used to transform a function into another function.

lapped tenons Two *tenons that overlap one another within a mortise.

lapping A process used to achieve a smooth polished surface.

larder A storeroom for food; historically, it was a room used for the storage of meat.

large-panel system A precast concrete panel cladding system where the panels span the full length between the structural columns of the building.

larrying (US larrying up) A method of constructing brick walling by sliding bricks into position, then filling the joints with a fluid mortar that is poured between the joints.

laser An acronym for light amplification by stimulated emission of radiation. Lasers are a coherent source of light, usually in the form of straight rays.

laser soldering A type of soldering process using laser light.

lashing The tying of cable or rope around the top of a ladder to ensure that it is secured in position.

latch A lock or catch with a sloped or bevelled tongue that is secured into a mortise as a door is pushed shut. The tongue is bevelled so that as it strikes the doorframe, the slope of the tongue pushes the catch back into the door. The bevel enables it to slide over the frame until it is over the mortise where a spring pushes the catch securely into place. The tongue is removed from the mortise and the door opened by twisting a handle that is connected to the tongue; this releases the catch allowing the door to be opened.

latchet A piece of metal that is used to secure or hold something in place. Generally latchets or tingles are made from folded metal strips that are fixed over the corner of a panel that it is holding. Part of the tingle is fixed to the substructure while the other part laps over the panel that it is securing in place. *See also* TINGLE.

latent heat The energy released when there is a change in state, for example, from solid to liquid.

lateral Sideways or at the side, as in **lateral movement**, referring to sideways movement rather than vertical or downwards movement.

lateral restraints 1. Structural ties, cables, walls, or members that provide resistance to sideways horizontal movement. **2.** Horizontal ties, rods, and cables used to tie the walls of the building together, stopping them splaying apart.

latest date In a logical or precedent network, the latest time an activity or event can be started or finished (latest start time or latest finish time) without delaying subsequent activities and the planned completion date.

latex A polymeric material with rubber-like mechanical properties. Latex refers generically to a stable dispersion (emulsion) of polymer microparticles in an aqueous medium. Latex may be natural or synthetic.

latitude 1. An imaginary line denoting points around the earth's surface of equal distance north or south of the equator; *see also* LONGITUDE. **2.** The distance from the origin in a traverse; *northings are positive, *southings are negative.

lattice 1. At a microscopic level in metals, the space arrangement of atoms in a crystal. **2.** A criss-cross framework.

lattice parameter At a microscopic level in metals, the length of any side in a crystal structure's unit cell.

lattice window A window containing small diamond-shaped panes of glass that are fixed in position using strips of metal.

laundry A place or room for the washing, cleaning, and drying of clothes and linen.

lava Molten rock that has erupted from a volcano or fissure within the ground.

lavatory 1. A room where the toilet, or WC (water closet), or urinal is installed. **2.** Another name for a toilet or the receptacle for foul water.

lay bar A horizontal *glazing bar.

layer board (lear board, gutter board) The board that forms the base onto which the felt or sheet metal for a *box gutter is laid.

laying-off Applying very fine, light strokes to a surface containing wet paint to remove any brush or roller marks, and provide a smooth surface.

laying-on trowel A plasterer's trowel, which is square in shape and used to place, smooth, and finish plaster on a wall.

laying trowel A bricklayer's trowel.

laying-up (laminating) To *laminate.

lay light (laylight) A horizontal window in a ceiling **(ceiling light)** or a roof **(roof light)**.

layout The plan positioning of a building. Layout drawings are produced that show the planned position of the building, walls, furniture, and other items that can be seen on a plan.

lay panel A panel that is longer horizontally than vertically.

leach To remove something slowly, typically relating to liquids. **Leachate** is the liquid from the decomposing waste in a *landfill.

leaching The process of liquid solder dissolving a metal coating.

lead A non-ferrous metal, chemical formula Pb, that is malleable, ductile, and dense. Lead is commonly used in construction in sheet form for roofing, flashings, and lead sheet cladding. Lead is dense and heavy (density 11,300 kg/m^3), but it is quite corrosion-resistant. It can be bent and folded repeatedly without failing, which is useful when forming flashings for roofing purposes. It melts at a very low temperature of 327°C. It is also alloyed with tin to form *solder; this is used mainly for making electrical and plumbing joints.

lead-caulked joint (lead joint) A type of joint used in cast-iron pipework where molten lead is poured in to seal and join the pipes.

lead-clothed glazing bar (lead glazing bar) A steel *glazing bar covered in lead.

leaded light (lead glazing) A decorative window comprising a number of small glass panes, usually rectangular or diamond shaped, that are held together with strips of lead. *Stained-glass windows are an example of leaded lights.

leader head (US) *See* RAINWATER HEAD.

lead flat *See* LEAD ROOF.

lead glazing *See* LEADED LIGHT.

lead glazing bar *See* LEAD-CLOTHED GLAZING BAR.

leading hand Person in charge of a group of tradespeople; a *chargehand.

lead joint *See* LEAD-CAULKED JOINT.

lead monoxide A poisonous insoluble oxide of lead existing in red and yellow forms; used in making glass, glazes, and cements, and as a pigment. Chemical formula PbO, also known as lead (II) oxide or litharge plumbous oxide.

lead plug A small hole drilled into a wall that is filled with lead and used as a fixing for a screw, nail, or other fastener.

lead roof (lead flat) A flat roof covered in sheet lead.

lead sheath A covering of lead around a steel bar. An example is a *lead-clothed glazing bar.

lead slate (leadslate) A lead flashing that provides a weatherproof joint where a pipe, such as a flue or soil stack, penetrates a roof.

lead time The time between the placing of an order and the delivery of materials, plant, units, or equipment to site. It is essential to be aware of how long it takes between placing an order and the item arriving on-site. Where changes are made during the construction phase and items are changed which have long lead times, considerable delays to the project may be experienced. Lead times can be affected by global demand, reduction in the availability of raw materials, and supply chain economics.

leaf 1. A *door leaf. **2.** One of the two main layers of a cavity wall. The main internal layer is known as the **inner leaf**, and the main outer layer is known as the **external leaf**.

leaf frame (leaf structure, framing) The structural components used to construct a *door leaf, namely the *door rails, *door stiles, and the *facings.

leakage coefficient The *air leakage rate of a building at an artificially induced pressure of 1Pa.

lean construction Use of production management principles, with no waste (time or resource). Current practice has the goal of better meeting customer needs while using less of everything, whereas lean construction takes a production approach to fully control the process and energy and resources that go into it. The result is a new project delivery system that can be applied to any kind of construction but is particularly suited for complex projects. Lean construction addresses uncertain and quick projects, has a clear set of objectives for the delivery process, is aimed at maximizing performance for the customer at the project level, designs concurrently product and process, and applies production control throughout the life of the project.

lean lime Lime with impurities such as magnesium and silica compounds contained within it.

lean production Optimization of performance against a standard of perfection to meet unique customer requirements with minimal resources and time. It is designed to make things efficiently with no waste, and is differentiated from mass and craft forms of production by the objectives and techniques applied on the shop floor, in design, and along the supply chains.

lean-to (lean-to roof, half-span roof) A mono-pitched roof that is supported at its highest point by a wall that is higher than the top of the roof.

leasehold Where the owner has bought the right to occupy or use a property or land for a given period of time; this is different to a freehold, where ownership is purchased outright.

least squares A mathematical method for plotting a best fit curve through a set of data points, by minimizing the sum of the squares of the deviations from the expected values, such that deviations are regarded as a continuous differentiable quantity.

least work A principle that applies to structures with a linear stress–strain response, and states that under an applied stress, deflections are such that the energy stored in elastic members is minimum.

leat A channel that diverts water to a watermill to provide power.

ledge 1. A horizontal platform. **2.** The horizontal platform at the foot of a window. **3.** The horizontal timber used to hold the panels in a matchboard door together.

ledged and braced door *See* BATTEN DOOR.

ledger 1. A horizontal beam fixed to a wall that supports the ends of the floor joists. **2.** A horizontal scaffolding pole used to support *putlogs.

Legionnaires' disease A severe and potentially fatal form of pneumonia that is caused by the bacteria *Legionella pneumophila*. It is contracted by inhaling tiny water droplets contaminated with the bacteria. The bacteria is commonly found in rivers and ponds, but can also grow in other water systems such as centralized air-conditioning systems, cooling towers, evaporative condensers, hot-water systems, ornamental fountains, room air humidifiers, showers, and whirlpool spas. The *Legionella* bacteria grows in water at temperatures from 20°C to 50°C and is killed by high temperatures.

legislation Parliamentary Acts that provide the legal rules that must be applied. The Health and Safety at Work Act, Building Act, and Factories Act are all examples of acts of Parliament that are applied to the industry. European Acts, legislative acts of the European Union, such as Construction Management Directives are also applied with interpretation coming from regulations, such as the Construction, Design, and Management Regulation.

lengthening joint (heading joint) A joint made between two materials that are joined end to end.

let in (sunk in) Fixed flush with the surface.

levée An embankment adjacent to a watercourse. An **artificial levée** is built to prevent flooding; a **natural levée** is formed from sediment during flooding.

level 1. An instrument used to provide a level line of sight so that the relative levels of the ground, parts of the structure, and buildings can be determined. There are different types of level: automatic level, tilting level, and laser level. **2.** A straight edge with a clear glass tube, containing liquid and an air bubble. The tube is set parallel with the straight edge, so that when the bubble floats in the centre of the tube, the straight edge is horizontal. **3.** When the ground is graded or moved by bulldozers so that it is horizontal and flat.

levelled finish A relatively smooth flat finish achieved using a float or trowel.

levelling. 1. Moving material around so that it is flattened. **2.** Moving earth or other material around so that it is horizontal. **3.** The act of taking readings using a tilting or automatic level to determine the height of objects and land, or to work out the profile of the land.

levelling rule A large straight edge, normally at least 3 m long, with a spirit level set in it, used for levelling screeds.

level of detail (LoDT) The description of the graphical contact of a model at each stage, i.e. schematic, concept, and defined. *See also* LEVEL OF INFORMATION.

level of information (LoI) The description of non-graphical information on a BIM model, such as spatial, performance, standard, workmanship, certification, etc. *See also* LEVEL OF DETAIL.

level-tube A tool to determine if an object is parallel to the ground, it has a transparent cylinder containing liquid with an air bubble trapped in it.

lever 1. A simple machine comprising of a rigid bar that is able to pivot about a fixed point. It is capable of multiplying the force that can be applied to an object. **2.** A flat metal tumbler in a lever lock.

lever arm The distance a force rotates about. *See also* MOMENT.

lever boards Adjustable wooden *louvres.

lever handle A horizontal lever attached to a door that operates a *latch. An alternative to a door knob.

lever lock A lock in which the key lifts a number of levers to operate the lock.

lever mixer (lever-operated thermostatic mixing valve) A mixer tap operated by a single lever. The water flow can be controlled by lifting the lever up or down, and the water temperature can be controlled by moving the lever left or right.

lever rule A rule utilized to calculate the percentage of a chemical phase in a metal at a specific temperature, when two different phases are present. This is calculated by reference to a binary alloy phase diagram.

levy A tax that is applied to construction contractors to fund training and education for building employees. The money is distributed through the Construction Industry Training Board (CITB) or Construction Skills.

lewis A mechanical device that fits into a dovetail hole and expands. Used for lifting stone or concrete blocks.

liability The consequences of failing to deliver legal obligations under contract or statutory instrument. Persons or organizations have legal responsibilities when operating under a contract or within the legislation that applies to operations undertaken; they will be liable for actions that breach any term or regulation.

library Room specifically for the storage and reading of books.

lid A cover placed on top of a vessel, bin, or lift shaft.

LiDAR (light detection and ranging) A remote-sensing method to measure distance by using light in the form of a pulse laser to illuminate a target and analysing the reflected light. The LiDAR instrument, which consists of a laser, scanner, and specialized GPS receiver, is normally attached to an aeroplane, helicopter, or *drone which is flown over the large areas of terrain to be surveyed. Precise three-dimensional information can then be generated from the data. Topographical surveys use a near-infrared laser to measure distances, whilst a bathymetric LiDAR employs water-penetrating green light to determine the depths of riverbeds or the seafloor.

life-cycle cost analysis (LCCA) A procedure to assess the cost-effectiveness of purchasing, owning, operating, maintaining, and disposing of an object or *asset.

lift (US elevator) A mechanical device that uses a *lift car to transport goods and people vertically from one floor to another.

lift and force pump A *displacement pump where the reciprocating action of the piston within the cylinder imparts movement to the water.

lift car The enclosed part of a lift that moves vertically from one floor to another.

lift controls (US elevonics) The electronic controls used by passengers to control and operate a lift.

lift drive The mechanical device that moves the *lift car. Various different types of drive are available, including electrical motors that operate counterweights, ropes, and hydraulic pumps.

lifting beam A straight steel beam or frame that is attached to a crane's hook, and has a number of fixings hanging from the length of the beam to attach onto the object being lifted. The beam is used for lifting long objects such as beams, wall panels, cladding panels, and so on.

lifting equipment Chains, hooks, slings, clips, and plant used by cranes in order to hoist, elevate, or lift items from the floor.

lifting eye A steel ring used to secure clips, hooks, and slings when performing lifting operations with a crane or other lifting devices.

lifting frame A piece of equipment that is attached to a crane's lifting hook to assist the lifting of long or wide objects. The frame is often built to lift certain objects, such as cladding panels, large beams, columns, formwork, and so on.

lift machine room (lift motor room, LMR) The *plantroom that contains all the equipment required to operate the lift. Usually located at the top of the *lift shaft.

lift-off hinge (loose butt hinge) A hinge where the top and bottom leaf can be slid apart. When the hinge is fixed to a door, the door can be simply lifted on and off the part of the hinge that is fixed to the door casing. When the hinge is coupled together, the hinge operates as normal.

lift pit (runby pit) The space at the bottom of the lift shaft that extends below the lowest floor level. In hydraulic lifts the pit will house the pump and the ram.

lift safety gear Safety equipment that is designed to ensure that the lift operates safely.

lift shaft (lift well, US hoistway) A vertical shaft within a building that houses a *lift. Usually incorporates a *lift machine room at the top of the lift shaft and a *lift pit at the bottom.

lift shaft module A prefabricated portion of a *lift shaft.

ligature 1. Reinforcement binders fixed around the main bars of reinforcement to form a cage. *See* STIRRUP. **2.** Slang term used to describe a wire, band, or other device that can close something off, for instance, restrict or close the movement of fluid along a pipe, like a tourniquet device.

light 1. A *window. If it can be opened, it is referred to as an **opening light**, otherwise it is known as a **fixed light**. **2.** Electromagnetic radiation that can be detected by the human eye. **3.** A source of illumination, either natural or artificial.

light alloys A low-density metal such as *aluminium, *magnesium, *titanium, beryllium, or their alloys. These type of alloys are extensively utilized as they combine high strength with low specific density.

light fitting *See* LUMINAIRE.

lighthouse A tall round tower building with a powerful flashing light, which is located on hazardous coastal regions to warn sailors of dangerous rocks.

lighting 1. The process of providing light, either artificial or natural, to a space. **2.** The electrical equipment used to provide light.

lighting bollard A low post that protrudes from the ground and incorporates a *luminaire. *See also* LIGHTING COLUMN.

lighting column A tall post that protrudes from the ground and incorporates a *luminaire. *See also* LIGHTING BOLLARD.

lighting control panel A *distribution board for all of the lighting circuits contained within a building. Can also be used to programme various lighting circuits.

lighting fitting *See* LUMINAIRE.

lighting installation All of the electrical cables and equipment required to provide artificial lighting within a building.

lighting point The point in a building where a *luminaire is to be installed.

light-loss factor (LLF, US dirt-depreciation factor) A factor used to calculate the luminance of a lamp at a particular period of time, usually just before relamping. It takes into account factors such as the age of the lamp, dirt accumulation on the luminaire, the frequency and effectiveness of cleaning, maintenance regime, and the age of the ballast. Formerly known as the **maintenance factor**.

lightness (value tone) The perceived intensity of light reflected by a colour.

lightning arrestor (surge protector) A device that is used to protect telecommunications and power lines from a lightning strike.

lightning conductor (US lightning rod) A thick strip of metal, usually copper, that runs from the highest point on a building to the ground. Protects the building from a lightning strike by transmitting any lightning that does strike the building to the ground.

lightning protection system A system installed on a building to protect it against a lightning strike. Usually comprises a series of *lightning conductors and *ground terminals.

lightning rod (US) *See* LIGHTNING CONDUCTOR.

light switch A switch, usually located on a wall, that is used to operate electric lights.

light well (lightwell, air shaft) An open space that usually extends several floors down a building that admits daylight and provides ventilation.

Lignacite An enterprise specializing in supplying concrete masonry products to the construction industry.

lignin An organic polymer that binds plant fibres together, giving them rigidity and resistance to decay.

lime The common name for calcium oxide (CaO). Lime is most commonly used in mortar along with sand and cement. In mortar, lime has a plasticizing effect, which improves the workability. Lime is also used as hydraulic lime in mortars, specifically termed 'lime mortar'. Hydraulic lime is a useful building material both as mortar and for conservation of historical properties.

lime putty A type of *putty made from lime (calcium carbonate or calcium oxide).

limestone A sedimentary rock rich in calcium carbonate. Used as dimension and aggregate stone, it is also burnt with clay to make cement. It can dissolve in rainwater to form *karst and sinkholes in the ground.

limit of works The point at which one contractor's or designer's work ends and another's begins. For example, where the main mechanical and electrical subcontractor's work package ends and the supplier of specialist electrical equipment begins, or where the builder's work ends and the plumber takes over. These points are particularly important as the contract's description must ensure that the interface between works covers all the work required. Where the gap between the limit of work is vague, claims and disputes over additional work are common.

limit state design (LSD) A design philosophy used in the *Eurocodes that considers the condition of a structure beyond which it can no longer fulfil the relevant design criteria, such as fit for use, structural integrity, durability, etc. There are two limit states that need to be considered, *ultimate limit state (ULS) and *serviceability limit state (SLS).

limpet washer A steel or plastic washer that is bent so that it can fit over the curve of a profiled metal or composite roof sheet. The bolt simply passes through the washer with the washer sitting in contact with the profiled roof.

line (bricklayer's line, builder's line, string line) Cord or string pulled between two points used for setting out straight lines in building works. The line can be securely fixed to profile boards, steel pins, dowels, or line pins, depending on the operation.

lineal measure (run) The length of something.

lineament A topographical feature, which is a result of an underlying geological occurrence, such as a fault.

line and level (line level) A spirit level suspended at the centre of a taut horizontal string line. With skilled use, the level can measure accurately to +/- 1 mm per metre.

linear Relating to a straight line.

linear cover The width of a tile minus any side lap.

linear diffuser (slot diffuser, strip diffuser) A *diffuser that comprises a number of slots through which air can be diffused.

linear polymers Polymer molecules containing a long chain of atoms, called the backbone. Normally, some of the atoms in the chain will have small chains (pendant groups) attached. The majority of polymers are linear polymers as this is the simplest and cheapest type of polymer to manufacture.

linear thermal bridge Discontinuities in the *thermal envelope that occur along a particular length of the envelope, such as at junctions between various elements. Examples include, at lintels, jambs, and sills, wind posts, and at junctions between various different elements, e.g. where external walls meet the floor. *See also* POINT THERMAL BRIDGE.

lined eaves (closed eaves) *Eaves that contain a *soffit board.

linen chute (laundry chute) A *chute that is used to collect dirty laundry.

line pins Steel pins, 80–100 mm long, which are inserted into mortar joints at each end of a wall and used to hold a *bricklayer's line. The bricklayer first builds up square and level corners of the wall. The pins are then inserted into the mortar joints and a builder's string line pulled taut between them. The bricks in the panel between the corners can then be laid level between the two corners.

line-tapped connection A direct connection that is made to an electrical cable without having to disconnect it.

lining 1. First cover or layer of paper, paint, or plaster that ensures a better ground for the finish. **2.** The internal surround of a door or window to cover the reveal.

lining paper The first cover of paper pasted on the wall to act as a base for the wallpaper.

link 1. Ring which connects to another ring to make a chain. **2.** Reinforcement bars that bend to the profile of the desired cage, spaced at regular intervals, and to which the main reinforcement is attached. The main reinforcement is normally attached on the inside of the link bars. The link bars appear to wrap around the main reinforcement.

linoleum (lino) A floor covering made from linseed oil (linoxyl). This type of covering is highly durable and is very commonly used in hospitals and clinics.

lintel (lintol) A horizontal beam that spans an opening to support the load from the structure above. Used above doors and windows. *See also* LINTEL DAMP-COURSE

lintel damp course (combined lintel) A lintel that is shaped to act as a *damp-proof course for a cavity wall.

lipping (banding, edging strip) A thin strip of timber fixed to the edge of a *door leaf to cover the *door face, and protect the door core.

lip-seal joint A joint used for push-fit pipes where the socket end of the pipe contains an inward-facing projection of rubber which seals the pipe when the spigot end of a pipe is inserted.

liquefacture Transforming into a liquid. Occurs in saturated cohesionless soils when the pore pressure is increased and the effective stress is reduced, resulting in overall reduction in shear stress. *See also* QUICKSAND.

liquefied gas A gas that has been converted into a liquid by pressurizing it and/or lowering its temperature. Examples of liquefied gas include butane and propane.

liquid A state of matter whereby a material is in fluid form.

liquidated damages An amount of money that is stated in the contract as the sum which would be payable in the event of a specified breach; the amount must be a realistic estimate of the loss that is likely to occur as a result of the said breach. The purpose of the term is to compensate and not act as a penalty. If the sum specified is excessive, then it is a penalty, and cannot be claimed.

liquidity index (LI) The ratio of the difference in *moisture content (w) and *plastic limit (PL) of a fine-grained soil to its *plasticity index (PI):

$$LI = (w - PL)/PI$$

liquid limit (LL) The point at which the water content in soil changes it from plastic to liquid behaviour. It is an *empirical boundary defined as the *moisture content at which the soil is assumed to flow under its own weight and is measured using a *cone penetrometer or Casagrande cup (see CASAGRANDE, ARTHUR).

liquid metal embrittlement A process whereby a metal becomes more *brittle when in contact with a liquid.

liquidus A line indicating the temperature at which a metal alloy of a specific composition will start to melt.

listed building A building that is considered to be of special architectural or historical interest is placed on a list by the Secretary of State for the Environment, and protected from alteration. Such buildings cannot be altered unless listed building consent has been obtained from the local planning authority.

lithification The processes by which sediment form *sedimentary rock. These include cementation, solution, repositioning, and compaction.

lithosphere The rigid outer layer of the earth's crust and mantle.

litigation The progressing of a dispute to a court of law. Litigation differs from adjudication in that the decision of the courts is final and binding, and evidence is given under oath; however, the court's process is slow and costly.

litre (l) The SI unit of fluid measurement. One thousand litres is equal to one cubic metre.

little joiner Small pellets, dowels, or pieces of wood that are used to fill, hide, and make good holes.

live (live wire, live conductor) The red sheathed conductor in an electrical cable that carries an electrical current.

live edge (wet edge) The wet edge of a fresh coat of paint.

live load (imposed load) A load that can be removed or replaced (for example, furniture in a building); see also DEAD LOADS.

live main A gas or water main that is transporting gas or water under pressure.

live tapping The process of connecting a pipe, such as a *communication pipe, to a *water main while it is live.

load A force or weight on a structure. See also LIVE LOAD and DEAD LOAD.

load-bearing Carrying load, for example, a *load-bearing wall.

load-bearing construction A building component or element that acts as a structural element, providing support or resistance, rather than providing an aesthetic function, divider, or other feature.

load-bearing wall A wall that offers support to other structural components such as floors and roofs, rather than just acting as a *partition. The load can be vertical, lateral, or as part of a bracing system to provide lateral restraint, preventing structural frames buckling. The wall is a longitudinal structure which supports other elements of the building, such as (and most commonly) the roof, upper wall, and floors.

loader shovel Tracked or wheeled earth-moving equipment, used for loading and transporting bulk materials such as soil, rock, and aggregate into dumper trucks.

loading 1. Something being added or carried by something else. **2.** A *filler or *pigment added to paint. **3.** The use of *ballast to prevent uplift. **4.** A pipe filled with sand to prevent distortion during bending. **5.** Forces imposed on a structure.

loading coat (loading slab) A concrete slab that is placed on top of asphalt or other tanking materials to resist hydrostatic forces or wind pressure which may otherwise force the waterproof material from its substrata.

load path The route a force takes through a structure, from the point of application to the point of exit; for example, a *live load could be transmitted through a floor slab, beams, and columns to the foundation, where it is then distributed into the ground.

load schedule A record to show the distribution of power across a number of electrical circuits or machines.

load shedding (automatic load limitation) The process where the power to certain customers or circuits is cut off due to a shortage in supply.

load smoothing (resource levelling) *See* RESOURCE SMOOTHING.

loam A material that contains roughly equal proportions of sand, clay, and silt.

lobby (vestibule) A short passage or room with entrance and exit doors at each end of the room used to control the movement of people or air. The opening and closing of the doors at each end of the room acts to seal and secure the passage of people or air. Where people are being controlled, the entrance door can be opened and the number of people passing through can be screened and controlled before exit, normally for security purposes. The room can also be used to prevent warm air escaping, or serve as an air trap that would restrict the development of fire.

lobster-back bend (cut and shut) A pipework bend that is formed by cutting the pipe into angled segments, which are then joined together to form the bend.

local area network (LAN) A local network of cables that enables computers and other devices to communicate with one another. In a wireless LAN, the computers can communicate with one another wirelessly.

local authority (council) The body that represents government and administers regional government control; the organization, as well as administering many other duties, is responsible for building control and enforcing the *Building Regulations.

local control A device that is capable of locally controlling remotely located equipment, such as an air-conditioning system.

location block Thin strips of resilient plastic that are used to locate and help hold the position of glass in a window frame.

location drawings (location plan, key plan) Plan drawings that show the placing of the building. The drawings are supported with dimensions to enable the positioning of the building and its components on-site.

location schedule A list of rooms with the finishings, fittings, and equipment to be located in the rooms identified. Normally, such schedules are used in buildings with many rooms, such as apartments or hotels.

lock A device fitted to a door or window, usually key-operated, to prevent entry.

lock block A block of timber in a hollow-core door that accommodates the mortise for the lock, and provides a fixing for any lock furniture.

locked fault A *fault which has not yet been displaced and is held in place solely by friction.

lock joint 1. An interlocking joint used to join two pieces of timber together at right angles. 2. A type of standing seam used in sheet metal roofing.

lock nut (locking nut) A nut that is prevented from becoming loose under load. This can be in the form of a nut being tightened onto an existing nut, or having a nut which contains a portion that deforms elastically when tightened to prevent it from coming loose.

lock rail (middle rail) A *door rail that accommodates the mortise for a lock.

lockset (lock set) The complete set of hardware required to lock a door. Includes the keys, handles, levers, lock mechanism, escutcheons, and so on.

lockshield valve (lock shield valve, LSV) A valve located on the return side of a radiator that controls the flow through the radiator to ensure that tall radiators receive enough hot water. Used to balance the system when radiators are first installed.

lock stile (closing stile) A *door stile that accommodates the mortise for a lock.

lodge 1. Dwelling set up and used by stonemasons; used as the living quarters and workshop when major medieval structures were built. 2. A small building offering accommodation for short stays. 3. A small gate house at the entrance to an estate or park.

loft (attic) 1. The space in the roof that can be used for storage, or converted to a room in the roof. 2. A floor of a commercial building, let unfurnished.

loft insulation Thermal insulation that is placed both between and over the ceiling joists in a dwelling to reduce heat loss. Can be applied in the form of a batt, a board, a quilt, or as loose-fill material.

loft ladder (disappearing stair) A telescopic or folding ladder that enables access to the loft.

logarithm The power to which a base must be raised in order to produce that number, e.g. the logarithm of 16 to the base 2 is 4, i.e. $2^4 = 16$. **Common logarithms** are noted to the base 10, whereas natural logarithms to the base e. **Logarithm scales** are used to present data over a large range or time period as increments are plotted exponentially.

log-construction A form of wall construction using tree trunks/logs, which are positioned horizontally on top of each other. The logs can be planed to give semicircles or flat sections.

London clay A stiff bluish/brownish *over-consolidated clay, which has been laid down under marine conditions and outcrops in the London area of the UK.

long and short work Corners of Saxon buildings where the large quoin and corner stones are laid on end and flat in alternate courses.

long arm Long pole with fixing on the end that allows it to be used to manoeuvre overhead equipment, such as roller shutter doors, Velux roof lights, roof vents, or other equipment that is too high to reach.

long float A flat trowel operated by two people for smoothing screed and concrete.

longhorn beetle (*Hylotrupes bajulus*) A wood-boring beetle that attacks only softwoods, its activity is restricted largely to the south of England. The insect causes rapid deterioration, leaving very large oval exit holes, around 10 mm long and 6 mm wide.

longitude The angular distance, east or west, of a point's meridian from the prime meridian, which runs through Greenwich, England. It is measured in degrees, minutes, and seconds.

longitudinal Running lengthwise, extending from top to bottom. **Longitudinal waves** propagate in the direction in which they were formed, for example, sound waves.

longscrew A pipe connector with threads that run parallel.

longstrip roofing Method of laying long strips of metal roof covering over a structure from end to end without the need for intermediate seams.

long sweep bend *See* SLOW BEND.

loop and loophole 1. A narrow slit in a castle or fortress wall through which missiles and arrows could be fired; often shaped in a crucifix with enlarged circular ends for better access. **2.** Series of access doors through which goods are delivered; used in warehouses, supermarkets, or superstores.

looped-pile carpet Woven carpet where the yarn is looped through the backing fabric.

loose-butt hinge *See* LIFT-OFF HINGE.

loose-fill insulation Thermal insulation material that is either blown or poured into a loft space or cavity wall. A number of loose-fill materials are available including mineral wool, cellulose fibre, mineral fibre, polyester beads, and vermiculite.

loose key 1. A tongue or *tenon that is not fixed to the main timber, hence it is loose and located into a mortise in each piece of timber that it is fixed to. **2.** A large T-shaped bar with a square recessed end that is used to turn stopcocks that are below the surface of the ground, often placed in the footpath or *pavement.

loose knot A knot in timber that is no longer held within the wood and may drop out.

loose material Particles of the main substrate, such as rust, scale, or flacking, or dust and debris on the surface of a material. The particles are removed by blast cleaning, scrubbing, or with a wire brush.

loose-pin butt (pin hinge) A butt hinge with a pin that can be removed, enabling the door to be quickly unhinged by removing the pin. This avoids the need to unscrew the hinge.

loose ring An O-ring used to seal pipes or joints. The O-ring is not fixed in place, but seals itself through its elasticity as it is stretched over the inside pipe and is compressed when the outer pipe is pushed into place.

loose side (slack side) The side of a veneer that was cut away from the main timber when it was formed. A timber veneer is formed by finely stripping timber with a cutting knife. As the knife peels the thin strip of a veneer from the timber, the side of the veneer that is in contact with the knife and pulled away from the main timber is called the loose side.

loose tongue (key slip tongue, slip feather, spline) An oblong strip of timber (*tenon or tongue) that fixes two pieces of timber together by locating the timber in two mortises formed in the timbers to be joined. The tenon is not fixed to the main timber, hence it is loose and located tightly into a mortise in each piece of timber that it is fixed to. *See* LOOSE KEY.

lopolith A saucer-shaped formation of intrusive igneous rock formed within the Earth's crust from slowly cooling magma.

LORAN A long-range navigation system which uses low-frequency signal pulses received from widely spaced radio transmitters to determine position.

lost-head nail A nail with a small slightly pointed head, which makes the head difficult to see when the head is driven just below the surface with a *punch.

loudness The perceived intensity of a sound.

louvre (US louver) A series of horizontal slats designed to admit air, light, and sound, but exclude rain. Used in doors, windows, lights, and ventilators.

Love wave A surface *seismic wave that travels in a horizontal direction, perpendicular to the direction of transmission.

low-angle light A light that impinges on a surface from a low angle.

low bid (US) The lowest priced tender.

low block A low-level podium.

low-carbon Producing a small net release of *carbon dioxide (CO_2) into the atmosphere. For example, a low-carbon power station would release less into the atmosphere than a traditional fossil fuel power station. Low-carbon power sources (stations) include wind, solar, hydro, and nuclear generation.

low-carbon economy A system of accounting for carbon (economical system) that follows a strategic approach to a minimal output of greenhouse gases and embodied carbon.

lower bound theorem A structure will not collapse under an external applied load if the distribution of *bending moments are in equilibrium and the plastic moment of resistance is not exceeded.

lowering wedges Formwork stays or wedges that are eased before the formwork is stripped.

lower yield point The lowest value of stress after the onset of *yield, for example, before the load starts to increase again. Particularly related to annealed carbon steels.

lowest tender The tender that has the lowest price out of all those that have tendered. *See also* LOW BID.

low-level suite A toilet basin with a matching attached close-coupled cistern.

low-pressure hot-water system (LPHW) A wet central heating system that operates at atmospheric pressure and the flow temperature of the water is at a maximum of 80°C. Used in the domestic sector and some small non-domestic buildings.

low-rise (low-rise building) A building with eight storeys or less.

low-velocity system A type of air-conditioning system that moves air around the building at low speed through large diameter ductwork.

low voltage Electric current greater than 50 volts but less than 250 volts. Usually supplied from the mains at 240 volts.

LPHW *See* LOW-PRESSURE HOT-WATER SYSTEM.

lubricant A material, usually liquid, used to reduce friction between two objects in contact.

lubrication A process whereby a *lubricant is applied.

luder bands Elongated surface markings or minor indentations in sheet metal caused by discontinuous yielding.

luffing (derricking) Reducing and increasing the radius of a crane by raising and lowering the jib. Luffing jib cranes are used on tight sites where over-sailing rights could not be obtained. As the crane passes the side of the site where it does not have the right to pass over, the crane jib is raised preventing any infringement of the neighbour's rights.

lug 1. A projecting piece of metal or plastic used for fixing. The lug projects out from the main unit and provides a fixing that enables the unit to be secured in position. Some fixing lugs have holes for screws or bolts, or fishtail ends that enable the lug to be bedded into mortar joints. **2.** A metal end fixed to a wire used to make an electrical connection. **3.** A plastic spacer used between the joints of tiles to accurately position floor or wall tiles.

lug end The end of the extensions on some window frames that are built into the wall reveal. Some windows have extensions on the sills that are longer than the window opening and are built into the wall; *see* LUG SILL.

lugged pipe bracket A wrap-around bracket with a *lug for fixing pipes to a wall.

lug sill A window sill that is longer than the window opening such that the ends of the sill are built into the wall at the jambs.

lumen (lm) The SI unit of *luminous flux.

lumen method A method of calculating the average illumination in a room.

luminaire (lighting fitting) A device that holds or contains a lamp, and controls the amount of *luminous flux emitted from a lamp. To achieve the required *illuminance within a space, the choice of lamp must be combined with the correct choice of luminaire.

luminance (L) A measure of the brightness of light emitted or reflected from a surface in a particular direction. Usually measured in *candelas per square metre.

luminous The process of emitting or reflecting light. A characteristic of a material or component.

luminous ceiling A suspended ceiling comprising back-lit translucent panels.

luminous flux The amount of visible energy emitted by a light source, measured in *lumens (lm).

luminous intensity (I) The amount of energy emitted from a light source in a particular direction. Measured in *candelas (cd).

lump hammer A hammer with a large heavy head used to break up objects such as concrete, drive in large pins or nails, or ease stubborn objects.

lump-sum contract (fixed price) A contract where the price is fixed and agreed with the client before the work starts. Cost-plus is a similar type of arrangement with most of the work being fixed price, but variations allow for work that is difficult to determine at the start of the contract.

lux (lx) The SI unit of *illuminance. One lux is equivalent to one *lumen per square metre.

luxmeter A device used to measure the amount of *illuminance.

Lyctus beetle (*Lyctus brunneus*, powder post beetle) A dark-brown, slightly reddish wood-boring beetle, 4–7 mm long, that attacks the sapwood of hardwoods such as oak, elm, ash, and eucalypt. Softwoods are not attacked by this beetle. Known to attack furniture and block or strip floors. Its exit holes are circular, 1–2 mm in diameter, and the bores tend to run parallel to the grain.

macadam In 1869 the first *flexible pavement road surface was built in London using tar mixed with aggregate and became known as '**tar macadam**' after its inventor John Loudon McAdam.

machine A device or tool containing several parts; it is normally driven by electricity to perform or undertake a certain task.

machine direction The direction in which something (e.g. a *geotextile) is manufactured. It is the direction it moves through and exits the machine and this defines the length of the product. *See also* CROSS-MACHINE DIRECTION.

machine learning A form of *artificial intelligence that enables computer systems to automatically learn and refine without being precisely programmed to do so.

machine special structural (MSS) Timber that has been graded by machine and is graded for special structural use.

(⊕) SEE WEB LINKS

• The Timber Research and Development Association website. Further information on stress grading is available on this site.

Mach number The ratio between the speed of an object travelling through the air or other fluid substance relative to the speed of sound as it is in that substance for a particular physical condition, e.g. temperature and pressure. An aircraft travelling at three times the speed of sound has a Mach number of 3.

$$Mach\ number = \frac{Object\ speed^*}{Speed\ of\ sound^*}$$

*For the same physical condition for the medium the object and sound is travelling through.

Mach-stem wave A wave that impacts a surface at an angle less than 90°, which produces a higher than normal wave.

macro 1. Something that can be viewed with the naked eye, without the use of a microscope; *see also* MICRO. **2.** A single instruction that provides a series of additional instructions for a computer to undertake a particular task.

macromolecule A large molecule comprising thousands of atoms.

made ground (made-up ground, built-up ground) Ground that has been formed by *fill.

maglev (magnetic levitation) A high-speed electrically operated train that uses a magnetic field to glide above the track.

magma Molten rock (at temperatures between 800 and 1200°C) that originates from below the surface of the earth; it cools and solidifies near or on the earth's surface to form *igneous rocks.

magnesium A light, *non-ferrous, moderately hard metallic element used in structural alloys and also used as an alloying element for other metal alloys. Magnesium is light and has excellent resistance to corrosion.

magnetic anomaly A disturbance in the earth's magnetic field due a deposit of magnetic minerals, such as iron ore, near to the surface of the ground.

magnetic bearing A location of a line in reference to the Magnetic North or South Pole.

magnetic field A force field surrounding a magnetized object (a magnet) or an area (the Magnetic North or South Pole).

magnetic polarity reversal When the earth's north and south magnetic fields change over. This typically occurs four to five times over a period of a million years.

magnetite A ferro-magnetic mineral with the chemical formula Fe_3O_4.

magnetization A measure of how easily a material can exhibit magnetic properties when subjected to a magnetic field.

magnetometer An instrument to detect the intensity and direction of a magnetic field.

magnitude The amount, degree, or size of something, for example, the size of an earthquake, which is measured on the *Richter scale.

mahogany A type of timber commonly used in furniture applications.

Maihak strain gauge An acoustic strain gauge.

main contractor (general contractor, prime contractor) The party who has the main contract with the client is responsible for the work on-site and the employment of subcontractors and specialists necessary to complete the construction work. Some contractual arrangements may result in contractors being employed directly by the client; in such cases the main contractor would be the party responsible for the majority of the construction work.

main distribution frame (building distribution frame) A place where telephone *terminals and exchange equipment is housed.

mainshock The largest *magnitude recorded in a series of earthquakes. Smaller earthquakes either side of the mainshock are referred to as foreshocks and *aftershocks.

mains-pressure cold-water supply The statutory service standard level of mains cold-water pressure in the UK is one bar or 7–10 metres/head (approx 14 psi). It is important to maintain a pressure in the pipe to ensure water reaches the consumer and to prevent other untreated water sources entering the property. In the UK the guaranteed standards scheme (GSS) (*see* Ofwat) sets out that water companies shall maintain a minimum pressure of water in the mains communication pipe serving the premises supplied with water of 7 metres static head.

(⊕) SEE WEB LINKS

• The Ofwat website gives information on water pressure in residential properties

mains voltage Low voltage power, defined as 240 volts in the UK and used, for example, to power household appliances.

maintainability The ability to keep something operational by regularly checking it and undertaking repairs as and when required.

maintenance The work necessary to keep things operating properly and in a good state of repair. To ensure that plant and equipment are always operational, a 'planned preventive maintenance' programme should be used. To ensure that machinery operates continually, parts that wear and have a limited life span need to be checked and replaced before they cease to function; regular maintenance should ensure that the parts are always in good working order. Other decorative items will also need to be checked and any necessary work undertaken to ensure that they are visually appealing. *See also* CONDITION-RESPONSIVE MAINTENANCE; PREDICTIVE MAINTENANCE; RELIABILITY-CENTRED MAINTENANCE; ROUTINE MAINTENANCE; SCHEDULED MAINTENANCE.

maintenance painting Repair and repainting to ensure the finish remains visually appealing or functional if protecting the substrate from deterioration, for example, protecting steel from rusting and timber from decay.

maintenance period The time that the supplier or contractor agrees to keep the property and equipment free from defects; in contractual terms this may be referred to as the defects liability period.

major axis An axis of an object (such as a beam) that has the largest second moment of area (*see* PRODUCT OF INERTIA). *See also* MINOR AXIS.

make good To repair an item so that it is in proper working order, and so that it appears as new.

make up To add material so that it is the full amount or finish something off so that the work is aesthetically pleasing and complete.

making good *See* MAKE GOOD.

makore An African mahogany.

malfunction Something that is not working correctly—perhaps due to a fault.

mall (maul) A large or heavy *mallet.

malleable How easy or difficult it is to shape or deform a material.

malleable cast iron White cast iron that has been subjected to different heat treatments to increase its graphite content, resulting in a relatively ductile and malleable material.

mallet A hammering tool with a large wooden or metal head.

management contract (US construction management) A standard form of contracting where the main contractor acts as the procurer and manager of the contract and sub-contract packages. The chief contractor acts as an advisor to the client, setting up and placing the contracts for the design and construction works, although this contractor never enters into a contract with the suppliers. The contracts are made directly with the client. The chief contractor organizes and manages the work packages.

managing contractor The chief contractor leading the management contract.

mandatory Something that must be undertaken, followed, or complied with by law or any other authority.

M & E (Mechanical and Electrical) Building consultants and contractors who are specialists in building services, including such activities as: *heating, cooling, air conditioning, *plumbing, *ventilation, fire control, lifts, escalators, security, and *renewable energy. As M & E works are now a significant part of most building projects, some consultants and contractors are now taking the lead role in building projects

man door (wicket door) A small gate or a door of sufficient height and width to allow a person to enter easily. Can be a door within a larger door; the larger door could allow plant and other mechanical equipment into a shed, however, when not in use the smaller door would allow people to enter and egress.

manganese A *non-ferrous metal with similar properties to iron.

manhole *See* INSPECTION CHAMBER.

man hour The quantity of work performed by an average worker during one hour.

manifold A chamber or pipe with a number of openings that branch into a series of smaller pipes.

man lock An air lock entrance to a pressurized working chamber, tunnel, or shaft.

Manning's formula An empirical formula to estimate the velocity of flow in an open channel or a pipeline driven by gravity. It is named after Irish engineer Robert Manning (1816–1897), who reworked the original formula proposed by French engineer Philippe Gauckler (1826–1905). Hence, it is also known as the **Gauckler–Manning formula**. It is defined as:

$$V = \frac{1}{n} . R^{\frac{2}{3}} . S^{\frac{1}{2}}$$

where:

V = Mean velocity (m/sec)
n = Manning's coefficient
R = Hydraulic radius (m)
S = Friction slope (m/m)

manometer An instrument for measuring fluid pressure, such as a *piezometer.

manor house The house which is the main administrative centre of a country estate or manor. The houses are sometimes fortified.

manpower The power provided by human effort; the number of people needed to physically lift something.

mansard 1. A roof that has two slopes on each face, with the lower at a steeper angle than the upper slope, also known as a French roof. **2.** The area directly under (enclosed by) a mansard roof.

mansion A very large well-appointed house often with more than ten bedrooms, a ballroom, library, and set in sizable grounds. In the US, a mansion is a building with over 740 m^2 of internal space set in extensive grounds.

manual handling Any transporting or supporting of a load (including the lifting, lowering, pushing, pulling, or carrying) by hand or bodily force; governed in a workplace setting by the **Manual Handling Operations Regulations 1992** (as amended in 2002).

manufacture To produce something, typically on an industrial scale.

manufactured aggregate Thermally or otherwise modified *aggregate.

manufactured gas A mixture of gases obtained by thermal decomposition of oil (such as petroleum gas), destructive distillation of coal (such as coal gas), or steam reaction.

map A geographical diagram, showing specific features of the land such as roads, contours, reservoirs, etc. **Mapping** is the process of making maps.

marble A *metamorphic rock formed from carbonate minerals, such as calcite or dolomite that have recrystallized under heat and pressure. Used as a decorative stone for tiles, worktops, and sculptures. **Marble cladding** is the use of a thin marble slab to face items such as columns to achieve a decorative effect. **Marbling** is representing the appearance of the crystalline structure of marble artificially—by the use of paints.

margin 1. The outer edge of an enclosed area. **2.** Boundary limit. **3.** Profit.

margin trowel A thin rectangular *trowel.

marigram A graph showing tide levels for a particular place.

marina A manmade harbour for yachts and small boats.

marine Relating to the sea.

marking drawing A drawing showing the position of building elements, often used to show the location of prefabricated modules and components. The drawing shows the location on a grid plan and identifies the position of each component. Each component is given a reference number in relation to its grid position.

marking gauge (joiner's gauge) A tool used to mark a line parallel to the edge of a piece of wood. The tool consists of a wooden block with a steel projecting pin, the block slides along a bar, which can be fixed at any point to define the depth of the line.

marking out *See* SETTING OUT.

marl A naturally occurring mixture of clay and calcium carbonate (lime). Used as a fertilizer and to soften water.

marsh An area of low-lying water-logged ground.

martensite An iron phase with a high carbon content that is the result of a diffusionless transformation in ferrous alloys, especially steel. The presence of martensite is desirable in steels where increased *hardness is required. To obtain martensite it is necessary to heat low carbon steel to about 800°C then rapidly cool it down to room temperature within 10 seconds maximum.

mash hammer *See* LUMP HAMMER.

mask angle In a global positioning system, it is the minimum elevation below which satellite signals are designed not to record signals.

masking The act of concealing the existence of something by obstructing the view of it.

mason A craftsperson skilled at working with stone and brickwork. Traditionally, the term was reserved for those who could cut, chisel, and work stone; in some areas it is now extended to include specialist bricklayers.

masonry A term used to refer to the construction of walls, dwellings, buildings, etc. using essentially brick, mortar, and concrete.

masonry cement *See* CEMENT.

masonry drill *See* TUNGSTEN-TIPPED DRILL.

masonry fixing Any fittings used to fix an object to brickwork, for example, an expansion bolt.

masonry nail A square twisted hardened steel nail that can be hammered into brickwork.

masonry paint Special type of paints which are typically smooth and textured, and are suitable for application to exterior walls of brick, block, concrete, stone, or renderings. Fine cracks in the substrate can often be hidden using the sand-textured material.

masonry primer A primer usually based on asphalt used to prepare masonry specimens for bonding with other asphalt-based materials.

masonry unit A building block, brick, or stone.

mason's putty A lime putty to which Portland cement and stone dust have been added; especially used in ashlar work.

mason's scaffold A free-standing *scaffold.

mass 1. A body of matter. **2.** The quantity of matter; mass (m) equals force (F) divided by acceleration (a), which is derived from Newton's second law of motion: $F = ma$.

mass concrete An amount of concrete, which does not contain reinforcement, but is sufficiently large to develop cracks from the heat of hydration if the amount of heat generated is not controlled.

mass haul diagram A graph to illustrate the volume of cut and fill in earthworks. The horizontal axis represents chainage; the vertical axis represents cut and fill—cut is shown as positive and fill as negative *ordinates.

mass timber construction A solid timber construction system that uses no glues or nails. Usually fabricated from softwood timber. Mass timber structures offer sustainable advantages over steel and concrete because wood is a naturally renewable resource with relatively low embodied energy.

mass wall *See* TROMBE WALL.

mast A vertical pole or slender tower.

Master of Engineering *See* MENG.

mastic A material of a non-hardening nature, which forms surface films when applied. Mastic must possess sufficient viscosity to resist sagging and flowing in the thickness required, but it must also be sufficiently ductile to flow plastically when joint movement occurs. Vegetable oils (including linseed oil), synthetic polymers, and bituminous materials may be used to make mastic materials.

mat 1. A US term for a *footing. **2.** A mesh of reinforcement. **3.** A *filament resin mixture used in *fibre-reinforced plastics.

matchboarding door *See* BATTEN DOOR.

mate A labourer or helper who assists with the organization and preparation of materials for the tradesperson.

material (building material) The matter or substance used to make things, particularly material used to construct building, such as aggregates, bricks, blocks, steel, and plaster.

mathematics The study of numbers, their relationships, and quantities.

matrix The predominant phase in a composite material in which another secondary phase or material is dispersed. For example, in a metal reinforced with carbon fibres, the matrix in this case will be the metal.

matt A non-gloss flat finish, particularly with regards to paintwork.

matter A *material substance with physical mass.

mattlock A type of pickaxe, with a long wooden handle containing a metal head that is pointed at one end and flattened (spade-like) at the other.

mattress 1. A large concrete ground slab used to support plant and equipment. **2.** A layer of blinding concrete to seal something. **3.** A layer of geotextile weighted down with rock to provide a *scour.

maul *See* MALL.

maximum The large value, amount, or level that can be obtained. **Maximum bending moment** is the largest *bending moment. **Maximum demand** is the largest quantity of something delivered or expected.

Maxwell diagram A diagram, constructed by using Bow's notation, that combines a series of *polygons of forces. It is used in the analysis of members that are all in the same plane and are connected together by *pin joints, such as a plane truss.

MCB (miniature circuit breaker) Automatic breaker positioned in an electrical circuit that breaks the circuit when there is an overload. The circuit can be reconnected once the fault is diagnosed and corrected.

MDF (medium density fibreboard) A timber product made of wood fibres utilized as a board.

mean The average, intermediate, or midway value. For example, the **mean sea level** is the average long-term sea level.

meander A bend in a sinuous watercourse. It is formed when moving water in a riverbed erodes the outer bank. Typically, meanders occur on a flood plain characterized by a series of big loops in the watercourse around which the water slowly flows.

mean radiant temperature The mean temperature of all the surfaces that surround an object.

measure A procedure to determine the size (length, area, or volume) or weight of something, by using a measuring device such as a measuring tape or balance.

measure and value contract A standard contract where the items of work are described in the *bill of quantities. The bills are priced by the contractor and submitted as part of the tender. Once the work is completed the prices are used to calculate the value of work. The work is remeasured to check the actual quantity of work completed.

measured item Part of works which is included within the descriptions in the *bill of quantities.

measured mile An exact mile marked on a road surface or an exact nautical mile that uses markers so that vessels can calibrate their speed.

measured quantities *See* MEASURED ITEM.

measured separately (ms) Aspect of the works which is quantified under a different section of the *bill of quantities, due to its size or unique qualities.

measurement The act of recording quantities, prices, and valuing work, the process of which is undertaken by a *quantity surveyor. The main part of the process is the *taking-off of the quantities from the drawings so that the work can be priced in accordance with the *Standard Method of Measurement.

measuring frame *See* BATCH BOX.

measuring tape A long, narrow strip of metal or plastic that contains marks at precise intervals (in millimetres, centimetres, metres or inches, feet, yards, or a combination of various units) to enable distances or lengths to be measured.

mechanic A skilled person who repairs, maintains, or operates machinery or engines.

mechanical adhesion Occurs when an adhesive flows into the microstructure of the surfaces to be bonded.

Mechanical and Electrical *See* M & E.

mechanical barrow *See* MOTORIZED BARROW.

mechanical computer-aided engineering (MCAE) The use of computer software packages to aid mechanical design and engineering.

mechanical degradation of timber A form of degradation of timber, it is the most common type of degradation that occurs in timber when subjected to continuous loading for long periods of time. For example, after 50 years the load that a piece of timber can withstand is less than half the stress it can carry at the onset of loading. Similarly, there is a reduction in stiffness (*Young's modulus) with time. Another form of mechanical degradation is the induction of compression failure in the cell walls of the timber. Timber overstressed in compression in the longitudinal direction forms kink bands and compression creases. Such defects can reduce the tensile strength of the wood by 10–15%, but the loss of toughness under impact conditions can be as high as 50%.

mechanical engineering A profession that deals with the discipline of designing, constructing, and operating machines, tools, and mechanical plant.

mechanical extract ventilation (MEV) A type of *mechanical ventilation system that is designed to operate continuously by extracting warm moist air from the wet areas of a building. Fresh air is then drawn into the building via background ventilators (i.e. trickle ventilators) and air leakage. It can comprise either a central extract system (centralized system) or individual room fans (localized system), or a combination of both. Systems typically have two speed settings. Continuous trickle ventilation is provided on the low speed setting and boost ventilation is provided on the high speed setting. The boost setting can be operated either manually or automatically (using a humidistat or PIR sensor).

mechanical fixing A joint or bond which relies on its interlocking form and shape to maintain the fixing interface, being different from chemical fixing, where glues or resins are used. A true chemical bond results from an attraction between atoms, ions, or molecules, resulting in the formation of chemical compounds.

mechanical floor (mechanical level, machine-levelled floor) A floor in a building that contains mechanical plant such as air-conditioning units. Some mechanical floors have to be designed to take large imposed loads from the plant and need to be perfectly level to ensure that the equipment runs smoothly.

mechanical properties The *properties of a material in relation to stress and strain.

mechanical room A room that houses mechanical plant.

mechanical ventilation A type of ventilation system where an electrically driven fan or fans are used to supply and move air around the building.

mechanical ventilation with heat recovery (MVHR) A type of mechanical ventilation system that is designed to operate continuously by extracting warm moist air from the wet areas of a building and supplying fresh air to the other rooms. Both airflows are ducted through the heat exchanger, where 70% or more of the heat from the outgoing air can be transferred to the incoming air. Systems typically have two speed settings: a low speed trickle ventilation setting and a high speed boost ventilation setting. A typical balanced MVHR system comprises an intake and an extract fan, a heat exchanger to transfer heat from the extract air to the inlet stream, ductwork to distribute air around the dwelling, and grilles to supply or extract air from individual rooms.

mechanics The study of motion and forces.

mechanic's lien The right over property if payment is not made for work undertaken, the supplier of materials, labour, or professional services can stop ownership of the property being transferred if a debt is not settled.

median 1. Middle point of a frequency distribution. **2.** A line that divides a *triangle or *trapezoid.

medium 1. The material substance that something (such as heat and sound) can be transmitted through. **2.** Something that is in the middle, i.e. between two extremes.

medium-density fibreboard *See* MDF.

medium-rise Buildings between four and six storeys high. Buildings tend to be defined as low rise, less than four storeys or high rise, more than six; however, buildings of four to six storeys are sometimes described as medium-rise.

medium voltage Electrical supply between 1 kV and 50 kV, medium voltage installations would include any builders' work, such as the laying of cables, which will host the supply.

meeting The bringing together of two or more people in a formal setting to discuss project issues. Types of meeting include **coordination meetings, management and design meetings,** and **progress meetings.** *See also* SITE MEETING.

megalithic A large stone that has been used to construct a structure or monument, either alone or together with other stones.

melamine A thermosetting polymeric material with a wide range of applications, chemical formula $C_3H_6N_6$. Melamine is mainly used for kitchen worktops.

meltdown Devastating collapse or breaking, for instance, when a nuclear reactor melts due to overheating of the fuel rods, resulting in radiation pollution.

melting A physical change from a solid to a liquid, for instance, when ice sheets melt and become water, known as **meltwater.**

member 1. A structural element in a building, for example, a beam. **2.** A person belonging to a specific group.

membrane A thin pliable sheet of material.

MEng A Master of Engineering degree award; typically a student would have successfully completed four years of full-time study at university.

meniscus The upper curved surface of a liquid when held in a tube, which is due to surface tension.

mensuration The calculation of lengths, areas, and volumes from measurements or dimensions.

mer The repeat unit of a polymer chain.

Mercator projection A map of the world that has been produced on a flat surface—*longitude (or *meridian) and *latitude lines are plotted as straight parallel lines that cross at right-angles to each other.

mercury A metallic element with the chemical symbol Hg, which is liquid at or near room temperature and pressure.

meridian A line of *longitude, defining an imaginary circle around the earth that passes through both the North and South Poles.

merlon The solid part of an embattlement parapet, the walled part of a fortified wall, between which there are gaps allowing those protecting the building to view through and fire upon those who may be attacking the building.

mesh An object consisting of semi-permeable barriers made of connected strands of metal, fibre, or other flexible/ductile materials. Mesh is similar to a web or net in that it has many attached or woven strands.

metabolic unit (MU) A physiological measure used to express the intensity of physical activity, defined as the ratio between the metabolic rate of a specific activity and a reference 'resting' metabolic rate equivalent to $58.2W/m^2$.

metal A material, usually with excellent mechanical properties, typified by ionic bonding. Most metals have excellent *tensile strength, *hardness, *Young's modulus, and *ductility.

metal arc welding (SMAW, MMA) A welding process utilizing an electrode.

metallic bond The bonding in metals involving the sharing of free electrons.

metallurgy The study of metals involving the physics, chemistry, and all engineering implications.

metal primer Paint used as a protective and first coat on iron or other metals.

metal-sheathed mineral-insulated cable Cables which are used primarily as heating units and power cables. Mineral-insulated cables typically comprise one or more wire conductors contained within a bendable metal sheath.

metal spraying (metallization) General term applied to the spraying of one of several metals onto a metal substrate.

metamorphic stone Rocks that have been transformed by a natural process involving intense heat or great pressure.

meteorology The study of the earth's atmosphere, particularly in relation to climatic conditions and the weather.

meter A device for measuring the amount of flow, e.g. a *water meter. A **meter bypass** allows the flow to bypass the meter such that the amount is not recorded. It can be used for fire-fighting equipment so that no charge is incurred by the owner of the property for the water used by the fire brigade.

methane A colourless, odourless gas with chemical formula CH_4, a major component of natural gas.

method of included angles A surveying method for a closed-loop traverse, where the theodolite is moved from station to station in a clockwise manner, such that angles obtained are always within the loop. Any closuring error is distributed evenly.

method of measurement The technique or system used for describing and measuring units of construction. *See also* STANDARD METHOD OF MEASUREMENT.

method of section A procedure for determining the internal forces within a structure, by making an imaginary cut at right angles through the structure. The internal and external forces can then be determined by resolving the forces.

method of slices A slope stability analysis where the slope is divided into parallel strips using prescribed guidelines. The sum of the weight of all the slices is then considered rotating about their lever arm, about the centre of the slip plane, against that of the shear strength of the soil along the potential slip plane. Guidelines regarding where to change a slice include: where the failure surface passes from one material to another (in terms of shear strength parameters); at the vertical through the centre of rotation; at a shape change in the slope angle (where it becomes flat at the top and bottom); use 5 to 8 slices approximately.

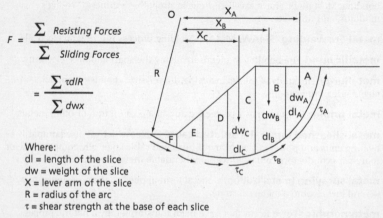

$$F = \frac{\sum \text{Resisting Forces}}{\sum \text{Sliding Forces}}$$

$$= \frac{\sum \tau dl R}{\sum dw x}$$

Where:
dl = length of the slice
dw = weight of the slice
X = lever arm of the slice
R = radius of the arc
τ = shear strength at the base of each slice

Method of slices

Methods of Assessment and Testing (MOAT) Performance conditions that need to be fulfilled prior to the issue of an Agrément Certificate.

method specification Detailed description of how to complete the task.

method statement A document that identifies how a particular job or task is to be undertaken; this is particularly relevant with regard to developing a safe working practice.

methyl methacrylate The *mer unit of Perspex*, chemical formula $CH_2=C(CH_3)$ CO_2CH_3.

metre An *SI unit of length, equivalent to 1000 millimetres or 39.37 inches.

metric A measurement decimal system based on weights and lengths, such as *kilogramme and *metre.

MEV *See* MECHANICAL EXTRACT VENTILATION.

micaceous iron oxide paint A high build finishing or undercoat containing micaceous iron oxide.

MICE Membership of the *Institution of Civil Engineers, which can either be as an *Incorporated Engineer (IEng) or *Chartered Engineer (CEng) level.

micra A silicate mineral with plate-like crystals forming thin flat cleavage planes. Colour is dependent on type, with **biotite micra** being black or brown, to **muscovite micra** being clear.

micro Very small, something that is viewed under a microscope; *see also* MACRO.

microbe A microscopic *organism, that can transmit disease; e.g. the bacterium *Vibrio cholerae* causes cholera.

microbiologically induced calcite precipitation (MICP) A form of *biocementation where bacteria are able to metabolize urea and water into calcium carbonate (calcite) as a result of possessing the catalysing enzyme urease.

microbiology The study of *microorganisms and their impact.

microbore (microbore pipework) Very small pipework diameter, typically 6, 8, 10 or 12 mm. Used in central heating systems to enable the pipework to run up through walls and under floors.

micro CHP *See* MICRO COMBINED HEAT AND POWER.

microclimate 1. A small area where the climate is different from the surrounding area. **2.** The local set of atmospheric conditions that differ from those in the surrounding area, often with a slight difference but sometimes with a substantial one. The term is relative to the broader context but when addressing urban and rural areas can often refer to areas as small as a few square metres or square feet (for example a garden bed or a cave) or as large as many square kilometres or square miles. Because climate is statistically described, implying spatial and temporal variation of the mean values of the describing parameters, within a region there can occur and persist over time sets of statistically distinct conditions, that is, microclimates. Microclimates can be found in most places. Microclimates exist, for example, near bodies of water which may cool the local atmosphere, or in heavy urban areas where brick, concrete, and asphalt absorb the sun's energy, heat up, and re-radiate that heat to the ambient air: the resulting urban heat island is a kind of microclimate. **3.** A building microclimate is a small area within the building structure, within voids, gaps, or spaces where the temperature, air, and moisture can be considered in detail, for example between and within building components, cavities in walls, voids in roofs, and under floors.

micro combined heat and power (micro CHP) A *combined heat and power (CHP) system that has an electrical capacity of less than 50 kW.

micron One millionth of a meter (0.000001) and a micrometer (10^{-6} m).

microorganism A microscopic *organism such as a virus, bacterium, or protozoan that can only be seen when viewed under a microscope.

microptic theodolite A surveying instrument that contains micrometers, such that the circle readings for horizontal and vertical angles are 180° apart.

microscopy Essentially the study of a material's surface under high magnification.

microstrainer A fine stainless-steel mesh drum that is used in the tertiary treatment stage of wastewater. The drum, which is partially submerged in effluent or raw water, slowly rotates and collects fairly large particles, such as alga, from the water. When the trapped particles reach the top of the drum they are dislodged by a spray into a discharge trough.

microstructure Essentially, the chemistry or physical composition of a material visible only at high magnifications.

microtunnel A very small tunnel, used for inserting pipelines. *See also* PIPE JACKING.

microvoid coalescence (MVC) The formation of very small and minute cracks or surface irregularities occurring in a material due to a change in chemistry.

microzonation A method of dividing an earthquake-prone area into zones, which are related to their geological and geophysical characteristics, to define potential risk.

MIDAS *See* MOTORWAY INCIDENT DETECTION AND AUTOMATIC SIGNALLING.

middle-third rule A rule that states there will be no tension in an unreinforced wall if the resultant force lies within the middle-third of the wall.

MIG (metal inert gas welding) A type of welding using an inert gas, primarily used on *non-ferrous metals.

mild steel A type of *steel containing no more than 2% carbon content and very few additional alloying elements. This type of steel is most widely used, especially for civil engineering (structural applications); although prone to corrosion, a protective coating (galvanizing) is used.

mile A unit of linear measurement, equivalent to 1760 yards or 1.6 kilometres.

milestone Significant point or event in a project. Key parts of the project may need to be delivered by this date or it may mark the point of key deliverables.

Millennium Development Goals (MDGs) A set of eight international goals, with eighteen targets, which were aimed at addressing humanitarian aid and *sustainable development in the developing world between 2000 and 2015. The goals were established at the Millennium Summit in 2000, and consisted of:
- Goal 1: Eradicate extreme poverty and hunger.
- Goal 2: Achieve universal primary education.
- Goal 3: Promote gender equality and empower women.
- Goal 4: Reduce child mortality.
- Goal 5: Improve maternal health.
- Goal 6: Combat HIV/AIDS, malaria, and other diseases.
- Goal 7: Ensure environmental sustainability.
- Goal 8: Develop a global partnership for development.

See also SUSTAINABLE DEVELOPMENT GOALS.

Miller indices A group of three digits used to determine the apparent location of atoms in a metal.

milli- A prefix relating to one thousandth (10^{-3}), e.g. a millimetre is one thousandth of a metre.

milling The action of grinding to cut or smooth metal objects.

mill scale The surface of hot rolled steel, usually flaky in texture.

mimic diagram Illuminated diagram showing floor plans, zones, and equipment. It may provide information on building use and equipment location.

mine A place and process where minerals are excavated from the ground.

mineral An inorganic substance that occurs naturally in the ground and is often excavated for its value, e.g. diamond.

mineral-insulated cable (mineral-insulated metal-sheathed cable, mineral-insulated copper covered, MICC) Electrical conducting cables made from copper.

mineral spirits A solvent based on petroleum used as a paint thinner.

mineral wool A light fibrous material used as an insulator.

miner's dip needle A compass-type of device, which indicates the dip of the earth's magnetic field and the occurrence of magnetic material (iron ore) within the ground.

Miner's rule (Palmgrem-Miner rule) A linear relationship to predict fatigue damage.

minibore *See* MICROBORE.

minimum The lowest possible number, point, quantity, standard, etc.

mining The extraction, deep boring, or digging of materials (minerals and other deposits) that are valuable.

minor axis An axis of an object (such as a beam) that has the smallest second moment of area (*see* PRODUCT OF INERTIA). *See also* MAJOR AXIS.

misclosure *See* CLOSING ERROR.

Mises yield envelope *See* VON MISES YIELD.

MIStructE Member of the Institution of Structural Engineers.

mitre Of matching ends. A **mitre joint** has been cut with a mitre saw to produce two ends to be joined at 45°.

mix A mixture of different materials to form something, e.g. a concrete mix contains cement, water, sand, and aggregate.

mixed construction (mixed use) Buildings with different uses, e.g. residential, commercial, and retail within one property.

mixed dislocation A type of dislocation in various forms. In other words, a combination of different types of *dislocations that exist in a metal.

mixed liquor A mixture of sewage effluent and organic material in an aeration tank for activated sludge treatment.

mixed-mode ventilation A combination of *natural and *mechanical ventilation.

mixer A device for mixing various components to produce a *mix, e.g. an *asphalt mixer.

mixing water (gauging water) The water used in a concrete or mortar mix. It should be free from contamination—typically drinking water is used.

mixture *See* MIX.

moat A broad ditch filled with water that surrounds a castle. Traditionally, the moat was part of the castle's defences; however, more modern buildings are surrounded by water to give an aesthetic appeal.

mobile concrete pump A vehicle containing a pump and a folding boom with a delivery pipe. Used to pump ready-mix concrete into hard-to-reach places.

mobile crane A crane with a telescopic boom that sits on a wheeled undercarriage, forming a vehicle that can be driven from site to site.

mobilization costs The cost of setting up a site, providing temporary accommodation, hoardings, and temporary power to the site.

mock-up A scaled or full-size model of the building, roads, component, or item of plant.

mode A value that occurs the most frequently. *See also* MEAN and MEDIAN.

Model Water Byelaws The UK's water regulations relating to preventing waste, undue consumption, misuse, and contamination of water.

modification A *variation or change to that specified or previously described.

modular construction The use of standardized building components to form a building. Factory-produced pre-engineered building units that are delivered to site and assembled as large volumetric components or as substantial elements of a building. The collection of discrete modular units usually forms a self-supporting structure in its own right or, for tall buildings, may rely on an independent structural framework.

The main sectors of application of modular construction are:

- Private housing
- Social housing
- Apartments and mixed-use buildings
- Educational sector and student residences
- Key-worker accommodation and sheltered housing
- Public sector buildings, such as prisons and MoD buildings
- Health sector buildings
- Hotels.

modular ratio The ratio of the *Young's modulus of steel to that of concrete, with regard to reinforced concrete.

modular system Building constructed out of *prefabricated components that are designed to fit together quickly and easily.

module (building module) Prefabricated unit of a building. Can be used within a modular system or could be a single unit, e.g. bathroom pod, or plant unit, that can be inserted into a more traditional form of construction.

modulus of incompressibility The ratio of pressure to volume change in a soil mass.

modulus of plasticity (plastic modulus) The ratio between the plastic moment of a beam and its yield stress.

modulus of rigidity The ratio between shear stress and shear strain; *see also* POISSON'S RATIO.

modulus of rupture Breaking strength in a non-ductile or *brittle solid as measured by bending.

modulus of volume change *Coefficient of volume compressibility divided by one plus the initial void ratio.

MOHO (Mohorovicic discontinuity) The boundary between the earth's crust and mantle.

Mohr's circle A diagram that represents normal stress on the x-axis, against shear stresses on the y-axis, such that any point in a circle, which is centred on the x-axis, represents the stress on a particular plane.

moisture Liquid, such as water vapour, held within a solid or condensed on a cool surface.

moisture content The amount of moisture held within something, e.g. wood or soil, divided by the weight of the dry wood or soil, expressed as a percentage. A **moisture meter** is a device that can measure the amount of moisture held at the surface of wood or concrete. It consists of two prongs that are pushed into the surface—a digital display then indicates the amount of moisture presented.

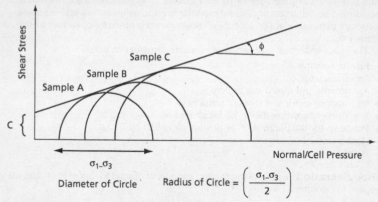

Mohr's circle

moisture-resistant adhesive (MR) An adhesive resistant to the ingress of water or moisture.

molar Relating to *mole—containing one mole per litre.

molding (US) *See* MOULD.

mole 1. SI unit of amount of a substance, defined at molecular level. **2.** A massive sea wall constructed of mass stone, used to protect a harbour. **3.** A tunnelling machine.

molecular Relating to a *molecule.

molecule A microscopic part of a substance that defines the properties of the material.

molten When a material has reached the liquid stage, passed its melting point.

molybdenum A non-ferrous metal with a very high melting point (2623°C), used as an alloying element in steel.

moment A force multiplied by a lever arm, producing a turning effect.

moment–area method Used in structural analysis to determine slope and deflection between two points on elastically deflected beams. It is based on the **moment–area theorems**. The **first moment–area theorem** states that the slope of the beam between two points can be determined from the area of the **M/EI** diagram between these two points. The **second moment–area theorem** states that the vertical deflection between two points on the beam is equal to the first moment of the **M/EI** diagram between these points multiplied by the distance where it is measured.

moment distribution A structural analysis method for calculating the bending moments in statically indeterminate beams and frames. Initially, every joint is fixed, so as to develop fixed-end moments. Joints are then released one at a time with the fixed-end moments being distributed to adjacent members until equilibrium is maintained.

moment of inertia A measure of the resistance to rotational change. It is the sum of the mass of every component in a body multiplied by the square of their distance from the axis.

moment of resistance The *bending moment that can be carried by a beam or element.

momentum The measurement of movement, the capability of progressing forwards.

monastery A building built to house those who wish to pray and also includes domestic quarters and workspaces.

monitor 1. To regularly check or take observations. **2.** An instrument for viewing data, such as a computer monitor.

monitoring well A groundwater well to allow water quality and/or level to be checked regularly.

monk bond Brick bond laid with two *stretcher faces between *headers; this is a variation on the *Flemish bond (header and stretcher).

monkey A mechanism to release a drop hammer on a pile-driving rig, so that it falls freely under its own weight.

monkey tail The scroll of the handrail at the bottom of a staircase.

monkey tail bolt *See* EXTENSION BOLT.

monochromatic 1. Having one wave length, e.g. a laser. **2.** In one colour, e.g. painted in one colour.

monolith A single great stone (often in the form of a column or obelisk).

monomer *See* MER.

monomictic Lakes that have a single mixing period each year, i.e. having a uniform temperature and density from top to bottom. There are two types, those that are covered with ice throughout the year and those that never freeze over.

monorail A single track railway, typically with the carriages hung from overhead.

Monte Carlo method A computer algorithm that uses repeated random sampling to obtain the result.

monthly certificate (interim certificate, progress certificate) A statement issued to the client recording the value of work completed to date, and recommending to the client the sum of money that should be issued to the contractor. The certificate is usually issued by the client's representative, quantity surveyor, architect, or engineer who has measured and valued the work. The certificate also states the retention, which is the amount of money that is to be withheld.

monthly instalments (interim payment, progress payment) Money given to the contractor for the work completed, the payment usually follows the valuation and issue of the *interim certificate.

monthly statement (monthly claim) Statement of work completed and value claimed.

monument A structure, building, or significant item created to commemorate an event or person of note.

Moody chart A dimensionless chart that relates Darcy friction factor, relative roughness, and *Reynolds number to determine pressure drop or flow rate along circular pipes, non-circular pipes, and open channels.

moraine The mass of earth and rock that has been carried and deposited at the front and side edges by a glacier as it retreats.

Morison equation An equation used to calculate the inline force on a body in an oscillating flow, e.g. the forces created due to waves on an offshore structure.

morning glory *See* SPILLWAY.

morphology The study of structure and form, particularly in relation to organisms.

mortality rate The ratio of deaths to the total number for a particular reason, e.g. the number of deaths due to water-related diseases in developing countries.

mortar A material used to bond bricks and blocks together in the construction of dwellings and buildings. Comprised mainly of sand, cement, and lime, although *lime mortars are made of lime, sand, and water.

mosaic A decorative wall or floor covering using small coloured cubes (*tesserae) stuck onto the surface.

motion The act of moving.

motive power circuit A circuit that is capable of providing a high start demand, e.g. to drive a lift.

motorized barrow (mechanical barrow, US buggy, concrete cart) A power-driven two-wheeled cart, used for transporting heavy goods such as fresh concrete.

motorized valve A valve that is opened and closed by an electric motor, used in central heating systems.

motorway A road with fast-moving traffic with limited access, typically with three lanes in each direction, separated by a central reservation barrier or grass verge, with hard shoulders on the outer edges.

motorway incident detection and automatic signalling (MIDAS) A network system of traffic sensors which uses induction loops, spaced at 500 m intervals, on motorway portal gantries to detect slow-moving traffic, queueing, stationary traffic, accidents, and other incidents. In connection with the *active traffic management (ATM) system, information is provided to other drivers approaching an incident to reduce speed or alert them of a queue ahead, i.e. to prevent further incidents from occurring. *See also* SMART MOTORWAY.

motte A raised mound, which usually has a stone or wooden structure on top known as a keep. The motte is historically linked to a motte-and-bailey castle, which is a structure situated on raised earth and surrounded by a fence. The raised earthworks and fence provided significant defences.

mottle Blotchy texture created by spots, different shades, streaks, or other random features.

mould (US mold) 1. A container for giving shape to things until it has set, e.g. concrete cube moulds. **2.** Fungus. **3.** Soil that is rich in humus (organic matter).

mouldboard A level board or curved blade on a *bulldozer or *grader used in *earthworks to push soil.

moulded brick A type of *brick shaped in accordance to desired or intended application.

mouth An opening or entrance.

movable Capable of being transported from one place to another with ease.

movement The change in dimension or position of something, *see* THERMAL MOVEMENT.

movement detector A device that detects movement due to the disturbance in low-power microwaves that it omits, used to control automatic doors, lights, etc.

movement joint A joint that is flexible, to allow expansion and contraction to occur.

MU *See* METABOLIC UNIT.

muck Excavated soil or mining waste.

mud Wet sticky soil.

mudsill *See* SOLE PLATE.

mulch A protective covering of organic material spread on soil or around plants to prevent erosion and moisture loss from the ground.

Mullen burst *See* BURST STRENGTH.

Muller-Breslau principle A method of drawing influence lines; it states that the influence line of an action (force or moment) assumes the scaled form of the deflected shape of the structure after removing the restraint.

mullion A vertical member that divides a window.

multimedia 1. A combination of two or more approaches to delivery or exposure, commonly referring to the broadcasting of information, programmes, and visualization software. **2.** A multiple-stage filtrations system using a combination of two or more different waste filters employed so that each medium and layer traps a specific range of particles or type of impurity. Choice of combination and type of media used depends on the impurities that have to be removed from influent water.

multi storey building A tall building having a number of floors.

municipal engineering Roads, sewers, water supply, etc., related to towns and cities.

mural A picture painted directly onto the face of a wall or ceiling.

murder clause (US) A contract clause that unfairly shifts the responsibility to another party who should not carry the burden.

Murphy's Law A saying that states that if something can go wrong, it will.

museum A building with artefacts, paintings, or collections of historical, scientific, or artistic importance, which are displayed and stored.

mushroom construction Reinforced concrete construction where there are no beams—the column has been enlarged at the top **(mushroom-headed column)** to take the load of the floor slab.

mushroom-headed push button A mushroom-shaped stop button, typically used to stop a machine in an emergency and so it is painted red.

MVHR *See* MECHANICAL VENTILATION WITH HEAT RECOVERY.

m

nanotech The use of science, engineering, and technology at an atomic or molecular level, which is between 1 and 100 nanometres. 1 billionth of a metre; $1 \text{ nm} = 10 \text{ m}^{-9}$

narrow strip foundation A length or longitudinal bed of concrete, sometimes reinforced concrete, of minimum width used on good load-bearing strata to support lengths of load such as a wall.

The width of the foundation is sufficient to take the load, typically the load of a wall.

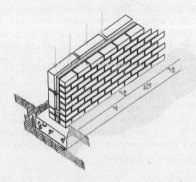

Narrow strip foundation

National Annex (NA) Documentation that supplements the *Eurocodes, containing specification and procedures that should be adopted in a particular member country. *See also* NATIONALLY DETERMINED PARAMETERS.

National Highway Traffic Safety Administration (NHTSA) A US agency with a mission to save lives, prevent injuries, and reduce economic costs due to road traffic incidents, via education, research, safety standards, and enforcement activity.

Nationally Determined Parameters (NDP) Design parameters which are left open in the *Eurocodes for national choice and are therefore obtained from the *National Annex. These values are to be used in the particular member country concerned, and relate to, for example:

- Country-specific data, e.g. geographical and climatic conditions.
- Values to be used where a symbol only is given a specific Eurocode.
- Alternative procedures given in the Eurocodes.

natural draught (US natural draft) The air flow that occurs due to the *stack effect.

natural greenhouse effect (greenhouse effect) A natural phenomenon in which short-wave radiation emitted from the sun penetrates the earth's atmosphere and warms the earth's surface. Infra-red (long-wave) radiation is then emitted from the earth's surface, some of which is absorbed by gases in the atmosphere (greenhouse gases). This warms the earth and the lower atmosphere. The natural greenhouse effect keeps the earth 33°C warmer than it would otherwise be, at an average of 15°C, allowing life as we know it on earth to exist.

natural rock (natural stone, rock stone, rock) Naturally occurring rock made of mineral deposits or natural processes deep within the ground. Rock can be sedimentary, metamorphic, or igneous, formed from different natural processes that give them different characteristics. Sedimentary rocks are formed from particles of sand, shells, and other fragments of natural material which settle and compress into hard rock over millions of years. Examples of this type of rock include conglomerate, breccia, sandstone, siltstone, and shale. Chemical sedimentary rocks include rock salt, iron ore, chert, flint, some limestones, and dolomites. Metamorphic rocks form under the Earth's surface as a result of intense heat and pressure (which are present deep within the Earth). The two basic classes of metamorphic rocks are (1) foliated metamorphic rocks, including gneiss, phyllite, slate, and schist, which are layered in appearance as a result of the pressure and heat, NS (2) non-foliated metamorphic rocks, including quartzite, hornfels, marble, and novaculite, which do not have the same characteristic appearance. Igneous rocks are formed from molten lava (magma) very deep in the Earth. The magma cools and hardens. When the magma cools slowly within the Earth, it has different characteristics and appearances; the crystals that form within the Earth have time to grow and are large. When the magma cools very quickly, crystals are very small and the resulting rock has a glasslike appearance. Types of igneous rock include quartz, feldspar, biotite, amphibole, and olivine.

natural ventilation The process of ventilating a space by opening windows and doors or purpose-provided ventilation openings. The ventilation is driven by the *wind effect and the *stack effect. Natural ventilation does impose restrictions on the plan form of the building, in that all of the ventilated spaces must be within a certain maximum distance from an open window or ventilation opening. However, this spatial constraint does permit the provision of natural lighting to much of the space, and allows the occupants good views of the outside world.

nave Area within a church where the congregation assembles.

Navier's hypothesis The assumption that plane strain sections remain plane. In the analysis of elastic bending, the stress and strain values in the cross-section of a beam are relative to the position of the neutral axis.

Navier Stokes equation Describes the dynamics of a fluid by a set of non-linear partial differential equations. They are derived from applying Newton's second law to fluid flow.

navigation The science of planning/plotting and following a course of travel, particularly used for directing ships and aircraft.

NAVSTAR (NAVigation Satellite Timing and Ranging) A system of 24 satellites that circles the earth every 12 hours to form the basis of the *global positioning system (GPS).

navvy A slang term for ground worker or builder's labourer.

navvy pick A double-pointed pickaxe or a pick with a point and chisel end.

NBS (National Bureau of Standards) *See* NIST.

NCAP *See* EURO NCAP SAFETY RATING.

NDT *See* NON-DESTRUCTIVE TESTING.

neap tide A tide with minimal difference between high and low water levels; occurs twice a month, shortly after the first and third quarters of the moon.

neat cement grout A grout made of cement and water.

neat size Timber or other materials cut to the sizes needed; *see* DRESSED TIMBER.

necking (metallurgy) Reduction of the cross-sectional area of a material when subjected to a tensile force. This phenomenon occurs exclusively in *ductile materials (or materials exhibiting *ductility). When a material is subjected to a tensile load to failure and no necking occurs, the material is said to be *brittle.

needle A support beam that is inserted through a hole in the wall and is either propped at both ends to support the wall above, or rests on the wall and supports parts of the building or scaffolding.

needle gun Power tool with 2.5 mm diameter steel rods, grouped together at the end of a gun which are vibrated against concrete and steel. The gun is used for cleaning surfaces and to roughen the surface.

needle scaffold Scaffolding or platform supported by beams that penetrate the wall (needles).

needling Inserting a beam through a wall to act as a *needle.

negative moment A sign convention to denote the direction of a *moment, typically in the anti-clockwise direction.

negative pressure Any pressure that is lower than atmospheric pressure.

negative skin friction An additional load that can develop on the *shaft of a *pile due to *consolidation of the soil adjacent to the pile, often assumed to be in the region of 10 kN/m^2.

negotiated contract Legally binding agreement where part or all of the contract terms are agreed by negotiation. The general terms of the contract are agreed and in place, with further discussions expected in order to arrive at an agreement for the remaining sections of the contract. The main contract may be a standard contract with parts, such as the contract sum, being negotiated, or the formation of the contract could be a process where the whole agreement is subject to negotiation.

Usually an estimate of the contract value is stated, but the actual contract sum is discussed and agreed. In a negotiated contract the client may have selected the contractor on the ability to perform the works rather than on a tender value, therefore the value of the works and the contract sum should be agreed by negotiation and inserted into the contract.

negotiation stage Following the appointment of the main contractor, the contract value or contract sum is negotiated; see *negotiated contract. The term can also refer to *tendering.

neoprene A polymeric material from the family of rubbers used primarily in construction for plumbing applications.

nephelometric turbidity unit (NTU) A standard unit to denote the amount of *turbidity (suspended solids) in a fluid. A **nephelometer** is typically used to measure turbidity. It measures the scatter of light caused by the suspended solids, when a light source is passed through a column (small sample bottle) of cloudy water.

nepotism A practice by those with power (for example a manager) hiring, promoting, or otherwise unfairly favouring family members or friends over other candidates, simply because they are part of the family or social network of those in power.

net A safety net used to prevent workers and materials from falling.

net load The load on the underlying soil from the structure and backfill; it is the difference in the loading conditions on the soil before excavation and when the structure is complete. Net values are used in settlement calculations; *see also* GROSS LOAD.

net pyranometer *See* NET RADIOMETER.

net radiometer A device that measures incoming and outgoing short-wave and long-wave radiation using a pair of *pyranometers and *pyrgeometers.

network 1. An arrow or precedent diagram that presents the logical connection between *activities, durations, and resources, and is used to plan activities and determine the *critical path. **2.** A grid of *cables, pipes, or *ducts used to distribute electricity, gas, water, or other similar utility services.

network analysis The method of determining the logical sequence of events for activities within projects, how long the project will take, where the *critical path lies, and the points of the project that have *float.

network polymer A three-dimensional polymer material made by crosslinking. A network polymer is composed of tri-functional mer units that form three-dimensional molecules.

Neumann's law A materials property rule stating that the molecular heat in compounds of analogous constitution is always the same. It is named after Franz Ernst Neumann.

neutral axis A location, in the cross-section of a beam, where the bending stresses and strains are zero. In a straight symmetrical beam, before loading, the neutral axis is located at the beam's geometric centre.

neutral conductor The neutral wire in an electrical circuit.

neutralization To counteract something to make it ineffective, for example, to make a liquid neither acidic nor alkaline, or to zero an electrical charge.

neutralization chamber A chamber provided in a wastewater treatment plant where the effluent from different streams combines and is neutralized. Often the neutralization is brought about by the addition of sulphuric acid (H_2SO_4) and caustic soda (NaOH), and the chamber is provided with an air blower at the bottom to ensure the mixing of added H_2SO_4 or NaOH while the pH is monitored. The amount of acid or base added should be sufficient to bring the pH of the tank contents within acceptable limits.

neutron A neutral particle that is present in the nucleus of an atom.

newel The vertical member of timber into which the diagonal strings of a stair can be fixed. The newel can form the base of the *newel post, which can be used to carry the stair *handrail.

newel cap The top of a newel post, often provides a turned decorative finish to the *newel post.

newel drop The bottom of the upper *newel post that projects below the ceiling level, making a decorative feature.

newel end An escalator balustrade end.

newel post The vertical timber at the end of a stair flight, used to carry the strings.

Newmark's charts A graphical method of computing the vertical stress beneath a loaded area. The shape of a loaded area is plotted to a set scale on the chart in relation to the depth and position where the vertical stress is required. The number of squares on the chart enclosed by the loaded area is counted and multiplied by the chart's influence factor and the load per unit area to obtain a vertical stress value.

newton (N) An *SI unit of force, corresponding to the force required to accelerate 1 kilogram mass at a velocity of 1 metre per second.

Newton's laws of motion Three physical laws that describe the fundamental motion of objects in relation to the forces acting on them:
1. Objects will remain at rest or in uniform motion unless acted on by the action of an external force.
2. The velocity of an object changes when acted on by an external force. $F = ma$; force equals mass times acceleration.
3. For every action there is an equal and opposite reaction.

NGOs *See* NON-GOVERNMENTAL ORGANIZATIONS.

NHTSA *See* NATIONAL HIGHWAY TRAFFIC SAFETY ADMINISTRATION.

nib A small projection from a surface.

nibbler A hand-tool that clamps together to cut or bite holes into thin sheets of metal. The tool can be powered or used with handles similar to pliers.

niche A recess in a wall, constructed to house an ornament or vase.

nickel A *non-ferrous metal (Ni), nickel's prime use is as a major alloying element for stainless steel. Nickel alloys have excellent thermomechanical properties—they maintain excellent mechanical properties even at temperatures greater than 500°C. As a result, nickel alloys are predominantly utilized in high temperature applications such as turbines, engines, heat exchangers, and so on.

nickel brass A copper alloy containing zinc and a small quantity of nickel.

nickel sulphide A compound that occurs in some glass-making processes. If trapped the compound can swell and crack the glass.

nicker A large flat mason's chisel used for cutting a groove into stone before splitting it.

nidged ashlar (nigged ashlar) Stone dressed with a pointed hammer, giving a roughened effect.

night lock (night latch) Cylinder lock or bolt.

night vent (night light) *See* FANLIGHT.

ninety-degree bend (90° bend) A *quarter bend.

nippers (steel-fixer's nips, tower pincers) Pair of wire cutters used to tighten and cut tie wire when fixing and binding steel reinforcement.

nipple 1. A rounded fixing pipe with a one-way valve used to fix a grease pump so that lubricant can be pumped into moving joints. **2.** A valve at high points of a hot-water system used to release air locks.

NIST (National Institute of Standards and Technology) A US federal agency measurement standards laboratory. Known as the **National Bureau of Standards (NBS)** between 1901 and 1988.

nitrate A type of salt based on nitric acid, chemical symbol NO_3^-.

nitrification A chemical process where a nitro group is added to a compound.

nitrite An intermediate oxidized ion of nitrogen. Nitrifying converts ammonia (NH_3) to nitrite (NO_2^-), to nitrate (NO_3^-) in the nitrogen cycle.

nitrogen A chemical element (N), present in the form of a gas at ambient temperatures. Nitrogen comprises approximately 80% of the earth's atmosphere by volume.

noble metal A type of metal renowned for its inertness, with a high resistance to chemical reactions.

node An intersection point; on a curve where it crosses itself, where lines converge on a chart or diagram, or where members meet on a *truss.

no-fines concrete A concrete made from coarse aggregate and cement, with no sand or fine aggregates present. It creates a lightweight concrete with voids between the coarse aggregates.

nog 1. A nog brick or block, a fixing brick made from wood or other malleable material. **2.** A *noggin.

nogging (noggin) Short horizontal timbers used to provide bracing and stiffen up stud work. Noggings are also used to provide grounds and fixings in timber walls to enable them to support radiators, shelves, and other fixtures and fixings.

noise Unwanted and undesirable sound.

noise absorption *See* SOUND ABSORPTION.

noise control on-site Noise from construction sites is monitored by the local authority. Guidance is contained within BS 5228–2:2009 Code of Practice for Noise Control on Construction and Open Sites and under Section 60 of the Control of Pollution Act 1974.

Noise Criteria (NC) A rating system used to design a maximum allowable noise in a given space. It uses a series of standard curves of sound pressure level against a range of frequency bands. The system is used to assess the noise produced from building services systems, such as ventilation systems, but it may be applied to other noise sources. *See also* NOISE RATING.

noise insulation *See* SOUND INSULATION.

Noise Rating (NR) The maximum noise level that can be tolerated from an item of equipment.

noise reduction coefficient (NRC) A rating system that produces a single figure for the amount of noise that is absorbed by a material, object, or construction. It is calculated by taking an average of the sound absorption coefficient at four separate frequencies: 250 Hz, 500 Hz, 1000 Hz, and 2000 Hz.

nominal dimension (nominal size) The sizes used to describe the approximate characteristics of the product rather than the exact dimensions. Most timber products are described in this way with the dimensions describing or naming the cross section and length that was close to the rough dimensions of the timber before it was cut, planed, or worked. The dimensions given are within a set tolerance ensuring the product resembles the description.

nominated Product, supplier, or contractor identified for a purpose or position.

nominated subcontractor Company selected by the client and identified in the contract to undertake a specific aspect of the works. The *subcontractor is normally identified in the *bill of quantities during the *tender process.

nominated supplier Company that supplies goods or specialist services selected by the client and identified as the supplier within the contract documents. The supplier is listed within the tender documents and bill of quantities. Within government contracts it is essential that the process allows for fair competition. Framework contracts, where a number of suppliers have qualified to be part of a contractual agreement, are becoming more common in government contracts.

nonagon A nine-sided *polygon.

non-bearing wall A wall that provides a partition but does not take any of the structural loads of the building. A non-load-bearing wall has to sustain its own loads, and loads of fittings placed on it, and remain rigid and stable, but does not carry any of the building loads.

non-biodegradable Materials that do not break down, for instance, plastics.

non-cohesive soil (cohesionless soil) *See* GRANULAR.

non-combustibility test A test undertaken in accordance with EN ISO 1182 where a material is exposed to a temperature of 750°C for a minimum of 30 minutes. Used to determine whether a material is combustible or not.

non-combustible A solid material that is not capable of burning. *See also* NON-FLAMMABLE.

non-concussive action A tap or valve that is self-closing.

non-contradictory complementary evidence (NCCI) Supplementary information and guidance as defined in a *National Annex.

non-crystalline A material that has a non-ordered atomic arrangement.

non-destructive testing (NDT) A process used to determine the properties of a material without inflicting any permanent damage or alteration in properties.

non-displacement pile *See* DISPLACEMENT PILE.

non-drip paint (drip-free paint) A thixotropic paint with a jelly-like consistency and used as an alternative to gloss paint.

non-ferrous A material that does not have iron as its prime constituent.

non-ferrous metal alloy A metal alloy not based on iron. A non-ferrous metal may contain iron in small amounts, however, a non-ferrous alloy is not based on iron. Examples include aluminium, copper, lead, titanium, and zinc. In construction, non-ferrous alloys are not used as extensively as *ferrous metal alloys.

non-flammable A liquid or gas that is not capable of burning. *See also* NON-COMBUSTIBLE.

non-governmental organizations (NGOs) Non-profit organizations that are cooperative rather than commercial. They work on a national or international level, within areas such as humanitarian, education, healthcare, political, social, human rights, and environmental protection.

non-load-bearing wall A partition wall that carries no load other than its own weight; *see also* LOAD-BEARING WALL.

non-manipulative joint A simple joint made by inserting pipes into a coupling tube. The jointing tube or sleeve has rings, olives, or glands within it that are compressed against the pipes inserted into the sleeve, making a watertight joint. Different compression joints are needed for light gauge pipes, plastic pipes, and copper tubes. Many of the fittings can be unscrewed and used again.

non-mortise hinge (surface-fixed hinge) A butt hinge that is cut so it is only as thick as one of its flaps when closed; the hinge may be cut so the flaps fit one inside the other.

non-performance When the contractor does not do the assigned work and is in default of the contract obligations. If the default goes to the root of the contract and is fundamental to the contract performance, the client may be entitled to bring the contract to an end; *see also* DETERMINATION.

non-renewable energy Energy that comes from a natural resource, which cannot be renewed, for example, coal, gas, and oil.

non-renewable resource *See* FINITE RESOURCE.

non-repeating thermal bridge A *thermal bridge that is intermittent and occurs at a specific point in the construction. Often caused by discontinuities in the thermal envelope which may be a result of the construction method used or may be due to changes in materials over the thermal envelope. They commonly occur around openings and other instances where materials of different thermal conductivities form part of the external envelope. Typical examples occur around windows, doors, and roof lights, around loft hatches, where internal walls or floors penetrate the thermal envelope and where steel I-beams have been used to support timber roofs. *See also* REPEATING THERMAL BRIDGE.

non-return valve A valve that allows a fluid or gas to flow in one direction only.

non-rising spindle A spindle that does not move up or down when turned; the spindle moves a threaded piston which controls the flow.

non-setting glazing compounds A flexible non-setting material used to bead glazing into the frame. Used in instances where the glazing may be liable to thermal or structural movement.

non-slip Describes floors, treads, or nosings with finishes designed to be slip-resistant even when wet, dusty, or greasy. The surfaces may be embossed with shapes and patterns, have recessed grooves, have abrasive materials inserted into the surface, rubber raised studs, or use other materials that are naturally slip-resistant.

non-splittable activity An activity on a network that must be completed before any other activities can start.

non-steady-state diffusion On a microscopic level in metals, the diffusion condition for which there is fluctuation of diffusing species through the metal. This usually refers to the transport of atoms through a metal, or from one metal to another.

non-trafficable roof A roof that is not designed to be walked on.

non-vision glass *See* TRANSLUCENT.

non-vision grille A grille that has overlapping horizontal louvres to limit vision through it.

non-woven A *geosynthetic structure that is bonded by a mechanical, chemical, thermal, or solvent process. Nonwoven structures are extensible, low in strength, thus have low modulus, but are more uniform (isotropic) in strength than woven structures as a result of the random orientation of the fibres during manufacture. They are also permeable, and have the ability to transmit water within their plane. Applications include *filtration, *drainage, and *separation.

normal A line that starts from the centre of a circle and extends beyond the circumference.

normal distribution (Gaussian distribution) A bell-shaped probability frequency distribution in which the mean, median, and mode are all of the same value, defining the mid-highest point on the curve.

normal fault A facture in rock that is characterized by the upper side of the fault displacing downwards.

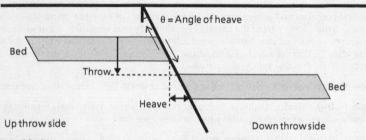

Ground surface

θ = Angle of heave

Bed

Throw

Bed

Heave

Up throw side

Down throw side

Normal fault

normalized leakage area *See* SPECIFIC LEAKAGE AREA.

normalizing (metals) For ferrous alloys (especially steels), heating up to or above a certain temperature (usually above 500°C), then air cooled. This type of treatment is very common in steels to improve their *ductility, which is vitally important for structural steel.

normal soluble salts A clay brick classification that means that the surface of the brick is liable to a white powdery efflorescence stain caused by soluble salts within the brick that leach out when the brick becomes wet.

normal stress *See* DIRECT STRESS.

normative In respect to the *Eurocodes, a requirement that has the force of a *European Standard.

northing A position or movement northwards in reference to latitude coordinates.

northlight roof *See* SAWTOOTH ROOF.

nose An overhang or prominent part of a projection.

nosing (nose) The overhang or front part of each tread.

nosing line (pitch line) The theoretical projection produced if a line is strung across the tip of each nose on the treads of a flight of stairs. The angle between the horizontal and the nosing line is the degree of pitch.

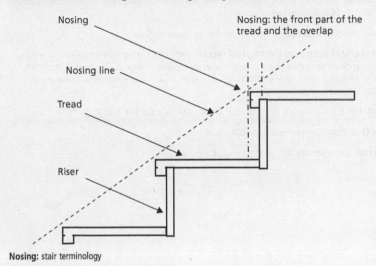

Nosing

Nosing: the front part of the tread and the overlap

Nosing line

Tread

Riser

Nosing: stair terminology

notch board A cut string.

notched joint A joint made of notching sections of timber to make an interlocking joint.

notched trowel A square trowel with rectangular notches cut out allowing adhesive to be pasted onto walls, floors, and ceilings. The rebates cut into the adhesive allow tiles and other fixtures to be placed and levelled. Placing tiles on adhesive without recesses proves much more difficult to level.

notch sensitivity A material susceptible to failure due to the presence of a crack or opening.

notice to bidders (US) Instruction that opens a contract to tenders.

notice to proceed Instruction from the client or client representative to commence work.

noxious Harmful to life/health—something that is poisonous.

N-truss *See* PRATT TRUSS.

nuclear Relating to the nucleus of an atom.

nucleation On a microscopic level in metals, the initiation or the initial stage of the formation of a new phase, which imparts certain mechanical properties to the metal. The mechanical properties of a metal, e.g. tensile strength, are strongly influenced by the presence of phases within the metal of a particular chemistry.

For a particular phase to be present it is necessary for it to nucleate (or form) at the first instance. This process is called nucleation.

nucleus The central or most important part of something, for example, the central positively charged region of an atom.

null No value, relating to zero, invalid, or meaningless.

numerical analysis (numerical modelling) The use of mathematical analysis to solve problems rather than using *experimental techniques. Typically, complex finite element and finite difference computer-modelling programs are used to solve interactive processes.

nutrient A substance that provides nourishment for life and growth.

NVQ National Vocational Qualification

nylon (polyamide ii) *See* POLYAMIDE.

n

OBE 1. *See* OPERATING BASIS EARTHQUAKE. **2.** Officer of the Most Excellent Order of the British Empire. There are many professionals in construction who have received honours for their professional and personal endeavours.

((())) SEE WEB LINKS
• Honours list

obelisk Tall column or shaft of stone with pyramid top, typically of Egyptian origin with hieroglyphics etched on each face.

objective lens The largest lens, nearest to the subject being viewed, in an optical instrument.

obligation What the parties to a contract agree to do for each other. A full list of the commitments and duties are contained in the conditions of contract. In a simple construction contract both the client and the contractor have obligations to each other; for example, a contractor has obligations to deliver a service and provide a building, while the client has obligations to allow the contractor access to the site and make payment for the works.

OBM (Ordnance benchmark) A level or datum the level of which has been officially fixed with reference to *ordnance datum. *See also* BM.

((())) SEE WEB LINKS
• Benchmark locator

observation panel A glass panel incorporated within a door that enables a view to be gained.

obsolescence The process of an item, building, structure, or service passing out of use so that it is redundant and no longer useful. The process can apply to old and new items, services, structures, or buildings that no longer serve a useful function.

obtuse 1. Slow to understand. **2.** An angle between 90° and 180°. **3.** A triangle with an interior angle greater than a right angle.

occupancy The number of people that a room or building is capable of accommodating while still adhering to fire and safety regulations. The number of people a room is capable of holding is important for determining how it will fulfil evacuation requirements and meet the standards set by the building regulations.

oceanic Related to the sea.

oceanography The scientific study of the sea.

octagon A closed shape with eight sides/angles.

octahedral position At a microscopic level in metals, a position an atom can take inside the metal. Atoms in several metals and some ceramic materials take the octahedral position.

octane number (octane rating) A measure of the quality of fuel in a combustion engine. The higher the number/rating, the less likely the cause of **knocking**, an explosion caused by premature burning in the combustion chamber.

OD *See* ORDNANCE DATUM.

odometry The use of data from motion sensors to predict change in position in respect to time. It is currently part of the technology used to develop *autonomous cars.

odour A smell, either pleasant or unpleasant.

oedometer A piece of laboratory equipment (test) to determine the magnitude and rate at which consolidation settlement will take place.

offcut 1. Any piece of material that remains when something is cut down to the required size. **2.** Wood remaining when a standard length of wood has been cut to the required size. These short, irregular pieces of wood are increasingly being used and recycled.

offer First part of a contract, an expression of interest by a party to be bound by an obligation to another party. If there is an offer, acceptance, and some form of consideration, a binding contract is formed.

offer up To hold a component in position against another to check that it fits or aligns with the other components.

off-form concrete The surface finish obtained when concrete is simply struck from its formwork or mould. The texture and shape of the mould provide the finish for the concrete.

offset A horizontal distance measured perpendicular to the survey line to obtain the position of a specific point or object.

offshore 1. Located in the sea a distance away from land, for instance, an offshore oil rig. **2.** Moving away from the shore, for instance, an offshore wind.

off-site Work carried out outside the site boundary: work undertaken in a joinery shop, or work that is *prefabricated.

off-site construction The fabrication, assembly, and completion of works away from the site location, i.e. *off-site. The components, elements, or whole buildings can be constructed (partially or almost fully assembled) and then delivered to the final place where the building will be assembled and remain. Off-site construction is becoming increasingly common as the construction industry transforms the way it builds, with an increasing focus on quality, performance, engineering, automation, off-site fabrication, and a skilled workforce tailored to modern methods of construction (MMC).

off-tamp finish (tamped finish) A textured concrete finish achieved by dabbing a straight edge of timber or metal up and down over the surface of the concrete, the tamping creates an undulating finish. The rough finish is often used for roads, ramps, and car parks to provide a skid-resistant surface.

Ofwat (Water Services Regulation Authority) The body that regulates water and sewage utilities in England and Wales.

(⊕) SEE WEB LINKS
Official Ofwat website

ogee (ogee joint, OG joint) 1. A rebated joint or half-lap joint. **2.** A spigot and socket joint that allows one tube to be located and connected within another. The widened neck of one end of one pipe allows the narrower end of another pipe to be housed within it. The narrowed neck is formed by the wall of the pipe being rebated to half its original thickness.

ohm The unit of electrical resistance. One ohm exists when a current of one ampere produces a potential difference of one volt between two points in a circuit.

Ohm's law A law that states that the current flowing between two points in a circuit is directly proportional to the voltage within that circuit.

O horizon An organic layer at the surface of the soil containing decaying plant and animal matter.

olive A brass ring that is slid over the narrow end of a coupling and then compressed using a nut to form a compression joint.

on-costs The overheads—the site, head office costs, and personnel that support the company operations. All of the costs must be covered within construction costs. Costs are added to contracts on a percentage or fixed fee basis.

one-and-a-half-brick wall (brick-and-a-half wall) A solid wall that has been constructed with a thickness of one and a half brick lengths (327 mm thick).

one-brick wall A wall that is constructed with the thickness of one brick length.

on edge *See* BRICK-ON-EDGE.

one-hole basin *See* SINGLE-HOLE BASIN.

one-line diagram A single line electrical diagram.

one-pipe system (single pipe, single pipe system) An above-ground drainage system that comprises a single vertical discharge stack, which conveys both soil and wastewater to the drain. An anti-siphon arrangement is required on most branch pipes to enable unrestricted layout of the appliances.

one-way switch A simple switch that either switches an appliance on or off.

on/off controls An automatic controller that enables equipment to be switched on or off.

on-site Work carried out within the site boundaries.

OPC (ordinary Portland cement) *See* CEMENT.

open assembly time The time allowed when adhesive is applied to bring two or more components together. Fixing the components after this time will result in a weaker bond.

open bidding (open tendering) Competitive tendering where the work is openly advertised and contractors are invited to submit tenders for the project. No restrictions are placed on the initial tendering process although there may be pre-qualification required to secure the work, for instance, good financial status and experience of certain types of work.

open BIM An open platform for collaborative design, realization, and operation of buildings, formed on open standards and workflows.

open boarding Planks of wood placed side by side with gaps between each board, often used to form a fence.

opencast mining (open cut mining, open pit mining) A method of excavating rocks and minerals at the surface of the ground where the mine is open to the atmosphere, rather than from a tunnel deep below ground.

open-cell ceiling A suspended ceiling with a cellular opening in the finish and through the tile. Although you can see through the open cells in the ceiling, one's sight is naturally drawn to the suspended ceiling finish and distracted from the structural ceiling behind it.

open cornice (open eaves) The part of the roof rafters that extend beyond the wall of the property creating an overhang. The eaves are classed as open when the soffit is not boxed in.

open cut An excavation that is open to the atmosphere. *See also* OPEN CAST MINING.

open defect Any surface features of wood such as knots, shakes, splits, and worm holes that are easily recognizable.

open drained joint (drain joint) A V-joint formed in the edge of concrete panels for drainage. The indented V is formed along the vertical edge of the panel. When adjacent concrete panels come together the two V joints form a square drainage channel. A neoprene gasket is placed in the centre of the channel to act as a baffle. Wind and water can be blown into the open joint but the baffle prevents the water passing through, allowing it to drip down the cavity and out of the open drain.

open eaves *See* OPEN CORNICE.

open floor An intermediate floor that has no ceiling.

open-graded aggregate An aggregate with particles of similar size that on compaction result in large air voids. *See also* GAP-GRADED AGGREGATE.

opening Structural gap in a wall, roof, or floor ready to receive a window, door, or panel.

opening face The side of a door or window, which when open, is furthest from the frame.

opening leaf The opening *leaf of a door.

opening light A window that opens; a window which does not open is called a *fixed light.

open joint Gap or rebate left between components when joined together to provide a feature.

open loop convective bypass A type of *convective loop bypass where the air located within the building component is replaced with external air, enabling the heat to transfer from one area to another.

open metal flooring A mesh type open flooring is formed by laying strips of metal on edge, or bars or expanded metal mesh on a steel frame. The metalwork provides a strong, slip-resistant floor structure with gaps between preventing dust and dirt build-up and allowing light to pass through the floor. Useful in warehouses where good vision at high level is needed.

open mortise (open mortice, slip mortise, slot mortise) Joint in wood used to fasten two pieces of timber at a 90° angle. A slot or groove is cut in the end of timber and made so that it is ready to receive the thin tenon of another timber to form a 90° angle joint.

open newel stair Geometrical stair without intermediate *newel posts.

open pit mining See OPEN CAST MINING.

open plan A room, normally an office, that is just divided by furniture and screens, no dividing walls.

open-plan stair A flight of stairs without any solid risers; open risers (see OPEN RISE) exist between each *tread.

open rise The vertical component of the staircase which is left open between the horizontal foot *treads. Where a staircase has treads mounted in strings but no back or vertical riser, it becomes an **open-rise staircase**.

open roof A roof without a ceiling, exposing the trusses to those inside the building.

open-source Describes software that is freely available for users to download from the Internet and use, i.e. without incurring any cost.

open stair A stairway which is open on one or both sides.

open string A string cut to match the profile of the treads, a zigzag string is produced as the string is cut around the profile of each tread and riser.

open tendering See OPEN BIDDING.

open traverse A *traverse that ends at a *station whose relative position is not previously known, it is not referenced back to the first station.

open valley An exposed valley. The tiles on the roof lead into the valley but do not cover it.

open-vented system A type of wet central heating system that has a feed and expansion, and is open to the atmosphere via an open-vent pipe. The water temperature in open-vented systems must be kept well below its boiling point in order to stop steam from being formed. *See also* SEALED SYSTEM.

open well A shallow *well that is hand-dug to depths between 1 m and 15 m, with a 1–2 m diameter.

open-well stair A stair flight with an open well and a gap between the flight of stairs and the wall.

operating basis earthquake (OBE) The amount of ground vibration a nuclear power station (or other such buildings) is designed to withstand from an earthquake without compromising health and safety.

operational research (operations research) The use of mathematical techniques to improve decision-making.

operative An on-site worker, usually refers to someone operating plant, a labourer, or a chain person rather than a professional or skilled worker.

operative temperature *See* DRY RESULTANT TEMPERATURE.

optical Relating to vision or visible light.

optical smoke detector A type of smoke detector that uses a photoelectric cell and an optical beam to detect the presence of smoke.

orangery Framed building with a large area of glass, much like a conservatory, which allows much natural light into the building.

orbit The path a celestial body (a planet, moon, or satellite) takes around another larger celestial body (for instance, the sun).

orbital sander A circular motorized sander with a flat surface turned in a circular motion to grind flat and smooth the surface on which it is placed.

ordering At a microscopic level in metals, the positioning of all types of ions within the material in an ordered, repetitious pattern, rather than in a random arrangement. Usually in metals, the ions are arranged in an ordered pattern.

orders Instructions given to undertake work.

ordinary Portland cement (OPC) *See* PORTLAND CEMENT.

ordinate The vertical or y-coordinate in *Cartesian coordinates. *See also* ABSCISSA.

ordnance benchmark (OBM) A permanent mark on a fixed object, denoting an elevation which has been related back to *ordnance datum. Used by surveyors to establish the initial elevation of a site.

ordnance datum (OD) The mean sea-level at Newlyn in Cornwall, UK, and used as a datum in Ordnance Surveys.

ore A naturally occurring mineral from which metals are extracted.

organic landscaping Soft landscaping of living matter: shrubs, bushes, trees, and other plants.

organic loading rate In waste (sewerage) management the term is used to express the availability of substrate (food) for microbes. It is expressed in the units of $kgCOD/m^3$ -day and has the formula COD applied/HRT (batch system); COD applied × flow rate/working volume (continuous system).

organic rendering 1. Use of masonry paint. **2.** Use of grass, ivy, and other wall-climbing plants to cover a wall.

organism A living thing; an animal or plant.

organizational culture The underlying beliefs, assumptions, values, and ways of interacting that contribute to the unique social and psychological environment of an organization; the way things are done.

oriel window A bay window that projects only from an upper floor.

orientation The position or direction of something.

orifice An opening (hole) in a body from which fluid is discharged.

origin 1. The starting point of something, the beginning. **2.** The point where the x- and y-axes intercept; *see* CARTESIAN COORDINATES.

original ground level The height of a site before *groundwork.

O ring (O-ring) A rubber ring, normally mounted in a groove, used as a seal against air, water, or oil.

ornament Statue or feature used to enhance landscape, building, or interior.

orogeny A mountain formed by folding, faulting, and uplifting of the Earth's crust.

orthogonal At right angles to each other; can be used to indicate the crystal structure in a mineral.

orthographic Made up of vertical lines.

oscillating Moving or swaying backwards and forwards. *See also* OSCILLATION.

oscillation Movement back and forth or regular variation in magnitude or position around a central point.

osmosis The diffusion of a liquid via a semi-permeable membrane from a less concentrated solution to a more concentrated one, until the concentrations are in equilibrium. Used in desalination.

ounce An imperial unit of weight equal to one-sixteenth of a pound (lb).

outcrop Strata exposed at the ground surface.

outer hardening A treatment used in metals to increase the *hardness or resistance to surface indentation. This treatment is undertaken at a microscopic level and involves moving the *dislocations within the metal; this alters the microstructure of the metal which is tailored to increase the hardness of the material.

outer shell The valence electrons that are very loosely bound to the nucleus of an atom. As the electrons are under a smaller force by the nucleus, so the outer shell's electrons are exchanged first in any reaction. Thus, the state of the outer shell is very important in determining the reactivity of a material.

outer string One of the two diagonal components of a staircase that carries the *treads and *risers; it is the *string which is furthest away from the wall.

outfall A place where a river (or drainage pipe) discharges into the sea (or a treatment works).

outfitting The construction work, fixing, finishes, and furniture needed to complete the internal aspects of a building.

out for tender The period between the advertisement of the contract and the time when the tender has to be submitted; this is also the maximum period the contractor has to prepare the contract.

outgassing The removal of excess or unwanted gases from a material.

outgo The pipe from a sanitary fitting that connects it to the drain. The final part of the pipework before the gulley or sewer connection.

outgoing circuit The wiring between the switchboard and connection points or outlets.

out of plumb Not vertical. Usually used to describe the degree that something is not vertical. Recorded using the horizontal measurement between the top and bottom of the item from a vertical line, or the degrees difference that the centre line is from the vertical.

out of the ground Used to describe the moment when the building structure starts to emerge above ground level. The first parts of a structure that extend above the ground are normally the structural frame or the walls.

out of wind Area that is sheltered from the wind.

outreach arm (bracket arm, carrying arm) The arm that extends from the main column or wall to carry a light or luminaire.

outrigger 1. A beam that extends from a building to carry scaffolding. The beam removes the need for the scaffolding to extend down to the ground. The beam is propped and secured against the structural ceiling and extends out of a building opening such as a window. Scaffolding, such as flying scaffolding, can be secured to the beam. **2.** Legs that extend out from a crane or excavator to provide extra stability

outside glazing Term used to describe external *glazing that has been installed from the outside of the building. *See also* INSIDE GLAZING.

outstanding works 1. Defects or work that need to be rectified. **2.** Work that is yet to be completed.

oval wire brad Nail with an elliptical shape that is positioned with the longer dimension in line with the grain to reduce the chance of the wood splitting.

overburden pressure The stress imposed at any point within the ground from the soil directly above. It is calculated by multiplying the unit weight of the overlying soil by the height of the soil at that point.

overcladding The process of applying an additional external cladding to a building as part of refurbishment.

overcloak Used in sheet metal roofing to describe the part of the sheet that overlaps the sheet beneath at a drip, roll, or seam.

overcoating The process of applying an additional coat of paint or varnish to an object.

over-consolidated clay A clay that has previously been subjected to a higher stress than its present-day overburden stress. The reduction in stress could have resulted from melting of ice sheets, erosion of overburden pressure, or a rise in the water table. Such a layer is more liable to swelling when wet.

overcurrent (excess current) The presence of a larger-than-expected current within a circuit. Usually occurs due to a *short circuit.

overdesign factor *See* DESIGN FACTOR.

over-fascia vent A hole cut into the top of the *fascia to allow airflow into the roof space at the eaves.

overflow (overflow pipe) A pipe connected to a basin, bath, sink, or tank that discharges excess water when it is full. The excess water will either be discharged to a *waste pipe or externally via a *warning pipe.

overhand work 1. Bricklaying from the inside of the building reaching out to the external leaf or the cavity to lay bricks or blocks. **2.** An awkward or complicated method of working.

overhang Projection from a structure, roof, or wall that extends beyond the face of the support.

overhanging eaves The lower-most edge of the roof that extends out and projects from the building, protecting the face of the building.

overhaul The haulage distance that exceeds the *free haul distance.

overhead crane (gantry crane) A crane that runs on a *gantry.

overhead door *See* SWING-UP DOOR.

overheads (administrative charges, establishment charge, on-costs) The operating costs of the company. The head office, management, administration charges, and the costs of offices that are used to operate the company and are charged proportionally to each contract, or at a flat rate per job.

overlap The degree that something extends over something else, the *lap. Roof tiles need to extend over each other to ensure rain does not pass through. Some work and management packages will need to extend into the next operation to ensure successful handover.

overlay 1. Something that is spread over the top of something else to provide protection or enhance performance. Insulation that is spread over the top of a ceiling is often called an overlay. **2.** Transparent drawing placed over the top of a plan to show the position of services or fixings to assist design coordination.

overload A load that exceeds the design load in structures and electrical circuits.

overpanel A infill panel placed above a door to close the gap between the ceiling and the top of the door.

override 1. Manual control switch that allows the user to take control away from an automatic system. **2.** To manage a group regardless of the system that is in place.

oversail To overhang or extend over something. A crane needs to ensure there is an oversailing agreement for it to be used within the airspace of another person's property.

oversailing course Brickwork course that extends out from the wall to act as a feature or to offer protection to the wall below.

oversite An operation, task, or feature that extends across the boundaries of the site.

oversite concrete Concrete placed across an excavation to seal the ground. The concrete normally covers the full area of the excavation.

oversite work Operations that take place within the full area of the construction site or up to a boundary.

oversize Piece of equipment or component that has larger-than-normal dimensions but has been specifically chosen for an operation.

overspray The excess paint outside the intended area to be sprayed. When paint is applied by a spray gun it will cover an area larger than intended. Masking of surfaces and covering of areas ensures that the spray does not damage the other areas.

over-tile 1. Decorative tiling laid over other tiles, also called **imbrex** or **Spanish tiling**. **2.** General tiling on top of existing tiles that have strong structural fixing to the substrate and are flat; this is not considered good practice but is sometimes undertaken in refit and refurbishment work.

overtime Additional working compared to standard or contracted hours. Sometimes paid at a higher rate.

overturning When something has the tendency to invert; retaining walls and other such structures are checked for overturning moments.

overvibration Vibration of concrete for longer than required leading to segregation of the aggregates and cement. Concrete is vibrated to compact it and remove air bubbles, but excessive vibration can lead to the large heavier aggregates dropping to the bottom of the concrete, and the finer aggregates being forced to the top, resulting in a defective concrete.

overvoltage *See* VOLTAGE OVERLOAD.

Owen tube A sampler used to measure the settling velocity of fine sediment.

owner (building owner, client) The person who owns the property and who often instructs work to be carried out. The contractors and architects in building projects undertake the works for the building owner.

oxbow U-shape bend in a river. An **oxbow lake** is formed when the river abandons its original meandering path by eroding a more direct path.

oxidation A reaction whereby a material undergoes a chemical interaction with oxygen.

oxide A compound formed by the reaction of an element with oxygen.

oxyacetylene A compound comprising of oxygen and acetylene used in *welding.

oxy-cutting Use of oxyacetylene flame to melt metal, then once metal is glowing, an increased supply of oxygen is used to react with the molten steel, producing greater heat which cuts the steel away.

oxygen A gas at ambient temperature comprising 20% of the Earth's atmosphere.

ozonation A water treatment technique. *Ozone is bubbled through the water to destroy bacteria, viruses, and traces of pesticides; it also helps to break down compounds that cause colour, taste, and odour in the water.

ozone A gas at ambient temperature usually present in the atmosphere which acts as a reflector for ultraviolet radiation.

ozone depletion The process whereby the concentration of *ozone in the Earth's atmosphere is reduced. The *ozone layer located over the Antarctic is particularly susceptible to depletion at certain times of the year due to climatic conditions. It is most severe above the Antarctic during the spring.

ozone hole An area of the stratospheric *ozone layer where the levels of *ozone have been depleted (*see* OZONE DEPLETION), resulting in a reduced concentration of ozone. This process is commonly referred to as thinning of the ozone layer.

ozone layer A term commonly used to refer to the layer of *ozone found in the stratosphere, a region of the Earth's atmosphere that begins approximately 10 km above the Earth's surface and extends up to approximately 50 km. Ozone is also found in the troposphere, the layer between the Earth's surface and the stratosphere. The majority of the ozone in the Earth's atmosphere is located in the stratosphere. In the stratosphere, the ozone layer is thinnest around the equator and denser towards the poles.

package deal A standard form of contract where both the designing and building of the project are contained within the same legally binding document. *See also* DESIGN AND BUILD.

packer Piece of rigid material used to lift, raise, or secure an item at the right height, level, or position. The hard resilient piece of material is placed under or between an object, ensuring it remains in the new position.

packing 1. Strip of *packers. **2.** Material used to fill a void.

packing gland The packing and seal around the stem of a tap, the packing within the gland makes the seal watertight. The *packing nut is tightened onto the fixing, securing the seal.

packing nut Gland nut used to secure a watertight seal on the tap.

pad 1. A block of stone or dense concrete placed under a concentrated point load such as a beam or column, the pad stone helps to distribute the point load over a wider area, reducing the load per unit area. **2.** A foundation that is square in plan, a point load such as a column sits in the middle of the pad, and the pad foundation distributes the load to the ground. **3.** A block or cushion.

padbolt A bolt that can be padlocked.

pad footing (base, pad foundation) An isolated *foundation that is square or rectangle on plan and used to distribute point loads, such as columns, to the ground. Most foundations are reinforced with steel rods. The columns that rest on the pad are secured and positioned using *holding-down bolts, which are cast into the *concrete. Other foundations may be connected into the *pad. To ensure there is a good link, steel reinforcement *starter bars are used to link the reinforcement in the pad foundation to the reinforcement in other foundations. Starter bars protrude from the concrete footing, enabling the adjoining footings reinforcement to be linked to it. *Ground beams are often connected into the side of pad foundations. Ground beams carry the longitudinal loads, such as walls and floors.

padsaw A pointed saw blade that is secured in a hand block.

padstone A grinding stone for sharpening cutting blades.

pagoda Frame used as an ornamental garden feature to train and support climbing plants.

paint (painting) A method of applying a thin coating of material to the surface of another solid or structure. The paint is applied in liquid form, which then becomes solid before being put into service. Thus paints are surface coatings, marketed in liquid form, and usually suitable for site use. They serve one or more of the following purposes:

- To protect the underlying surface by excluding the atmosphere, moisture, chemicals, fungi, insects, etc.
- To provide a decorative, easily maintained surface.
- To provide light and/or heat reflecting surfaces.
- To give special effects, for instance, inhibitive paints for protecting metals, electrically conductive paints to provide a source of heat, condensation-resisting paints.

painter Skilled tradesperson who applies paint to the surface of materials.

painter's labourer Normally an apprentice painter who assists the painter by mixing paint, ensuring materials are ready, and also assists with painting that requires less skill.

painting The application of undercoat and paint to a surface to achieve the desired finish. The substrate or surface to which the paint is applied is cleaned and prepared ready to receive the undercoat and final finish.

paint removal The stripping of paint from a material's surface; for best results the removal of the finish should be down to the substrate. Chemicals can be used to break down the paint, which can then be removed with a *paint scraper. Alternatively, a blow torch or hot air blower can be used on oil-based paints to heat the paint, causing it to bubble and then, once soft, it can be scraped off the surface. Paint can also be removed by shot blasting.

paint scraper Flat tool used to remove paint from the substrate such as wood or metal. The paint is first treated with chemicals or a hot air blower or torch. The treatment helps to lift the paint away from the surface making it easier to remove the paint with the scraper.

paint stripping The removal of paint from the surface of a material. *See* PAINT REMOVAL.

paint system The prescribed application, preparation, and build-up of layers of paint to provide a robust surface finish.

paintwork The finish achieved when paint is applied to a surface.

pair Two identical or matching items.

palaeoseismology The study of past earthquakes by looking at geological sediments and rocks.

pale (pales, paling) Vertical metal or wooden struts, stakes, or boards, forming a palisade fence.

palisade Vertical struts driven into the ground or fixed together to form a fence.

Palladian window A window that has three vertical sections; the top of the middle section is arched in shape and wider than the two side sections, which are horizontal along their tops.

pallet 1. A thin strip of wood used as a fixing fillet. **2.** A timber frame used to stack, store, and transport materials such as bricks, tiles, and stone. The frame is built so that it can be lifted easily by a forklift and transported on a truck.

Palmgren-Miner rule *See* MINER'S RULE.

pan 1. Shallow tray used for catching liquids or debris, placed underneath an object to catch and contain the substance. **2.** The bowl-shaped part of the toilet (*water closet or WC) that contains the water and receives human excreta and urine.

pan connector (sanitary connector, WC connector) A flexible or rigid connector, usually plastic, that connects the *water closet *pan to the soil pipe.

pane A framed sheet of glass installed within a window or a door.

panel A distinct sheet of material or sections of infill (brick, concrete, or stone) that are placed between a structural frame.

panel form Small formwork panels, 600 mm x 1200 mm, that lock together to create larger sections of permanent formwork.

panel heating A type of space-heating system comprising a panel that is heated using hot-water pipes or, more commonly, electric conductors. The panels can be ceiling-, floor-, or wall-mounted.

panelled door (panel door) A door comprising panels between the *rails and stiles.

panelling Regular units used to clad or line walls, ceilings, or floors, providing a modular decorative finish.

panel pin Small wire nail with a small head, which makes the nail less visible when embedded in the timber.

panel planer A piece of equipment that reduces lengths of timber down to a set thickness.

panel products Long units of building materials used for cladding and lining walls, ceilings, and floors.

panel saw A small handsaw, with short narrow-gauge cutting teeth, used for making precise cuts.

panel wall A non-load-bearing wall made up of regular units of timber, brick, or block.

pan form (waffle form) Void formerly used within in-situ concrete floors. The box-shaped *formwork creates regular rectangular indents in the soffit of the floor resembling a waffle pattern.

panic bolt A device that is capable of operating the cremona bolt of a fire or emergency door from the inside.

panic hardware *Door furniture that can be used to open a fire or emergency door from the inside when required.

panic latch A device that is capable of operating the *latch of a fire or emergency door from the inside.

pantile An elongated S-shaped clay or concrete interlocking roof tile.

pan wash sink 1. A *sink located in a kitchen and used to wash cooking pans. **2.** A bedpan sink.

paperhanger Decorator skilled at hanging wallpaper.

parabola A curve formed such that any point along it is equidistant from both a fixed point, termed the focus, and a fixed straight line, termed the directrix.

paraboloid A three-dimensional shape in the form of an ellipse or hyperbola, which has been created by turning a parabola around an axis of symmetry.

parallactic angle The angle made between the differences in line of sight due to *parallax.

parallax The apparent change in position when an object is firstly viewed with one eye and then with the other, or when the person viewing the object changes position.

parallel Two objects or lines running consistently at an equal distance apart.

parallel coping A *coping used to cover the sloping *parapet wall at a gable end.

parallel gutter A *box gutter.

parallelogram Four-sided figure or shape with opposite sides parallel.

parallel thread A *thread that has a constant diameter. *See also* TAPER THREAD.

paramagnetism A type of magnetic behaviour exhibited by certain materials.

parameter A measurable quantity of a material, which can be altered to vary the results or outcome; *see also* PROPERTIES.

paramount partition Prefabricated non-load-bearing partition wall.

parapet A low wall along the edge of a roof, balcony, or terrace. Used to protect people against a sudden drop and for decorative purposes.

parapet gutter A concealed gutter, rectangular in section, located behind a *parapet wall, between the wall and the roof.

parapet wall An extension of a wall at the edge of a balcony, roof, terrace, etc. creating a wall that acts as a barrier to an edge or boundary. The origin is Italian, from the word *parapetto* (from *parare* 'to cover or defend').

parge (pargeting, parging) 1. A thin coat of *plaster or *mortar that is used to coat walls. Traditionally used as a lining material for chimney flues or to waterproof external walls. Currently used to improve the acoustic performance of masonry aggregate block party walls and to improve the *airtightness of masonry aggregate block party walls. **2.** Decorative ornamental plasterwork.

paring chisel Long narrow cutting tool used for clearing out mortise joints, the sharp bladed tool is worked by hand without the need of a mallet.

Parker truss A *Pratt truss that has an inclined (arch shape) top chord.

parliament hinge (butterfly hinge, H-hinge, shutter hinge) A hinge in the shape of the letter H.

parlour A room that was traditionally set aside for conversations.

partial factor (γ) A value greater than or equal to 1 applied (by multiplication or division) to the *representative value to take into account any uncertainty within the derivation of that value, from which the *design value is obtained.

partial fill (partial fill insulation, partial fill walls) Describes a cavity wall where the void is only part-full of insulation, being different from a fully filled cavity, where the insulation fills the whole cavity.

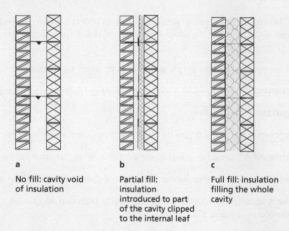

a
No fill: cavity void of insulation

b
Partial fill: insulation introduced to part of the cavity clipped to the internal leaf

c
Full fill: insulation filling the whole cavity

Partial fill

partially separate system A combination of the combined and separate below-ground drainage systems, where the majority of the surface water is discharged via a surface water drain to the surface water sewer. The remaining surface water, usually at the rear of the building, is discharged via the foul water drain to the foul water sewer, which flushes the foul water drain. Cheaper to install than a separate system. *See also* COMBINED SYSTEM and SEPARATE SYSTEM.

particle Small part or fragment of a material.

partition An internal wall used to separate one space from another.

partition block A hollow clay bock used to construct *partitions.

party floor A shared dividing floor between two properties. *See also* PARTY WALL.

party wall (parting wall) A shared dividing wall between two properties. *See also* PARTY FLOOR.

PAS 1192 *See* SEVEN (7) PILLARS OF BIM.

pascal (Pa) The international standard (SI) unit of pressure. One pascal equals one newton per square metre.

pass door *See* WICKET DOOR.

passive earth pressure (Pp) The maximum horizontal stress that is exerted from the soil on a retaining wall (typically from the soil in front of the wall) as the wall moves towards the soil. *See also* ACTIVE EARTH PRESSURE; COEFFICIENT.

passive fire protection A method of fire protection where the form, layout, and fabric of the building is designed in such a way as to protect the building and its occupants from a fire. *See also* ACTIVE FIRE PROTECTION.

passive solar heating Where the form and fabric of a building are designed in such a way as to maximize the collection of solar radiation that is received directly from the sun, the sky, and the ground. Any solar radiation incident on the building is transmitted indoors through windows or other glazed elements, and converted into heat by absorption on opaque elements of the building. Nearly all UK buildings benefit from some passive solar heating, except where special measures have been taken to exclude solar radiation.

passive stack ventilation A ventilation strategy achieved using the natural buoyancy of the warm internal air within a building. The *natural ventilation system consists of a series of vents located around the building at the outer extremes and in the 'wet' areas of a dwelling which are connected via natural buoyant streams of air moving in horizontal and vertical directions towards the ridge outlets or other roof terminals. The vents are usually fitted to the ceiling. Fresh air is drawn into the dwelling via background ventilators (i.e. trickle ventilators), lower inlets, and *air leakage. The building is constructed so that the internal air within the building ventilates through a central chimney (terminal) or inlets. As the building air warms and becomes more buoyant, it rises and circulates, naturally passing to the central highest point. As the air passes through the building, it draws the outside air in through openings and ducts in the lower parts of the building. The method of ventilation is an effective natural ventilation strategy, as it uses a combination of cross ventilation, buoyancy (warm air rising), and the venturi (wind passing over the terminals, causing suction) effect. The system avoids the requirement for mechanical ventilation and is thus considered passive and a zero-energy approach to ventilation and improved air quality (as long as the inlets draw in clean fresh air).

Passivhaus Certification (Passive House Certification) A certification process that confirms that the dwelling has been built to the *Passivhaus Standard.

Passivhaus Planning Package (PHPP) A design tool that is used to assess whether a dwelling meets the requirements of the *Passivhaus Standard.

Passivhaus Standard (Passive House Standard) A voluntary energy performance standard that is designed to reduce the space heating and cooling requirement whilst maintaining thermal comfort and internal air quality. Passivhaus Standard is of German origin, although the Passivhaus brand is internationally recognized for its ability to deliver low-energy buildings, with minimal energy required for space heating and cooling.

patching The process where minor defects in concrete are repaired.

patent A licence granting exclusive rights to an inventor to manufacture, use, or sell an invention for a set period.

patent axe (comb hammer) A hammer with nine sharp points on the striking part of the hammer head. Used for scabbling concrete to provide a good key to which new concrete will bond.

patent glazing Transparent or translucent materials, usually glass, used to clad a building.

pathogen A microorganism that causes disease, typically used to describe a wide range of viral, bacterial, and protozoan waterborne diseases such as cholera, dysentery, rotavirus, and typhoid.

patio Paved or flagged surface of a garden.

patio door A horizontally sliding external door, usually comprising two panes of glass. *See also* CASEMENT DOOR.

pattern 1. A design, instruction, plan, model, or *template, which is used as a guide to make something. **2.** Something that is repeated over and over, e.g. certain shapes on wallpaper, or a trend in a data series.

patterned glass Glass that has a pattern incorporated into one or both sides of the glass. Used for privacy and decorative purposes. Also known as **rolled** or **figured glass**.

pattern staining Discolouration of plaster due to the substrate to which it is fixed. Where the plasterboard is in contact with plaster dabs, timber, or other fixing, it is usually colder or warmer at this point. If colder, due to a cold bridge (*see* THERMAL BRIDGE, it is more likely that condensation will form on the surface of the plaster. If the surface is warmer due to the substrate, the discolouration is often due to the dust which is trapped in air, and is more likely to circulate over the warm area of the plaster and deposit on the surface.

pattress box (back box, conduit box, socket box) A box used to mount electrical items, such as a switch or *socket outlet.

pavement Hard flat surface laid for pedestrians either as a footpath on its own or adjacent to a road used for mechanical vehicles, such as cars and lorries. Equally, pavements can be classed as any hard surface laid for use by either vehicles or foot traffic.

pavement lens (pavement prism) A glass block or lens installed within a *pavement light.

pavement light (vault light) A window, usually constructed from glass blocks, that is located within the pavement to provide daylight to the space below.

paver (paving brick, paviour) Building units, bricks, and blocks used for external hard standings.

pavilion A small structure set aside for an ornamental summerhouse in the garden or a building on a cricket or sports field used for viewing the activities.

pavilion roof (polygonal roof) A roof that is hipped equally on all sides and is of a regular polygon shape in plan.

paving 1. The surface of a *pavement. **2.** The process of laying a *pavement. **3.** The material used to form the pavement, such as concrete slabs, *pavers, or stone.

paving slab (flag, precast flag) A large precast concrete or stone *slab used for *paving.

paviour *See* PAVER.

PC sum *See* PRIME COST.

pea gravel Course rounded gravel used to surround buried pipes.

pearlite (steels) A phase found in steels and cast irons. Pearlite consists of alternating layers of alpha-ferrite and cementite (both are chemical constituents of iron).

peat Fibrous soil of organic origin, which forms boggy ground.

pebble A small stone that has been rounded by the action of water, sand, or wind.

pebbledash A rendered wall finish containing pebbles that have been pushed into the surface of the render. It is a cement or line render applied to the external face of a building and contains gravel, pebbles, or shells that adhere to the external surface of the render, providing a rough surface. The render is placed smoothly over the surface of the wall through a pump, gun, or trowel, made level, and then the stones, gravel, or shells are thrown at the render surface with a trowel, scoop, or by hand. The stones stick to the surface, providing the rough finish.

pedestal A base or support usually for a column or statue.

pedestal basin A *basin mounted on a *pedestal.

pedestal WC A *water closet where the bowl is mounted above the floor on a *pedestal.

pedestrian A person travelling by foot in an area used by vehicles.

pedestrian crossing (zebra crossing) A designated place on a road for pedestrians to cross on. It is characterized by alternating dark and light strips. It is a driving offence in the UK not to give precedence to pedestrians on the crossing; hence pedestrians have the right of way.

pedology The study of soil science in its natural environment, concerning its properties and classification.

peelable (strippable) Wallpaper fixed with adhesive that is made to be easily removed.

peeling Finish such as paint that has not properly adhered to the surface and has started to come away from the substrate.

peen hammer Striking tool with a rounded end, used for hammering and working metal.

peening The working of metal using a *peen hammer.

peg 1. A post fixed into the ground, marking building gridlines and levels used for setting out the building. **2.** A timber dowel, normally oak, fixing nail, or galvanized steel dowel used to locate and hold a tile in place.

peg tile A plain tile with a hole at the top of the tile for a peg or nail to be inserted. The peg fixes over the top of the batten and holds the tile in place.

pellet (stud) A small cylindrical-shaped piece of wood that is used to cover the head of a screw that has been countersunk into a piece of timber.

pelmet A thin horizontal panel or curtain placed above the head of a window to conceal the curtain rail and fixings.

Pelton wheel A water turbine wheel, where a jet of water impacts on blades or buckets around the perimeter of the wheel causing it to turn; *see also* REACTION TURBINE.

penalty (penalty clause) A condition within the contract that requires payment in excess of losses suffered. Terms that require payment in excess of loss anticipated are not legally binding in the UK.

pencil arris (pencil round) A corner or angle of timber or plaster rounded to a radius of approximately 3 mm.

pencil bar A thin reinforcing bar, 6–8 mm in diameter.

pencilling The process where the mortar joints in a brick wall are painted white.

pendant (pendant fitting, droplight) A light fitting that hangs from the ceiling.

pendulum A mass that swings freely under the influence of gravity from a fixed point.

penetration An opening made in a material or component.

penetrometer An instrument used to undertake a **penetration test**. The instrument consists of a cone or plunger, which is fixed to a series of connecting rods. The penetrometer is pushed or hammered into the ground. From the amount of resistance offered from the ground to penetration, an indication of bearing capacity, shear strength, and an indication of the amount of settlement can be obtained.

penning gate A *sluice gate, rectangular in shape that moves upwards to open.

Pennsylvania truss (Petit truss) A variation of the *Parker truss, developed by the Pennsylvanian Railroad in the 1870s, that contains sub-struts/ties to resist/transmit stresses; *see also* BALTIMORE TRUSS.

pen-on-section-test A test that is undertaken to check the continuity of the *primary air barrier. It involves using a line to mark the location of the primary air barrier on a set of general arrangement drawings. The line should be continuous and separate the heated (conditioned) spaces from the unheated (unconditioned) spaces.

penstock 1. A valve or *sluice gate used to control the flow of water or the discharge of sewage. **2.** A pipe or channel to supply water, typically under pressure, to something, for example, to a *hydroelectric plant.

pentagon A five-sided polygon, with each side of equal length.

penthouse The most prestigious apartment in an apartment block located on the uppermost floor, may occupy the whole floor and have terraces.

penthouse roof A single pitch roof which does not abut a wall.

people in construction (PiC) A human resource used in the realization of physical production activities in construction. They are 'boots on the ground' managers, professionals, craftspersons, and general workers in the front line of production.

peptizing agent A product that increases dispersion of a substance into colloidal form (*see* COLLOID) by depolymerization, or reducing *flocculation.

percentage A proportion of a whole expressed as an amount out of a hundred, for example, half of an amount would be expressed as 50%.

perched groundwater When the groundwater lies above the surrounding groundwater due to an isolated body of impervious soil, such as a clay lens.

percolation The slow movement of gas or water passing through a porous substance, such as rainwater percolating downwards through the soil to the water table.

percussion boring Piles (columns) of precast concrete, steel, or timber are driven into the ground using a drop or vibrating hammer. When using a drop hammer, a large weight is lifted vertically in the piling rig frame and drops under gravitational force onto the head of the pile (column). The weight is repeatedly lifted and dropped until the set point is reached. The set point is a predetermined distance of pile movement; when the distance is equal to or less than the drop of the pile, the required resistance is reached.

percussive drilling A process used to break through rock using repeated hammering action.

perforated baffle (distribution baffle) A baffle added in a sewerage waste channel where homogenization (mixing/equal distribution of constituents) of the influent is required prior to the application of a treatment process.

perforated brick A brick that incorporates a number of vertical holes. It is lighter than solid brick and has better thermal insulation properties.

performance Implies the satisfactory carrying out of the works in accordance with the contract.

performance bond (completion bond) A bank guarantee provided by the contractor that secures the client with ensured remuneration in the event of the contractor's default.

performance management A holistic set of management processes which bring together training and development, performance measurement, and human resource planning. It is an activity aimed at setting expectations and establishing objectives through which individuals and teams can see their part in the organization's overall mission and strategy. It is focused on improving performance and holding people to account for their performance.

An appraisal, or performance review, has traditionally been an integral part of performance management, offering a means for managers and their employees to review and discuss the latter's performance. Its purpose has been twofold: (i) to inform administrative decisions on contractual aspects of employment, such as pay, bonuses, promotions, or termination, and (ii) to identify areas for growth and improvement and inform suitable development plans.

However, in recent years there has been much debate around the usefulness of the traditional appraisal and whether they are 'past their sell-by date'. Current thinking suggests that, instead of the annual appraisal, more regular monitoring of progress towards goals, immediate feedback, and an ongoing focus on improvement better achieve performance improvements (CIPD, 2018).

(((((⊕)))) SEE WEB LINKS
• CIPD performance appraisals factsheet

performance specification Description that specifies the requirements of the building, structure, and materials rather than specifying or naming the materials themselves.

pergola An arch formed by a double standard wooden frame that allows plants to be trailed over it to provide a walkway in a garden.

pericline 1. A small elongated dome-shaped fold in sedimentary rock. **2.** A variety of the mineral albite, characterized by long white crystals.

periglacial Land that borders a glacier or ice caps. Also used to refer to an area of land where there is *permafrost.

perimeter 1. A boundary line around an area. **2.** The length of the boundary line around an area or shape.

perimeter angle The angled trim installed around the perimeter of a suspended ceiling at the junction where it meets the wall.

perimeter beam The most external beam or beams that run around the circumference of a building or structure.

perimeter diffuser A *diffuser that is located along the perimeter of a floor or ceiling.

peritectic A reversible reaction in metals occurring at a microscopic level where a liquid and solid react together to produce one different solid phase when a metal is being cooled. This type of reaction is common in many metals (including steels) when cooled from the liquid state.

peritectoid A reversible reaction occurring in metals at a microscopic level where two solid phases react together to form a completely different solid phase when a metal is being cooled from elevated temperatures.

permafrost An area of land (for example the polar regions), or subsurface layer of the ground that remains permanently frozen throughout the year.

permanent formwork (permanent shuttering, sacrificial form, absorptive formwork) Formwork that remains in place once the concrete has set and becomes part of the structure.

permeability The rate at which water under pressure can flow through the interconnected voids (or pore spaces) within a material, such as soil.

permeameter A laboratory instrument used to determine the *coefficient of permeability of a soil sample.

permissible deviation (tolerance) Differences in moisture contents, weights, mixtures, ingredients, strengths, and sizes from that specified, that is considered acceptable. It is also the distance or movement from the required position that is allowed. Such allowances are necessary to accommodate inaccuracies in instruments, casting, variability of materials, and ability of humans to accurately cut, weigh, set out, and measure.

permissible stress The maximum allowable stress that can be specified under certain conditions in elastic design.

permit (permit to work) Form that provides authorization to undertake work in a controlled or hazardous area. The permit would normally state the area controlled, risks and hazards, equipment, and procedures that must be undertaken, persons authorized to work, periods of work allowed, period for which the permit is valid, and signature of the person authorizing the work.

permutation An ordered arrangement of elements from a set.

perpend (cross-joint, perp, perp joint, US head joint) The vertical joint in masonry construction. Commonly referred to as **perps**.

perpendicular 1. At right angles to another object, for example, a perpendicular joint is at 90° to the horizontal mortar bed in a wall. **2.** Something that stands vertical to the ground, is at 90° to the ground.

perpendicular style Gothic architecture that is exaggerated by vertical straight and slender aspects such as windows and panelling.

perpend stone (bond stone, parpend, perpender, perpent stone) A large, long stone that extends through a wall from the inner to the outer face. Used to bind the wall together.

perps *See* PERPEND.

Perry-Robertson formula An empirical formula used to derive the buckling loads for long, slender beams and axial loaded struts.

personal protective equipment (PPE) Clothing or items worn to prevent a person coming into direct contact with a physical, electrical, heat, chemical, biohazard, or airborne particulate matter which could do harm. PPE includes such items as *hard hats, goggles (for *eye protection), earplugs (for *hearing protection), and face masks (such as a *respirator).

personnel door A door to an area of a building that has restricted access, such as a staff area.

Perspex (Plexiglas) *See* POLYMETHYLMETHACRYLATE.

PERT (Project Evaluation Review Technique) Planning technique for analysing logical networks to determine earliest start, latest start, earliest finish, latest finish, and float of activities to determine the *critical path.

pest An unwanted organism that causes damage to livestock, crops, or humans.

pesticide A chemical compound (such as a herbicide or insecticide) used to kill *pests.

pet cock *See* AIR-RELEASE VALVE.

petrol intercepting chamber (petrol interceptor, petrol intercepting trap) A trap used to remove petrol from surface water runoff.

phase Section of works normally linked to an aspect of the works that has distinct characteristics, for instance, a design or construction phase.

phase change materials Materials which absorb and release latent heat as they go through a change in their physical state, e.g. moving from liquid to solid. In construction such materials can be embedded within building components to help store and release heat energy into the building. The materials are used to help manage internal comfort conditions.

phase diagram A graphical representation of the relationships between environmental constraints (for instance, temperature and sometimes pressure), composition, and regions of phase stability, ordinarily under conditions of equilibrium for metal alloys. A phase diagram shows the limiting conditions of temperature and pressure, or solubility and temperature under which two phases are in equilibrium with one another.

phase transformation When a phase in a metal undergoes a change to another or two other phases, which are chemically different.

phenol A specific class of compounds consisting of at least one hydroxyl group attached to an aromatic hydrocarbon ring. These are derived from benzene and are commonly used in resins, plastics, pharmaceuticals, and in dilute form, as a disinfectant and antiseptic. A number of reproductive disorders in humans and wildlife are linked to these compounds as they are carcinogenic and have the capacity to disrupt the body's endocrine (hormone) system.

phenolic foam A foamed insulating board made from phenolic resin. Phenolic foam is light with a density of only about 48 kg/m^3, which has the advantage over many other expanded plastics foam boards in that they burn only with considerable difficulty, with very little smoke or toxic gas, and are usable at temperatures up to 150°C. The boards are not strong enough to carry the feet of a ladder, but covered with quarry tiles their performance is greatly improved.

Phillips-head screw A fixing device with helical thread and a crosshead rebate in which a crosshead (Phillips) screwdriver is inserted to drive the fixing into the timber.

phon A unit used to measure perceived loudness.

phosphate dosing A method of adding small amounts of phosphate into the drinking water to reduce lead becoming dissolved into the water from lead pipework, which can be found within some old domestic properties.

photic Relating to light, particularly the upper layer **(photic zone)** in lakes and seas where there is sufficient penetrated light for photosynthesis to occur.

photochemical degradation of timber A form of deterioration of timber material caused exclusively due to the presence of the sun. Exposure to sunlight causes the colouration of the heartwood of most timbers to lighten. This degradation can be very appreciable and can be slowed down by careful application of various finishes to the wood. It is important that degraded surface layers are removed before applying any protective surface coatings, otherwise they will not adhere to the surface.

photogrammetry The use of photographs to create representations, particularly the use of aerial photography to produce maps.

photographic survey Images of the site, structure, or building recorded and logged with notes for future reference. The survey may be conducted prior to development taking place showing the condition of the land and surrounding buildings. Surveys can also be undertaken during the construction period to record progress and the actual construction of components.

photosynthesis A process by which plants and other organisms convert carbon dioxide and hydrogen into simple carbohydrates (sugars) by using energy that chlorophyll absorbs from the sunlight.

photovoltaic cell (PV cell) A device that converts solar energy into electricity. It generates electricity using layers of semiconductor materials with special electronic properties.

PHPP *See* PASSIVHAUS PLANNING PACKAGE.

phreatic *See* GROUNDWATER.

pH value A number indicating if a solution is acid (1–6) or base (8–14). Pure water has a pH value of 7, which is neutral.

physics The scientific study of matter, energy, force, and motion, and how they interact with each other.

phytoremediation The use of plants to absorb or degrade contamination from soil or water.

pi Represented by the symbol π, it is the number of times the diameter of a circle will wrap around a circle's circumference. The ratio $\pi = 3.142$, hence 1 diameter fits around a circle 3.142 times.

piano hinge (continuous hinge) A long butt hinge that often runs along the full length of the door that is opening and closing. The hinge comes in long strips which can be cut to length.

piano nobile The first or main floor of a building, containing the living and reception rooms.

piano wire Long thin wire that can be attached to a plumb bob or used for setting out. Wire can be pulled much tighter than string and is often considered more accurate for setting out; however, in most cases laser lines have superseded both string and wire.

piazza Open square surrounded by buildings. Many piazzas earn a reputation because of the historic or eminent buildings that surround them.

pickaxe A double-ended pick with a point on one end and chisel head on the other, used for breaking up concrete and loose stiff materials.

picked stock facings *Stock bricks that have been selected to be used as *facing bricks.

picket A pointed piece of timber that is driven into the ground. Can also be used to form part of a fence (a **picket fence**).

pick hammer A small hammer used by roofers for making holes in slates and for driving in nails.

picking (stugging, Scotland colouring) Using a pointed pick to create a pitted effect on the surface of stone.

picking up Painting next to wet paint and joining without any visual signs of the two edges being painted at slightly different times, sometimes referred to joining the live edges.

pico- A prefix in the metric system that denotes 10^{-12}.

pictorial projection Three-dimensional drawings used for marketing purposes, includes axonometric, isometric, and oblique projections.

picture gallery Room in a building used for displaying pictures.

piecework (piece work) Work where the employee is paid for each unit (piece) of work that is completed.

pie chart A circular graphical representation of survey, tests, or market research results, with the results presented as segments that represent the percentage of the whole.

piecing-in Repairing a damaged portion of a material or surface by inserting a replacement piece that is the same size as the damaged piece.

piedmont A region at the base of a mountain or mountain range.

piend (Scotland) A *hip.

pier 1. A load-bearing buttress or brickwork column between two openings. **2.** A structure made out of wood, steel, or concrete that provides a platform that extends out into the sea, river, or lake. Used as a landing stage for boats or as an attraction to walk along to admire the coastal view.

piezometer An instrument for measuring pressure, for instance, the compressibility of a material or the amount of fluid pressure.

pig Brickwork that has been built out of gauge, so that one corner has one course more than the other. As the brickwork travels to the other course it often tapers up or down meaning that the brickwork is not level. Such mistakes are made when string lines are pulled to the wrong course at a corner and the brickwork is built out of line.

pigeonhole wall A wall constructed using a *honeycomb bond.

pig lug A box corner formed by folding a piece of metal or metal angle without making any cuts.

pigment A powder used to colour, especially paints.

pile 1. Concrete column that is driven or bored into the ground; piles are either end-bearing with the load being transferred to the end of the pile that rests on good load-bearing strata, or friction piles that transfer the loads by friction along the sides of the piles. **2.** The fibres that project from a woven carpet.

pile cap A steel plate or reinforced concrete slab that is placed on top of a *pile to distribute the load from the superstructure evenly over the pile or pile group.

pile driver A machine, consisting of a hoist and a leader, used to drive (hammer) piles into the ground. Other pile-driving techniques include jacking, jetting, screwing, and vibrating.

pile foundation A structural building element (*foundation) used to transfer the loads of the building's *superstructure deep into the ground, via structural columns of timber, steel, or concrete. The piles are bored, hammered, or turned into the ground. Piles can be end-bearing, where the loads are transferred to the end of the pile down to load-bearing strata (see figure), or friction piles, where the loads are distributed along the length of the pile.

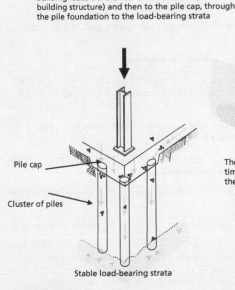

Building loads transferred to the columns (or building structure) and then to the pile cap, through the pile foundation to the load-bearing strata

Pile cap

Cluster of piles

Stable load-bearing strata

Stable load-bearing strata

The pile foundation (columns of concrete, timber or steel) transfer the loads deep in the ground to load-bearing strata.

Pile foundations

pile helmet A temporary steel cap that is fitted to the top of a *pile to prevent damage to the pile during driving.

pile shoe A cast-iron or steel point, which can be fitted to the foot of driven piles to facilitate penetration and provide protection.

piling The boring or driving of concrete, steel, or timber *pile foundations into the ground.

pillar A non-circular free-standing vertical pier. *See also* PIER.

pillar tap A tap that stands proud of the basin or bath on a pillar.

pilot hole A small hole drilled in a material that acts as a guide for a nail or screw, or for a larger drill bit. The pilot hole prevents the material from cracking or splitting when the nail, screw, or larger drill bit is inserted.

pilot light A small continuous flame in an appliance, such as a boiler, used to automatically ignite a much larger burner when required.

pilot nail A nail that is driven in a material to hold it in place temporarily until the main nails are installed.

pin 1. A flexible joint that is held together by bolts or rivets. **2.** A slender wire nail, wooden dowel, or peg.

pinch bar (case opener, claw bar, jemmy, wrecking bar) Small hexagon-shaped rod 14–16 mm diameter with one chisel end and a hooked chisel on the other end. A smaller version of the *pry bar, which can be used in one hand.

pinch rod Measuring and checking rod, cut to measure the height between floors, windows, or doors. Used to check heights quickly.

pine Fast-growing coniferous tree that produces softwood used for structural timber components and finishes.

pin hinge A butt hinge with a pin that can be removed.

pinhole A very small hole on the surface of a material usually caused by surface imperfection or trapped air. The very small holes caused by trapped air on a film of paint are referred to as **pinholing**.

pin joint A connection in a structure where members can rotate with respect to each other—such joints do not transmit moments.

pinnacle 1. The natural peak or top of something. **2.** An ornamental turret on the top of a spire, buttress, cone, or pyramid-shaped roof.

pinning 1. Use of panel pins to fix pieces of timber together. **2.** Use of dowels to fix timber together.

pinnings Different coloured stones set in rubble walling to provide a chequered effect.

pinning up Filling the gap between an underpinning foundation and existing foundations by inserting and ramming in dry or semi-dry mortar. Non-shrinkable grout or expanding grout is often used to avoid shrinkage and settlement.

pin tumbler lock A cylinder lock with spring-loaded pins offering many different key combinations, ensuring good security.

pipe A long hollow cylindrical tube, usually constructed from metal, plastic, clay, or concrete.

pipe bracket A *bracket attached to a wall, floor, or ceiling to support a pipe.

pipe clip A *clip used for fastening pipes to walls, floors, or ceilings.

pipe closer (fire-stop sleeve) A *fire stop that fits over a pipe where it penetrates a wall, floor, or ceiling.

pipe cutter Tool for cutting pipes. Pipes are rotated in a vice causing two cutting discs to cut a V-groove into the pipe. Clamps holding the cutting discs are tightened to cause the discs to cut through the pipe.

pipe duct 1. A pipe that is used to draw cables through from one position to another. **2.** A *duct that only contains pipe runs.

pipe fitter Skilled person who fits water, gas, oil, and steam pipes.

pipe fitting A *fitting that is used to join pipes together. A number of different fittings are available such as *bends, *elbows, and *tees.

pipe flashing (vent soaker) A *flashing installed around a pipe where it penetrates through a roof.

pipe hook 1. A hook used for lifting and moving large pipe sections.
2. A hook-shaped device, usually metal or plastic, used to support a pipe from a *rafter or a *joist.

pipe jacking A trenchless technique for installing underground small-diameter pipelines, ducts, and culverts. From a launch pit or drive shaft, hydraulic jacks push specially designed pipes through the ground behind a shield, as excavation is taking place within the shield.

pipelayer *See* DRAIN LAYER.

pipeline (pipe line) A long pipe formed from lengths of pipe joined together. Used to convey fluids or gases.

pipeline failure Mechanical and/or chemical degradation of gas and oil pipelines. In general, most pipelines are made from corrosion-resistant steels, and the failure mechanism is a combination of creep, fatigue, weld defects, and other forms of corrosion.

pipelining A process of driving a length of steel pipe into the ground to form a shaft, which is then filled with concrete to form a *pile. The steel pipe can be open-ended or fitted with a steel shoe.

pipe ramming A trenchless technique of driving a pipeline through a sort distance under the ground, for example, below a road. Similar to *pipe jacking, but employs a percussion hammer rather than jacks.

pipe ring A cylindrical bracket used to support a pipe. Can comprise either one or two pieces.

pipe sleeve A short section of pipe that fits around a smaller-diameter pipe as it passes through a floor, ceiling, or wall.

pipe tail 1. The open end of a section of pipe that is installed prior to the connection of additional pipework. **2.** A short length of pipe that connects a sanitary fitting to a branch pipe.

pipework A collective term used to describe a collection of pipes and their associated fittings.

pipe wrap (pipewrap, wrapping tape) Collective name given to various tapes that are wound around pipes. Tapes are used to provide protection, thermal insulation, acoustics insulation, or for identification purposes.

pipe wrench (cylinder wrench, Stillson) A wrench for gripping circular objects, threaded bars, threaded pipes, and pipes.

piping 1. Instability due to seepage as a result of the pore pressure exceeding the weight of soil and water above it. It can occur on the downstream (dry) side of *dams and *cofferdams; it appears as if the soil is boiling. **2.** A section of pipe. **3.** Cavities in metal. **4.** A high-pitched noise.

pit A large hole in the ground.

pitch 1. The angle of a sloping roof to the horizontal, measured in degrees. **2.** A perceptual quality of a sound that is dependent upon the frequency of the sound source. **3.** Name given to petroleum-derived bitumens used for waterproofing. **4.** Term used to refer to the distance between items that are spaced equally apart, such as reinforcement bars in concrete or nails in wood.

pitch board (gauge board, pitch block) A triangular-shaped template used when constructing stairs. The sides of the template are the same size as the rise, the going, and the pitch of the stairs.

pitched roof A roof that has a pitch greater than or equal to 10°. *See also* FLAT ROOF.

pitch-fibre pipe (US bituminized fibre pipe) A type of pipe commonly used for sewers that was manufactured from wood or asbestos fibre and was heavily impregnated with pitch. No longer used.

pitching The action of positioning a pile and runners ready for driving it into the ground.

pitch line *See* NOSING LINE.

pitch mastic A jointless floor made from pitch with limestone or silica sand aggregate, fluid when hot, spread to a thickness of 16–25 mm. Oils and fats affect it less than they do asphalt, and it can have a polished or matt finish.

pith The core of a tree, containing weak parenchyma and most of the log's defects. It is contained in boxed heart or a centre plank.

pitot tube A device for measuring fluid (air or water) velocity. One end of a small-bore tube is placed, pointing upstream, into the flow while the other end is connected to a manometer. The value obtained is the sum of the pressure head and velocity head, i.e. the **pitot pressure**. If another manometer is connected to the system, which is not in line with the flow (e.g. at the side of the pipe), just the pressure head can be recorded. Hence, the velocity head can be deducted and the velocity calculated.

pitting corrosion A type of corrosion resulting in the formation of holes or pits on the surface of metals.

PIV *See* POSITIVE INPUT VENTILATION.

pivot A device used to hang swing doors or a rotating window.

pivot window A window that opens by rotating on horizontally or vertically located pivots.

pixel An individual dot of light on a computer or television screen, many of which form images by illuminating in different colours.

placing of concrete (pouring) The process of laying wet concrete.

plafond A ceiling or soffit.

plain bar A smooth-surfaced steel reinforcement bar.

plain tile A small rectangular roofing tile, usually clay or concrete, that has a slight camber.

plan Drawing that depicts the view if the person was to look down on an object or building, it provides an outline of the horizontal plane. Plans are often provided to scale and with dimensions, and are used for setting out the building and components.

plancier piece (US) A *soffit board.

plane A handtool for smoothing and levelling the surface of timber.

plane element An element of the *thermal envelope, such as a wall, floor, roof, window, or door.

planed timber Wood with its surface dressed with a plane.

planer (rotary planer) A portable handheld power tool for smoothing, shaping, and levelling the surface of timber. Rotating cutting blades cut into the surface of the timber.

planing machine (planer) A static power tool for smoothing, shaping, cutting, and levelling the surface of timber, metal, and stone.

planing mill (US) A workplace (saw mill) used to cut and shape timber into planks of matchboard, floorboards, and other *planks.

plank A long, flat, solid timber board, usually installed face down in parallel rows.

planking and strutting Braced planks laid vertically at the side of an excavation to provide temporary support.

plank-on-edge floor (US) See SOLID WOOD FLOOR.

planned maintenance A regular maintenance programme for a building and its services.

planner 1. A professional responsible for the organizing and sequencing of operations on-site, who produces a programme and networks, and examines resource and cost implications. **2.** Person employed by the local authority responsible for controlling the development of land.

planning 1. Local authority controlling the development of land **2.** Organizing, scheduling, and sequencing activities.

planning and scheduling The sequencing of works and resources for efficient project delivery.

plant Mechanical equipment and services. Can be mobile such as cranes, dumpers, and excavators, or can be stationary and also include buildings, such as substations and refineries.

planted A surface-mounted moulding bead, for example, an architrave.

plant level (mechanical floor, mechanical level) A floor in a building that houses mechanical equipment, such as boilers and air-conditioning systems. *See also* PLANTROOM.

plantroom (US mechanical room) A room in a building that houses mechanical equipment, such as boilers and air-conditioning systems. *See also* PLANT LEVEL.

plasma Ionized gas containing roughly equal numbers of ions and electrons. Used in welding, cutting, and metal spraying.

plaster base A continuous surface to which plaster will adhere.

plaster bead *See* ANGLE BEAD.

plasterboard A board comprised of gypsum, fibreboard, or paper, and used in the construction of internal walls; also known as **gypsum board** or **wallboard**.

plasterboard nail A galvanized or zinc-coated flat, round-headed nail used to fasten plasterboard to a wall or ceiling. *See also* PLASTERBOARD SCREW.

plasterboard screw (drywall screw) A screw used to fasten plasterboard to a wall or ceiling. Generally, more expensive than *plasterboard nails.

plasterboard trowel Plaster trowel with a thin flexible blade used to smooth over tapered joint plasterboard.

plaster dabs Small amounts of plaster or gypsum-based adhesive that are used to fix plasterboard to walls.

plasterer Skilled worker capable of applying a smooth jointless surface of plaster.

plasterer's float A large rectangular trowel used for levelling plaster.

plasterer's labourer Unskilled operative used for cleaning, sealing, and preparing surfaces, mixing plaster, organizing materials, and cleaning tools. Often an apprentice or trainee *plasterer.

plasterer's small tool A small tool with a flat blade at one end and spoon blade at the other, it is used to shape and finish small areas that are difficult to access with a standard float. Often used to access tight corners.

plasterer's trowel A large rectangular trowel used for levelling and smoothing plaster.

plastering The process of applying plaster to a ceiling or a wall.

plastering machine Piece of plant that mixes and pumps plaster onto the walls ready to be finished and trowelled by hand.

plaster stop *See* STOP BEAD.

plasterwork Any work undertaken using plaster.

plastic A type of polymeric material (*see* POLYMERS).

plastic analysis The analysis of a structure using *plastic theory, mainly used to analyse statistically indeterminate structures.

plastic deformation Permanent or non-recoverable deformation after release of the applied load. To effect plastic deformation the material must be subjected to a force higher than the *yield stress.

plastic failure A failure mode characterized by large deformation and no brittle failure.

plastic hinge In *plastic theory, it is a point in a structural section that has lost all stiffness, the stress in the material is at or above its yield point.

plasticity The ability of a material to deform under load, without facture, and when the load is removed the deformation remains; *see also* ELASTICITY.

plasticity index (PI) The range of moisture content over which the soil remains in a plastic condition.

plasticizer A substance added to organic compounds to create a more flexible finished product.

plasticizer migration The loss of *plasticizer from plasticized plastics in contact with other materials, making the plastics brittle.

plastic limit The moisture content at which a fine-grained soil becomes plastic. It is the moisture content at which a 3 mm diameter thread of soil can be rolled by hand without breaking up.

plastic modulus *See* MODULUS OF PLASTICITY.

plastic moment (Mp) The theoretical maximum *bending moment that a structural section can resist, and at this point a *plastic hinge will form.

plastics *See* PLASTIC.

plastic theory The analysis and design of structures based on the idealization of elastic-perfectly plastic material behaviour.

plastisol A protective of *polyvinyl chloride coating applied to galvanized steel to provide further resistance to corrosion.

plate A horizontal structural timber member, normally 100 mm x 50 mm, that is used to support studs and roof trusses, for instance, a **sole plate** or a **wall plate**.

plate-bearing test An in-situ test used to determine an approximate value for the bearing capacity of the soil. A steel plate is loaded against the ground, typically at the bottom of a trial pit, until failure of soil occurs, or until a specified amount of settlement has been reached.

plate cut *See* FOOT CUT.

plate exchanger A type of *heat exchanger comprising a series of metal plates that transfer heat between two fluids or gases. Due to the large surface area of the plates, they are more efficient than a conventional heat exchanger.

plate girder A very large steel section, typically an I-section, with flanges that have been welded to the web; in older girders, rivets and bolts were used rather than welds.

platen A smooth metal plate that loads or holds an item in place, e.g. the loading plates that are in contact with the surface of a concrete cube during a compression test.

platen-press A hydraulically or pneumatically operated press that presses a component between two rigid metal plates. Used to manufacture laminated board.

plate tectonics Large-scale movement of the earth's *lithosphere. There are seven major plates on the globe, with smaller plates or platelets between some of these. Although the plate movements are slow, a few centimetres a year, over geological periods (millions of years) this adds up to thousands of kilometres. The movement on plate boundaries is not continuous, but intermittent, with energy being gradually increased until the forces within the ground can no longer be contained, whereupon earthquakes may occur.

platform A raised level area, for example, to keep things clear of the ground, or for people to stand on so that they can be visible to the audience.

platform floor *See* RAISED FLOOR.

platform frame A type of timber-frame construction where the structure is built up one floor at a time. Once each floor is complete, it is used as a working platform to construct the next floor. The majority of timber-frame construction in the UK is platform frame. *See also* BALLOON FRAME.

platform roof (Scotland) A flat roof.

playa A dry desert lake with no outlet that periodically fills with water.

Pleistocene A geological period from 2,588,000 to 12,000 years ago, denoting the earlier part of the Quaternary period and characterized by retracting continental ice sheets and the arrival of humans.

plenum An enclosed space that is filled with air, such as the space under a raised floor or above a suspended ceiling. Often used as an air duct for an air-conditioning system.

plenum system A type of air-conditioning system where air from a plenum is distributed through ducts.

pliable conduit *See* FLEXIBLE METAL CONDUIT.

plinth 1. The projecting base at the bottom of a *pedestal. **2.** The projecting base of a wall. **3.** The base of a cupboard. Also known as a *kick plate.

plinth block *See* ARCHITRAVE BLOCK.

plinth course The projecting course of masonry at the base of a wall that forms the plinth.

plinth return A special brick that has an angled face along one of the long edges and one of the short edges. Used to form a right-angled corner on a plinth.

plinth stretcher A special brick that has an angled face along one of the long edges.

plot 1. A small area of land, for example, where a building is constructed. **2.** To establish a graph from data points.

plot ratio (US floor area ratio) The maximum floor area of the building compared with the land area; this limits the number of storeys that can be placed on a building.

plough (US plow) A plane used for making a groove in timber.

ploughed and tongued joint Two rebates or mortises formed in timber with a strip of wood (loose tongue) inserted into the rebate forming a joint.

plug 1. A cylindrical device, usually plastic, that is inserted into a pre-drilled hole to hold a screw. **2.** An electrical device that is inserted into a socket to provide an electrical connection. **3.** A device that is inserted into the base of a bath or basin to prevent water escaping. **4.** Any object that is designed to tightly fill a hole.

plug cock (plug tap) A valve comprising a tapered plug that is used to stop the flow of a fluid or gas. Operated by turning the valve a quarter turn.

plugging Insertion of material, plastic or wood, into a hole to provide a fixing for nails or screws. As the screws or nails are inserted into the timber or plastic, the material is pushed against the surrounding material making a tight fixing.

plugging chisel (plugging, drill US star drill) A steel bar with a cross-shaped tip that is struck by a hammer to make a small hole in masonry or concrete walls.

plug-in connector (flex cock, plug-and-socket gas connector) A type of connector for a gas pipe that incorporates a socket.

plug-in switchgear (withdrawable switchgear) *Switchgear that is plugged in rather than screwed into the *switchboard.

plugmold (US) Surface-mounted trunking providing power outlets.

plug tenon (spur tenon) A *stub tenon that has a four-shouldered tongue.

plumb Vertical. Something which is plumb stands in an upright position, being exactly vertical.

plumb bob (plummet) Small steel weight placed on the end of a string line to hold the line tight and vertical. Used for checking that something is plumb or checking the position of something over the ground, grid reference, or in relation to other objects.

plumb cut (US) The cutting of a *birdsmouth on a roofing rafter where the rafter sits on the *eaves or *wall plate.

plumber Skilled tradesperson who works with pipes, sanitary fittings, and lead work on roofs. As pipes used to be made out of lead, the original meaning of the word plumber, meant a person who works with lead.

plumber's labourer (plumber's mate) Operative with skills to assist a *plumber by measuring and cutting pipes, organizing fittings and equipment, and clearing away rubbish and debris.

plumber's metalwork Sheet metal work undertaken by a plumber, includes dressing roofs, fitting flashings, and gutters.

plumbing 1. Working with pipes, sanitary fittings, and lead work. **2.** Setting something vertical and *plumb.

plumbing unit Prefabricated bathroom pod or module, craned and plumbed in position.

plumb level Spirit level fitted with two bubbles, one set for horizontal work and the other setting out vertical work and checking for *plumb.

plumb line String line with *plumb bob attached, which holds the line vertical, used for checking and setting out vertical work.

plummet *See* PLUMB BOB.

plunger Flexible or rubber sucker stuck to the end of a wooden rod. The tool is placed over openings such as sink drains and *water closets, then pressed to cause suction and create negative pressure drawing water in the opposite direction to its normal flow. Used to unblock pipes.

pluton A major intrusion of igneous rock formed at one time and contained within a single boundary within the Earth's crust from slowly cooling magma. *See also* BATHOLITH; STOCK.

plutonic rocks Igneous rocks that have formed deep within the Earth's crust from slowly cooling magma. They are characterized by coarse texture crystals which are visible to the human eye.

ply To join together, as by moulding or twisting.

plymetal *Plywood covered on one or both sides with sheet metal.

plywood Layers of timber material glued together to form a whole structure; for additional strength and durability the timber is often formed with the grains of adjacent layers at right angles to each other.

PMV *See* PREDICTED MEAN VOTE.

pneumatic Relating to the use of compressed air, particularly to operate tools.

pneumatic structure An **air-inflated** or **air-supported structure**.

pneumatic tools (air tools) Equipment driven by compressed air; can be used in wet areas without risk of shock.

pocket (box out, US blockout) 1. A hole that is cast in concrete using a *former as the concrete is laid. **2.** A hole in a wall that is constructed to support the end of a beam.

pod A volumetric part of a building, such as a bathroom or a bedroom pod.

podium (low block) A platform that is slightly raised above its surroundings.

pod urinal *See* BOWL URINAL.

point A sharp V-shaped end of an object, such as the point of a nail.

point cloud An assembly of data points (x, y, and z) used by a coordinate system to create a 3D parametric model.

point detector A fire detector that is designed to detect the presence of a fire at a particular point within a building.

pointed arch An *arch formed from two curves creating a point at the apex.

pointed vault A type of ribbed vault where the ribs meet at a point.

pointing 1. The external mortar joint between individual masonry units. **2.** The process where the joints between masonry units in a wall are filled with mortar after the wall has been constructed. The mortar can be coloured or finished in a number of ways to improve the appearance of the joint, or to provide greater weather protection. *See also* JOINTING and REPOINTING.

point of articulation A link that allows movement between two or more joined components.

point thermal bridge Discontinuities in the *thermal envelope that occur at a particular point in the construction. Examples include balcony and supports. *See also* LINEAR THERMAL BRIDGE.

Poisson distribution A probability distribution that expresses the number of random events occurring in a fixed period of time.

Poisson's ratio A constant that relates longitudinal strain in the direction of the load to lateral strain perpendicular to the load, for example, it defines the ratio of how a material will increase lengthwise and contract widthwise when stretched.

polar 1. A region or location around the North or South Poles. **2.** Something that has an electrical or magnetic field.

polariscope An instrument used to study polarized light.

polarization The direction of oscillation waves, e.g. light waves oscillate perpendicular to the path they travel.

polar second moment of area Defines the ability of a section (e.g. a beam) to resist torsion about an axis.

polder Low-lying land that has been reclaimed from a body of water (sea, lake, or flood plain) by constructing *embankments or *dykes, and *dewatered. Polders have a high risk of flooding.

pole A long, usually circular in section, piece of a material.

pole-frame construction (post-frame construction) A construction method where the building is supported on vertical poles or posts that can either be driven into the ground or supported on a slab or footing.

pollute To cause *pollution.

pollution The introduction of any contaminants into an environment.

polyamide (nylon) A polymer containing monomers of amides. Polyamides are renowned for retaining excellent mechanical properties at elevated temperatures.

polycarbonate (PC) A *thermoplastic polymer material made from carbonate. PCs are commonly used as an unbreakable glass substitute especially as bullet-proof or vandal-proof windows utilized in banks, public transport stations, high-risk (federal government) buildings, etc. It has half the density of glass and transmits more light in comparison.

polychlorinated biphenyls (PCBs) A polymeric material based on chlorine. PCBs have a variety of uses, most common applications are in electrical wiring and components.

polyethylene (PE) A widely utilized thermoplastic polymeric material. In construction applications it is used for interior plumbing pipes, and waterproof sheets for damp-proof applications as polyethylene is impermeable to the passage of water (waterproof). It is available in many different densities, which determines its application.

polyfusing Heat fusing of plastics.

polygon A closed figure that is contained by closed straight lines.

polygonal roof *See* PAVILION ROOF.

polygon of forces If more than three forces act through one point and are in one plane (termed coplanar forces) they can be represented in magnitude and direction by the sides of a *polygon. If the polygon closes, the forces are in equilibrium.

polymer A material usually composed of hydrocarbon compounds with extensive applications and usage. Polymers are widely utilized in construction, and are generally low-density materials used mainly in non-load-bearing applications. Approximately 20% of polymers produced in the UK go into construction. The most commonly used polymer is *polyvinyl chloride (PVC), and this material finds use as pipe materials for rainwater, waste and sewage systems, electrical cable sheathing, cladding, window frames and doors, and flooring applications. Polymers are generally classified into three categories based on their mechanical properties, they are *thermoplastics, *thermosets, and *elastomers.

polymerization The process whereby polymers are formed due to a chain reaction of *mers. The process of polymerization is a complex chemical reaction but in simplistic terms involves the repetition of the basic chemical block of a polymer 10,000–100,000 times depending on the desired density and application of the polymer.

polymethylmethacrylate (PMMA, perspex, acrylic resin) A polymer material based on methyl methacrylate. Perspex® (PMMA) is a transparent plastic, most commonly used as a shatterproof replacement for glass, especially in municipal buildings and windows in bus stops. PMMA has excellent transmission properties and toughness. In comparison with glass, it has half the density and transmits more light.

polynomial A mathematical expression for the sum of a number of terms, each of which is a constant multiplied by at least one variable raised to a positive integral power, e.g. $2x^2 + 4y^2 - 7$.

polypropylene (PP) A *thermoplastic polymeric material. PP has properties which are similar to low-density and high-density polypropylene. In construction it is used for sewer pipes or lavatory seats.

polystyrene (polystyrene foam) A polymeric material based on styrene. Polystyrene foam has low thermal conductivity making it ideal for cavity-wall insulation; however, it needs protection due to its flammability.

polytetrafluoroethylene (PTFE, Teflon) A polymeric material with a very low coefficient of friction, PTFE is thus commonly used in containers and pipework for reactive (corrosive or hazardous) substances.

polythene *See* POLYETHYLENE.

polyurethane (PU) A polymeric material made from urethane, usually utilized in foam form. Polyurethane is widely used in the construction industry, particularly as thermal insulation panels in cavity wall insulation. Polyurethane-based glue is also used as an adhesive. Polyurethane-based coatings and varnishes are used in carpentry or woodworking as this results in a hard, inflexible coat that is especially popular for protecting floors. Another common use of polyurethane foams is in commercial and domestic furniture.

polyvinyl acetate (PVA or PVAc) A polymer material based on vinyl acetate monomer commonly used as wood glue, and in paints and industrial coatings.

polyvinyl butyral (PVB) A polymer resin used to bind glass; this further improves the toughness of glass panels.

polyvinyl chloride (PVC, uPVC) A common thermoplastic polymer based on vinyl used extensively in the construction industry. UPVC (unplasticized polyvinylchloride) is used as window frames, doors, and pipes in most dwellings and buildings; unplasticized refers to the fact that the material contains no *plasticizers.

polyvinyl fluoride (PVF) A polymeric material mainly used in flammability lowering coating applications, particularly in the construction industry. Polyvinyl fluoride is a *thermoplastic fluoropolymer and has similar properties to *polyvinyl chloride (PVC).

polyvinylidene chloride (PVdC) A polymer material based on vinylidene chloride. PVdC has excellent resistance to chemical attack and is also impermeable to the passage of moisture.

polyvinylidene fluoride (PVDF) A polymeric material used in metal paints.

pommel A globe-shaped ornament used at the top of a *pinnacle.

ponding The accumulation of shallow pools of water on a horizontal surface such as a flat roof or flat surface.

pontoon A floating platform.

pony truss A bridge truss that contains a deck supported by two side trusses. It is characterized by having no lateral bracing between the top *chords of the two side trusses. This is because the side trusses are relatively low and, having lateral bracing, would prevent traffic passing beneath.

poplar A type of tree from the genus *Populus*.

popout (knockout) A partially cut or thin section of a component that is designed to be knocked out to provide an access hole.

popping Where plaster looses and comes away from its background. Commonly occurs at the heads of plasterboard nails.

pop rivet A type of permanent fastener used to join two sheets of metal together. Comprises a hollow cylindrical shaft with a flat head and a pin that extends through the centre of the shaft and head. It is inserted into a pre-drilled hole, and a specially designed tool is then used to draw the pin through the head and shaft. The pin expands the shaft, securing it in the hole, and then is broken off, creating a 'popping' sound.

population 1. A collection of an interbreeding species (e.g. people, animals, or plants) present in one place (e.g. world, country, or region). **2.** A group of items from which a statistical sample may be obtained.

pop-up waste A drain plug at the bottom of a sink or basin that is operated by a lever.

porcelain enamel An inorganic coating bonded to metal by the fusion process.

porch A covered approach to a doorway or a small roof placed over the entrance to a house, the sides of which can be closed in or left open.

pore A tiny hole on the surface of a material that allows the passage of a liquid or gas. These holes are present in many construction materials—timber, blocks, bricks, concrete, etc.

pore treatments The process where a chemical barrier, usually silicone, is used to reduce the amount of rainwater that penetrates masonry walls. Although the treatment will reduce the penetration of rainwater into the wall, it can also inhibit the flow of water vapour out of the wall.

porosity The presence of holes, space, or gaps inside a solid. A *porous material is thus not fully dense. Porosity is generally undesirable as pores can act as crack nucleation points; however, a porous material is a very good insulator.

porous A material which is not fully dense containing holes or voids. Examples of porous building materials include *aircrete, *mortar, and *bricks.

porous pipe A pipe, which lets in water, used for subsoil drains.

portable Capable of being transported (moved) with ease.

portal frame A simple structural frame comprising two vertical columns and two sloping roof beams that join in the centre. Typically used where a large internal free span is required, such as in factories and warehouses.

portcullis Historically, this was a defensive gate at the entrance to castles. The metal reinforcement grid is lowered to close and protect the castle or fortress and raised to allow people through.

Portland cement (ordinary Portland cement) A material with adhesive and *cohesive properties capable of bonding mineral fragments (sand, bricks, stone, etc.) together. It is capable of reacting with water to give a hard, strong mass. The main ingredients of Portland cement are: calcium carbonate (from *chalk or *limestone), silica (from *clay/shale), and alumina (from clay/shale). It is manufactured by heating limestone and clay together to form *clinker rich in calcium silicates. The main stages in the manufacture process are:

- Chalk and clay are mixed together either in a slurry (known as the wet process), or blended and transported in an air stream (known as the dry process).
- The mixture then moves down a kiln undergoing a number of changes, as it becomes hotter:
  - Water evaporates at 100°C.
  - *Carbon dioxide is given off at 850°C
  - Fusion takes place at 1400°C, *calcium silicates and aluminates form in the resulting clinker (partly glassy, partly *crystalline material).
- The resulting clinker is ground to form a fine powder and *gypsum is added to 'control' the rate of setting of the concrete.

The end product is Portland cement and consists of four compounds:

1) Tricalcium silicate $3CaOSiO_2$ C_3S
2) Diacalcium silicate $2CaOSiO_2$ C_2S
3) Tricalcium Aluminate $3CaOAl_2O_3$ C_3A
4) Teracalcium alumino ferrite $4CaOAl_2O_3$ C_4AF

By varying the properties and types of these compounds and changing the fineness of the cement particles during manufacture, the four main types of Portland cements can be produced: ordinary Portland cement (OPC); rapid-hardening Portland cement (RHPC); low-heat Portland cement (LHPC); and sulphate-resisting cement. The reaction between water and the chemical compounds of cement is *exothermic (known as the *heat of hydration) and produces a largely crystalline structure referred to as cement gel. The product of these chemical reactions are calcium silicate (CSH), calcium hydroxide $Ca(OH)_2$, and calcium aluminate hydrates (CAH).

Portland stone A limestone from the Isle of Portland, Dorset. Portland stone has been used in a number of famous buildings in the UK.

positive input ventilation (PIV) A *mechanical ventilation system, typically located in a cold, well-ventilated loft space, which supplies a constant supply of fresh air to the dwelling via a central diffuser. Wall-mounted systems are also available. Systems can incorporate heating elements to preheat the colder incoming air.

possession Indicates which party has control of and rights to use a site. Once a contractor is in control of the site, the contract can commence. Without possession, the contractor is prevented from starting. Once the works are complete, the contractor hands back the site and building. The control of the site is transferred back to the owner. Even if there is no statement within the contract terms about possession of the site, it is taken that the contractor must have control (possession) of the site in sufficient time to complete the works.

post A vertical member that may be used to provide support for a structure. *See also* COLUMN.

post-and-beam construction A traditional method of timber-frame construction where a skeletal frame of vertical posts and horizontal beams supports the floors and the roof.

post-contract stage Building work has finished, *possession of the site has been handed back to the client, the *final account can be prepared, and the final period for retention monies starts.

post-crash intervention system *See* COLLISION AVOIDANCE SYSTEM.

post-frame construction *See* POLE-FRAME CONSTRUCTION.

potable water *See* DRINKING WATER.

potboard The lowest shelf of a cupboard on which the pots are placed.

potential difference The different voltage between two ends of an electrical path that allows voltage to move through it.

potential energy The energy that is stored in a body or system due to its position within an electric, magnetic, or gravitational field, e.g. a mass at an elevated height due to gravity; *see also* KINETIC ENERGY.

pot floor A suspended floor formed with hollow clay blocks spanning between the concrete beams with a *screed or concrete topping placed on top.

pothole 1. A hole formed in the surface of a road. **2.** A deep vertical hole in rock (particular limestone), or in a river bed.

pot life The working life of glue, paint, and other chemicals stored in containers.

poultice attack The corrosion of aluminium kept in contact with a wet, porous material. It can destroy aluminium foil in insulation, for example.

pound (lb) An *imperial unit of mass containing 16 ounces, which is equivalent to 0.45 kg.

pour The placing of concrete, normally associated with large operations involving the placing of large quantities of concrete, such as concrete walls, frames, foundations, and floors.

pour and roll The pouring of hot liquid bonding compound placed prior to the rolling out of asphalt or bitumen roofing felt.

pouring rope Non-combustible rope (asbestos rope or similar) wrapped around cast iron pipes while lead caulking is poured.

powder coating A continuous film/coating formed from fine, dry powder using the electrostatic process.

powder post beetle Wood-boring beetle that leaves exit holes around 1 mm to 2 mm diameter. The **lyctus powder post beetle** or *Lyctus brunneus* attacks only hardwoods, including oak, ash, and particularly elm. The beetle will eat the sapwood of hardwoods with high starch contents and large pores. As timbers become older they are less susceptible to attack from the powder post beetle. It is also common for blockboard and plywood to be attacked. The beetle's exit holes are circular and the dust from boring is very fine, like talcum powder, distinguishing it from that of the deathwatch beetle, which produces a much coarser dust.

power 1. Authority to act and control. **2.** Low-voltage temporary electric supply. **3.** Capacity for exerting physical or mechanical force. **4.** In physics the rate of energy output from an object.

power-activated tool A mechanical gun that fires nails into timber.

power and lighting installation The components and services within a building that provide power for the appliances and provide artificial lighting.

power float (machine trowel, rotary float, helicopter) Mechanical piece of plant with rotating blades (floats) used for smoothing concrete.

power panel A *distribution board that only contains the circuits that provide power to the building.

power point A *socket outlet.

power tool Portable hand-held electrical tools used on-site, such as mechanical drills, breakers, sanders, and saws.

pozzolan A powdery siliceous material used to make *Portland cement. Pozzolans react with hydrated lime to form calcium silicate and silicoaluminate hydrates. Pozzolans commonly used in modern concrete construction include fly ash, which is also known as pulverized fuel ash (PFA), rice husk ash, ground granulated blast furnace slag (ggbs), silica fume, maize cob, and metakaolin (calcined clay).

pozzolana *See* POZZOLAN.

PPD *See* PREDICTED PERCENTAGE DISSATISFIED.

PPE *See* PERSONAL PROTECTIVE EQUIPMENT.

practical completion (substantial completion) Works are accepted to be sufficiently complete to hand over to the client, although minor defects noticed are still required to be rectified. In some contracts it is the point at which the defects liability period begins, the contractor's liability for insurance ends, plus it is the trigger for other contractual machinery to begin.

Pratt truss A *truss with vertical web members in compression and diagonal web members sloping down towards the centre so that they are in tension—the reverse of a *Howe truss. It is a statically determinate structure and as such can have long spans, up to 75 m. Variations of the Pratt truss include the *Baltimore truss and *Pennsylvania truss.

preamble Information provided at the start of each description in the bills of quantities. The introduction will provide rules of measurement, departures from standard methods of measurement, and description of the trade.

pre-boring 1. Putting a *pilot hole in a material. **2.** Drilling a hole for a pile.

pre-camber A beam that has an unloaded upward deflection so that when loaded it avoids the appearance of sagging.

precast Previously made material or structure which is transported to a site for installation.

precedence diagram (precedence network) Logical network that calculates the *earliest start and *latest start for each task, the amount of *float available, and the *critical path.

precedence of documents The hierarchy of documents and the recognition of those documents which are to be accepted as correct if two or more documents disagree. Provides a legal and contractual position as to which documents are of most importance in determining any dispute.

pre-chlorination The chlorination of drinking water prior to another treatment stage, for instance, filtration or chemical process.

precipitate 1. Water in the form of condensate falling to the ground. **2.** The separation of a solid from a liquid.

precipitation The formation of a new phase in metal alloys, usually from another different liquid phase.

precipitation hardening An increase in *hardness of a metal by influencing the chemistry of the material, usually resulting in the formation of a new chemical phase within the metal.

precipitation heat treatment Metals subjected to artificial aging in which the chemistry of the material changes, usually resulting in enhanced mechanical properties.

precision The exactness of detail.

pre-concentration (wastewater) Enhancement of the organic strength of wastewater before applying to the actual wastewater treatment system. This novel concept relates to the anaerobic treatment where a high organic concentration of influent wastewater is required by the anaerobic microorganisms for optimal treatment performance.

preconstruction process The tendering, pre-contract planning, and coordination meetings prior to construction commencing.

pre-contract stage Tendering, contract negotiations, and other meetings prior to the contract being signed and work commencing on site.

predicted mean vote (PMV) A *thermal comfort metric based upon a seven-point thermal sensation scale. The recommended acceptable value for an internal space is between −0.5 and +0.5.

Value	Sensation
3	Hot
2	Warm
1	Slightly warm
0	Neutral
−1	Slightly cool
−2	Cool
−3	Cold

predicted percentage dissatisfied (PPD) A *thermal comfort metric used to predict the percentage of occupants that will be dissatisfied with the internal thermal comfort conditions. The recommended acceptable value for an internal space is less than 10%.

predicted performance The forecasted performance of a building based upon what was built.

predictive maintenance Action that is scheduled as a result of analysis of data about a structure or system to identify its potential to fail at a particular time in order that suitable maintenance can be undertaken to prevent the failure from occurring.

prefabricated Describes building components or elements manufactured in a factory and then assembled on-site.

prefabricated building Structure with components that have been manufactured in a factory, delivered to site, and assembled. The building may be delivered in flat-pack form with panels lifted up and bolted together, or produced in full volumetric modules that have walls, roofs, and floors.

prefabricated modules (modularization, prefabricated units) Components of a building that have been manufactured in a factory to be assembled on-site. The process of off-site manufacturing reduces the build time on-site.

prefabricated scaffold (frame scaffold) Scaffolding system with fixing lugs and clips built into the poles and boards making the scaffolding easier to assemble.

pre-filter (dust collector, sand filter) A *filter installed before the main filter in an air-conditioning system that is designed to remove larger particles such as dust from the fresh air.

prefinished Components that have received their final finish (paint, sealant, or stain) in the factory and are delivered to site in a finished state.

preliminaries 1. General description of the works and list of the contractor obligations and restrictions imposed by the employer. **2.** In the US mobilization costs.

preliminary work Proprietary work undertaken by others prior to works taking place on-site.

preloading (prestressing, pre-tensioning) A method of introducing load, stress, or tension to a material before it is in service to counteract undesirable effects. A **preload** may be applied to highly compressible soil prior to the construction of a structure so that the final settlement of the structure is reduced. A **prestress** may be introduced to structural members of a building to offset in-service loads. A reinforcing bar may be **pre-tensioned** before the concrete is cast to produce a prestressed concrete beam—a beam that is in compression before it is loaded.

premixed plaster Plaster supplied factory-mixed, either wet in drums, ready-mixed, or dry in bags, and usually containing lightweight aggregate. A dry mix may be based on retarded hemihydrate gypsum or *Portland cement, so that on-site only water is added. The high thermal insulation of these plasters makes them useful in cold climates, and for fire encasement of structural steelwork. Thin-wall plaster is one type.

prepainting Priming a surface prior to painting.

preparation (surface preparation) The process of preparing a surface prior to it being painted or coated.

prequalification In order to enter the bid process, the client may request minimum criteria that must be met, including financial stability, experience of undertaking similar work, demonstration of competence, and relevant insurance. If such factors are not met, the contractor would not be qualified to enter the tender.

preservation The process of maintaining and protecting buildings and natural resources.

preservative An element or compound used to extend the longevity of an element or object. This usually involves the addition of a chemical to inhibit spoilage.

pressed brick A brick that has been subjected to moulding under mechanical pressure.

pressed glass Glass units, such as pavement units or glass block, that are pressed into shape.

pressure The force that is applied to an object over a particular area. Measured in newtons per square metre or *pascals (SI unit).

pressure bulb A stress or settlement contour beneath a loaded area (foundation). The size of the bulb is proportional to the area of foundation and the magnitude of load it supports.

pressure circulation Circulating air or water through a system using fans or a pump.

pressure cut-out A sensor that operates a control circuit when a preset pressure is reached.

pressure-equalized joint A type of joint used in cladding systems and windows that minimizes rainwater penetration by having a cavity within the frame or behind the cladding that is ventilated to outside air. Since the air pressure within the cavity can equalize with the external air pressure, there is no force to push the rainwater through the joint.

pressure gun A sealing or caulking gun, used to force the fluid mixture out of a nozzle enabling easy application.

pressure head The pressure exerted by a column of fluid above a certain point.

pressure-maintenance vessel A *diaphragm tank installed on a boosted cold-water supply system.

pressurization test A *fan pressurization test.

pressurized escape route A fire escape route within a building that is slightly pressurized to prevent smoke entering the escape route in the event of a fire.

pressurized structure An *air-inflated or air-supported structure.

pre-standard European standard for materials and processes.

prestressed concrete A type of concrete, pre-conditioned to improve the tensile properties of concrete. This involves using high-tensile steel rods, which can absorb high-tensile loads. As a result prestressed concrete is commonly used in buildings as floors, and also in bridges.

pretreatment 1. A preparatory process to the main intervention, ensuring that the main works or remedial action can take place. **2.** A process employed to alter or prepare influent (sewage) according to influent characteristics that can be accommodated by a unit downstream or following on in the treatment plant. For example, some membrane processes in treatment systems are very sensitive, and without changing the nature of the sewage through pretreatment, fouling of the membrane can quickly occur.

priced bill (priced bill of quantities) Document submitted as part of the tender which has rates fixed against the work described and measured. The descriptions and measurement of the works are normally distributed to those tendering for the contract. The contractor, within the tender, places prices and rates against the descriptions, and submits these with their tender documents. If the contractor is successful, the priced bill becomes part of the contract documents.

price index List of prices of materials and building works published and updated at monthly intervals.

price work Work based on measured rates and outputs rather than hourly rates. Prices are fixed against the work to be performed; at the end of the week the work is measured and payment is made for the amount of work performed.

pricking up The first of three coats of plaster applied to laths. Once dry, the surface of the plaster is scratched to form a key for the second coat. *See also* SCRATCH COAT.

primary air barrier The elements in the building that act as the main barrier to air leakage.

primary beam The main beam in a structure that supports the main loads and may support secondary beams. *See also* SECONDARY BEAM.

primary cell An electrochemical cell where an irreversible chemical reaction produces an electric current.

primary element A main structural or supporting element of building. The primary elements include the foundations, walls, floors, and roof. *See also* SECONDARY ELEMENT.

primary fixing 1. A fixing built into the structure of a building to support a secondary fixing, for instance, a *fixing channel. *See* SECONDARY FIXING. **2.** A type of fastener used to secure cladding panels or profiled metal sheeting to walls and roofs.

primary flow and return pipes The main pipes from which water flows to and from an appliance.

primary lining A permanent lining, typically precast or prefabricated segments, used to provide initial support in soft ground tunnelling.

primary treatment The initial stage in a water treatment plant involving coagulation and sedimentation, followed by *secondary treatment, and *tertiary treatment.

prime cost (PC) Description and fixed cost within the bill of quantities for work or goods which cannot be properly measured. It is often viewed as a fixed allowance for the price or the work; however, even when the price is stated, the courts have determined that, unless otherwise expressly worded in the contract, the figure is only an estimate. If the contractor is unable to obtain the goods and services within the stated figure the costs could increase. The supplier of the service or product is often nominated. Some terms of contract may provide the contractor with a percentage for managing the supplier and providing the goods.

primer A component of paint which must adhere well to the substrate, seal the bare surface, offer protection against deterioration or corrosion, and provide a good base for the undercoat. To ensure good adhesion, the surface must be free from loose or degraded material.

priming An initial coat of paint applied to a surface.

principal A traditional roof truss, such as a king and queen post truss; *see also* PRINCIPAL RAFTER.

principal axes The main axes through the centroid of a cross-section in which the product of the second moment of area is zero.

principal rafter The main *rafter in a traditional trussed roof that provides support for the *purlins. The purlins then provide support for the *common joists.

principal stress The normal stress perpendicular to the plane on which the shear stress is zero. In an element of soil there are three mutually orthogonal planes where the shear stresses are zero: the largest of these is known as the **major principal stress**, and the smallest of these is known as the **minor principal stress**.

principles In respect to the *Eurocodes, these are general statements, definitions, requirements, and analytical models for which there is no alternative permitted, unless specifically stated. Principles clauses are preceded by the letter P in the Eurocodes. *See also* APPLICATION RULE.

priority circuit An electric circuit that supplies power to essential items of equipment.

prism 1. A solid object contained by plane surfaces. Different prisms include square prism, octagonal prism, triangular prism. **2.** A transparent optical element with polished flat surfaces that refracts light.

prismatic Relating to a prism—a transparent polygonal solid, typically triangular in cross-section, used for dispersing light. A **prismatic girder** is a triangular-shaped roof beam constructed from three steel tubes.

prismoidal formula An equation used in earthworks to calculate volume; *see also* SIMPSON'S RULE and TRAPEZOIDAL RULE.

prison Secure building for keeping people in captivity, often used to contain those convicted of criminal acts.

privacy latch A latch that enables a door to be locked from the inside. Used on bathroom and *water closet doors.

private branch exchange (PBX) A telephone exchange that serves a building or office. It serves as the central point for all internal and external calls and connects together all of the internal telephone lines. Automated electronic systems are known as a **private automatic branch exchange (PABX)** or an **electronic private automatic branch exchange (EPABX)**.

privatization The transfer or sale of public utilities and buildings to private companies. The gas, water, electric, and telecoms industries in the UK used to be owned by the state, but are now owned by private companies.

processed shakes *Shakes that have been modified to look as though the timber has split.

Proctor compaction test A standard laboratory test to determine the compacted state of a soil sample, by varying the moisture content of different compacted samples. It is possible to determine the maximum compacted state (at optimum moisture content) a soil can attain.

product (building product) A component, element, or material that has been designed and manufactured to be used in a building.

product of inertia (product second moment of area) A cross-sectional area property about a pair of perpendicular axes. It is the sum of the products of mass (or area) of each element in that cross-section multiplied by the product of the distance of each element corresponding to the given axes.

proeutectoid steel A type of *steel subjected to a specific heat treatment to influence its chemistry (microstructure) and hence its mechanical properties. This type of treatment involves heating the steel to the *austenite phase and then cooling the material.

profile The outline, edge, or vertical section through an object.

profiled sheeting (metal decking, tray decking, troughed sheet) Metal sheeting, steel, or aluminium, that is folded into a sinusoidal or trapezoidal shape in order to improve its stiffness. Used for cladding walls and roofs, and to construct a *composite floor.

program A computer software program; a list of instructions, written in a programming language, that tells a computer to execute a given task.

programme (US progress chart) Schedule or plan of work, showing the activities, start times, durations, and progress (work completed) in a horizontal bar chart. The activities are normally linked by a network of tasks. Resources may also be built into the network calculating resource demand. By distributing and levelling the resources, they may be used more effectively and efficiently.

programmer Person who builds and constructs the *programme.

progress certificate A monthly certificate that shows a measure of the work done to date.

progress chart *Programme of works prepared by the contractor that shows the percentage of work carried out and completed for each task. Normally presented in a horizontal bar chart with the bars shaded to denote the amount of work completed.

progressive failure 1. Something that is failing gradually, for example, the slow movement of soil along a slip plane, failure could have occurred at one point along the plane, while further down the plane the soil has yet to fail. **2.** Something that causes something else to fail (having a knock-on effect) with the consequence of the first failure being quite different to the end result of the final failure.

progress payment Monthly payments based on the amount of work completed.

progress report The progress chart and a summary report provided to show the planned work and the actual work undertaken. The report also identifies suggested activities and areas of work to be rescheduled.

prohibition notice A notice given by the factory inspector, a representative of the Health and Safety Executive, which requires dangerous activities to stop immediately or situations to be made safe. For the notice to be given, the activities are considered to be a health and safety risk to the public or workers that could result in serious injury.

project The work or tasks to be undertaken. The process of *project management is used to deliver the outcome.

projecting hinge A hinge that allows a door to open 180° beyond an architrave or any other projection.

projecting scaffold A cantilevered scaffolding that is built out of one of the upper storeys of the building, the scaffolding does not extend to the ground

projection 1. An estimate of the rate of something. **2.** Something that protrudes. **3.** A drawing method to represent a three-dimensional object on a two-dimensional plane.

projection plastering (mechanical plastering, spray plastering) Applying plaster via a spray plastering machine and a nozzle.

project management The management of a set of tasks, events, and resources to deliver one significant outcome. The management of multiple projects with related outcomes is often referred to as programme management.

project manager Person responsible for the management of the building *project.

project network *See* NETWORK.

project representative 1. *Clerk of works. **2.** Person who acts on behalf of the main contractor and liaises with local residents, the general public, and other interested parties.

proof stress The amount of stress needed to cause a definite small permanent extension (usually 0.2%) of a sample of material. This value is roughly equal to the *yield stress in materials that do not exhibit a definite yield point.

prop A temporary support that is placed under or against an object to prevent it from falling down.

propane An alkane material, a gas at room temperature, commonly used as fuel.

propeller A rotating shaft fitted with spiral blades. Used to push a ship (or aircraft) through water (or air).

propeller fan (axial flow fan) A fan that uses a propeller-shaped rotor.

properties The inherent characteristics of a material; *see also* PARAMETERS.

proportion *See* RATIO.

protected membrane roof An *inverted roof.

protected opening An opening in a fire-resistant construction that is constructed from fire-resistant materials so that the fire resistance of the construction is not compromised.

protected route A fire-resistant stairway or corridor in a building that is used by the occupants as an escape route in case of a fire.

protected shaft A fire-resistant vertical shaft.

protected stair (enclosed stair, US fire tower) A fire-resistant stairway used as an escape route in case of a fire.

protection The process of taking measures to ensure that a building, building works, building occupants, and operatives are kept safe.

protection fan *See* DEBRIS-COLLECTION FAN.

protection gear Electrical devices or equipment that are designed to protect large electric circuits. *See also* PROTECTIVE DEVICE.

protection of finishes The process of temporarily covering finishes during construction to prevent them from being damaged. Once the construction is complete, the coverings are removed.

protective device Electrical device that is designed to protect small electrical circuits, such as a *MCB. *See also* PROTECTION GEAR.

protective finishes to metals Finishes, usually in the form of coatings, that are applied to metals to protect them from corrosion.

protective rail A rail mounted on a wall to protect it from impact damage.

protein An organic matter comprised of carbon, hydrogen, oxygen, nitrogen, sulphur, and amino acids. Protein is essential to living organisms.

proton A nucleon and part of an *atom.

protractor Instrument used for measuring and projecting angles. Liner graduations show the degrees; normally graduated to 90°, 180°, or 360° degrees.

proud Protruding from a surface.

provisional sum (PS) A figure (sum) included in the tender or *bill of quantities for work that cannot be sufficiently measured at the outset of the contract; included for items that cannot be properly specified.

provision and use of work equipment regulations (PUWER) Regulations aimed at keeping people safe whenever equipment or machinery is used in a workplace setting.

pry bar (crowbar, jemmy bar, jimmy bar, prise bar) A tool used for opening gaps, separating timbers that have been previously joined together, or forcing components apart. Where two pieces of timber are nailed or fixed together, and they can be prised apart, the bar is worked into a gap and moved backwards and forwards forcing the components apart. Where the bar has a flattened and forked end, the bar may also be used for removing nails.

pseudorange The approximate (uncorrected) measured distance between a satellite and a receiver in a *global positioning system.

psi value (linear thermal transmittance) The rate of heat flow per Kelvin temperature difference per unit length of a *bridge. Measured in W/K.

PSV *See* PASSIVE STACK VENTILATION.

psychological contract The subjective, unwritten, and continually changing expectations, beliefs, and obligations, as perceived by the employer and the worker (CIPD, 2018).

((⊕)) SEE WEB LINKS
• CIPD performance appraisals factsheet

psychrometer An instrument used to measure relative humidity—also known as the *wet and dry bulb thermometer.

psychrometric chart A chart that graphically illustrates the relationship between various parameters relating to the moisture content of air. The parameters include *dew point temperature, enthalpy, dry bulb temperature, *moisture content, *relative humidity, specific volume, and *wet bulb temperature.

P-trap A P-shaped *trap used on the waste outlet of a sink or toilet to prevent sewer gas entering into the building. *See also* U-BEND and S-TRAP.

public service Work undertaken by someone or an organization that benefits the general public, such as the emergency services.

puddle Clay and sand that has been mixed with water and tamped to form a nonporous material, used as waterproof lining.

puddle flange A ring, usually rubber, that is placed around a pipe that passes through an external wall or shaft in a basement. It prevents water penetrating through the wall or shaft.

pugging A material that is placed between the joists on a timber floor to increase the floor's mass and so improve the airborne sound insulation of the floor. Materials used include mineral wool, plasterboard, and sand. If the material is laid on a board, the boards are known as **pugging boards** or **sound boarding**.

pugging boards (sound boarding) *See* PUGGING.

pull *See* DOOR PULL.

pull box *See* DRAW-IN BOX.

pull-cord switch (cord switch, pull switch) A ceiling mounted switch for a light or fan that is operated by pulling on a cord that hangs from the switch. Usually found in bathrooms.

pulley A simple type of machine comprising a grooved wheel that a rope or chain runs through. Used for lifting heavy objects.

pulley stile The jambs of sash windows that incorporate the pulleys.

pulp A soft, moist, shapeless mass of matter.

pulse technique *See* AIR PULSE TEST.

pulverized Consisting of fine particles, normally a solid in powder form, usually prepared by grinding.

pumice A material used as an abrasive.

pump A mechanical device that can move fluids and gases.

pumpability The ease with which a mix of a material, such as concrete or plaster, can be pumped from one place to another.

pumping main The main pipe that sewage is pumped through under considerable pressure.

punch Pointed hand-held tool, with a sharp or flattened head, used to make indents in metal or drive the heads of nails deeper into the timber so that they are hidden. When used in metal, the pointed punch is used to create an indent in the surface of the metal. The indent created provides a temporary housing for the tip of a drill so that the metal can be drilled without the drill bit slipping over the surface. When nails are to be driven below the surface of the timber, the flat head punch is used.

puncheon 1. A tool used for piercing or punching. **2.** A short wooden vertical post.

punching shear A type of failure that can occur in reinforced concrete slabs or highly compressible soils (such as loose sand), which is due to a high localized load.

punch list (US) A list of defective works that are required to be undertaken.

purchasing officer Historical title of a buyer.

purge 1. The process of cleaning or flushing a pipe or system. **2.** The preheating or pre-pressurizing of a building to stabilize its condition for a subsequent condition, action, or test, or the preheating of a building to store heat energy. **3.** Purging of natural gas includes a process in which nitrogen gas is pumped into a reactor/tank with pressure while the outlet is opened to let the entrapped air/oxygen move outside with the introduction of nitrogen gas. The nitrogen gas also restabilizes the environmental conditions inside the reactor/tank.

purlin A horizontal roof member that runs parallel to the ridge and spans between the roof trusses. Used to support the roof covering.

purlin roof A double roof.

purpose groups The division of buildings according to their main uses, for example, residential and social housing, commercial, industrial, and non-residential.

purpose-provided ventilation The designed and controllable air exchange between the inside and outside of a building by means of a range of natural and/or mechanical devices.

push bar A horizontal bar on a door that is used to open a fire door in an emergency.

push button A button that operates an electrical circuit when pushed. For example, the button that is pushed to call a lift.

push-button control A simple type of control system for a lift.

push-fit joint (gasketed joint, push-on joint) A simple type of pipe joint that is made by pushing a spigot into a socket. A rubber seal within the joint is compressed forming an air and watertight seal.

push plate A *finger plate.

push-pull prop A type of telescopic steel prop.

putlog A short horizontal scaffold pole with one flat bladed end and one tubular end. The flat end is built into the wall while the tubular end is supported on a *ledger.

putlog adaptor A flat blade that is attached to the end of a scaffold tube to convert it to a *putlog.

putlog clip A clip that holds a *putlog to a *ledger.

putlog hole A hole in a masonry wall used to take the flat end of a*putlog.

putlog scaffold (bricklayers' scaffold) Load-bearing scaffolding that is built in the wall as the building is constructed.

putties Materials mainly used for glazing purposes that are designed to act as a bedding material, as well as a filler and a sealant against rainwater. The most common type is linseed oil putty, which consists of vegetable oil (usually linseed oil) and whiting or filler (usually powdered chalk). When used, a hard skin forms by oxidation within six to eight weeks, depending on the temperature. When used in wooden frames, stiffening of the putty is aided by soaking the linseed oil into the wood; this does not happen with metal window frames. Long-term, putties fail by shrinking and cracking brought on by excessive oxidation, and as a result the putty loses its sealing and bonding properties.

putty knife Flat knife with curved and flat edge used for applying putty to glazed doors and windows.

PUWER *See* PROVISION AND USE OF WORK EQUIPMENT REGULATIONS.

PV *See* PHOTOVOLTAIC.

pycnometer A glass jar with a special top that is used for measuring *specific gravity of soil particles.

pylon 1. A tall steel lattice structure, used for carrying high-voltage cables. **2.** A tall vertical structure, which has been used historically to form a decorative entrance or approach to a building or bridge.

pyramid A solid shape with triangular sides that meet at the top, typically with a square base.

Pyran (Pyrostop) Fire-resisting glazing.

pyranometer A device used to measure short-wave radiation.

pyrethroid An insecticide.

pyrgeometer A device used to measure long-wave radiation.

pyrite An iron sulphide mineral with a shiney metallic lustre, it is the most important source of sulphur after native sulphur. It is also known as fool's gold.

pyro Reinforced and insulated electrical cable.

pyroclastic Relating to fragments of material rock blown out of a *volcano.

pyrolysis A process whereby waste material is subjected to thermal degradation. As a result char, syngas, and pyrolysis oil are produced. An example of pyrolysis is the conversion of wood to charcoal.

pyrometer An instrument for measuring very high temperatures; it measures the temperature indirectly by converting radiation that is given off, so does not need to directly touch the very hot material.

Pythagoras' theorem A mathematical formula that enables the length of one side of a right-angled triangle to be calculated by knowing the lengths of the other two—it states that the square of the hypotenuse is equal to the sum of the other two sides squared.

p

Q-bop A version of the basic oxygen process in a furnace. A Q-bop furnace has no overhead oxygen lance; instead, the oxygen is blown in through tuyeres (pipes) at the bottom of the furnace. Q-bop is a more efficient process for making steel than other processes (more efficient than the basic oxygen process).

quad (quadrant moulding, quarter round) A decorative moulding in the shape of a quarter circle, used with wooden flooring to cover the uneven gap between the floor and the *skirting board.

quadrangle 1. A four-sided geometrical shape, for example, a square or a rectangle. 2. A courtyard surrounded on all four sides by tall buildings.

quadrant 1. A quarter of a circle or a 90° segment. 2. A device for measuring angles used in astronomy and navigation.

quadratic Relating to the second degree. A **quadratic equation** contains a *variable raised to the power of two but no higher.

quadrature 1. Making something square or dividing into squares. 2. A technique employing mathematics to make an area equal to a given surface. 3. The position of two celestial bodies 90° from each other when viewed from a third position, for example, the position of the sun from the moon when viewed from the earth.

quadrilateral A two-dimensional figure with four sides.

qualification 1. An award given to show that a person or body has met a standard. The standard is normally met by achieving specific and predefined criteria. 2. A condition that must be met before the right can be gained to enter into a certain group. For example, preconditions or set criteria may need to be demonstrated before a company can tender for a specific contract. The process of checks ensures that companies have the right attributes to be included on the list of companies that will be used for future work; also called the prequalification process. Prior to a company being included on a tender list, it may be asked to demonstrate that it is financially sound, has relevant experience, and has the suitable expertise to be considered for future work.

qualitative analysis The analysis of variables collected from research. The analysis makes use of links, associations, and trends without heavy reliance on quantities and statistics.

quality 1. A standard of service or a product. 2. The degree that the characteristics of a product fulfil its requirements. 3. The product's fitness for use or purpose. 4. The expected characteristics of a product or service that are entirely defined by the end user, evolving and developing with customer requirements and expectations.

quality assurance (QA) An administrative and procedural control system used to ensure that products and services are delivered to a required standard and meet the clients' and customers' needs. The system is used to standardize processes so that they are consistent and easily replicated. By recording and specifying correct procedures, other workers are able to follow the description of the process, ensuring that the end product or deliverable is consistent with that expected. The formal procedures are normally broken down into stages or parts that are checked and signed off by the person undertaking the task and by others, normally more senior, or a nominated quality assessor. The checks validate that the standard has been achieved.

quality audit The observations, assessments, and checks used to ensure that participants are following the specified procedures. The audit ensures that procedures have been defined, are in place, and are being followed.

quality control (QC) A system of checks and tests that ensures components are produced within accepted tolerances and meet with a set criteria. Products that do not comply with the standards set will be rejected. The checks and associated procedures rigidly enforce standardization and consistency—specification and tolerance is of key importance. Where products fall outside the acceptable tolerance, they cannot be used. Products that are rejected normally trigger investigations that aim to identify the reason for the defects. If necessary, modifications are made to the process to prevent unacceptable products.

quality culture A working culture that embraces quality procedures and strives for improved quality. The culture of an organization exists in the attitudes of the company workers and staff. With modern systems of management, the quality culture of an organization would also include the attributes and attitudes of subcontractors and supply chains that deliver materials and services. For quality to be achieved, all stakeholders must have an interest in the service and product, communicate potential problems and areas of possible improvement, and suggest changes where required. Without employee commitment to the processes, documents, checks, and communication practice that make up quality management systems, the true potential of a quality system will not be achieved.

quality management (QM) A management system used to ensure operations are carried out consistently, reliably, and to the reasonable satisfaction of all the parties that have an interest or stake in the services. The system makes use of quality checks, *quality control, *quality assurance, *quality management systems, and meetings and workshops to deliver an overall approach to the management of quality. Quality is managed in ways that are clear, identifiable, documented, efficient, and controlled.

quality management system (QMS) All the administration, management, and support operations that are used in the management of quality. The systems are based on documentation and clear communication that inform processes to ensure that procedures are correctly followed so that expected standards of work are achieved. Ensuring that correct procedures are known and followed requires the implementation of training, reporting systems, and audits. Effective systems are based on good communication systems, clear lines of responsibility, efficient and clear documentation, training, auditing procedures, and a working culture that is interested in the quality of service which it delivers.

quality manual A key document in *quality assurance, the manual is specific to each organization and describes the organization, its operating procedures, standards of work, work undertaken, previous, potential, and existing clients. The production of such information is expensive and should be written so that it benefits all areas of the organization and avoids duplication. Production should use the document as part of its training and auditing work. Standards and specifications contained within the document should be live, being adjusted as necessary so that the information is always up to date. A superseded or out-of-date document is potentially dangerous. Control of the quality depends on the document being realistic, achievable, and usable.

quality-price balance The cost of the controls needed to achieve desired quality compared with the cost of the direct consequences due to not conforming to procedures. If quality systems were not in place then errors may occur. The associated costs of these errors are measured by their estimated cost. With all management systems there is a cost. To assess whether the cost of a quality system is appropriate, an assessment of the costs that would be incurred if such systems were not in place can be made.

quantitative analysis The analysis of research data using inferential statistics. Variables are coded so that numerical values can be applied and statistics used.

quantitative risk analysis The analysis of possible consequences of an action using numerical and statistical methods of assessment such as the *Monte Carlo simulation technique.

quantitative safety assessment The use of numbers to rate the risk or hazards. Tables can be used to compile and compare the hazards and risks. Each item in the table or matrix is given a risk weighting, e.g. 1—low risk, 2—medium risk, 3—high risk.

quantities The amount of work, materials, labour, and services required to undertake a task. The building quantities are listed in the *bill of quantities. The bills break work down into measurable units of labour, materials, and services. Products are described using the *Standard Method of Measurement for building works (SMM), and Civil Engineering * Standard Method of Measurement (CESMM). Using standard systems of measurement, the bills can be priced up by different parties and prices compared. The standardization helps to ensure that each party involved in the bidding process is pricing a comparable unit or volume of work.

quantity surveying The practice of a *quantity surveyor.

quantity surveyor A construction professional whose work lies in the field of measurement, valuations of work, cost, and contractual advice. Work can be varied and include preparing cost plans, estimating the value of work as a project progresses, preparing final accounts, administering contracts, and assessing the value of properties. Most quantity surveyors are chartered by the *RICS (Royal Institute of Chartered Surveyors).

quantum The quantity of something.

quantum meruit A reasonable sum for work provided; the term is normally used when there was no expressed agreement of price. The price relates to as much as a person deserves for the work, or as much as the item is worth. In practice, it normally relates to a payment for work, fair remuneration according to the amount and quality of work provided.

quarantine anchorage A docking place where vessels anchor while being granted permission (a clean bill of health) to enter the port.

quarrel A square or diamond-shaped pane of *glass within a leaded panel.

quarry An open-pit mine used for the extraction of rock or minerals, such as stone and other building aggregates.

quarry sap The moisture that can be found in stone that is newly quarried. The moisture quickly dries out, having a case-hardening effect.

quarter bend A pipe fitting that turns through 90°.

quarter point Two points on a beam set a quarter of the way in from either end. These points are used by two-point lifting equipment to minimize bending stresses in the beam during installation.

quarter round (quadrant moulding) *See* QUAD.

quarter-sawn The cut pattern of timber boards from a felled tree trunk (log). The log is first cut into quarters along its *axis, after which a series of cuts are made perpendicular to the tree's *annual ring. This produces a reasonably consistent *grain.

quarter-space landing A small square *landing where the sides of the landing are equal to the width of the stairs.

quartz A *crystalline mineral, typically transparent, with a hardness of 7 (on the Mohs scale of hardness) and density of 2.65 Mg/m^3. Common to *granite, and the main element found in sand, gravel, and *sandstone.

quartzite A *metamorphic rock which originates from *sandstone, however, it is changed due to heat and pressure, normally due to *tectonic compression. Quartzite varies in colour, from white/grey in its pure form, to other colours due to the level of impurities present such as iron oxide producing pink/red shades. Due to its hardness (Mohs' scale of hardness = 7) it is suitable for aggregates and railway ballast.

quasi-permanent actions *Actions that are applied for more than 50% of the *design life.

quasi-steady state The condition where the internal environment is held constant, whilst the external conditions are allowed to vary over time.

quaternary 1. Occurring in fours or consisting of four parts. **2.** The present geological period, just over the last 2 million years to the present day.

quatrefoil An architectural feature with four leaves, it is used to describe elements of a building which have four overlapping circles. Typically found at the top of gothic church windows.

quay (wharf, jetty) A structure or strip of land adjacent to the edge of a port/harbour, where ships dock to load and unload cargo or passengers.

queen closer A brick cut in half along its length to form a half *header. Used on each course of *Flemish and *English bonds to complete the pattern.

queen post The vertical members of a *queen post truss.

queen post truss A traditional timber *roof truss that has two vertical *queen posts, positioned a third of the way across the span of the *tie beam, that extend up to the sloping sides of the roof truss.

quench Rapidly cooling a material to achieve a desired property in the material, for instance, rapid quenching of low carbon steel from approximately 750°C to room temperature achieved within a few seconds results in increased hardness.

Quetta bond A type of *Flemish bond used to construct reinforced brickwork. Comprises a one-and-a-half-brick-thick Flemish bond wall where the resulting half-brick-thick vertical spaces (pockets) in the centre of the wall are filled with steel reinforcement and concrete infill.

queuing theory A mathematical theory used in operational research for assessing waiting time, storage time, and average time taken to reach the front of the queue. Different systems of queuing can be used and analysed to see which performs the best. The theory can be used to assess traffic flow, material distribution, and waiting time.

quick bend A sharp or tight bend in a pipe.

quicklime (hydrated lime, calcium oxide, CaO) A white caustic alkaline crystalline material formed by heating (>825°C) limestone to drive off carbon dioxide. The chemical compound lime is widely used in the construction industry. Applications include mortar, plaster, water treatment, and soil stabilization to create more desirable properties. For example, in mortar and plaster it is used to increase the rate of hardening as well as to improve adhesion. In soil it has the following advantages: moisture is absorbed, optimum moisture content is increased, *plasticity index drops, linear shrinkage and swelling decreases, unconfined compressive strength and CBR values increase, disintegration of clay lumps during pulverization is accelerated.

quicksand An unstable condition in fine-graded *granular soils (sands and silts), where the upward *seepage force due to rising water produces a force equal to the effective weight of the soil particles. The *direct stress becomes zero and as a consequence no *shear strength is developed between the soil particles. The soil can carry no weight and behaves like a *liquid. *See also* PIPING.

quickset level A levelling instrument, similar to a *dumpy level, but where the three levelling screws have been replaced by a ball and socket to provide quick initial setup.

quick step A decorative circular moulding of small radius.

quire Another name for choir area of a church which houses the altar, normally located in the western part of the chancel between the sanctuary and nave.

quirk A V-groove cut into timber, the edges of the groove are rounded, giving the appearance of an open book; often used as a decorative feature in timber beading.

quirk bead Rounded edge of timber (creating a bullnose or bead) with a rebate cut directly next to the rounded edge. The rebate (*quirk) is set in from the edge of the timber and next to the bead. The quirk, in a quirk bead, often has one straight edge and the other rounded.

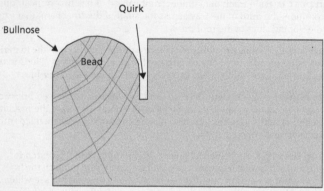

Quirk bead

quoin The exterior corner of a building or a wall.

quoin block An L-shaped block used to form the corner of a wall.

quoins The brick or stonework used to form the *quoin. Commonly used for decorative purposes when they project slightly from the face of the wall.

quotation (quote) The price given to undertake work.

quotient The result of a division, for example, 8 divided by 4 gives a quotient of 2, where 8 is the **dividend** and 4 is the **divisor**.

rabbet A rebate or recess.

racking back (raking back) The stepping of bricks at the ends of the wall to be laid/constructed. The *bricklayer uses the stepped bricks as a guide for level and line to complete the wall.

racking course A layer of aggregate that is used to fill voids on the foundation layer of a road before it is surfaced.

radial circuit An electrical power circuit for lights and appliances that starts at the consumer unit and finishes at the last socket or lighting point. There is no connection from the last socket or lighting point back to the consumer unit, unlike a *ring main circuit, which is connected to the consumer unit at both ends.

radial-sett paving A paving set laid in semicircular patterns, often constructed to a radius within which the paver's arm will reach.

radian (rad) A unit of angular measurement. One radian is the angle that is made by an arc on the circumference of a circle equal to its *radius. There are 2π radians in $360°$.

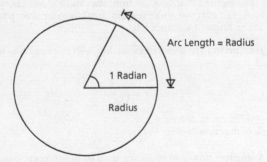

Arc Length = Radius

1 Radian

Radius

Radian

radiant heating A type of space heating system that heats a space using *radiation and *convection. Usually comprises a series of pipes that are embedded in a floor, ceiling, or wall—*underfloor heating is a type of radiant heating system although emits much of its heat through convection.

radiation The process where heat is emitted from a body and is transmitted through a space as energy.

radiation detector *See* FLAME DETECTOR.

radiator 1. Any heater that warms a space by emitting heat energy primarily by convection and a small amount by radiation. **2.** A metal heater that comprises a series of pipes and metal fins to transfer the heat from hot water to a space.

radiator key A small key used to open the air vent located on the top of a radiator. This process is known as **bleeding**.

radii The plural of radius.

radioactive dating Also known as **radiometric dating**. Various methods are used to determine the age of something by assessing the decay rates of its radioactive components. The most common dating techniques include **radiocarbon dating, potassium-argon dating**, and **uranium-lead dating**.

radioactive testing A *non-destructive test using *X-rays or *gamma rays to inspect for any sub-surface flaws in concrete and welded joints.

radioactivity Decay of nuclei in various isotopes, usually occurring in a spontaneous nature.

radio-frequency heating The process of using electromagnetic radiation to heat materials.

radiography A *non-destructive test that uses *X-rays to inspect the composition of the materials. Used to locate sub-surface defects in concrete and welded joints.

radioisotope An *isotope that is radioactive.

radius (r) 1. The length of a straight line from the centre of a circle to its circumference. **2.** The effective work range (or area) from a centre point, for example, the lifting range of a crane's *jib.

radius of gyration (r_x) The area a cross-section is distributed about its centroidal axis:

$$r_x = \sqrt{\frac{I_x}{A}}$$

where I_x = second moment of area
   A = area of the cross-section

radius rod A length of timber or moulding used by plasterers to create radial patterns in wall finishes. The rod will have a nail or point at one end, which is used as the centre of the radius.

radon A radioactive inert gas.

rafter The sloping beam that spans from the ridge to the eaves of a roof. Intermediate rafters span the full length from eaves to ridge, whereas jack rafters span part of the roof linking between hips and valleys. Also describes the sloping metal sections of a portal frame roof and used as an abbreviation for *truss rafter.

rafter filling (beam filling, wind filling) The infilling between roof *rafters above the *wall plate. Bricks, blocks, or insulation are laid on top of the wall plate to fill the void between joists, reducing air flow.

raft foundation Support underneath a building that transfers and disperses the building loads over a large area, often the total plan area of the building. By dispersing the load over such a large surface area, the load transferred per unit to the ground is significantly reduced. Due to its ability to reduce the loads on the ground, the raft foundation is often used on poor ground. Raft foundations are normally heavily reinforced to ensure the loads are spread over the total area.

ragbolt Bolt with barbs or protruding metal fixings that help lock the bolt into the substrate into which it is fixed.

ragging A paint finish created by dabbing a scrunched-up cloth, with paint on it, on the surface of a wall, or created using a roller made from loose flapping rags.

raglet (raggle, raglan, reglet) The rebate made in a wall or the recess in brickwork joints to receive the *flashing. The flashing is turned into the recess and secured using mortar. The recess and the inbuilt flashing prevents water getting behind the face of the flashing.

ragwork Using a rag roller or scrunched-up rag to create a pattern and texture on a wall.

rail 1. A horizontal piece of timber or steel from which objects can be hung or secured. Objects typically hung from rails include curtains, kitchen utensils, and tools, and objects secured include fence panels and boarding. **2.** The top part of a stair balustrade that provides a continuous pole, which allows those using the stairs to assist movement by holding the rail. **3.** The horizontal member that subdivides a doorframe. **4.** A *handrail. **5.** A horizontal member on a fence.

rail chair *See* DADO RAIL.

railing A fence constructed from rails and vertical posts.

railway signal A mechanical or electrical device that is situated adjacent to railway lines and is used to inform the train driver about the line ahead; whether to proceed or not, at what speed to progress, etc.

rainwater goods Collective name given to all the components required to convey water from a roof. The components will include gutters and downpipes.

rainwater head (rainwater header, rainwater hopper, US conductor hopper, US leader hopper) A box-shaped container at the top of the downpipe that collects rainwater from the gutters. Usually constructed in cast-iron or lead.

rainwater outlet An outlet that takes rainwater from a roof and discharges it into a *gutter.

rainwater plumbing (external plumbing) The process of installing *rainwater goods.

rainwater shoe A short curved fitting at the bottom of a downpipe that discharges rainwater into a *gully.

rainwater spout A duct used to discharge rainwater from a roof. *See also* GARGOYLE.

raised and fielded panel A panel that is raised in the centre and has a sunken margin. Commonly used in panelled doors.

raised countersunk head screw A wood screw with a domed head.

raised fibres Timber fibres that stand proud of the surface. As timber dries out, the softer wood between the fibres shrink, leaving the fibres proud. This characteristic also remains after painting or varnishing; by lightly sanding and reapplying the finish, the raised appearance of the grain can be reduced.

raised floor A floor mounted on pedestals or battens, allowing services to be installed below the floor. *See also* ACCESS FLOOR.

raised panel Access panel elevated from the surface onto which it is fixed.

raising piece A packing material used to lift the unit to the correct position so that it is fixed in line and level.

rake (batter) A sloping angle from the horizontal. Earth banks are often landscaped to a set slope; this is the rake.

raked-out joint Mortar joints that are recessed out to approximately 10 mm in depth. The joint, which is recessed along the full length of the brickwork bed, exaggerates and enhances the appearance of the joint. To improve the water resistance of the joint, the mortar should be compressed and compacted into the joint when *pointing.

raking cutting 1. The grading of soil or earth at an angle. **2.** The cutting back of existing brickwork to provide an exposed edge ready to receive new brickwork.

raking flashing The stepped cutting of flashing so that it follows the line of the roof and is rebated into the mortar joints of the abutting brickwork.

raking out The removing of mortar at the surface of brickwork to create a recessed or raked-out joint.

raking riser A vertical riser of a stair that is positioned at an angle to increase the tread area of a step.

raking shore (inclined shore, shoring raker) Prop or strut that is placed at an angle and secured to a wall to provide lateral stability. The rake refers to the slope or angle of the prop. Shores can be made out of steel or timber.

rammed-earth construction Building of walls out of raw materials, such as earth, chalk, sand, gravel, clay, and lime. The materials are normally sourced locally, making the form of construction sustainable. With the right selection of materials little energy is demanded to extract and use the materials. *Formwork is used to provide a mould for the materials, and mechanical plant may be used to compress the materials into the forms.

ramp Inclined platform used to get from one level to another. An incline of 1 in 12 or less is often used to enable wheelchair users to travel up and down the incline safely.

rampant arch An arch that has one support higher than the other.

rampart A defensive wall.

ram pump A very simple displacement pump containing two moving parts. For the pump to operate, an elevated water source is required. Water flows into the lower chamber causing the air in the upper chamber to compress. Once the pressure is sufficient, water is pushed up the outlet pipe. Various valves open and shut to control the flow of the water.

random ashlar (random bond, irregular bond) Wall made from cut stone of different heights that prevents the stone from being laid in regular courses.

random courses Walling constructed from courses that differ in height.

random pattern Paving or walling constructed from units of different size, that do not fall into a regular arrangement.

random paving Arrangement of paving slabs that do not follow a regular pattern.

random rubble (random rubble walls) Stone of different shapes and sizes placed together without any attempt to build in regular courses.

random slates Slates that have been split in random form and are rugged in texture.

random variable A variable that can have a range of values that occur randomly, but has an associated probability.

range 1. Allowable set of values or items. **2.** Large cooker with double oven and hob.

range masonry (coursed ashlar, range work) Stones cut to equal heights so that they can be laid in regular courses.

ranging A method to establish a straight line **(ranging line)** using *ranging rods.

ranging rod A pole that is used in surveying for *setting out. Typically the pole is 2 m long and is painted in alternative red and white sections of 0.5 m.

Rankine's method of tangential angles A surveying method for *setting out a transition curve.

Rankine theory A method to predict the *active and *passive earth pressures on a smooth (friction is neglected) vertical back retaining wall in cohesionless soil. It accounts for the vertical and lateral pressures within the soil adjacent to the wall. *See also* COEFFICIENT.

rapid-hardening Portland cement A type of *Portland cement that reaches optimum strength at an accelerated rate.

rapids A fast-flowing turbulent part of a river, where the flow is over rocks and boulders near to the surface.

rapid sand filter Used to filter water in the tertiary stage of a water treatment plant. The filter contains coarse sand and is primarily used to remove impurities that have been trapped in the *floc. It is cleaned by backwashing; *see also* SLOW SAND FILTER.

rasp A tool for shaping wood, which consists of a long steel bar with cutting teeth that protrude from the flat surface.

rate 1. Output, quantity of work performed per hour. **2.** Price per unit of work, or price per hour.

rate of growth How quickly an element or object expands or increases in size.

rate of pay (wage rate) Payment per hour or per unit of work performed.

rate of rise detector A *fire detector that responds more rapidly to an abrupt rise in temperature than a fixed-temperature heat detector.

rating 1. The performance of an electric motor. **2.** The level in a watercourse (a stream or river) and the discharge from it. **3.** The amount of water that can be taken from a watercourse without causing damage.

rating flume A tank that is used to calibrate current meters and pitot tubes.

ratio A number divided by another number—undertaken to present a proportional relationship.

rat-trap bond Similar bond to a Flemish bond (alternate *header and *stretcher), but with the bricks placed on their side. The bond uses fewer bricks, making the construction method more economical. Rather than header and stretcher, when a brick is placed on its side the header end of the brick is called the **rowlock** and the top face of the brick, which is also exposed, is called the **shiner**. In a one-brick wall the two shiners are placed between the rowlocks. As the bricks are on their side there is a gap in the centre of the wall between the two shiners. The name 'rat trap' is given due to the gap created within the centre of the wall.

rawbolt A bolt contained within a sleeve. When inserted into a drilled hole and tightened, the bolt is drawn into the sleeve causing it to expand (tighten) against the side of the hole.

ray (wood ray, xylem ray) A wooden part of a plant consisting of tissues, which helps transport water.

Raykin fender A *fender that has a sandwich-type construction, being made from a series of rubber layers bonded between steel plates. Used to protect docklands from the impact of mooring ships.

Rayleigh wave A surface earthquake wave that moves in an ellipse fashion in planes normal to the ground surface and parallel to the direction of propagation.

Raymond pile *See* STEP-TAPERED PILE.

razor socket Plug point created for electric two-pronged shaver.

RC *See* REINFORCED CONCRETE.

reach A stretch of open water.

reaction 1. An equal but opposite force acting on a body. **2.** A chemical change, for instance, when two substances are mixed together to form a new substance.

reaction turbine (Francis turbine) A *turbine that is driven by moving water entering a wheel round the circumference, it combines both radial and axial flow components. Principally used for electrical power production.

reactive Will produce a chemical *reaction. Reactive components can be used in resins and paints to enable them to harden/set.

reactor 1. A place (container, vessel, structure, etc.) where a chemical or nuclear reaction occurs. **2.** An electrical component in a circuit that produces reactance, for example, a capacitor or an inductor. **3.** Something that produces a reaction.

readout The information obtained from a monitoring device.

ready-mixed concrete (ready mix, ready mix concrete) A type of concrete that has been proportioned or batched at a plant and mixed at the plant in a static mixer or truck mixer prior to transport to site.

ready-mixed mortar A type of mortar that has been proportioned or batched at a plant and mixed at the plant in a static mixer or truck mixer prior to transport to site.

reaeration A method of introducing air bubbles into a lower layer of a reservoir, lake, or docklands. This causes the lower water to rise to the surface where oxygen is absorbed through surface *aeration.

realignment The adjustment made to alter the path/route of a road or railway line.

real time The actual time it takes for something to happens, for example, the time it takes a computer to read data from an external source, process, and upload it.

rebar A *reinforcing bar.

rebate (Scotland rabbet) 1. A cut into a material. The rebate is usually used to house another material such as a glass window. **2.** A rectangular recess cut on the edge of a material.

rebated joint A joint created by forming a rebate in a material, usually formed in wood capable of receiving the end of another material.

rebated weatherboarding *Weatherboarding that incorporates a rebate along the thickest edge of the board.

rebound hammer (Schmidt hammer) A device to determine the strength of concrete non-destructively. The device consists of a spring-loaded plunger that is pressed against the surface of the concrete to load the plunger. The plunger is then released to give an indication of the surface hardness and penetration resistance; with reference to the conversion chart an indication of the strength of the concrete can be obtained.

receptacle (US) A *socket outlet.

recess An indent in a wall or surface, sometimes built-in as a design feature, other times to provide a storage area for folding doors or other building furniture.

recessed joint Brickwork mortar joint that is raked out to a consistent depth, between 5–10 mm below the surface of the brickwork. The joint tends to hold water that falls on the surface of the wall. Porous brick walls with recessed joints are more susceptible to frost attack than *flush mortar joints.

recessed luminaire A *luminaire that is fixed flush to a ceiling, wall, or other surface.

reciprocal The result of dividing 1 by the number in question.

reciprocal lattice The arrangement of a group of points arranged in a metal at a microscopic level which describes the positioning of the atoms inside the metal.

reciprocal levelling A levelling procedure used when long sights are required (over rivers, ravines, or obstacles) to minimize *collimation, *curvature, and *refraction errors.

recirculated air Air extracted from the conditioned space that is sent to the air-conditioning or ventilation system and mixed with fresh air. The mixed air is then supplied to the conditioned space.

reclamation 1. Bringing unsuitable land (waste land, marsh land, or desert land) back to use, for construction or farming. **2.** The extraction of useful substances from waste.

reconditioning To restore an item to a condition resembling that of the original, both in functional performance and aesthetics.

reconfigure The ability to alter or change.

recovery 1. To return to a normal state. **2.** The reclamation of useful substances from waste.

recreational accommodation Buildings constructed for pursuits of leisure: sports, theatre, cinema, concerts, and other leisure activities.

recruitment and selection The process of attracting appropriately qualified candidates to apply to work for an organization. This may be directed towards filling a vacant position, or proactively, seeking to establish interest in the labour market for possible future openings in the organization. Advertising in the press and online, open days, and use of recruitment agencies are some common methods of recruitment. Selecting candidates is focused on assessing the applicants to decide who should be given a job offer. A range of different methods can be used to assess candidates, such as interview, psychometric tests, group work, role play, and assessment centres. Arguably, an assessment centre is the highest level of selection methods as these benefit from a relatively high predictive validity rating. Assessment centres require candidates to complete a number of different tasks and they often combine behavioural ratings and cognitive and personality assessments obtained from multiple sources. Fairness is important in selection decision-making (CIPD, 2018).

 SEE WEB LINKS
• CIPD selection methods factsheet

rectangle A four-sided plane figure, having four right angles, adjacent sides of different length, and opposite sides of equal length; *see also* SQUARE.

rectangular on plan Building that is a regular quadrilateral when viewed from above.

rectilinear Consisting of straight lines or in a straight line direction.

recycle To reuse something.

redevelopment To clear an area of buildings and structures so that the area can be replanned and developed as if the original development had not existed.

red lead A type of lead oxide, red in colour used as pigments in paints.

red oxide Iron (III) oxide, one of the three main oxides of iron; a paint primer used to prevent rust. It has generally replaced the toxic red lead.

reduced bearing An angle that lies between 0 and 90°, in a north or south direction of the survey line; *see also* WHOLE-CIRCLE BEARING.

reduced level The elevation of a point to a reference datum, e.g. *ordnance datum.

reducer 1. A pipe fitting used to connect a large diameter pipe to one of a smaller diameter. **2.** A substance that is capable of reducing the concentration of another substance.

redundancy 1. Dismissal from work because the employee is no longer required. **2.** Duplication or superfluity, for example, additional members within a structure that are not necessary but have been included to increase robustness and rigidity. **3.** Termination of the contract of employment for economic reasons when an employer needs to reduce the size of their workforce. One of the following conditions must be satisfied:
- the employer has ceased, or intends to cease, continuing the business, or
- the requirements for the employee(s) to perform work of a specific type, or to conduct it at the location in which they are employed, has ceased or diminished, or is expected to do so (CIPD, 2018).

(⊕) **SEE WEB LINKS**
- CIPD introduction to redundancy

redwood (European redwood, *Pinus sylvestris*) A type of tree known as Scots pine found in Europe and Asia.

reed bed An area of marsh land where tall slender grass plants grow. Used to treat wastewater by acting as a filter and providing a habitat where microorganisms can thrive and break down contaminants.

reeded Convex or beaded moulds that sit together in parallel rows giving a fluted appearance.

reed thatching Roof covering constructed with best-quality reeds to make durable thatching. This roof-covering technique mainly uses *Arundo phragmites* reed on most of the roof and *Cladium mariscus*, which is a more flexible reed, for the ridges.

reef 1. A ridge of coral or rock just below a water surface. **2.** A mining vein containing ore.

refectory Room for eating or dining.

reference The use of something as a source of information.

reference panel (sample panel) Unit of work constructed to provide an example of the materials to be used, quality of workmanship, and the finish to be expected.

reference specification Description of standards, codes of practice, and functions that the contracor will perform rather than describing or listing in detail the products to be used.

reflectance A measure of the proportion of illumination that rebounds from a surface; the illumination given from the surface.

reflected Describes something that has been projected back towards the source, particularly light, sound, or heat.

reflected radiation Solar radiation that has been reflected off buildings and surfaces.

reflective glass Glass that incorporates a coating or film to reflect solar radiation from the sun.

reflective insulation Insulation materials that have a reflective surface to reduce heat loss by *radiation.

reflector A device incorporated behind the lamp in a luminaire to redirect light back into the space.

reflow soldering A type of *soldering whereby solder fillets are formed in metallic areas.

reflux valve *See* CHECK VALVE.

refraction The change in direction when a wave, such as light, passes through a material such as glass or water.

refractory A heat-resistant material. The term refractory refers to the quality of a material to retain its strength at high temperatures. Refractory materials are used to make crucibles and linings for furnaces, kilns, and incinerators. The best example of a refractory material used in construction is *brick that has remarkable strength retention at elevated temperatures. There is no clearly established boundary between refractory and non-refractory materials, though a practical requirement often cited is the ability of the material to withstand temperatures above 1100°C without softening. Refractory materials must be strong at high temperatures, resistant to thermal shock, chemically inert, and have low *thermal conductivities and coefficients of expansion.

refractory mortar A type of mortar with excellent retention of properties and performance at elevated temperatures.

refrigerant A fluid that provides cooling by changing state from a fluid to a gas. Used in air-conditioning systems, heat pumps, refrigerators, and freezers.

refrigerated storage A storage area that is refrigerated. *See also* REFRIGERATOR.

refrigeration unit (US chiller, refrigeration machine) A mechanical device that produces large amounts of cooled water for air-conditioning systems.

refrigerator A mechanical device that is designed to store goods at a low temperature.

refuge 1. A traffic island in the middle of a road. **2.** A shelter to provide protection from something, for example, a chamber cut into the side of a tunnel for pedestrians to shelter from passing traffic.

refurbishment (rehabilitation) The improvement of a structure or building so that it returns to being a functional property that is aesthetically pleasing and structurally sound.

refuse Rubbish or waste.

refuse chute A vertical *chute used in multistorey buildings to dispose of rubbish.

region An area with a defined boundary, which would indicate geographical, political, or cultural characteristics.

register An outlet or terminal unit for air supply from an air-conditioning unit. The outlet is usually fitted with a damper and diffuser.

reglet A narrow groove chased into a wall to receive lead flashing. *See also* RAGLET.

regrating Cleaning of stone walls and other masonry.

regression A statistical relationship between a random variable and one or more independent variables. Used to determine the value of the random variable.

regular (regular coursed rubble, US coursed ashlar, range masonry, range work) Stone rubble that is laid in regular courses, but due to the irregular nature of the stone, the depth of separate courses varies.

regulating reservoir A reservoir designed to control the supply, whether to prevent the supply running dry or flooding.

rehabilitation *See* REFURBISHMENT.

reheat unit (terminal reheat unit) An *air terminal unit that incorporates a heater to increase the supply air temperature.

reinforce To strengthen something by providing internal or external support.

reinforced beam (reinforced column) A concrete beam or column that contains reinforcement, typically containing a collection of *reinforcing bars that have been assembled to form a cage of reinforcement, around which concrete is poured and sets.

reinforced brickwork Brickwork that has been reinforced by wire mesh.

reinforced concrete Also known as ferroconcrete, a type of concrete in which *reinforcing bars (rebars) or fibres have been incorporated into the concrete matrix to strengthen the material that would otherwise be brittle. Reinforced concrete is widely used in civil engineering.

reinforced concrete degradation Material degradation of reinforced concrete either by rusting of the steel components or freeze–thawing of the concrete matrix.

reinforced soil wall A retaining earth wall that has been constructed in layers, with a tensile inclusion (*geosynthetic or steel strips) between each layer, which are attached to the excavated face.

reinforcement *See* REINFORCED CONCRETE.

reinforcement bar *See* REINFORCING BAR.

reinforcement schedule A list with quantities of all of the reinforcement needed for a job, the list specifies the type, size, and shape of the bar, also known as the bending bar schedule.

reinforcing bar (rebar) A steel bar that is placed in concrete to increase its tensile strength. The bar is normally manufactured from carbon steel and may be smooth or ridged—the latter to increase bond strength.

rejointing The pointing or repointing of brickwork.

relamping The process of replacing the lamps within a *luminaire.

relative density (specific gravity) The density ratio of a substance to a specific reference material. The term 'specific gravity' is normally used when the reference material is water at 4°C. For sand the relative density is its in-situ density relative to its minimum and maximum compacted states, i.e.:

$$D_r = \left(\frac{\gamma_{dmax}}{\gamma_d}\right)\left(\frac{\gamma_d - \gamma_{dmin}}{\gamma_{dmax} - \gamma_{dmin}}\right)$$

where: D_r = relative density; γ_{dmax} = the maximum dry unit weight; γ_{dmin} = the minimum dry unit weight; γ_d = the in-situ dry unit weight.

relative humidity (RH) The ratio of the amount of moisture contained within a given sample of air at a particular temperature to the maximum amount of moisture that the air could hold at that temperature, which is expressed as a percentage. The relative humidity within buildings should lie between 40% and 70% for human comfort.

relative navigation A method to establish the relative position of two positions, where one or both points may be moving.

relaxation The reduction in tensile stress with time when a constant strain is applied.

reliability-centred maintenance (RCM) A process used to identify the policies that must be implemented to manage the failure modes which could cause the functional failure of any physical asset within a specific given operating context. Assessments of reliability, which are part of good design practice, are normally based on analysis of data, such as failure rate versus time.

(((●))) SEE WEB LINKS
• SAE evaluation criteria for RCM processes.

relieving arch (safety arch) An arch built into a wall above a lintel. The arch reduces the load that is imposed directly on the lintel.

religious accommodation Building used for spiritual needs or place of worship, and includes mosque, church, synagogue, cathedral, and temple.

reloading To restore power to a circuit.

relocatable partition Movable and demountable partition wall.

remeasurement The measurement of the quantity of work after the work has been completed. The measurement is then presented against the list of agreed rates for calculation of the value of the work.

remedial work Procedures, activity, and work to improve the condition of something, make good for a specific purpose, or restore to a former condition. Remedial work to walls and floors often involves structural work; such work on the ground can include improving its stability and/or removing or sealing contamination.

remediation A technique used either to remove a contaminated source from the ground, modify contaminant pathways without necessarily removing it, or to destroy/modify the contamination. Such techniques include: excavation and disposal, thermal treatment (desorption, incineration, and vitrification), bio-remediation, soil washing, capping, vapour extraction, and stabilization/solidification.

remote-entry system A system where the occupant of a building can permit or deny individuals access to the building by electronically activating the door.

remote gain *See* ISOLATED GAIN.

remoulding The action of altering the internal particle structure of a clay or silt from its in-situ state. Particles tend to become aligned and consequently a loss in shear strength and increase in compressibility is obtained.

Renaissance A period of rebirth or re-emergence of a culture or learned style which unfolded from the 14th century to the 17th century. In architecture it can be seen in the re-emergence of the Classical style, featuring symmetry, order, proportion, and relationships. Some of the most important work and thought of the Renaissance was created by Leonardo da Vinci and Michelangelo. While da Vinci was an architect, he was also a painter, sculptor, musician, scientist, engineer, mathematician, inventor, geologist, botanist, and writer; much of his work was typified by relationships and was described in its mathematical order and proportionality.

render (rendering) **1.** A sand, cement, and lime or resin mix, which is applied to walls to provide a smooth or textured finish. Can be used to improve water resistance and durability of a wall. **2.** Layer of sand, cement, and lime, or a resin mix that is applied to give a smooth or textured finish to a wall, and is called a **rendering coat**. **3.** Base coat or first coat on three-coat plaster work.

renewable energy Energy obtained from a renewable resource, such as wind, solar, hydro, geothermal, and biomass, rather than from a *finite resource.

renovation The improvement of the structural and aesthetic appearance of a building, or item of furniture, or equipment; could also include restoring the item to resemble its original appearance and quality.

renovation plaster Lightweight plaster that is often used in damp conditions, it tends to be stronger and more adhesive than other plasters.

repainting (maintenance painting, overpainting) Painting over existing paint, which has been prepared ready to receive the paint. All cracked and flaky paint is removed and the surface is sanded down ready to receive the new coats.

repair To fix or restore something that is broken, to a good/working condition.

repeating thermal bridge A *thermal bridge that follows a regular pattern and is evenly distributed over an area of the *thermal envelope. Its frequency is usually known and consistent. Typical examples include ceiling joists in cold pitched roofs that are insulated at ceiling level, ground floor joists in an insulated suspended timber ground floor, timber studwork and I-beams in timber-frame construction, mortar joints in an insulating block inner leaf, and steel wall ties in masonry cavity external wall construction. *See also* NON-REPEATING THERMAL BRIDGE.

representative action (F_{rep}) A value of the action used in the *Eurocodes for the confirmation of a limit state. (*see also* SERVICEABILITY LIMIT STATE; ULTIMATE LIMIT STATE). It represents either the *characteristic action or the assembly of suitable combinations of *characteristic values F_k together with a *combination factor (ψ).

repointing Grinding or raking out of existing mortar and replacing with new mortar and pointing as required.

repressed brick A wire-cut brick; once it is cut, it is also pressed to create a more regular finish with true sides.

resawing The sawing of large pieces of timber (flitches) into smaller pieces, during the conversion from large logs to usable sections of timber.

rescheduling The updating and reconfiguring of the project programme in order that the project deliverables are managed. The project programme will be updated with the actual progress, which once added to the project logic may affect the delivery of key activities and deliverables. If activities have slipped causing a delay to the project, extra resources may be added to reduce the time required to complete critical activities.

resealing trap A trap that allows air to pass through the water seal under negative pressure.

research A systematic study to investigate a particular subject area in order to determine facts, figures, relationships, test/revise theories, draw conclusions, and make recommendations.

resection A surveying technique used to establish the position of a point by observing from that point other points of known location.

reservoir 1. A lake or container for storing water. The lake can be either naturally or artificially formed. The latter is typically formed by constructing a *dam across a river. **2.** Water being stored for human consumption or agricultural use.

reshoring Back propping, adding additional support to formwork or other temporary works. Back propping is often used under concrete floors slabs while setting to ensure that the concrete has temporary support until it reaches full maturity.

resident engineer (RE, client's engineer) The engineer who represents the client and is based on the site so that he or she can observe and check the work. Normally the engineer will have an office on-site.

residential Property built for people to live and sleep in.

residual The amount remaining after the main component has gone. **Residual shear strength (τ_r)** is the ultimate shear strength on a failure plane, after the peak, occurring at large displacement. **Residual soil** is soil that has formed in-situ from weathered bedrock.

residual current device (RCD, residual current circuit-breaker, current-balance) A safety device that disconnects an electric circuit if any current leakage is detected. Modern consumer units incorporate RCDs and they can also be plugged into a socket that powers appliances.

residual stress The level of *stress resultant on a material when there is no external force or load. This type of stress persists in a material even when there is no external force or load in the material. Residual stresses can be either *tensile or *compressive, and is an undesirable property as it can accelerate failure of a material.

resilience 1. The ease with which a material returns to its original shape after an *elastic deformation. In physics, it relates to the amount of potential energy stored in an elastic material when deformed that can be released to enable the body to return to its former shape. **2.** The ability of a physical body, system, or organic material to re-establish itself. **3.** The ability of an ecosystem to return to its original state after being disturbed, damaged, or attacked. *See also* INFRASTRUCTURE RESILIENCE.

resilient mounting *See* ANTI-VIBRATION MOUNTING.

resin A polymer material, usually a viscous fluid at ambient temperature.

resin flux A *resin and small amounts of organic activators in an organic *solvent.

resistance 1. Hindrance caused to flow, for example, the force water experiences as it flows through a pipe due to wall friction. **2.** The capacity of something (e.g. a member or component of a structure, or the ground; *see also* BASE RESISTANCE; SHAFT RESISTANCE) to withstand the *actions imposed upon it without failing. The term is used throughout the *Eurocodes and can refer to resistance of the ground, bending resistance, buckling resistance, and tensile resistance.

resistance welding A type of *welding which utilizes an electric current.

resistivity The ability of a material to oppose the flow of electricity—the inverse of *conductivity.

resolution of force (resolving of forces) A method to represent a force of given magnitude and direction into coordinate directions.

resorcinol formaldehyde (RF) A resin based on phenol-formaldehyde, used in adhesives for wood gluing.

resource smoothing (load smoothing, resource levelling) The logical rearranging of activities so that the resource use is more evenly distributed over the project or programme activities. When the project logic alone is taken into account, the resource usage will vary considerably, with resource peaks and gaps in resource use. By altering the timing of activities, without affecting the project logic, the resource use can be more evenly distributed. In programme management, where more than one project is running concurrently, the use of a single resource pool across all projects helps to ensure the most effective and continuous use of resources.

respect for people The practice of treating people with dignity, respect, fairness, and understanding in the workplace and beyond. People should never be treated as a means to an end in a workplace.
  Seven priorities have been identified for respect for people in construction:
- Diversity: Promotion of workplace diversity;
- Facilities: Adequate and good site facilities and the site working environment;
- Heath: A good framework and support system for healthy working and living;
- Safety: High standards of H&S
- Career development and lifelong learning;
- The off-site working environment: Consideration for others to take place away from the work environment; activities to encourage greater integration and/or consideration for others;
- Behavioural issues: Conduct that is considerate of others' feelings, beliefs, and cultures.

(⊕) SEE WEB LINKS
- Detailed information on the Respect for People programme.

respirator Personal protective equipment in the form of a mask, which one breathes through to filter out toxic airborne particles.

responsibility The contractual or project commitment to obligations and duties.

responsibility matrix A table showing a list of resources, and the activities and duties that each resource has been allocated.

rest bend *See* DUCKFOOT BEND.

restoration The repair, cleaning, and structural improvement of an item to bring it back to as close to its original condition as possible.

restraint To hold back or prevent movement, such as with a fixed end connection (*see* FIXED BEAM).

restricted tendering Process that allows only those contractors that have met the prequalification criteria, are part of the framework agreement, or those that have been selected to submit bids for the contract.

resultant A *vector that is the sum of two or more vectors added together.

retaining wall Structural encasement (wall) constructed to hold back soil, water, or materials. Retaining walls are used to increase the amount of level usable building area, retaining soil at a higher level, and preventing it from encroaching into the building or another useable area.

retarder (retarding admixture, retarding additive) A chemical used to slow down a chemical reaction.

retempering (knocking up) The addition of water and remixing of concrete that has started to stiffen. Retempering significantly reduces the strength of concrete.

retention Sum of money held back by the client at each intermediate payment, often 5% of the payment certificate value, with half of the sum being released back to the contractor at *practical completion, and the other half being paid to the contractor at the end of the defects liability period.

reticule (reticle) A grid of fine lines in the eyepiece of an optical instrument (telescope or microscope) to determine the scale or position of the object being viewed.

retrofit The strengthening, upgrading, or fitting of extra equipment to a building once the building is completed.

return 1. A short section of wall at right angles to the main wall. **2.** A pipe that carries a fluid or gas back to an appliance.

return air Air that is extracted from a space and returned to the air-conditioning or ventilation system. *See also* RECIRCULATED AIR.

return fill *see* BACKFILL.

return latch (catch bolt, spring latch) The tongue of a door latch.

return pipe A pipe that carries a fluid or gas back to an item of equipment from which it originated.

return wall A part of a wall with a change in direction.

reuse of formwork Temporary forms that are used as moulds for concrete, then stripped, cleaned, and used again.

reveal The area of wall at a jamb that is not covered by the frame.

reveal lining A finish applied to the inside of a window or door *reveal.

reveal pin An adjustable clamp that is used to secure a horizontal piece of scaffolding across a *reveal (a **reveal tie**). The reveal tie is used to tie scaffolding to the building.

reverberation The repeated reflection of a sound from hard surfaces within a room.

reverberation time The time taken for a sound within a space to reduce to a millionth of its starting level (i.e. by 60 dB). The reverberation time of the space can be reduced by the insertion of sound absorption materials.

reverse fault A facture in rock that is characterized by the lower side of the fault displacing downward; *see also* NORMAL FAULT.

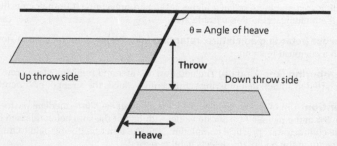

Reverse fault

reversible lock (double-handed lock) A lock that can be installed on either jamb of a door.

reversible movement The expansion and then contraction of materials, building elements, and building components.

revision The sequential notation attached to a document, drawing, or schedule showing whether the information is current or a previously updated document.

revolving door (tambour) A cylindrical door, normally comprising three or four leaves, that rotates around a central axis. Used to reduce heat loss through the doors and reduce draughts. Usually located at the entrance to office buildings and supermarkets.

rewards All financial provisions made to employees, including cash pay and the wider benefits package (pensions, paid leave, etc.). It can also include non-pay benefits, such as recognition, development, etc. (CIPD, 2018).

(⊕) SEE WEB LINKS
• CIPD reward and pay factsheet.

rewirable fuse (semi-enclosed fuse) *See* FUSE.

rework 1. To improve, make good, or put right something that has a defect. **2.** The remixing of concrete that has started to hydrate. This incorrect use of concrete, which may involve the addition of water to make the concrete more workable, produces an extremely defective concrete mix. Once concrete has started to hydrate (set) it should not be disturbed, since remixing concrete breaks bonds that have been created during the initial hydration, which will not be restored. Water added to concrete once the hydration has commenced is not part of the chemical reaction and reduces the concrete's strength. Water that does not take part in the chemical reaction eventually evaporates leaving voids where water previously existed. *See also* RETEMPERING.

Reynolds number A dimensionless number that gives the ratio of inertial forces ($\rho V^2/L$) to viscous forces ($\mu V/L^2$): where ρ = density; V = velocity; μ = viscosity relative to some length L, e.g. the radius of the pipe. **Reynolds critical flow velocity** defines whether the flow is *laminar or turbulent. If the number is above 2100, inertial forces will prevail and the flow will be turbulent; however, if the number is less than 1100 viscous forces will then prevail, and consequently the flow will be laminar.

rheology The study of how solid and semi-solid matter flows, for example, the creep of steel, polymers, and asphalt.

rhombus (rhomb) A *parallelogram with four equal sides and oblique angles, for example, a diamond shape.

R horizon A partially weathered layer of bedrock that underlies a soil profile.

rib A longitudinal projection from the surface of a wall or ceiling; the projections can be structural beams or columns, but often they are ornamental projections adding a feature to the ceiling. The projections are often seen on the surface of vaulted ceilings at the intersection of arcs.

rib and block suspended floor A *floor made from strengthening beams and concrete, clay, or hollow blocks. The floor can be cast *in-situ, but mostly the floor is constructed from precast concrete beams with concrete blocks laid between them. Modern suspended floors of this nature are now using insulation board laid between the structural beams and topped off with structural *screed. Such floors are quick to assemble and construct.

RIBA (Royal Institute of British Architects) The professional awarding body of British architects. For a person to be called an architect in the UK they must be a member of RIBA.

ribbed-sheet roofing Corrugated or profiled sheet roofing, the profiles add strength to the roofing material, reducing sag between the roofing spars or beams.

ribbon Length of flat tape.

ribbon board (ledger) A horizontal beam that is fixed to a wall or onto timber studs to provide support for floor or ceiling beams to rest on.

ribbon cable Flat electric cable that can be used underneath carpets or floor coverings without causing a large undulation, often used for telephone connections or distributing data.

ribbon course A course of bricks or tiles of a different colour to the main brickwork to create an aesthetic feature. To enhance the appearance, the brickwork may project out from the main course and tiles may have a different shape or texture.

ribs and lagging A temporary support while tunnelling in unstable rock. Ribs are steel I-sections in the shape of the tunnel's profile. Timber lagging then spans between the ribs; however nowadays *shotcrete is sometimes used in place of timber.

Richter scale A logarithmic scale from 1 to 10 used to measure the *magnitude of an *earthquake.

RICS (Royal Institute of Chartered Surveyors) The recognized professional body of quantity surveyors, building surveyors, and other professions associated with land, construction, and property. More recently, the institute has also become a professional representative of facility managers and estate agents. RICS was founded in London in 1868, and granted a Royal Charter by Queen Victoria in 1881. The Charter requires the RICS 'to maintain and promote the usefulness of the profession for the public advantage'. The institute has become a body that builds and provides data on the sale of houses and increases and decreases in the value of land and property, and provides information on environmental issues.

(⊕) SEE WEB LINKS

• The Royal Institute of Chartered Surveyors web page provides information on how to join the institute, governance, and information on property, land, and the environment.

RIDDOR (Reporting of Injuries, Diseases, and Dangerous Occurrences Regulations), A set of regulations that places a legal duty on UK employers, self-employed people, and people in control of premises to report work-related deaths, major injuries, or over-three-day injuries, work related diseases, and dangerous occurrences (near-miss accidents).

(⊕) SEE WEB LINKS

• The website is the source of the above definition and includes details on how to report an accident.

ride A door that scrapes the floor as it is opened or closes is said to ride; it should be adjusted by trimming the underside of the door or slightly adjusting the hinges. Where possible the floor should be levelled.

rider shore A short shore that does not extend all of the way to the ground, but relies on a secondary longer shore for support. It is the upper part of a raking shore that is connected to the wall and back shore.

ridge The top horizontal part of a roof structure where the two sloping parts of a roof meet.

ridge board (ridge piece) Horizontal timber board to which the diagonal rafters are secured.

ridge capping The covering over the top of the ridge of a roof. The ridge is often covered with capping tiles, ridge tiles, folded sheet metal such as lead or copper, or impermeable fabric such as bitumen or felt.

ridge course The uppermost course of tiles that is next to the ridge. Ridge tiles and the ridge course need to be fixed firmly in position as they are exposed, and are prone to damage in high winds if not fixed properly.

ridge end A special tile used at the ends of the ridge to give a desired appearance.

ridge stop A flashing to cover the intersection of a ridge that meets a wall. The flashing is sunk into the *raglet or *chase in the wall, then dressed over the wall and abutting ridge.

ridge terminal An outlet for a soil and vent pipe or duct that exits through or close to the ridge.

ridge tile A roof tile suitable for covering a ridge or a hip.

ridge vent A raised unit that allows air to escape from the roof space, ensuring that stagnant moist air does not build up in the roof.

rift A gap or break—something that splits apart. A broad central steep-sided valley formed by two parallel faults is a **rift valley**. An area that has been subject to a large number of faults is a **rift zone**.

right angle A 90° turn or corner.

right solid (right angle triangle) A triangle with a 90° angle in it.

rigid Firm or stiff.

rigid damp-proof course A course of bricks or slates that provides an impermeable barrier to rising moisture or damp.

rigidity Resistance to deformation from shear forces. *See also* MODULUS OF RIGIDITY.

rigid pavement A *pavement that is constructed with a concrete slab laid on the *sub-base rather than *asphalt or *tarmac; *see also* FLEXIBLE PAVEMENT.

rill 1. A small stream. **2.** A groove or channel within the ground.

rim latch A latch that is fixed to the surface of the door and the locating locking style is fixed on the surface of the doorframe.

rim lock Lockable *rim latch.

ring A cylindrical strip of metal used to join two pipes together.

ring beam (edge beam) 1. A horizontal beam placed at the top of a wall to tie the wall together. **2.** A horizontal beam placed around the internal face of an enclosure to provide support for the floor. **3.** A circular beam at the base of a dome.

ring main circuit (ring circuit, ring main) An electrical power circuit for lights and appliances that links a number of sockets or lighting points together. It is connected to the consumer unit at both ends.

ring-shanked nail (improved nail) Nail with raised rings on the shank that increase the hold and withdrawal resistance of the nail.

rip The sawing of wood in the same direction as the grain, parallel to the grain.

riparian Relating to the riverbank.

ripper A large pointed steel tool that is used to penetrate and spilt open firm ground and thinly bedded rock. The tool is fitted to the back of a bulldozer or tractor and is hydraulically forced into the ground. The bulldozer/tractor will have sufficient traction to pull the ripping tool through the ground.

ripple A small surface wave, for example, the waves created when a stone is thrown into a pond.

rip-rap Large stones placed on a riverbank or on the upstream side of a dam to prevent erosion from the wave movement.

ripsaw Saw with coarse chisel-cutting teeth with each tooth being set at alternate angles, enabling the saw to *rip wood.

rise The vertical distance from one *tread to the next on a *flight of *stairs. The rise of each step should be equal, and for private stairways in the UK, should be less than 220 mm high.

rise and fall Fluctuations in height, up-and-down movement of an item.

rise and fall method A surveying method for levelling, where reduced levels are calculated as a rise or fall from the previous reading or known reduced level; *see also* HEIGHT-OF-COLLIMATION METHOD.

riser 1. The vertical face at the back of a step. **2.** A vertical service duct, pipe, or cable used to take water, gas, or electricity to an upper floor. **3.** A water service pipe that runs up a building, used to supply water for fire-fighting. For fire-fighting, pipes can be dry when not needed and connected to fire hydrants or mobile water supplies or can be wet and filled with water under pressure so that they are available when needed. The dry risers are then filled under pressure with water, providing water to the upper storeys to fight the fire. Wet risers remain filled with water under pressure so that water can be drawn off instantly when required.

rising damp Water that rises from the ground through masonry, timber, or concrete by capillary action causing dampness in the lower parts of a building. Rising damp leads to mould growth, creates damp conditions that lead to insect infestation, and causes the decay of adjoining materials. The use of damp-proof courses (DPCs), with the exception of where the DPC has been bridged, has largely alleviated the problem of rising damp.

rising main (rising pipe) The cold-water service pipe that rises vertically from the floor of a building. It is required to have an internal *isolating valve and drain-off valve provided within it at the lowest point.

rising spindle A threaded spindle that rises up and down when a tap or valve is turned on or off. *See also* NON-RISING SPINDLE.

risk The likelihood of an unwanted event occurring, such as variation, an accident, additional costs, delay, price fluctuations, etc. In a contract, items described as contractors' risk are those which the contractor takes on.

risk analysis The assessment of the potential risk associated with a specific event.

risk assessment A procedure to assess the possible risk involved, the magnitude of the risk, and the likelihood of it occurring.

( ) SEE WEB LINKS
• Details information on how to undertake a risk assessment.

risk management The identification of potential risk and the procedures to take to avoid or minimize impact. Processes are associated with the identification, analysis, evaluation, and treatment or mitigation of the identified factors with assigned risk (the possibility of losing something that has a value).

riven slate Slate that is reduced to the desired thickness by splitting it rather than sawing it to size. Due to the separation of the layers, the slate has a natural mottled surface.

rivet A short metal rod with a head at one end. When the shaft of the rod is inserted through aligned holes on sheets of material, the projecting end is then flattened to form a head on the other end, thus holding the sheets of material together. Used in aircraft, steel frameworks to buildings, and in bridge construction.

road A stabilized length of earth, typically containing a hard surface, on which vehicles can be driven.

robotics The mechanization of tasks that would normally be undertaken by a human. Robots can undertake work in difficult and hazardous places, reducing risks to operatives; they can also perform repetitive tasks, often quicker and more accurately than humans. Machines are now being used to produce concrete walls and floors, spray and level plaster, and lay bricks and blocks.

rock A hard material consisting of a collection of different minerals, which forms the majority of the relatively thin outer curst of the surface of the earth. Rock falls into three broad groups based on its origin, namely: *igneous, *metamorphic, and *sedimentary.

rock anchor (rock bolt) A steel cable or rod that is used to stabilize rock. A hole is initially drilled into the rock, after which the rod is inserted. An expansion device or *grouting is then used to secure the rod in place. Typically used when tunnelling through rock and on highway cuttings through rock to stabilize steep side slopes.

rocker shovel A mechanical loading device that is used where headroom is limited, such as in a tunnel or a mine. The shovel lifts spoil in a rocking action from the front of the machine and deposits it at the back of the machine.

rock excavation Contractual term used to describe excavation that is too difficult to undertake without mechanical plant such as breakers or by means of blasting. When excavation can only be undertaken by such means, an extra cost for the dig can be claimed.

rock face Stones and rocks with the natural surface exposed and used for walling.

rock mechanics The study of the behaviour of rock and rock masses, it includes drilling and tunnelling through rock as well as mining and excavating it.

rock pocket (US) Air bubbles or holes in rocks caused by loss of fines in the concrete aggregate, insufficient compaction and vibration, or using a poor concrete mix.

Rockwell hardness test An experimental procedure utilised to determine the *hardness of a material. The hardness is verified by measuring the depth of penetration of load impacted by a diamond cone.

rock wool (rockwool) A type of insulation made from glass fibre.

rod A measuring timber cut to a predetermined length with increments marked, used for setting out elements of construction that have consistent dimensions or spacing, for example, brickwork courses, windows openings, floor to ceiling heights, etc.

rodding A method using a drain rod to unblock a pipe. A **rodding eye** is a small hatch at ground level that allows access to the drain for rodding. A **rodding point** is a section of pipe at 45°, which connects the drain to the surface of the ground so that it can be rodded.

rod float A weighted wooden rod that floats upright such that it is subjected to varying stream velocity.

roll 1. Sheeting material that has been wound around itself to form a compact cylindrical shape. **2.** A semi-cylinder ornamental and partially functional shape used to round off the rim of sinks and baths; it can be totally aesthetic when used at the top of columns. **3.** A joint used in sheet-metal roofing where the edges of two metal coverings are rapped over a semi-cylinder length of wood. The raised semi-cylinder joint, which runs down the slope of the roof preventing moisture entering the joint. **4.** In Roman architecture, a semicircular architectural feature found at the head of columns.

roll-capped ridge tile A ridge tile with a roll feature added to the top of the tile.

roller An absorbent cylindrical painting tool, used to apply paint to a wall in long even strips.

roller door (rolling shutter) A vertically revolving door or shutter comprising a series of vertical slats that are rolled up over a horizontal roller located at the top of the opening when open and are rolled down when closed. *See also* ROLLING GRILLE.

rolling grille A type of security grille that operates in the same way as a *roller door.

Roman cement A type of cement obtained by burning septaria, found in clay deposits.

Romanesque Medieval European architecture characterized by semicircular and pointed arches. In England, Romanesque is better known as Norman architecture.

Roman tile A type of *single-lap tile. Both single and double versions are available. Single Roman tiles have one central channel while double Roman tiles have two channels.

roof The uppermost part of a building that protects the structure from the rain and other external elements.

roof abutment A part of the structure that is in direct contact with the roof.

roof boarding Decking or boards placed adjacent to each other on the top of the roof enabling the roof to carry flexible sheeting material. The boarding provides a firm base on which insulation and impervious weather protection can sit.

roof cladding The external finish or material that is placed on the roof structure.

roof conductor A copper tape lightning conductor fixed to the highest part of the roof or parapets to provide the air terminal through which lightning can be safely conducted to the ground and earthed.

roof covering The external surface of the roof, normally the part that protects the structure from the rain and other external elements.

roof cutting The skill of setting out and cutting of timber rafters to form a traditional framed roof.

roof decking The covering of a roof using boarding or timber planks to provide a solid surface on which the flexible sheeting or liquid (asphalt) roof covering can be placed.

roof drain The outlet from a roof used to collect rainwater (*surface water) and discharge it through a downpipe.

roof drip A point of rainwater discharge at the eaves of the roof.

roofed-in stage The phase of construction where the roof is erected and provides weather protection to the workers and structure below.

roofer (roofing specialist) A tradesperson who constructs or repairs roofs.

roof extract unit (roof extractor) A extractor fan or series of fans that are located on the roof of a building.

roof guard 1. A protective handrail or fence placed around the perimeter of a roof to prevent personnel falling from the roof. **2.** A snowboard placed on the roof to prevent large volumes of snow sliding off the roof and injuring people below or damaging property.

roofing (roof covering) Uppermost part of the building used to provide a waterproof structure and protection from the weather. The weatherproof skin can be made from a membrane, slates, or sheeted material.

roofing felt Weatherproofing sheet material that contains bitumen or asphalt.

roofing insulation Sheet, wool, or loose fill material that has good thermal resistivity, reducing heat flow through the roof space. Rigid insulation is often used on top of the roof rafters with the weatherproof covering being fixed over the insulation; wool and loose-fill insulation is often used between the ceiling joists providing a thermal barrier between the *roof space and main rooms within the house.

roofing nail Any nail used for fixing roof tiles, battens, or roof coverings to the roof.

roofing punch A steel rod with a sharp point used for making holes in sheeting or corrugated roofing material.

roofing screw Long threaded screw used for fixing corrugated and profiled roofing sheets.

roof light (skylight) A window located in a roof.

roof light sheet Transparent sheet of roofing material used to allow natural light in through the roof.

roof pond An *indirect-gain passive solar system where the thermal mass comprises water located on the roof of a building.

roof space The attic or void contained within the roof structure, in many cases the attic is converted to form a room in the roof.

roof tile (roofing tile) Clay, slate, plastic, or concrete unit in a flat form that is laid on a roof, overlapping and interlocking into other tiles to enable rainwater to be discharged from the roof.

rooftop unit An air-conditioning, heating, or ventilation system that has been designed to be located on a roof.

roof truss Prefabricated timber roofing unit that utilizes triangulation to maintain its rigid form and transfer roof forces. The timbers are held together using pressed gangnail plates.

roof waterproofing (roof weatherproofing) The membrane or covering that prevents the ingress of rain, snow, and wind through the roof.

room air conditioner A type of wall-mounted air-conditioning unit that is designed to condition only the room in which it is situated. Usually consists of a small refrigeration unit with an integral air circulation fan. Air is drawn from the room, cooled, and returned. They can be noisy, are generally not very efficient, and normally only provide comfort cooling, although filtration and heating can also be provided. Humidification is not available.

room-heater A heating appliance that is designed to heat only the room in which it is situated.

room-sealed appliance An appliance that obtains fresh air for combustion from outside and discharges the flue gases externally.

root The section of a tenon where it widens out at the shoulders

root mean square value (RMS value) A statistical measure of a varying quantity, calculated by square rooting the mean of the squares of the values.

rope A length of cord made by twisting strands of fibre or wire together.

ropiness (ropy finish) 1. A paint surface where the brush marks have not flowed out **2.** Rough or untidy finish.

rose 1. A **ceiling rose** is a decorative surround placed where the light fitting passes through the ceiling. **2.** A *shower rose. **3.** The decorative part of a door handle that the lever passes through.

rostrum A plinth or stand for holding documents, used by presenters while making an address to an audience.

rot Decay in wood, caused by moisture or mould growth.

rotary cutting Method of spinning logs on a lathe and cutting with a blade so that a continuous veneer is cut from the log. Prior to cutting, the log is soaked in hot water making it easier to cut the veneer.

rotary drill A method of drilling where the drill bit moves in a rotational movement.

rotary float A power float for levelling and smoothing concrete floors.

rotary planer A powered planer with a rotating blade used to produce a flat smooth surface.

rotary veneer A thin sheet of timber veneer produced from *rotary cutting.

rotational A circle movement around an axis, e.g. a rotational slope stability slide moves in a circular motion around the centre of gravity of the slip plane.

rotunda Any building that is circular on plan. The buildings are often covered by a dome.

rough arch An arch that is constructed of rectangular blocks, stone, or bricks that are not cut to the voussoir (wedge) shape.

rough ashlar Stone that is not cut neatly, having been roughly broken, hewn, or chiselled out of the quarry.

rough-axed Tiles or bricks that have been cut or shaped using an chisel-shaped hammer or axe.

roughcast (slap dash, wet dash) Plaster or render that is sprayed or roughly thrown and levelled on the surface of a wall.

roughcast glass Translucent, rolled sheet glass, one face of which has a slightly rippled texture.

roughcast machine (plastering machine, roughcast applicator) Piece of plant used to quickly apply plaster to the surface of walls. The plaster is pumped through the machine and sprayed onto the wall. Once applied it is levelled and finished.

rough cutting Cutting or breaking of bricks, tiles, or slates to approximate size with a *hammer, *bolster, or *trowel edge.

rough floor A sub-floor constructed from timber to carry floorboards.

rough grounds Timber battens fixed to provide a level base upon which to nail plasterboard.

roughing filter A graded filter used in the pretreatment process to separate out suspended solids.

roughing in (roughing out) The first fix, first level, or general installation of work that will be unseen. The levelling out of plaster or the first fix in plumbing.

roughness The level to which something is not smooth.

rough string 1. A beam that runs under the centre of a flight of stairs, shaped to the same profile as the stairs, also known as the carriage of a stair. **2.** The sides of a stair, otherwise known as strings, that are shaped to the profile of the stairs rather than having parallel edges. The stair *string is the diagonal beam that is used to carry the *treads and *risers of a stairway.

rough-terrain forklift Truck designed for lifting pallets and packages, and carrying them over building sites and other rough terrain. The trucks are either two- or four-wheel drive with tractor tyres.

rough work Trade work that will not be seen so does not have to be aesthetically pleasing. In brickwork this would be *common brickwork that is not seen once the finishes are applied; in plasterwork it is the first coat of plaster used to dub out or build up the plaster to provide a level surface to work on; in joinery it is the carcassing or the first cut of timber that provides the approximate shape before finishing; and in plumbing it is the first fix work that will be hidden behind finishes.

roundabout A more or less circular intersection (junction) on a road which allows traffic to move round it in one direction.

roundel A circular-shaped object.

rout To cut out a groove in wood, stone, or metal.

router Tool for cutting grooves in timber using a rotating cutter.

routine maintenance Regular maintenance undertaken, normally using a regular maintenance schedule, in order to preserve the condition and service levels of buildings or infrastructure. Such maintenance is normally aimed at preventing deterioration or failure. It may also be used to correct minor issues.

rowlock cavity wall (all-rowlock wall) A wall constructed in *rat-trap bond.

rowlock course A course of bricks constructed with gaps between using *rat-trap bond.

Royal Institute of British Architects See RIBA.

Royal Institute of Chartered Surveyors See RICS.

rubbed arch A flat arch formed of bricks that are cut and shaped (rubbed) to provide angles that extend from the *key stone and hold the arch in place. The mortar joints are very thin to ensure good contact and compression between the bricks.

rubbed bricks (rubbers) Bricks made of a soft clay, often without a frog, that are easily cut, ground, or sanded into a desired shape.

rubbed finish Concrete rubbed down with an abrasive carborundum stone to create a very smooth finish.

rubbed joint Masonry joints simply rubbed smooth with a sack to create a flat mortar joint.

rubber A resilient flexible polymeric material made from natural and synthetic elastomers.

rubber buffer (rubber silencer) A resilient stopper used to prevent doors slamming open on walls and other building furniture.

rubber cork tile A resilient tile made from a mixture of latex, rubber, and cork.

rubber flooring Resilient latex flooring that provides a smooth, non-slip surface.

rubber ring *See* JOINT RING.

rubbing down Removing rough spots, high spots, and other undulations by sanding the surface ready to receive paint, wax, or varnish.

rubbing stone An abrasive block used for grinding concrete, stone, and other ceramic materials.

rubbish (builder's rubbish, building rubbish) Debris created from the works on-site including: off cuts, trimmings of timber, brick, rubble from demolition, excess material, waste, and building debris from site-based work.

rubbish bucket (kibble skip) A large steel bucket lifted by a crane, and used for removing rubbish.

rubbish container A steel skip used for collecting, storing, and removing rubbish from building sites.

rubbish pulley Hoist used for lowering building debris to the ground.

rubble (rubble ashlar, rubble masonry) A type of stone with an irregular shape utilized in walls and foundations of buildings.

rubble wall Wall made from randomly cut stone.

rule off To smooth off plaster or concrete by drawing a smooth float that has two handles for drawing the float over the surface. Excess material is drawn off and the surface is levelled.

run 1. A straight or horizontal dimension. **2.** A stretch of cables or pipes. **3.** A length of bricks laid end to end, also known as a course of bricks. **4.** A defect in paintwork where excessive paint has been allowed to drip, forming a thick bead of paint that has flowed down the surface.

runby pit *See* LIFT PIT.

run moulding (horsed mould) A fibrous moulding or template that is pulled through the wet plaster to provide the desired shape.

runner The rail that guides a drawer, sliding door, or sliding window.

running bond Walling made entirely of stretcher bricks—bricks laid end to end with the longest face exposed. The joint of the upper course is positioned over the centre of the brick below, ensuring that the joints on different courses are staggered.

running mould *See* RUN MOULDING.

running tile Patterned or coloured tiles that run a continuous length providing a long border.

run-off The amount of surface water flow from precipitation when the infiltration of the ground is at full capacity.

runs *See* RUN.

rupture Something that has broken suddenly.

rural Areas within the countryside, depicted by woodland, grassland, fields, lakes, and other natural environments or estates that have a low population density rather than the densely populated built up *urban areas.

rust A form of material degradation highly prevalent in metals, especially *ferrous metals. Rust is usually initiated by the presence of oxygen and/or water.

rusticated stone Masonry block that has a rough textured (unfinished) face and wide joints to highlight the edge of each block.

rustication 1. Regularly cut *ashlar stone with blocks of stone that are left angular and protruding while other blocks have smooth faces. The smooth- and rough-faced blocks are randomly distributed to give a rough undulating face. **2.** The irregular feature created by stonework that is chiselled, drilled or hammered to provide a rough texture.

rustic joint A recessed joint set in rough-faced *ashlar stone.

rutting The sideway deformation of a road surface to form a sunken track due to the passage of vehicles, exaggerated by heavy vehicles and hot weather.

R-value *See* THERMAL RESISTANCE.

Sabin The unit of sound absorption. One Sabin is equivalent to one square metre of 100% absorbing material.

sacrificial anode A metal plate, which is electrically connected, used in cathodic protection (a method to control corrosion) of piping or other equipment. The metal plate must be more corrodible than the material to which it is attached.

sacrificial coating (sacrificial layer) 1. A thin coating of metal that oxidizes to protect the main metal from corrosion. A good example of this is galvanized steel (steel coated with zinc). The sacrificial layer oxidizes and protects the main metal from further corrosion. The coating protects the underlying metal so long as a the layer remains in place. **2.** A layer of paint, which can easily be covered or removed if defaced with graffiti.

sacrificial form Permanent formwork that provides a surface for the concrete to be cast on or cast against and remains as an integral part of the structure. Galvanized profiled steel, often called wriggly tin, is commonly used as floor formwork and remains in place after the concrete is poured.

saddle A piece of metalwork that is dressed and shaped so that it can fit underneath roof tiles and other roofing material to provide an extra layer of protection against rain. The piece of metalwork, often cut into squares of 460 mm can be used where a sloping roof meets an abutting wall, or a ridge joins a sloping roof.

saddleback board A strip of timber, chamfered along its length and placed on the floor between the jambs of a door. The raised strip of timber provides a neat strip that the door can be fitted to. The raised surface allows the door to be able to swing open clear of carpets and other finishes, while still fitting close to the board.

saddleback coping A coping stone (or shaped tile) that is cut or formed with a raised central point (apex) and overhangs the sides of the walls which it covers. The coping is shaped so that water runs away from the apex and is cast off away from the face of the wall due to the overhang.

saddle clip A pipe fitting that is U-shaped with fixing lugs on it, so that it can be positioned over a pipe and clamped to a bracket or fitted to a wall.

saddle-jib crane A *tower crane with a central mast and a horizontal jib that extends out from the central mast and is used as the lifting arm. A counter jib shorter than the main lifting arm carries a counter-weight to counter-balance objects that the crane lifts. Large tower cranes normally have the cab mounted on top of the mast, giving the operator a central position with the best possible views of the site around which objects are being lifted and moved.

saddle piece A piece of plumber's metalwork that is shaped or fitted over something. The saddle may be clamped to a pipe to provide a seal, or positioned over a ridge to provide weatherproofing.

saddle scaffold Scaffolding that is built over the apex of the roof.

saddle stone The stone used at the apex of a *gable end of a roof.

saddle trusses Small roof trusses that sit on top of the main roof trusses to enable the roof to change direction or angle. These are used where a roof turns a corner and sit on the top of existing trusses at right angles to the ones below, ensuring the structure remains stable.

SAE international A global professional society of automotive engineers.

SAE levels of automation (0–5 levels) A range defining the levels of human intervention in *autonomous car development, defined as:
- **Level 0:** Fully manual vehicles.
- **Level 1:** Single level of automation that assists the driver, e.g. steering, speed, or braking control.
- **Level 2:** Automation in respect to helping the driver steer and accelerate/ decelerate, e.g. lane departure system and self-parking features.
- **Level 3:** Environment detection, which is the lowest class in an automated drive system. For example, overtaking slower-moving vehicles but still requiring human override to execute certain tasks.
- **Level 4:** No human intervention required but is provided as an option, if required for manually overriding.
- **Level 5:** Human driving is completely eliminated; hence there is no need for a steering wheel or braking system for a human to interact with.

safe 1. Not in danger, secure. **2.** A lockable box used to keep valuables secure.

safety The feeling of being secure from risk of unwanted events or injury, a condition of being safe. Being safe is the control of oneself, others, and environments such that risks of unwanted events are suitably managed to avoid occurrence. The *Health and Safety at Work Act provides the power and the *Health and Safety Executive is empowered to ensure that the work environment and the acts of workers and employers are suitably controlled to avoid injury and maintain the health and welfare of those at work and those who enter into the work area.

safety arch An arch that is used to add strength to the structure and relieve the load placed on smaller arches or lintels below; also known as a relieving arch.

safety factor *See* FACTOR OF SAFETY.

safety glass (toughened glass) A reinforced laminated glass that is designed to break into small blunt fragments when broken. *See also* TEMPERED GLASS.

safety helmet *See* HARD HAT.

safety ladder (jacket ladder) A steel ladder, permanently fixed to the building with protective hoops that allows a person to pass through and offers additional protection from falling off the ladder.

safety net A net placed over open areas within partially constructed buildings to prevent workers, materials, and equipment falling through or over the side of buildings.

safety officer Person appointed by a company to ensure that the statutory provisions of the Health and Safety Act are followed. Under the Health and Safety Act, employers are required to appoint a Health and Safety Officer to meet company legal obligations under Health and Safety Law.

safety railing A *guard rail to prevent people falling from heights.

safety signs Signs required on construction sites to warn of dangers and risks. These make employees aware of their legal duties, inform them of protection needed, and ensure correct guidance and management of the site is maintained to ensure the health, safety, and welfare of employees. The legal requirements are available from the *Health and Safety Executive.

(⊕) SEE WEB LINKS

• The Health and Safety Executive website provides essential guidance on the use and format of safety signs.

safety valve A valve that opens to ensure the pressure (or flow) is not exceeded (or lost) from a system.

safe working load A load that a structure or foundation can carry safely—normally factored (divided by a *factor of safety) to accommodate a margin of error to ensure failure does not occur.

sag (sagging) Bending downwards. A **sagging bending moment** is a *bending moment with tensile along the bottom and compression along the top; *see also* HOG.

Saint-Venant principle A principle that states that when a set of locally applied forces (in equilibrium) are imposed on a section of the surface of a body, strains are negligible at distances that are large compared with the dimensions of that part. It was developed by a French mechanician and mathematician Adhémar Jean Claude Barré de Saint-Venant (1797–1886) and is based on the theory of elasticity.

saline Containing *salt.

salt A mineral made up of sodium chloride (NaCl), which can exist in the form of other metal chlorides, such as lithium chloride.

saltation The jumping/bouncing motion of sand and soil particles being carried/transported by the wind or moving water.

salvage To save something from destruction.

sample A part of a whole, a selection of a few articles where there are many, taken for inspection, reference, or determining, and predicting the nature of the larger sample.

sample panel (reference panel) A section of brickwork, cladding, roofing, or flooring that is constructed in the required format. It is inspected and improved by the architect or client's representative, and is used as a benchmark of the standard expected throughout the construction.

sampler A device to enable a sample of material to be collected for testing. *See also* SOIL SAMPLER.

sampling error A statistical error recorded from a sample rather than from the whole population.

sampling rate The speed at which values are obtained, e.g. twenty readings per second.

sand A *granular material of fine grains (between 0.06 and 2 mm) made up of rock fragments or mineral particles, usually of quartz origin. It has zero cohesion when dry and an apparent cohesion when wet due to the surface tension of the water within the pore structure.

sand bedding A 40–50 mm layer of sand, placed on an excavated area to provide a level surface on which the floor structure can be built. Where the sand is used to create a separation from the substrata below, it may also be called sand blinding.

sand blasting The propelling of sand or other siliceous material from a gun or hose, fired under high pressure and used to clean, remove the surface or contaminant, or reshape a surface. There are many different types of blasting, including hydro-blasting, micro-blasting, bead-blasting, and wet abrasive blasting, which are all variations of the sand-blasting process.

sand boil *See* LIQUEFACTURE.

sand box A box or tube that can be filled with sand to act as a support for a scaffold *prop.

sand catcher A device to collect and monitor the amount of sand in suspension in a flowing river or wind stream.

sand drain (sandwich drain) A drainage channel filled with *sand.

sander (sanding machine) Mechanical equipment that is fitted with *sandpaper and rotates on a belt, moves backwards and forwards, or moves in a circular motion, and is used to smooth and clean surfaces. Sanders can be belt, disc, or orbital, and may be hand-held or floor-standing bench tools.

sand filter A filter made from sand. The sand is placed within a container through which water runs, the impurities and salts are caught within the sand, and the water which passes through is removed of a high proportion of impurities.

sanding The action or process of rubbing sandpaper over a surface to remove high spots, paint, irregularities, or to clean the surface. The abrasive material removes a thin layer of the surface material.

sandpaper Strong reinforced paper or cloth coated with abrasive silicone material or metals that are graded from fine to coarse. The abrasive paper is used for removing a thin layer of surface material from an object—levelling, shaping, or cleaning. Different abrasive materials include sand, emery, aluminium oxide, silicon carbide, aluminium zirconia, chromium oxide, and ceramic aluminium oxide. The different materials are used for different purposes.

sand pile A heavy weight is dropped on silty ground to form an indentation, which is filled with sand. The process is then repeated and a column of compacted sand is formed within the ground.

sandstone A sedimentary rock with sand grains mostly of quartz, but also includes calcite, clay, or other minerals.

sand streak A line of exposed aggregate in the surface of concrete, caused by bleeding.

sand trap A portion in a channel where flow is slowed down (for example, where the channel has widened), and the suspended material is dropped from the flow.

sandwich beam An H- or I-beam, made of different composite materials, improving the strength-to-weight ratio, therefore increasing the performance.

sand wick A type of *sand drain consisting of a small diameter vertical column. It is typically used under embankments constructed on soft compressible clays, to reduce the drainage path the excess pore pressure has to travel.

sanitary accommodation A room that contains a sanitary *appliance.

sanitary appliance (sanitary fitting) An appliance that receives either soil or wastewater. For instance a *water closet, a *bath, a *basin, or a *sink.

sanitary connector *See* PAN CONNECTOR.

sanitary cove 1. A curved tile used as the junction between the wall and the floor to prevent water penetration. **2.** A metal *cove installed on a stair between the *riser and the *tread.

sanitary fitting *See* SANITARY APPLIANCE.

sanitary pipework (sanitary plumbing) All of the pipework associated with a sanitary appliance.

sanitation The provision of clean water and sewage disposal in relation to human health.

SAP *See* STANDARD ASSESSMENT PROCEDURE.

sapele (*Entandrophragma cylindricum*) An African tree that grows to approximately 45 m high. The wood has a significant commercial value and is similar in appearance to mahogany; it is particularly sought-after for use as flooring due to its durability and graining.

saponifier An alkaline chemical added to water to further dissolve different types of residues.

sapwood (alburnum) The younger wood of a tree, pale in colour and sat just under the bark. It functions as a water conductor from the roots to the leaves.

sarcophagus 1. Carved or decorated stone box receptacle or coffin made to accommodate a corpse. **2.** The word is occasionally adopted in construction to mean something that contains something, therefore being a heavy structure for containment of an object or a person.

sarking A flexible *roofing felt or layer of roof boards that are placed under the *roof covering.

sash (window sash) The frame that supports the glazing in a window frame.

sash balance A coiled spring integrated into the jambs of a *sash window. Used instead of a cord, pulleys, and weights to operate the window.

sash bar (sash astragal) The vertical glazing bar in a *sash window.

sash clamp A small clamp used to hold components together when they are being glued.

sash cord (sash line) A cord attached to each *jamb of a *sash window that supports the *sash weights.

sash fastener (sash lock) A fastener attached to the sash of a window used to keep the window closed.

sash fillister A *plane used to cut rebates for glazing.

sash haunch A joint traditionally used in the framing of a sash window, where a full tenon would weaken the frame. The size and position of the tenon is altered to maintain strength.

sash lift A handle on the inside of a *sash window used to open or close it.

sash pulley A pulley set within the jamb of a *sash window. The *sash cord passes over the pulley and supports the *sash weights.

sash ribbon A thin band of steel used to support the *sash balance.

sash stop A moulding attached to the inside face of a jamb that keeps the sliding portion of the *sash window in place.

sash weight A weight set within the jamb of a *sash window and hung from the *sash cord. Used to counterbalance the weight of the sliding portion of a *sash window.

sash window (double-hung sash window, hanging sash, vertical sliding window) A window comprising two sashes (window casements) that opens by sliding one of the sashes vertically. *See also* SLIDING WINDOW.

satellite A celestial object (such as a telecommunication device) that orbits the earth or any other planet, and used in *NAVSTAR and *global positioning systems.

satin paint A type of paint with a delicate shiny finish, known for its durability and stain resistance.

saturate Something that is completely full, for instance, the pore spaces in soil with water.

saturated Describes a material, liquid, or container that has absorbed or contains the maximum quantity of another fluid or material that it can hold.

saturated solution A liquid that has dissolved or suspended the maximum amount of solid material without separation or segregation.

saturated vapour pressure The pressure associated with a given sample of air when it is completely saturated.

saturation chroma The intensity of a hue when describing a colour compared with a neutral grey that is of a similar lightness.

saturation coefficient The volume of water that can be absorbed by a brick. Different coefficients and absorption tests exist for the UK and US.

saturation temperature The temperature at which a liquid boils.

saturation point The point at which a given sample of air contains the maximum amount of water vapour that it can contain.

saucer dome A *domelight.

savannah A flat grassland with scattered tress found in tropical and subtropical climates.

Savonius rotor A type of *vertical axis wind-turbine, designed by Finnish engineer Sigurd Savonius in 1922.

saw A toothed tool for cutting wood, steel, or plastic depending on the blade and type of saw. As the toothed edge of the saw is run backwards and forwards across a material it cuts into it making a groove known as a kerf. The teeth of the saw are angled with adjacent teeth set in alternate directions. The angle of the teeth is known as the set, with the outward difference between adjacent teeth creating the width of cut or kerf. Saws can be hand-held or bench, powered or manually operated.

saw bench A table or platform with a circular saw mounted within it. The blade of the saw protrudes up into the table allowing wood to be passed over the table and cut by the protruding blade.

saw doctor A professional skilled at resetting and sharpening the teeth of *saws.

sawfile A small file, circular in section, measuring approximately 1.5 mm in diameter. The blade is held in place by a jig or hacksaw frame. The file is used for cutting intricate shapes in timber, plastic, tile, clay, metals, and other hard materials.

saw horse Bench, trestle, or stool with four legs providing a long thin platform approximately 1 m long and 100 mm wide that can be used alone for working small lengths of timber, or more commonly used in pairs to support pieces of timber that are to be cut or worked.

sawing The cutting of timber using a *saw.

sawn damp course The cutting of a masonry course after the wall is constructed to allow the insertion of a *damp-proof course. The cut is made into the horizontal joint in small sections and the damp-proof course is inserted. Slight settlement of the wall above between 1–3 mm is to be expected even when the joint is refilled with mortar or expanding grout. In remedial and refurbishment work it is more common to have the masonry drilled and injected to create a dense impermeable damp-proof course.

sawn stone Stone that has been cut to shape for walling, *cladding, and *ashlar stonework.

sawn veneer A veneer cut with a thin saw as opposed to being sliced or rotary-cut.

saw set (swage) A tool used to give the correct set for saw teeth.

sawtooth roof (northlight roof) A roof comprising a series of triangular sections that are placed in parallel with one another. The result is a roof with a profile similar to that found on the teeth of a saw. The steeper face of each triangular section is usually glazed and orientated to the north.

SBEM *See* SIMPLIFIED BUILDING ENERGY MODEL.

scabbler A power tool used for breaking up and removing the upper surface of concrete. The rough surface then provides a mechanical key for new concrete to adhere to. The tool scabbles the surface, removing any weak finish that may exist on the surface. This ensures new concrete cast against the surface can form a good mechanical fix.

scabbling (hacking) The removing of a weak concrete surface and roughening up of the surface in order to provide a strong rough surface and key for new concrete. When concrete is poured and vibrated, excess water carrying the aggregate fines and cement comes to the surface of the concrete, creating a weak film called laitance. When new concrete is cast against existing concrete, the laitance should be removed. The surface of the concrete can be hacked by a *scabbler, a pointed tool, brush hammer, or comb. The scabbler breaks off the surface creating a rough undulating surface, exposing the concrete that is strong and not affected by the laitance. The rough concrete provides a good mechanical key for the adjoining concrete; this is not as strong as the chemical bond that would be formed between two elements of newly laid concrete, but does provide a strong bond.

scaffold (scaffolding) A temporary structure that is erected to enable work to be carried out at an elevated level. Scaffoldings are used to provide access at heights so that walls can be erected, and ceilings, floors, and roofs can be constructed, decorated, and repaired. The framework provides access for a worker, a place to temporarily store materials, and if providing temporary support, it can act as *falsework. Various scaffolding systems exist, including quick assembly patent systems, tower scaffolds, independent scaffolds, platform scaffolds such as birdcage assemblies, and putlog scaffolds that are partially supported by the building. Scaffolding was traditionally made from timber tied or bolted together. The tubes, poles, and standards of today are more commonly aluminium and steel, though in some countries, bamboo remains a material that is commonly used.

scaffolder (scaffold hand) A competent and skilled worker capable of safely erecting, changing, and dismantling scaffolding. Scaffolders should be properly trained and must be competent in the erection of scaffolding. The Construction Industry Training Board provides training courses in the erection and checking of scaffolding.

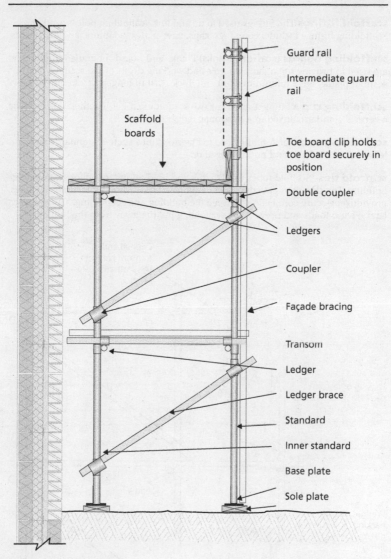

Guard rail

Intermediate guard rail

Scaffold boards

Toe board clip holds toe board securely in position

Double coupler

Ledgers

Coupler

Façade bracing

Transom

Ledger

Ledger brace

Standard

Inner standard

Base plate

Sole plate

Scaffold: Independent scaffolding

scaffold fittings The fittings used to tie and link scaffolding poles together. Scaffolding fittings include: couplings, clips, base-plates, adaptors, and toe boards.

scaffolding boards (scaffold planks) Planks and boards of graded timber that are used to provide the main platforms and working decks of the scaffold. The scaffold planks are a minimum of 40 mm thick, 3.70 m long, and 230 mm wide.

scaffolding clip A fitting used to link two scaffolding tubes together; this could be a vertical standard, linked to a horizontal ledger or brace.

scaffolding coupler A fitting used to link two tubes such as standards and ledgers, or standards and putlogs together.

scaffold ties *Scaffold tubes used to link and hold an independent scaffold to a building. The ties often pass through or are braced into *window or *door openings, providing a secure connection between the building and scaffolding. The ties resist lateral wind loads and prevent the scaffolding pulling away from the building.

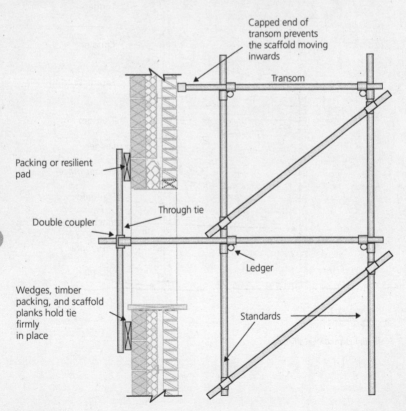

Capped end of transom prevents the scaffold moving inwards

Transom

Packing or resilient pad

Through tie

Double coupler

Ledger

Wedges, timber packing, and scaffold planks hold tie firmly in place

Standards

Scaffold ties

scaffold tube Any of the tubes used in a *scaffold, such as a putlog, brace, ledger, standard, transom, etc.

scalar A quantity that has magnitude but no direction, e.g. mass and time; *see also* VECTOR.

scale The ratio between a model or drawing, and the actual building or structure that it represents; for example, 1:20 means that the actual building is twenty times greater in size than the model or drawing.

scaled dimensions Measurements taken from a drawing using a scale rule rather than dimensions. Scaling dimensions from drawings or models are less reliable than using dimensions that provide the true size.

scale of chords A rule that is used to set out angles for a radius and chord length.

scallop A carved or casted ornament, fixed to a building or structure, that resembles a scallop shell.

scalloping A decorative pattern made by a row of half circles that join or link together.

scallops Flexible sticks, also known as withes, used to hold the thatch at the *verge of the roof; they are held down by the spars.

scantling Wood of non-standard size but between 47–100 mm thick and 50–125 mm wide.

scarf (scarf joint, scarfed joint) A joint made in laminated timber, formed by bevelling each strip of veneer at an angle of 1:6 to 1:24 and gluing the angled edges together.

Fingers of veneer or laminates cut at an angle
and glued against each other

Scarf

scatter Irregular spreading out.

schedule 1. A document listing the doors, ironmongery, finishes, reinforcement, windows, or other fittings, finishes, or components within a building. The document lists the component or finish, quantity, special instructions or quality, and the place to be used. **2.** A *programme of events, showing the time, duration, and sequence of activities; the chart may also provide notes on the associated resources.

scheduled maintenance (programmed maintenance) A longer-term approach than routine maintenance, would normally include works that are more extensive than minor maintenance. It may be planned as part of a whole of life asset management strategy.

schedule of defects (schedule of outstanding works, snagging list, punch list) A list of defects identified or works still outstanding that need to be completed before the certificate of practical completion is awarded.

schedule of prices (schedule of rates) List of prices placed against materials, plant, and labour, used where the nature or quantity of work is unknown. Where the exact quantities of work are unknown, a list of prices, plant, and labour rates can be made available, and the work quantified and costed once complete.

scission The process of cutting is termed scission.

scissors lift A Mobile Elevated Working Platform (MEWP) with a concertina lifting system. The scissor support structure contracts and extends to raise the working platform. The lifts come in different sizes with large lifts capable of lifting multiple workers and substantial materials. The lifts come with different wheels and tyres making them suitable for navigating in different conditions, e.g. suitable for smooth flat indoor, or external site terrain.

SCOOT *See* SPLIT CYCLE OFFSET AND OPTIMIZATION TECHNIQUE.

scope of works A description of the extent and nature of work to be undertaken. The specification gives details of the works, limitations, and any work that is excluded.

score (scoring and snap) To mark a line, by using a sharp instrument (knife or point) to cut a rebate into a material. The indent provides a permanent line or mark that can be used to guide a deeper cut. Scoring provides a precise mark and guides the next cut with greater precision than would be achieved by cutting alone. The term is also used to describe a mark which has been unintentionally left on a surface causing a permanent mark and damage.

Scots pine A species of pine (*Pinus sylvestris*), native to northern Europe and now extinct in England. It is distinguished by its blue-green leaves and orange-red bark.

scour The erosion of a river, a seabed, or river bank by moving water.

scrap Waste material that is of no use, particularly metal that is to be re-melted for casting.

scraper 1. Essentially a large blade, spade, or bucket on wheels, which can be either self-propelled or towed. The front edge of the bucket has a cutting edge, which is lowered (200 mm) into the ground and then dragged through the ground thus loading the bucket. Used in *earthworks. **2.** A bladed tool for removing paint from surfaces. The paint may be softened using heat or chemicals, and the blade is run across the surface forcing itself between the layers of paint and the substrate, causing the paint to peel away.

scratch coat The coat of plaster with an etched or keyed surface that provides a good bond before the finishing coat is applied. The scratch coat helps level undulations in the substrate and provides a good key for subsequent layers of plaster.

scratcher comb A plasterer's float with nails projecting from the surface, also called a devil float, used to etch the surface of plaster to help provide a key.

scratching The making of a shallow mark on a surface.

scratch tool Plasterer's tool used for etching decorative features and enrichments into the surface of the plaster.

screed The top layer poured over insulation or concrete to which floor finishes can be applied.

screed board (rule, straight edge) A straight board used for levelling the surface of screed. The board is longer than the screed dabs, guides, or rails so that it can level between them.

screeding The laying of a fine aggregate, sand and cement concrete screed.

screed pump Mechanical pump used for distributing concrete screed to the place where it is needed.

screed rail Boards, rails, or tubes that are set level and parallel in bays so that screed can be levelled between them using a *screed board.

screed to falls A screed which is laid to a gradient falling towards a drain or gulley, allowing water to be discharged from the surface.

screen 1. A frame with a mesh grid built into it. **2.** A framed mesh sieve used for separating fine and coarse aggregate **3.** A door with a mesh panel that allows natural ventilation, and prevents insects and other animals entering the dwelling. **4.** A framed mesh ventilation hole used to allow ventilation into roof eaves, under suspended floors, and above windows.

screens passage A passage at the end of a medieval hall, where screens conceal the main entrance.

screw A fixing with a helical threaded shank and pointed end that self-taps into wood as it turns and penetrates the timber. Screws can have a single flat groove in the head for a flat-bladed screw driver to be inserted or can have a crossgroove to receive a Phillips head screw driver.

screw cap 1. Screw fixing for lightbulbs. **2.** A decorative dome head that can be screwed or fixed on top of the screw once the screw has been inserted into the timber.

screw cup A decorative brass or steel cup with a central hole, positioned under a screw so that the underside of the countersunk screw fits into the cup.

screw-down valve A valve that is operated using a threaded spindle.

screwdriver A hand-held manual or powered tool used for turning and driving *screws into materials. The screwdriver can have a flat-bladed or Phillips cross-head for inserting into the screws.

screwdriver bit A replacement tip for a screwdriver. Some screwdrivers come with detachable heads, the shafts of such screwdrivers having a hexagon recess into which the hexagonal bits can be inserted. Different sized and shaped bits can be inserted.

screwed and glued joint A joint which is rebated, glued, and screwed to make a strong reliable joint.

screw eye A loop fixed to the head of a screw, used as an anchor or fixing for ropes or hooks once the screw is in position.

screw gun A power tool used to drive screws, also called an electric screwdriver.

screw plug Two discs with a rubber seal between them used for sealing drains. The discs are engineered so that they can be threaded inside a threaded pipe, creating a sealed access to the drain.

scribe To cut a light groove into the surface with a scriber. The scriber is used to mark the surface of the material.

scribed joint (cut and fit) Joint made by offering the item to be cut or joined against the surface to which it is to be fitted. For an exact fit the joint position is scribed while in place and then removed for cutting. When returned to the desired position the material should provide a snug fit.

scriber A sharp pointed tool for accurately marking the surface of materials.

scrim Woven cloth used to provide a reinforcement and help bridge the joints of plasterboards. The material is used with tapered edge plaster board, plaster, or adhesive, and placed across the joint. The joint is then skimmed with fine plaster. The joint tape comes in rolls and can be used for flush and corner joints. The scrim is normally cloth or hessian, but can be cotton or metal—all being used to provide reinforcement to plaster.

scrimming The applying of *scrim over joints.

scroll A decorative spiral providing a feature to the end of a shelf or under a *handrail in a stairway or hall.

scroll saw An electrically powered *fretsaw.

scrubber An apparatus for removing impurities in gas.

Scruton number An aerodynamic number—the higher the number the less the oscillations will be; used to assess the pedestrian excitation of bridges.

scullery A room set aside where cooking utensils and dishes were washed.

scum A fine thin layer of dirt on the surface of a liquid.

scumble glaze (scumble stain) Transparent paints used for modifying the colour, texture, or appearance of the finish.

scumbling The removing of paint or texturing to expose part of a substrate below to give a broken finish to the paintwork.

scupper An opening in a *parapet wall that enables the rainwater collected in a *box gutter to be discharged to a downpipe.

scutch A bricklayer's hammer with a chiselled cross-peen on both ends of the head, used for fair and rough cutting.

SDGs *See* SUSTAINABLE DEVELOPMENT GOALS.

sea gate 1. A long rolling swell of the sea. **2.** An access channel to the sea. **3.** An adjustable barrier providing protection against the sea.

seal 1. A material that is used between two components to prevent leakage. **2.** The process of filling a gap or space with material to prevent leakage. **3.** The water found in a *trap that prevents foul air from the drainage system entering the building.

sealants A substance (such as an *epoxy resin) used to seal something. A material used in small gaps or openings to prevent the passage of water, air, and noise, i.e. silicone.

sealed system A type of central heating system where the expansion of the water occurs within a sealed expansion vessel.

sealed unit A multiple-glazed unit where the panes of glass have been sealed together in a factory.

sealer 1. A *sealer coat that prevents the passage of water and air or gases as well as protects the base onto which it is applied.

sealer coat Finish used to seal the surface. Sealer coats can be used on concrete, paint, wood, plaster, and other materials. They are often used to seal porous surfaces, but may be used to protect or enhance a finish.

sea level The average level of the sea, halfway between low and high tide.

sealing compound A viscous liquid that is used to fill and seal joints. The material fixes to the sides of the joint and bridges and fills the gap, preventing the ingress of water, debris, dust, and wind. Sealants can be two-part epoxy or elastomeric products or one-part mastic sealants.

sealing ring A elastomeric ring used to provide a water- and air-tight seal around adjoining pipes. The ring is fitted over the sprocket or end of one pipe so that the collar of the coupling pipe or coupling ring can slide over the pipe with the sealing ring trapped and compressed between the pipes.

sealing strip Compressible foam or elastomeric strip applied to movement joints.

seam (welted seam, welt, lock joint) A joint between two sheets of roof metal where the joining edges of the sheet metal are placed back to back, up standing, and then folded over to create a seal. A single fold is called a single lock and double fold a double lock. Seams that slope down the roof may be left upstanding, whereas horizontal seams are folded down the roof to help improve the seal and allow rain water to flow over the seam.

seamer Electrically powered pliers used for rolling and folding sheet-metal roof seams.

seamless tube A pipe made by extruding metal from a solid.

seam roll A *hollow roll where the edges of sheet metal are placed against each other in a standing seam and simply rolled together to create a water resistant roof or wall joint.

sea puss A seaward undercurrent formed when the waves have broken on shore.

seasoning The drying out of timber by natural drying or air drying.

seating The surface that an object, equipment, or structure rests on.

sea wall A structure to divide the sea from the land and designed to prevent flooding and erosion.

secant A straight line that cuts a curve. The **secant modulus** is often employed to define a modulus value when the stress–strain response of a material is non-linear; values are typically given for different percentage strains, e.g. 5 and 10%.

secant piling A below-ground retaining wall that has been constructed from overlapping cast in-situ board piles. Once the piles have been cast the ground can be excavated from one side.

SECED Society of Earthquake and Civil Engineering Dynamics.

second 1. A unit of time that is equal to a 60th of a minute. **2.** A unit of angular measurement equal to 360th of a degree.

secondary beam A beam that spans and is supported by other beams.

secondary element A non-essential part of a building's structure; includes fittings and finishes.

secondary fixing A fixing that is housed by another bracket, frame, or bolt. The primary fitting is fixed directly to the building or structure in order to carry the secondary unit. Bolts may be fitted to carry the chains to hold air handling units and lights. Primary frames and cables are fitted to carry the frame of a suspended ceiling.

secondary glazing A type of double glazing that is formed by fixing a second sheet of glass or translucent material to the inside face of single *glazing.

secondary lining The finishing coat over primary lining in tunnelling.

secondary reinforcement Reinforcement perpendicular to the main reinforcement, often used to prevent cracking.

secondary sash glazing (applied sash glazing) A single glazed *sash that has been secondary glazed.

secondary treatment The middle stage in a *water treatment, typically involves rapid or slow filters in drinking water treatment plants and *trickling filters or activated sludge processes in *wastewater treatment plants.

second fixing Fittings that are fitted after plastering and at the same time as the internal finishes are applied. Such fixings include electrical sockets, switches, light fittings, water taps, and other sanitary units.

second foot *See* CUSEC.

second moment of area *See* PRODUCT OF INERTIA.

seconds Bricks with slight defects that can still be used in common brickwork where the face of the brick is hidden.

second seasoning The loose framing of high-quality joinery items allowing the units to adjust to the humidity levels when in place. Units can also be stacked in the new environment to allow movement before fitting. If any excessive movement or defects occur before fitting the units can easily be replaced.

secret dovetail A dovetail joint that is formed within and hidden in a mitre joint.

secret fixing Any method of fixing and jointing that cannot be seen and is hidden from view once complete. Examples would include *secret dovetail and *secret nailing.

secret gutter A gutter that is hidden or largely concealed from view. The gutter is almost hidden by the tiles or roof covering. Due to the tiles covering the gutter, hidden gutters are susceptible to blockages.

secret nailing (blind nail, edge nail) Nailing that is hidden from view. The nails penetrate into the edge of the floor boarding at an angle, the abutting boards hide the nails. Nails can also be hidden in tongue and groove and rebated timber. The nails are inserted into the groove in tongue and groove timber or into the rebate of lapped timber.

secret screwing Screw heads are inserted into keyholes with slot-shaped fixings that are used to hold panelling and trim in place.

secret wedging Wedges fixed into the end of a tenon that are hidden from view. The wedges are inserted and loosely tapped into saw cuts in the end of a stub *tenon before being inserted into the mortise. As the tenon is inserted into the mortise the pressure of the mortise bed (also known as the blind end of the mortise) forces the wedges into the tenon. The tenon then opens out against the sides of the mortise.

section 1. A unit of work, e.g. a bay of concrete or an area of construction. **2.** A group of workers.

sectional insulation Moulded insulation that is specifically designed and manufactured to fit around services, pipes that service and distribute heat or are part of a refrigeration system, and other irregular objects. The insulation is delivered in sections that effectively insulate the specific component.

sectional tank (bolted tank) A storage tank that has been manufactured from standardized panels.

section manager (section engineer) A manager or engineer responsible for or in charge of an area of work and all of the professionals and operatives that work on or within the *section.

section modulus The ratio of the *moment of inertia compared with the distance from the *neutral axis to the top or bottom edge of the object. Asymmetrical sections have maximum and minimum section values.

section mould A template used for provide the outline or guide for cutting and shaping material.

section property Geometrical properties of an object, for example, area and *section modulus.

secular Occurring over very long periods, particularly geological changes over centuries.

security Measures taken to reduce the risk of theft, unauthorized entry, and vandalism, as well as controlling and monitoring personnel on the site.

security fence Hording or barrier placed around the perimeter of the site to prevent unwanted access, protect the public from site-based activities and risks, and control entry.

security glazing A type of glass or polycarbonate material that is designed to provide security and withstand attack.

security lock (thief-resistant lock) A lock that is designed to prevent unauthorized entry.

sediment 1. Eroded material that has been transported by wind and water and then deposited. **2.** Fine material that has settled at the bottom of a liquid.

sedimentary rock (sedimentary stone) Rock that has resulted from external forces on the earth's crust and is formed of particles deposited by rivers, glaciers, the wind, the sea, or by chemical deposition from lakes or the sea. Common examples include *limestone, conglomerate, and *sandstone.

sedimentation 1. The process of suspended particles settling in liquid to *sediment. **2.** A wastewater treatment stage where the solid components to sewage settle out.

S E duct A type of shared flue found in multistorey buildings, where fresh air enters the flue at the bottom and the flue gases are discharged at the top of the flue.

seep For a liquid to slowly pass through or leak out of something.

seepage The passage of a liquid through a substance, for example, water flowing through a soil due to a *hydraulic gradient.

seepage pit A porous walled pit that is connected to a *septic tank to allow the liquid waste from the tank to slowly drain into the surrounding ground.

segment Part of an object, often a regular division of an item.

segmentation Dividing or splitting into segments, particularly in relation to a fault along its length, as a result of other faults crossing it.

segregated sewerage system *See* SEPARATE SYSTEM.

segregation The separation of fine aggregate in a concrete mix due to excess water in the mix or incorrect compaction and placing; for example, over-vibration or dropping the fresh concrete from a height through the reinforcement.

seiche The movement on the surface of an enclosed body of water (a lake), caused by the disturbance of the wind or an earthquake.

seismic Relating to *earthquakes; tectonic movement of the earth's crust.

seismic constant A steady acceleration, used in building codes dealing with *earthquake design, that buildings must withstand.

seismic fragility The maximum load equipment can withstand while still remaining in service when subjected to an earthquake.

seismic gap Where a segment of an active *fault has not slipped for a considerably long period of time.

seismic hazard Potential danger due to an *earthquake.

seismicity The frequency of earthquakes based on geographical and historical information.

seismic moment The size of an earthquake based on the area of slip and the force that caused it.

seismic reflection (seismic refraction) A geophysical surveying technique that uses sound waves to penetrate the ground. The sound can be generated by setting off explosives or by hitting a steel plate placed on the ground with a *sledgehammerr. The sound waves are refracted back to the surface at an interface between soil/rock layers of different velocity. Geophones are placed along the surface of the ground to detect the arrival of the refracted sound waves, which are connected to a *seismometer to record the data.

seismic wave A shock wave that travels through the earth from either an earthquake or an explosion. Such waves can travel along the surface as *Love waves, or through the earth as P waves and *S-waves.

seismic zonation Designating areas, locally, regionally, or nationally, that have differing potential risk and damage due to earthquake activity.

seismogram A record produced by a *seismometer to show the ground response to movement caused by an earthquake or explosion.

seismology The scientific study of earthquakes and the structure of the earth by the use of *seismic waves.

seismometer (seismograph) An instrument for measuring ground movement caused by an earthquake or an explosion.

seismotectonic zone (seismotectonic province) A geographical area that has similar earthquake characteristics due to the underlying geology.

selected tender (US invited bidder) Different from an open tender (*see* OPEN BIDDING), a contractor is nominated or previously listed and allowed or qualified to *bid. A party may be invited to bid or has met prequalification criteria and is part of a framework agreement.

selective tendering The process of asking those who have been *selected tender to bid. Only those listed or included in the framework can take part in the bidding process; this is different from an open-tendering process where any party is allowed to provide a *bid.

selenium A non-metallic element with the chemical symbol Se, used heavily in glassmaking (to give a red colour), chemicals, and pigments. It rarely occurs in its own right and is obtained as a side-product from other elements. It also functions as a semiconductor, which oddly conducts electricity better in the light than in the dark.

self-centring formwork Moulds or shuttering for concrete floors, that sit on beams that are extendable (self-centring), allowing the beam to be adjusted to fit within the floor and support the formwork. The term is a little misleading as neither the beam nor the floor formwork is self-centring. The beams used simply open out to fit different widths and are adjustable.

self-cleansing Surfaces and services that are designed not to retain dirt, dust, debris, or other solids, therefore being low-maintenance and not requiring regular cleaning. Self-cleansing pipes, drains, and traps allow solids to be flushed away, façades that are self-cleansing are cleaned by the rain and the wind. The surfaces are designed to be non-stick, avoiding adhesion where possible.

self-climbing tower crane A tower crane that is fixed to the shaft of a lift, central core, or building, climbing the building as the construction takes place. The crane achieves the climbing by fixing to the new solid parts of the structure as the building rises, with the lower attachments of the crane being removed once upper sections are attached and secure.

self-curing The process of curing without heat application.

self-drilling screw A *screw with a tapered thread that is turned into a pre-drilled metal sheet, creating a thread within the material as it is screwed in.

self-embedding screw A screw with a drill bit point that cuts into the metal, forming its own hole, and taping a thread into the material as it is rotated by a power drill.

self-extinguishing Property of a material that renders it capable of catching fire; however, the act of initial combustion causes it to extinguish itself very quickly.

self-finished (factory finished) Material or component that has a surface prepared for final application needing no further treatment. The component needs to be handled with care during transportation and fitting so as not to scratch or damage the surface.

self-finished felt Bitumen felt with a factory-applied finish, often metal foil, that provides a protective surface.

self-healing concrete A concrete that will biologically produce limestone in the presence of air and water when cracks appear. Bacteria, such as *Bacillus pseudofirmus* or *Sporosarcina pasteurii*, can be added to the concrete mix during manufacture. The bacteria can then lie dormant for up to a few hundred years as long as conditions are appropriate. When a crack then appears, presenting moisture and air to the bacteria, the bacteria will start to produce limestone, which will eventually repair the crack itself. *See also* BIOCEMENTATION; MICROBIOLOGICALLY INDUCED CALCITE PRECIPITATION.

self-illuminating exit sign (self-contained exit sign, emergency exit indicator lighting) An illuminated exit sign that operates in an emergency. It has its own power source, so can operate even if there is a power failure.

self-levelling screed (self-smoothing screed) Highly workable screed mixture that is simply spread over an area in sufficient quantity such that the fluid properties of the screed cause it to find its own level across an area.

self-protective material 1. Material that protects itself against oxidation or corrosion. **2.** Material, such as felt or plastic, with additives that protect it from the ultraviolet rays of the sun. The fillers or covering material reduce breakdown of the material and bonds.

self-siphonage The removal of the water seal in a *trap by the discharge from a *sanitary appliance.

self-supporting Material, component, or element that requires no additional support to maintain its shape and form when in position. *Profiled sheeting, *cladding, and some roof structures such as *shell structures need no additional support to function.

self-tapping screw A screw that cuts a thread into the material to which it is inserted.

selvedge To reclaim from waste, transform from damaged, or make good a material, component, or element of construction so that it can serve a function.

semi- Prefix meaning half of something; for example, *semi-detached means that, from a front elevation, half the building is detached from another property, with the other part being attached to another *dwelling, *semicircle means half a circle.

semi-bonded screed The structural concrete is cast in place and sets, the screed is laid over the top of the matured concrete. The screed makes a mechanical fix, locking into the pores of the concrete, but does not make a chemical bond with the concrete.

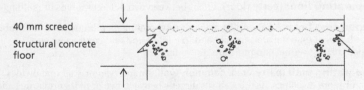

Structural concrete laid and allowed to set

The screed interlocks into concrete pores but does not make a chemical fix

40 mm screed

Structural concrete floor

Semi-bonded screed, laid over the top of a structural concrete floor

semicircle Half a circle.

semiconductor A material, such as silicon, that has electrical conductance between that of a conductor and an insulator.

semi-detached house (US duplex) A pair of houses that are joined together by a common wall.

semi-diurnal Having a half-day cycle, such as the tide.

semi-enclosed fuse *See* FUSE.

semi-gantry crane An *overhead crane that has one of its sides supported by a leg that runs on a track at ground level.

semi-skilled worker An operative who has experience and training sufficient enough to ensure that the activities requiring general construction knowledge, skill, and understanding can be competently undertaken. It does not include operatives who require training to master a skill such that they can be described as a *tradesperson.

semi-solid-core door A door where part of the core is filled with a material such as particleboard or wooden blocks.

sensible heat The heat that is applied to an object to change its temperature but not its state.

sensitive ratio The ratio between the shear strength of an undisturbed sample and a disturbed sample of soil.

sensor A device that is used to measure some physical quantity—temperature, humidity, movement, or pressure.

separated flow When the flow of a liquid passes round either side of an object, which can result in eddies forming on the downstream side of the object.

separate screed (bonded screed) A screed which has a mechanical bond to the substrate that it rests on. Generally the concrete floor is scabbled (*see* SCABBLING) to ensure the screed can lock into the surface below, creating a strong mechanical bond. A bonding agent may also be used to create a chemical link from one material to another.

separate system A below-ground drainage system that utilizes one drain and sewer to discharge foul water, and one drain and sewer to discharge surface water. Commonly used but installation costs are higher than a combined or partially separate system. *See also* COMBINED SYSTEM and PARTIALLY SEPARATE SYSTEM.

separating floor (party floor) A floor between two different rooms or buildings.

separating layer A debonding agent or a separating membrane used to ensure that layers of material are separate and free to expand and contract when experiencing thermal and moisture movement.

separating wall (party wall, common wall) An intervening wall that divides two or more dwellings. As the wall sits directly between two or more properties, the owners of each property have rights on alterations to the wall under the *Party Wall Act. The party wall in habitable buildings should restrict the passage of sound, moisture, and heat transfer as described by the *Building Regulations.

separation The process of keeping things apart. This can be achieved by the use of *geosynthetics which separate course and fine soil materials and/or man-made aggregates, while allowing the free flow of water across the geosynthetic. An example of their use is in the construction of *pavements, where a geotextile placed between the *subsoil and the *aggregate *sub-base prevents the aggregate being forced down into the soil by compaction during construction and through the weight of vehicles.

septage The *septic contents of a *septic tank.

septic An anaerobic bacterial condition, used to decompose raw sewage, specifically in a *septic tank.

septic tank A large tank used for the collection and treatment of sewage. Found in buildings that do not have a connection to the main sewer system.

sequence arrow An arrow linking tasks in a network diagram showing the logical flow of work for the project. The arrow is used to link tasks together showing the preceding and subsequent tasks linked to each activity. Sequence arrows are used to show the links in *arrow, *precedence, and other network diagrams.

sequence of operations The order that activities and tasks take place in a project. The order of tasks can be seen in the logic mapping of activity *networks and *Gantt charts. The sequence arrows show the links and order between activities.

sequence of trades The order in which skilled workers undertake their work on a traditional development.

sequester 1. To isolate, confiscate, impound, hold or commit something for safe keeping. **2.** The natural or artificial method of capturing carbon (carbon sequestration) and other emissions, removing them from the atmosphere. Forests and timber is one natural form of carbon sequester used to capture and store carbon.

series A progression of one thing after another; for instance, the succession of rock strata being formed from different layers of rock.

Serpula lacrymans Fungus that causes *dry rot in the UK.

servant key A key that opens one lock.

service Facilities for public needs—water, electricity, toilets.

serviceability limit state (SLS) A state that corresponds to conditions beyond which specified service requirements are no longer met (e.g. excessive deflection, vibration, and damage which can affect appearance, comfort, and function). SLS together with *ultimate limit state (ULS) forms the two subsets which are considered in the *Eurocodes.

service cable The cable that supplies electricity to individual buildings. It runs from the electricity supply main in the street to the buildings electricity meter.

service clamp A pipe saddle fitting that produces a watertight clamp around a pipe and provides a pipe outlet from the main pipe producing a tee junction.

service life The timescale for which something is designed to remain operational.

service lift (goods lift) A lift used to transport goods from one level to another.

service pipe 1. The water pipe that runs from the external stop valve (located just outside the boundary of the property) to the inside of the property, where it terminates at the internal stop valve. **2.** The gas pipe that runs from the service main in the street to the property, where it terminates at the customer's meter control valve.

service reservoir (distribution reservoir) An underground or covered reservoir that holds drinking water to feed into the water distribution network.

service road A minor road next to a main road, which can be used for access to business premises to deliver goods.

services The collective name given to the electrical and mechanical services within a building.

services core (mechanical core) The grouping of building services into a series of vertical shafts in a multistorey building. The shafts are usually positioned beside the lifts.

services duct A duct used to convey building services.

set 1. Components, tools, or furniture that share a common link and purpose providing the full compendium of items for the purpose. **2.** The chemical hardening of a material. **3.** The angle of a saw-tooth blade to make the *kerf, being just wider than the saw blade, enabling the blade to cut and easily run through the narrow gauge of the timber cut. **4.** The frame, rebate, and door for a door opening.

set back The distance the building line is positioned from the centre of the road. The line is positioned by the building authorities and is there to prevent developers encroaching on the street line.

set coat The *finishing coat of plaster.

setdown A recess in a floor to take a door mat or carpet.

set-off An agreed amount that has been deducted for work that has not been carried out according to specification.

set screw A screw with a rectangular head that can be tightened by a spanner to create friction and fix another object into place. Headless screws, also known as *grub screws, are available with recessed hexagon heads allowing an Allen key to be inserted into the screw and tightened, locking other units into place.

set square A drawing instrument in the shape of a right-angle triangle.

sett A cube of granite that is used for creating roads, paved areas, and hard-standing. A traditional surface for road construction.

setting The initial hardening of plaster, concrete, or mortar after which the material should not be disturbed as it will affect the chemical bonds.

setting block *See* GLAZING BLOCK.

setting coat The final coat creating the finish, *see* FINISHING COAT.

setting out (marking out) The establishment of temporary marks, steel pins, and wooden pegs to show where the building or structure is to be positioned.

setting shrinkage The shrinkage that takes place in concrete between its placement, and during initial set.

settlement The vertical downwards movement (compression) of soil due to an imposed load (from a structure), or a lowering of the *groundwater level. The compression of the soil can be due to several causes: elastic deformation of the soil grains (known as **immediate settlement**), which occurs immediately on loading and is recoverable; decrease in the volume of voids, known as *consolidation and is only recoverable to an extent by a reduction in overburden pressure; lateral flow of soil particles, known as plastic deformation of secondary consolidation; and collapse.

settling When fine suspended material in water sinks to the bottom and forms *sediment.

set up To erect or put something in position.

seven (7) pillars of BIM The key principles of BIM as defined by the UK Government, which are:

1. **PAS 1192-2**[+] *Specification for information management for the capital/delivery phase of assets using building information modelling.* A publicly available specification (PAS) to be used during the operational phase, which covers employer's information requirements (EIR) and BIM execution plans (BEP).
2. **PAS 1192-3**[+] *Specification for information management for the operational phase of assets using building information modelling.* A specification covering information management in the delivery phase of construction projects, which has led to the introduction of new concepts such as the organizational information requirement (OIR), the asset information requirement (AIR), and the asset information model (AIM).
3. **BS 1192-4** *Collaborative production of information. Part 4: Fulfilling employers' information exchange requirements using COBie—Code of practice.* A specification defining the transfer of information between parties throughout the life cycle of an asset.
4. **building information model (BIM) protocol** A contractual document which categorizes building information models that are required by the project team. It sets certain obligations, liabilities, and associated limitations for the use of such models. For example, it can be used to ensure specific ways of working.
5. **government soft landing (GSL)** Where the project team still watches over the project after construction, for up to three years after completion.
6. **digital plan of work (dPoW)** A digital plan that enables an employer to specify the deliverables expected at each stage of the design, construction, maintenance, and operation of built assets. The plan should be available digitally for members on the project so that they know what and when information is expected of them.
7. **classification system** A standard, structured, digitally enabled classification system to ensure all data is available in a common format so that it can be integrated into the dPoW.

[+] These documents can be downloaded from the British Standards Institution (BSi) website.

Seveso III Directive EU legislation concerning the control of onshore major accident hazards involving dangerous substances. The directive came into effect as a result of the **Seveso disaster** in 1976, when a small chemical plant in Italy exploded, releasing toxic fumes into the atmosphere, which adversely affected thousands of local residents together with the surrounding environment.

sewage The waste liquids and solids that are carried away for treatment by drains and *sewers

sewage disposal network A network consisting of sewage pumping stations and outfalls which convey the collected wastewater to the point of final disposal.

sewage system (sewer system) A system consisting of pipes (*sewer) used to collect sewage from the point of generation (usually households, industrial settings, surface run-off, including storm water). The sewers can be of different diameters and embedded with different slopes based on the design considerations of the respective system.

sewer A large underground system of pipes and tunnels used to transport *foul and *surface water. *See also* FOUL DRAINAGE; SURFACE WATER SEWER.

sewer chimney (US) *See* BACKDROP.

sextant A navigational instrument used to work out latitude and longitude.

shaft A vertical duct for building services.

shaft resistance (R_s) The load transmitted to the soil along the length of the pile, as defined in Eurocode 7. *See also* BASE RESISTANCE.

shake A separation of wood fibres that occurs when the tree is felled or the wood is seasoned. A shake is a defect that is not normally acceptable in structural wood. *Gross defects such as shakes are not an *allowable defect when grading. Shakes can be *heart, *radial, *ring, or star shakes.

shaking test A *dilatancy test to determine if the material is a silt or clay; the water will disappear when the pat is pressed if it is a silt.

shale A sedimentary type of rock, formed from thin layers of compressed clay, silt, and mud.

shallow Not very deep, e.g. **shallow foundations**, which include strip, raft, and pad types of foundations.

shanked drill A drill bit that has a reduced diameter shaft, allowing it to be held by a small *chuck.

shape code A standard numbering system to define different shapes to which reinforcing bars are bent.

shape factor A value that can be incorporated into a *bearing capacity calculation to accommodate three-dimensional shearing at the corners of square or rectangular *foundations.

shaping The second stage of sharpening the teeth of a saw, where a file is used to ensure the teeth are true and uniform in their profile and size.

sharp arris The edge of a corner or angle that has been left with a sharp edge or point. Some arrises are rounded, creating a rounded arris that removes the hard edge.

sharp-crested weir A *weir with a thin cross-section and a pointed upper edge; used for measuring flow.

sharpening Grinding and filling blades and teeth to restore their cutting edge.

sharp sand A type of sand that has coarse angular grains, suitable for use in concrete mixes.

shave hook A T-bladed scraper. The blade is drawn towards the user with the handle fixed to the end of the T-blade.

shear (shearing, shear action) To displace something relative to something else, for example, the upper portion of soil against that of the lower portion of soil in a *shear box.

shear box This is the simplest form of laboratory shear strength test. It is often referred to as the **direct shear test** as it relates the shear stress at failure directly to the normal stress, therefore, the failure envelope may be plotted directly from the results. The apparatus comprises a square box construction in two separate pieces, an upper piece and a lower piece. The vertical normal load is applied directly through the upper pressure plate and is divided by the plan area of the box to give the normal stress, σ_n. The sample is caused to shear along the plane dividing the upper and lower pieces by applying a horizontal load to the upper piece while the lower piece is held in position. The load is generally applied via a proving ring, hence the load causing the sample to shear can be read directly and the shear stress, τ, is the load causing shear divided by the plan area of the box. The test is repeated several times on different specimens of the same sample using different normal loads. The results are then plotted, to give the shear strength envelope, from which the shear strength parameters (c and ϕ) may be obtained. The volumetric behaviour of the soil can also be determined during the test by measuring the amount of horizontal displacement and vertical displacement using dial gauges.

shear centre A point in the cross-section of a structural member where the resultant shear force forms *bending without *torsion.

shear cone Failure of an anchor bolt, caused by a cone-shaped piece of concrete or brickwork failing (being pulled out) from around the bolt.

shear connector A connector between, for example, a beam and a slab to prevent longitudinal shear.

shear lag A slow response to the development of shear stress in a material or structural member, e.g. where a non-uniform stress distribution occurs.

shear load The load perpendicular to the normal load, causing a material or element to *shear apart.

shear modulus *See* MODULUS OF RIGIDITY.

shear reinforcement *Reinforcing bars designed to resist *shear loads.

shear slide A *landslide.

shear test A test to determine shear strength parameters of a soil, such as a *shear box and *triaxial test.

shear wall A structural wall designed to resist swaying forces.

sheathed wiring Electrical wiring that is covered by a protective, usually plastic, coating.

sheathing 1. A layer of boards applied to the studs of a timber-frame wall or to the rafters of a pitched roof to strengthen the construction. **2.** The outer protective covering of electrical wiring.

sheathing boards Large boarding sheets used for packing, decking, walling, roofing, and *shuttering.

sheathing felt A type of bitumen roofing felt.

sheathing paper Building paper.

sheave A grooved pulley wheel, for a rope, cable, or belt to run through.

she bolt A metal tie made out of ribbed reinforcement bar, over which winged or large nuts are used to create a nut and bolt tie. The tie bar is often used on wall shuttering placed through the two faces of the wall shuttering, with the bolts either side, enabling the shuttering to maintain the correct width once it is filled with concrete. The tie bar resists the horizontal hydrostatic loads created by the wet concrete.

shedding *See* LOAD SHEDDING.

shed dormer A dormer window with a roof that slopes in the same direction as the roof in which the dormer is located.

sheep's foot roller A self-propelled or towed soil compacting roller, which is characterized by having projections like sheep's feet around its *perimeter.

sheet Large thin flat material formed by rolling, pressing, extruding, or cutting down to size. *Sheet glass and *sheet metal are examples of sheet materials.

sheet flow A thin layer of liquid flowing over a surface.

sheet glass Thin glass commonly used for windows, doors, and wall applications.

sheeting Sheet cladding material.

sheeting rail A *cladding rail.

sheet metal A thin metal plate; in construction sheet metal is primarily used for roofs in buildings.

sheet pile A section of steel, concrete, or timber, typically in a U, W or Z shape, that when sunk into the ground, side-by-side, form an underground barrier, such as a *cofferdam.

shelf Horizontal platform used as a store for items such as books, tools, utensils, or as a display for ornaments and trophies.

shelf angle Length of galvanized steel bracket, attached to the structural frame, used to provide support for brickwork cladding.

shelf life The time that something can be stored in a usable state, for example, the time adhesives, varnishes, and paints can be stored and used.

shell The framework of a building.

shell bedding The usual method of bedding hollow blocks with the mortar pasted along the edge of the blocks and the perpendiculars. Thin strips of mortar are pasted along the edges of the block leaving a gap where the hollow section of the block occurs.

shell door canopy A shell-shaped canopy located above a door.

shield A protective guard.

shield bolt An anchor bolt made from a threaded tapered bolt and steel sheath. As the anchor bolt is tightened, the tapered bolt pulls back compressing the sleeve against the sides of the masonry or concrete.

shim (packing shim) Strips of metal, plastic, or timber used to lift, space, and hold components so that they can be fixed and secured in the correct position.

shingle 1. Wooden or slate roof or wall titles, placed in overlapping rows. **2.** Small rounded peddles found on a beach. **3.** A process in the manufacture of wrought iron, where slag is removed.

shingling (US) The process of placing overlapping layers of roofing felt.

shiplap (shiplap boards, shiplap boarding, US shiplap siding) Overlapping boards of rectangular cross-section that have a rebate on one side of the board and a tongue on the other.

ship spike (boat spike) A square section bar with a wedge point used for fixing and jointing large timbers.

shoal 1. Shallow water. **2.** An underwater sandbank.

shoddy Slang term meaning poor quality, substandard, or rushed.

shoe 1. A short bend or angled section of pipe attached to the bottom of a downpipe to convey the water away from the building and into a drain. **2.** A metal plate used to support the end of a rafter. *See also* PILE SHOE.

shoe mould (US shoe mold, base shoe) A quarter rounded bead or strip of timber, used to provide a finish and hide the junction between the skirting board and floor.

shoot To remove the rough or sharp edge of a mitre joint, removing the rough cut of a saw by planing.

shop A workshop used as a factory.

shop drawing A drawing used in a factory rather than on-site.

shopfitter Craftsperson or carpenter who is skilled in assembling and fitting out offices, shops, and stores. The fitter provides and fits the fittings and furniture.

shop priming Priming that is applied in the factory, generally ensures better adhesion.

shop work Work undertaken in a joiners' shop or factory, not on the construction site.

shore A prop or support to a building, such as a *flying shore *prop, or a *dead shore.

shoring Providing temporary support to or propping up a building, such as underpinning, placing *shores against the building.

short circuit A fault in an electrical circuit that occurs when an accidental connection is made between the live and neutral conductors.

short-grain timber Also known as brash timber, which breaks with little resistance.

short oil A characteristic of paint or varnish with a low-oil base.

shot blasting Firing small pellets at a surface to clean it before painting.

shotcrete A type of concrete material commonly known as sprayed concrete, whereby layers of concrete are extruded from a gun or hose to form the required thickness.

shotfired fixing An explosive fixing using a nail gun to fire nails into timber, concrete, or steel. Often used for fixing sheet material to steel and concrete.

shothole A hole left by a wood-boring beetle or grub, usually 2 to 4 mm in size.

shovel 1. A hand-held tool, consisting of a long wooden or plastic handle and a metal flat end, which is used to move loose material. **2.** A bucket-like device that is attached to an excavator and used to excavate material.

shower diverter A mixing *valve for a shower.

shower enclosure A cubicle in which to shower.

shower head The perforated water outlet that is used to direct the spray of water in a shower.

shower room A room that contains a shower.

shower rose A large shower head that sprays the water out at low pressure.

shower tray (shower base, shower receiver) The prefabricated base of a shower that collects the water and directs it towards the drain.

shrinkage A reduction in size of a material usually due to the loss of water or change in temperature. Excessive shrinkage can lead to cracking or separation from *substrate material.

shrinkage limit *See* ATTERBERG LIMITS.

shrinkage ratio The rate at which the volume of a fine-grained soil decreases in respect to the decrease in its moisture content.

shuffle glazing A type of glazing system formed by inserting the glass into angled sections at the head and the sill. The angled head section will be of a deeper profile than the sill section, enabling the glass to be manoeuvred into position.

shutter 1. A protective panel that is pulled over a window or a door to provide security, reduce heat loss at night, or provide privacy. **2.** A piece of *shuttering.

shutter bar A metal bar used to secure window or door shutters in place.

shutter bolt *See* ESPAGNOLETTE BOLT.

shutter hinge *See* PARLIAMENT HINGE.

shuttering Concrete formwork used as a temporary mould and support for wet concrete.

shuttering hand A shuttering joiner's labourer or assistant. Assists the joiner, carrying and loading out materials.

shutting jamb (close or post) The jamb that a window or door is closed against. *See also* HANGING JAMB.

shutting stile The stile that a window or door is closed against. *See also* HANGING STILE.

sick building (sick building syndrome) A building that causes its occupants to feel irritated, uncomfortable, and even become ill. The symptoms are normally a result of poor ventilation, mould growth in the building and services, lack of natural or adequate light, irritants within the water system, and poor drainage. Common symptoms suffered include sore eyes, sore throat, skin rashes, uncomfortably hot or cold, headaches, and stomach irritation.

side A face of a board or section of square sawn timber.

sidefill Gravel that is laid at the sides of a pipe in a trench.

side form The *formwork that is used to provide the edge *shuttering or edge boards for a concrete beam or box.

side gutter A small gutter used at the junction where a vertical surface penetrates the roof slope.

side-hung door A door that is hinged along one side.

side-hung window A window that is hinged along one side.

sidelap The amount by which the edge of one component overlaps the edge of another component. Used in roofing and cladding to provide a weather-tight joint.

sidelight (wing light) A fixed window placed at the side of a door.

siding *See* CLADDING.

S

sieve analysis The determination of the proportion of different-sized aggregates in a sample of mixed aggregate. The aggregate sample is weighed, placed in the top pan (sieve) on a nest of sieves. Each sieve has a different-sized mesh with the top sieve allowing all but the largest grains of aggregate to pass through, the lower level sieves catch the very small particles, and the pan at the bottom of the sieve captures the dust and silt. The sample of aggregate contained within the sieves is then vibrated causing it to drop, filtered out into different sizes, and be captured by the sieve that it cannot pass through. The contents of each of the different-sized sieves are weighed to determine the proportion of the size of aggregate in the aggregate mix.

sight A *foresight, *intermediate sight, or *backsight.

sighting distance The furthest distance an object can still be viewed.

sight size (daylight size) The size of glazing that is available to admit daylight.

signal Transmitted information by an electronic device such as a telephone, radio, or television.

sign convention A way of representing if something is in compression or tensile, or displacing upwards or downwards, by assigning a positive or negative symbol.

silencer A device that reduces the noise of machinery or engine.

silica (silicon dioxide) A silicon oxide, used as the primary constituent in glasses, formula SiO_2, which occurs naturally in sand and quartz. One of the most abundant materials within the earth's crust.

silicates Group of minerals based on silica.

silicon A material found in abundance in the earth's crust; an important constituent of glass as an oxide. Silicon is a semiconductor, meaning that depending on the inclusion of different atoms, it can be used as a conductor or non-conductor.

silicone A polymeric material comprised of silicon and oxygen, used for its flexibility and heat resistance.

silicone paint A type of paint based on *silicone with excellent heat resistance properties.

sill 1. The horizontal bottom member of a window or external doorframe.
2. A horizontal layer of *igneous rock that has been forced between layers of sedimentary rock due to a *volcanic eruption.

sillboard (Scotland) *See* WINDOW BOARD.

sill height The distance from the top of the sill to the finished floor.

silo A large cylindrical container on a supporting framework, used to store granular material such as cement, grain, and animal feed.

silt A fine-grained soil that lies between sand and clay, between 0.06 and 0.002 mm. It can be subdivided into coarse silt: 0.06–0.02 mm; medium silt: 0.02–0.006 mm, and fine silt: 0.006–0.002 mm.

silting The settling of *sediment on a river bed.

silver brazing (silver soldering) A *brazing or *soldering process whereby the principal filler metal is a silver alloy.

silver sand Fine rounded sand used for mortar and filling dry joints in block paving.

simazine A herbicide that is monitored in drinking water.

simplex concrete pile A *displacement pile that has been cast in-situ, by firstly driving a steel tube with a sealed base plate in the ground. Reinforcement is then inserted into the tube and concrete is poured around the reinforcement. The steel tube is then withdrawn from the ground using a vibrator, and the base plate is left within the ground.

simplex control A control system used to operate a single lift.

Simplified Building Energy Model (SBEM) A steady-state calculation method used to determine the energy use and CO_2 emissions associated with non-domestic buildings in the UK.

Simpson's rule A formula for approximating the integral of function by dividing an area up into parabolic arcs; *see also* TRAPEZOIDAL RULE.

simulation The use of a computer to model the behaviour or performance of a material, structure, and so on.

sine A trigonometric function for a right-angled triangle, which is the ratio of the length of the side opposite the given angle, to the length of the hypotenuse.

single-coat plaster Plaster that is applied in one coat only.

single door A door that has one leaf.

single glazing Glazing that comprises a single sheet of glass.

single-hole basin (one-hole basin, single-taphole basin, single-taphole sink) A *basin that contains a single tap.

single-hung window A sash window where only one of the sashes, usually the bottom one, can move.

single-lap tile (interlocking tile) A tile that overlaps at the bottom, and sides of the tile, and interlocks at the edges; at the centre of the tile there is only one thickness of tile covering the roof. A double-lap tile will have up to three layers of tiles at the top and bottom edges, and a minimum of two overlapping tiles at the centre of the tile. Single-lap tiles have interlocking edges that prevent the rain penetrating and being blown into the joints.

single-lever mixer A *lever mixer.

single-lock welt A sheet steel joint made by folding one sheet of metal over the other.

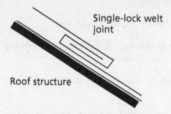

Single-lock welt joint

Roof structure

Single-lock welt

single medium (*plural* single media) 1. One channel, method, or delivery. **2.** A sewer system that employs a single type of medium (or mixture of media at one point in the system) in the filter, i.e. the medium is consistent in chemical composition and size throughout the filter.

single-outlet combination tap assembly A tap that has a single outlet for both hot and cold water; also known as a **mixer tap**.

single-phase (single-phase supply) The distribution of alternating current through two *conductors. *See also* THREE-PHASE SUPPLY.

single pipe (single pipe system) *See* ONE-PIPE SYSTEM.

single-pitch roof A roof with just one pitch to it; also called a mono-pitch, and if the top of the roof abuts an adjacent wall it may also be called a lean-to roof.

single-ply roof A roof with a single covering of waterproof membrane, the membrane is high-performance made from a synthetic rubber material such as ethylene propylene diene rubber, polyvinyl chloride, polyisobutylene, and other modern synthetic coverings, which are then laid on a plywood sheeting. Single-ply roofs have an expected lifetime of 30 years.

single-point heater *See* INSTANTANEOUS HOT-WATER HEATER.

single-pole switch A switch that simply breaks or connects one circuit. The switch only has one pole.

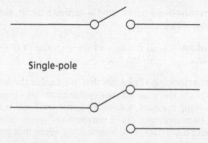

Single-pole

Double-pole

Single-pole switch and double-pole switch

single-sided ventilation A type of *natural ventilation strategy where ventilation openings are provided on one side of a space only. Where a single opening is provided, this is termed 'single opening'. Where two ventilation openings are provided, this is termed 'double opening'.

single-stack system An above-ground drainage system that comprises a large-diameter single vertical-discharge stack that conveys both soil and wastewater to the drain, as well as providing ventilation. All of the appliances connected to the stack require deep seal traps. The length of the soil and the waste pipe connections from the stack are restricted. Extensively used nowadays in domestic housing. *See also* ONE-PIPE SYSTEM and TWO-PIPE SYSTEM.

sink (kitchen sink) A shallow bowl used for cleaning and the preparation of food.

sinker drill A hand-held rock drill, used in shaft sinking.

sink grinder *See* GARBAGE DISPOSAL SINK.

sinking A hole, recess, or rebate that allows for a door mat, hinge, or plug to sit neatly and flush with the top surface.

sinking-in The effect of the substrate or undercoat absorbing the top layer of paint. With gloss and sheen paint this results in the loss of the desired finish.

sink plunger (force cup) A hand-held device used to unblock sinks. Comprises a rubber cup attached to a long pole. The flexible cup is pushed over a plug hole and pressed down then released creating suction which draws out the blockage.

sink unit A joinery fitting that is designed to house a sink.

sintering A process utilized to increase the density (reduce the air content) particularly in ceramic materials; usually conducted at elevated temperatures.

siphonage The suction of fluid up or down a pipe. A siphonic drain uses the downward force of gravity to help draw excess water and control the flow of water from the top of a pipe or cistern. Excessive movement of water in some drain or sewer pipes can lead to siphonage of water from traps and seals, which is undesirable.

siphonic closet (siphonic water closet) A *water closet that contains a double trap.

site accommodation The temporary cabins, toilets, storage, and welfare facilities provided on construction sites.

site agent (site manager) The main contractor's lead representative on-site, responsible for control of site-based operations and activities.

site boundary The perimeter of the land and building that is under the control of the contractor during the construction and development period. The boundary forms the outer edge of where construction operations take place without permission of neighbouring landowners.

site cabin *See* JACK-LEG CABIN.

site clerk (clerk of works) Acts as one of the client's representatives on-site, checking operations and ensuring the quality of work is carried out to the standard specified. The clerk of works will make reference to standards, specifications, and legal requirements, ensuring the contractor is undertaking the works in the manner expected and specified in the contract documents. Where there are any variations costed on the work undertaken, the clerk will monitor activities, man-hours, and operations required to complete the variation. Due to the way the clerk oversees and checks activities, the site staff and operatives on-site may also be referred to the clerk of works as the checker or timekeeper.

site constraints Aspects or features in and around the site, such as services, adjoining properties, railway lines, and watercourses, which restrict the operations that can be carried out. Buildings and structures near to the site may need supporting, underground culverts and drains may need to be bridged, or there may be restrictions preventing construction operations directly over the top of them. Cranes may need permits and agreement to use the airspace above neighbouring properties, and other aspects such as noise and visual pollution may restrict operations.

site diary A document kept on-site by the *site agent or site manager to record the daily activities, operations, and labour; incidents, weather conditions, and other relevant site activities are entered into the book or digital document. It is a source and mode of recording the activities, progress, and other issues related to site activities on a regular daily basis. The diary is an important document, often referred to in the event of disputes, accidents, or other events where there is a need to check or cross-check site operations.

site engineer (US field superintendent) The engineer on-site, employed by the contractor to control and check operations. The site engineer may be required to establish the position of structures, set out roads and buildings, determine the levels of the land and structures, calculate and check the volume of materials removed and required. Operations will vary depending on the nature of the site. The engineer's duties will vary depending on whether the site encompasses mainly *civil works or construction. Most of the engineering design work is carried out off-site by the structural or *civil engineering contractor.

site instruction (US field order) Written order issued by the client's representative (client's architect, engineer, or clerk of works), identifying work that should be done; the order normally indicates something that has immediate effect, unless stated. Such instructions can be a written order stating a variation to the work, clarification of works described under contract, order to comply with safe working practices, order to stop work, or carry out operations at specified times or in certain conditions.

site investigation (ground investigation, site survey) 1. The collection of information through visual inspection, observations, and tests on the ground, to determine site characteristics and ground conditions, and assess the suitability for development. There are various types of site investigation, including desk-top study, site reconnaissance, and soil investigation. **2.** Desk-top study, collection of all relevant reports, and historic information on the site. **3.** Site reconnaissance, exploration of the site, recording visual features, walking and travelling around the site collecting samples, and making observations. **4.** Soil investigation, the sampling and testing of soil, looking at the strengths, chemical characteristics, and make-up of the ground at different levels. Tests may be carried out on-site and in a soils laboratory. **5.** A process that can be undertaken as part of the tendering process and prior to commencement of the contract or project where the contractor or interested parties ascertain the viability and stability of the site through site visits, inspections, and collection of samples (for example soil samples).

site manager *See* SITE AGENT.

site meeting (progress meeting, management and design team meeting) An on-site meeting that gathers the relevant professionals and subcontractors to ensure that operations can be properly coordinated, scheduled, and managed safely, and problems foreseen, or those that have occurred, can be dealt with. Those invited will depend on the agenda of the meeting and the matters to be discussed. Meetings may be internal for those employed directly by the contractor or can involve all of the main parties connected to the site with representatives for the client, contractor, architect, structural and civil engineers, mechanical, electrical, and plant engineers, and all specialist subcontractors. The meeting will look at the progress, operations of the different parties, and the overall plan of operation with a view to coordinating activities and managing changes.

site practice The operations that take place on-site. Good site practice is the sequence of events, operations, and activities required in order to produce the goods to the right quality, and in a safe and professional manner.

site roads The designated through-passes for vehicles and pedestrians on-site. On some sites these may be made-up roads with hardcore, or constructed to highways standards, or may be simple designated and protected ways through and around the site. The safe movement of vehicles and pedestrians around the site should be planned, and the roads and footpaths fenced or guarded to prevent risk of injury.

site security The use of *hoardings, gateways, checkpoints, and security personnel to ensure that movement into, out of, and around the site is observed and controlled, to reduce the risk of theft, and control personnel entering the site. Security of the site contributes to controlling the health and safety of those within the site and general members of the public in the vicinity of the site that may be affected by operations. Controlling the flow of people helps to ensure that materials and equipment coming into and leaving the site are as expected, and theft does not occur.

site services Services that are temporarily connected to the site to enable the work to be undertaken include electricity, gas, water, drainage, and telephone.

sitework Work undertaken on site, including alterations to prefabricated components and adjustment to works.

SI units The international system of units, with the standard convention being the metric system, which is based on decimalization (the number 10). The SI system is the most common system of measurement in the world.

size A sealant used to seal and reduce porosity of timber, plaster, and concrete, making it suitable to have finishes applied to it. The liquid sealer can be made from a thinned-out paint or varnish or watered-down glue, adhesive, or paint.

sized slates Slates that are of a consistent size. *See also* RANDOM SLATES.

sizing To use size to seal a surface, making it less porous and suitable to receive finishes.

skeiling (skilling, skeeling) A flat sloping part of a ceiling. It can be a small section sloping upwards as the eaves meet the roof or, where the roof space is used to make a habitable room, the sloping section (skeiling can be considerable). Skeiling is the flat part on the underside of a pitched roof beneath the rafters, whereas the ceiling is the flat horizontal part beneath the flat members of the roof joists (ties).

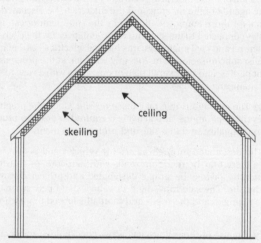

Skeiling (and ceiling)

skeleton construction A method of construction where all of the dead and live loads are supported by a structural frame of beams and columns. The external walls are non-load-bearing.

skeleton core door A type of hollow-core *door.

skeleton stair An *open stair with *treads and no *risers.

skelp A piece of metal used to manufacture pipe or tubing.

sketch A rough freehand *drawing, produced to illustrate something quickly.

skew Positioned at an angle to the main component.

skewback The upper surface of a stone, brick, block, or top of a *springer that carries an arch. It is the starting point of the arch stones.

skewback saw A saw with a slightly curved and concaved back.

skew corbel An overhanging element of brick or stone at the foot of a gable that is used to carry the brickwork and stone of the extended gable roof.

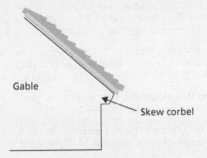

Skew corbel

skewnail (skew nailing, toe nailing, tusk nailing) Inserting nails at alternating angles to securely fix two or more pieces of timber together.

skid-mounted equipment Plant and equipment that is mounted on skids to allow the equipment to be pushed or pulled into position.

skids Ski-like rails placed under heavy plant and equipment, that enable the objects being moved to slide into position.

skilled operative (craft operative) A skilled worker on-site with a training in a craft such as plastering, joinery, electrics, plumbing, ground-works, and bricklaying.

skim coat (skimming coat) The final coat of plaster, maximum 3 mm thick, made of a fine plaster, is applied with a trowel filling perforations and undulations in the substrate below. It can be applied as a finish to cover over plasterboard joints, or can be applied over the whole area of plasterboard, or to cover a plaster basecoat.

skintle A way of placing bricks such that they are stacked out-of-line, for example, to allow air flow during drying.

skip 1. An area where the final coat of paint has been accidently missed. Undercoats of paints may have a slightly different texture or shade to help identify where the final coat has been applied and areas missed. **2.** A flat-bottomed container, made of steel, that is capable of being lifted onto the back of a flat-bottomed lorry. The container is used to collect and deposit rubbish and other building materials. The skip provides temporary storage; once full the container is removed from the site. The lifting eyes attached to the container enable it to be quickly lifted onto a lorry and removed from site. **3.** A container that can be attached to a crane for the moving concrete, mud, or materials across the site. Mud and rubbish skips are basic buckets that can be loaded and simply tipped upside down to remove the materials. A concrete skip is slightly more sophisticated with a chute attached to the bottom of the skip for discharging the concrete. The skip is loaded with concrete from a *concrete mixer, hoisted, and moved into position using a crane and when positioned where required, a wheel is used to open a flap that discharges the concrete down a chute, allowing the concrete to be poured steadily into formwork or discharged where it is required.

skirt 1. Material placed around the lower edge of an object, and used to dress or protect the lower face. **2.** A downstand on a foundation that penetrates into the ground, and used to prevent the ground being washed from under the foundation.

skirting (skirting board) A timber or PVC board that is run along the lower edge of a wall to provide a neat aesthetic finish between the plaster and the floor. The board can be fixed with nails or adhesive, or be plugged and screwed.

skirting trunking (US plugmold, wireway) A box section panel positioned along the base of a wall and used to run electrical cables through it.

sky component (sky factor) The light received directly from the sky. One of the three components required to calculate the *daylight factor.

skylight *See* ROOFLIGHT.

skyscraper A habitable high-rise building typically greater than 50 storeys high, exceeding 100 m, although this is low in comparison with most buildings that are classed as skyscrapers. Iconic skyscrapers include the Chrysler Building (319 m), Sears Tower (443 m), and Burj Khalifa in Dubai (828 m) amongst others.
    The first building considered to be a skyscraper was a steel frame building built in Chicago in 1884, although it was only 10 storeys high (42 m). The Empire State Building in New York was the first building to have more than 100 floors and is 381 m high, with the mast added it stands at a notable 443 m. The North tower of the World Trade Centre stood at 417 m and had 110 floors, but, with the South tower, was tragically destroyed in a terrorist attack on 11 September 2001. In 2010 the Burj Khalifa was built at a height of 828 m and 160 storeys tall, and was the tallest building in the world at that date.

slab 1. A flat block of concrete, slate, or stone that is used for paving footpaths and roads. **2.** A *solid concrete floor.

slab-on-grade A concrete floor slab resting directly onto the ground or a sub-base of hardcore.

slag (blast furnace slag) A by-product of steel and copper production that has glassy characteristics. The material can be used as *pozzolans as a partial cement replacement in some concretes and mortars. The introduction of the slag gives concretes and mortars different characteristics; depending on the mixes, the addition can improve the strength and setting qualities. Slag can also be used to make mineral wool with acoustic and thermal properties. The slag residues from the production of iron and copper have slightly different characteristics and properties.

slag wool Mineral wool made form fine filaments of blast furnace slag. The dense properties of the mineral give the wool good sound absorption properties, as the wool material traps and holds still air; it also resists the flow of heat.

slaked lime (calcium hydroxide, Ca(OH)$_2$) Hydrated lime.

slam buffer (US mute, rubber silencer) A nail or screw with a rubber dome that is fixed into the jamb or rebate of a door to reduce the noise of the impact if the door is slammed shut. The dome sits just proud of the surface, providing a cushion to the impact of the door.

slamming strip (slamming stile) A vertical strip of wood attached to the shutting *stile of a door that the door closes against.

slant The inclined distance between a base point on a pyramid or cone, and the vertex.

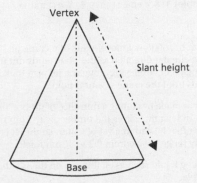

Slant: Slant height of a cone

slat A thin strip of wood used in fencing and on louvred windows. On louvred windows the slats are held in a frame, placed at an angle to provide shade from the sun, and allow light to penetrate through the gaps.

slate A fine-grained material derived from sedimentary rock and commonly used as a roofing material. The sedimentary rock is composed of volcanic ash or clay, which is compressed and has metamorphosed in layers of the deposits. Due to the cleavage and grain of the slate (lines of separation) it can be split into thin sheets and used for roofing. Slate is very durable, with minimal maintenance, and slate roofs can exceed the lifetime of the occupants.

slate boarding *Close boarding installed on a roof.

slate hanging (weather slating) Slates hung vertically.

slating Roofing with slate tiles.

sledge A flat platform-like vehicle on runners, used to pull equipment behind tunnelling machines.

sledgehammer A large heavy *hammer used to break up rock, stone, and hardened concrete.

sleeper 1. A timber with a large cross-sectional area that is used to support the rails of a railway track. **2.** A horizontal strip of wood placed in or over concrete to support the floor covering above.

sleeper clip *See* FLOOR CLIP.

sleeper plate The wall *plate on a *sleeper wall.

sleeper wall (basement wall) Wall laid to carry the timber floor joists or concrete beams. For timber floors the wall is usually laid in *honeycomb bond with spaces between the bricks allowing air to circulate under the *floor. *See also* DWARF WALL.

sleeve (expansion sleeve) A *pipe sleeve.

sleeve coupling (sleeve connector, sleeved joint, socket coupling) A small hollow cylinder used to join the ends of two pipes together.

sleeve piece (thimble) 1. *See* PIPE SLEEVE **2.** *See* FERRULE.

slender Tall and thin.

slewing The circular or rotational movements of a crane jib. Most of the crane's movements tend in a circular direction as the jib extends out from a central point. The lifting of a *luffing jib or *trolleying along the jib arm makes the movement outwards and inwards from the crane's central point.

sliced veneer Veneer made by cutting a thin slice of timber, using a machine with a knife blade up to 5 m long that slices the timber away from the main timber log (fitch). The timber is often heated in a vat of water, so that it is durable when cut. The veneers generally range in size from 0.2 to 1 mm thick.

slick A film or polluting liquid such as oil floating on the sea that has resulted from an oil spill, for example.

slickenside A polished, smooth rock surface caused by friction when two rock faces pass over each other, such as on a *fault plane.

slide bolt (thumb slide) A small *barrel bolt usually found at the top and bottom of double doors.

sliding door A door that is opened by sliding it from side to side.

sliding-door lock A lock for a *sliding door.

sliding-folding door See FOLDING DOOR.

sliding window (sliding sash, slider) A horizontally opening *sash window.

slime A slippery, thick, unpleasant liquid.

slip A thin piece of material.

slip brick (brick slip) A thin strip of brick used as a tile for cladding, giving the impression of a solid brick wall.

slip feather (sliptongue) A thin strip of timber that fits into grooves located at the edge of panels or boards to keep the boards aligned.

slipform A type of formwork that is designed to slowly move in the direction of the pour, continuously or at short intervals, to prevent the formwork from constantly being struck and erected. Used to form concrete shaft linings, towers, and pavements.

slip joint A joint made between pipes by sliding the end of one pipe into the end of the other.

slip mortise (slip mortice) A mortise, groove, or chase ready to receive a *slip tongue or feather.

slipper bend A *channel bend.

slip-resistant (non-slip) Describes floors with perforations, embossed patterns, ribs, studs, and inlays of abrasive material that resist slips.

slip sill A *sill that is the same length as the width of the door or window. *See also* LUG SILL.

slope 1. A side of a hill. **2.** A surface that is inclined to the horizontal.

slope deflection method An analysis method used to define the stiffness of a structure.

slop hopper A large sanitary fitting used in hospitals to dispose of human waste.

slop sink A large deep sink used to dispose of spilled or splashed liquid.

slot A groove or thin rebate cut into a surface. Some screws have a slot that allows a flat-bladed screwdriver to be located and the screw to be rotated and driven into the wood.

slot diffuser *See* LINEAR DIFFUSER.

slot drain A drain that is hidden beneath a narrow gap in the surface that is being drained.

slough 1. An area of low-lying muddy ground. **2.** A mud hole.

slow bend (long sweep bend) A long shallow bend. *See also* QUICK BEND.

slow sand filter A graded sand filter for treating drinking water that requires little or no power. The filter involves a bed of fine sand (0.1–0.3 mm diameter; typically 0.6–1.2 m deep), which overlies a thick gravel bed. Sometimes a coarse layer of charcoal or low-grade coal is used between these two layers to aid colour adsorption. A constant head of water (1.0–1.5 m) is allowed to flow through the bed at rates of 0.2–0.4 m/hr. The slow filtration rate allows a thin layer of microorganisms (*Schmutzdecke*) to form within the sand bed, which provides microbiological treatment and helps to remove colloidal material. The *Schmutzdecke* layer remains active, providing the sand bed stays wet. Periodically, the *Schmutzdecke* layer becomes overgrown, allowing colloidal material to build up and eventually block the filter. The top region of the sand layer is replaced or cleaned and the *Schmutzdecke* layer allowed to regrow. *See also* RAPID SAND FILTER.

sludge The solid waste that has settled out during sewage treatment.

sludge acclimatization The preservation of sludge in specific environmental conditions, generally aerobic and/or anaerobic, where a substrate (organic food) is regularly provided to help microbes nurture and grow. The acclimatization can take one month in case of an aerobic environment and two to three months in an anaerobic environment.

sludge blanket clarifier A treatment unit where three treatment processes take place, namely mixing, flocculation, and sedimentation, which are conducted within same unit without the use of mechanical mixers. Influent is homogenized through hydraulic mixing, followed by screening/filtration/contact with settled sludge in a cone-shaped depression and with the skimming off of clarified water at the upper periphery.

sludge dewatering A technique used to draw water from settled sludge before applying it to the actual treatment system. It reduces the sludge volume, as most of the sludge composition is water (99%) and only 1% is solids, and reduces the capital and operational cost of sludge treatment systems.

sludge drying beds The sand beds usually used to dry the settled sludge from the settling units of a treatment plant. They consist of layers of sand, while a uniform gravel layer is also applied underneath for support. The drying of sludge involves natural ways and the water percolates into the layer of sand below under gravity. The sand beds are designed in accordance with the sequence of usage.

sludger *See* BAILER.

slug 1. A small density volume of liquid that remains separate to the main volume of the liquid. **2.** A unit of mass equal to the acceleration 1 ft/sec^2 when acted on by a 1 lb mass.

sluice An artificial water channel or floodgate, used to control the flow of water.

slum A very poor and overcrowded urban area where living conditions are inadequate, consisting of decrepit housing units with an insufficient infrastructure.

slump The collapse or downwards movement of a body under its own weight, e.g. the vertical downwards movement of concrete in a *slump test.

slumping The movement of a substance such as a sealant or concrete downwards; the material flows downwards under the force of gravity. Most modern sealants applied by gun are designed to be non-slumping. Concrete slumps when piled in its wet state; once the process of hydration has taken place the concrete has set and should not slump.

slump test Concrete test used to determine the *workability of concrete. The test uses concrete compacted in three layers within a cone that is upturned and the difference between the cone and slumped concrete is measured. The steel cone is 300 mm tall with a 100 mm closed bottom and 200 mm diameter open top. The concrete is loaded into the cone in 100 mm layers, tamped, and compacted at each layer, and once the upper layer of the cone is compacted and levelled, the cone is upturned and the concrete discharged to form an inverted cone. The *slump recorded is the measure of the distance from the top of the steel cone to the top of the slumped concrete.

slurry A liquid mix of cement and water, or cement, fine aggregate, and water.

small-bore system A type of central heating system that uses small diameter pipework, usually around 15 mm, to feed the heat emitters.

small-bore unit A boiler that can be used to feed a *small-bore system.

smart city An *urban area that utilizes digital data to manage and optimize the *built environment efficiently for *sustainable development.

smart device An electronic device that connects to other devices or to a network, typically wirelessly, to remotely or autonomously control that device. Examples include smart thermostats, security cameras, and various kitchen appliances.

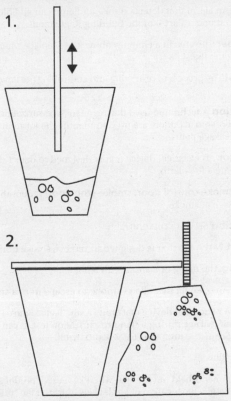

Slump test: 1. The cone is filled in three 100 mm layers, and each layer is tamped and compacted. 2. The cone is inverted and discharged, and the level between the top of the cone and the slumped concrete is measured.

smart dust A collection of tiny **microelectromechanical sensors (MEMS)**, robots, or other devices that are typically only a few millimetres in size. They can be used to detect light, temperature, vibration, magnetism, and chemicals, and are operated by wireless communication technology.

smart motorway A section of a motorway that uses *active traffic management (ATM) to improve flow capacity, typically at busy times, by the use of variable speed limits and the use of hard shoulder running.

SMM *See* STANDARD METHOD OF MEASUREMENT.

smog A mixture of fog and smoke. Smog is often used to refer to a type of air pollution resulting from vehicle fumes.

smoke A collection or grouping of particles present and visible in the air; usually occurs as a by-product of fires.

smoke alarm An alarm that detects smoke and emits an audible alarm when activated; required under Part B of the Building Regulations.

smoke chamber The area in a chimney above the fireplace where smoke gathers prior to entering the flue.

smoke control The process of restricting and controlling the flow of smoke within a building.

smoke detection A technique used during a fan *pressurization test where *smoke pencils or *smoke puffers are used to identify the location of *air leakage points and *air leakage paths.

smoke detector An electrical device that is designed to detect the presence of smoke and activate a fire alarm.

smoke door (smoke control door, smoke stop door) A door that is designed to prevent the passage of smoke.

smoke explosion *See* BACKDRAUGHT.

smoke extract fan A fan that is designed to remove smoke from a building.

smoke-logging The filling of a space with smoke.

smoke outlet An opening that allows smoke to escape from a space.

smoke pencil A hand-held device that emits a small jet of smoke. Used to identify the location of *air leakage during a *fan pressurization test. It can be battery- or hand-operated. *See also* SMOKE PUFFER; WIZARD STICK.

smoke pipe A *flue.

smoke puffer A hand-held device that, when squeezed, produces a small jet of smoke. Used to identify the location of *air leakage during a fan *pressurization test. *See also* SMOKE PENCIL; WIZARD STICK.

smoke rocket (rocket tester) A rocket-shaped canister filled with smoke used to undertake a *smoke test.

smoke shelf A horizontal ledge located above the fireplace and below the smoke chamber. It prevents down draughts and collects any rain entering the chimney.

smoke test A test that can be undertaken on drainage systems to identify leaks. Would be undertaken after an air or water test.

smoke venting *See* FIRE VENTING.

smoothing (resource smoothing) The balancing out of human and mechanical resources on a project by the relocating of tasks and activities, ensuring that resources are used to maximum effect with minimum breaks in activity.

smoothing compound A liquid grout or self-levelling compound.

snagging Ensuring all of the jobs, minor defects, and finishes are checked and rectified before handing the building over.

snagging list A document or schedule containing a list of all the jobs that need finishing, defects that need correcting, and adjustments made to ensure the building and structure meet with the specified standard.

snake (electric eel, plumbing snake, toilet jack) A flexible *auger that is used to unblock drains.

snap header (half bat) A brick cut to approximately half its length or 100 mm long that is positioned in a wall with its uncut header exposed, giving the impression that the header brick penetrates fully into the wall.

snapping line (snap line) A chalk line used for marking a straight line over a surface. A string line is pulled through chalk dust, the chalked string is then pulled taut over the surface, from point to point, where the straight line is required. Gripped by the worker's fingers the string is pulled slightly off the surface and then allowed to snap back into position and onto the surface causing the chalk to leave the string and mark the wall, floor, or board.

snap-ring joint *See* O-RING.

snap tie A tie for formwork, made of notched flat steel, enabling the exposed end of the tie to be broken off once the concrete has been cast and the formwork is removed.

sneck In stone rubble walls, a small stone less than 75 mm high, distinctly smaller than normally cut stone.

snecked rubble A rubble wall built of irregular-sized stones or small cut stones.

snib (check lock) 1. A hand-operated catch used to alter the position of the latch within a lock. **2.** (Scotland) A bolt, catch, or lock for a door or window.

snib latch A *latch that is operated using a snib.

snots Excess mortar on the face of masonry that has been allowed to drip or drop onto the horizontal and vertical surfaces of the masonry.

snow board (snow guard, snow slats, (Scotland) snow cradling) A device installed on a pitched roof above the eaves designed to prevent any damage that may be caused by large accumulations of snow sliding off the roof. The board retains the snow, allowing it to either drop off in small amounts or melt completely. Various types of snow board are available including horizontal boards, horizontal slats, wire meshes, horizontal pipes, or a series of individual pads.

soakaway A hole in the ground, either open or filled with stone, designed to temporarily hold excess surface water until it can drain away into the surrounding ground.

soaker A small section of flexible sheet metal, usually lead, used to make a watertight joint between a pitched roof and an abutment. The soakers are overlapped and are held in place by turning the top edge of the soaker over the tile on the pitched roof. A stepped *flashing is then used to hold the upstand in place and provide weather tightness.

soap 1. Release mould oil painted onto formwork to reduce adhesion and improve striking **2.** A brick of standard length but with its width equal to its height. The standard dimensions of a soap brick in the UK are 215 x 46 x 46 mm.

social value (social impact) The contribution and impact that any [construction] organization, project, or programme makes to the lives of internal and external stakeholders affected by its activities, including those working in the industry and in the communities in which it operates (Raiden, Loosemore, King and Gorse, 2018).

sociotechnical system (STS) A system that appreciates the interaction of people and technology in a workplace setting.

socket 1. The enlarged end of a section of pipe into which a *spigot is inserted. **2.** A hollow cylindrical pipe fitting that contains a thread on the inside. **3.** A *socket outlet. **4.** An electrical outlet into which a lightbulb is inserted **(light socket)**.

socket coupling *See* SLEEVE COUPLING.

socket former A hand-held tool used to join copper pipe to form a socket that can be jointed using a capillary joint (*see* CAPILLARITY).

socket inlet The electric plug that is inserted into a *socket outlet.

socket iron pipe A cast-iron pipe with a spigot and *socket joint.

socket joint *See* SPIGOT AND SOCKET JOINT.

socketless pipe *See* UNSOCKETED PIPE.

socket outlet (US receptacle, power outlet, socket) An electrical outlet into which the electric plug from an appliance is inserted.

sodium A metallic element, symbol Na, sometimes used as an alloying element in steel and other commercial metals. Sodium is a soft white metal, has a white silvery colour, oxidizes in air, and reacts violently with water.

Sod's Law *See* MURPHY'S LAW.

soffit (soffite) The underpart of an overhang or overarching element of a building or structure.

soffit board (eaves lining, US plancier piece) A flat board or lining placed under the overhanging part of the eaves.

soffit form (soffit shutter) Formwork used for the underside of a floor or ceiling deck, beam, or overhang.

softboard A soft, porous particleboard used for insulation.

soft landscaping (organic landscaping) Live plants, trees, bushes, and shrubs within the external works.

soft sand Sand that has rounded, rather than sharp aggregate, and is used for render and mortar as it is easier to work. It is rarely used in concrete.

software A computer program or operating system.

soft water Water that has a low concentration of calcium carbonate, a hardness value of less than 60 mg/l, and as such, it is easy to form a lather with soap.

softwood The description of timbers that belong to the gymnosperms group, being different from hardwoods. Softwoods are generally faster-growing and cheaper than hardwoods, they are considered less decorative, and generally are less prone to *moisture movement.

soil branch A branch *pipe that is connected to the *soil stack.

soil cement The binding clay used in *cob construction.

soil classification *See* AIRFIELD SOIL CLASSIFICATION.

soil drain A drain that conveys the waste from the *soil pipe to the *sewer.

soil fabric friction The amount of shear resistance developed at the interface between a soil and a fabric, such as a *geotextile.

soil fitment (soil appliance) A *sanitary appliance that is connected to the *soil pipe.

soil mechanics The investigation, classification, and determination of a soil's properties and parameters, together with its behaviour.

soil mixing The mixing of a soil with another material, e.g. lime, to improve its behaviour.

soil nail A steel rod that has been inserted into the ground, typically on a *slope, to increase stability.

soil pipe A pipe that conveys the liquid and solid waste from a *sanitary appliance.

soil profile The vertical layers of soil at a particular location.

soil-reinforced wall A mass of soil that has been retained internally by the use of embedded *geotextile layers (or steel strips). Cladding is used to provide an aesthetic finish and prevent soil loss rather than to provide structural support.

soil sampler A tube-like device that is inserted into the ground to collect an undisturbed soil sample, e.g. a **split barrel sampler** is used to obtain soil samples from a borehole.

soil stabilization The use of another material, as in *soil mixing, to improve certain characteristics of the in-situ soil, e.g. shrinkage, optimum dry density and shear strength.

soil stack A vertical *soil pipe that conveys the waste to the *soil drain. It is usually vented at its highest point.

soil-structure interaction The study of how soil and structure will perform together, for example, the amount and effects of differential settlement.

soil vent A *vent pipe used to ventilate a *soil stack.

soil water Water that is discharged from a *soil fitment.

solar collector (solar panel) A device designed to collect solar radiation.

solar control film A thin film of material that is applied to a window to convert it into *solar control glazing.

solar control glazing (anti-sun glass) Glazing that filters out certain wavelengths of infrared radiation emitted from the sun, but still allows the transmission of visible light. Used to reduce solar heat gain and control glare.

solar dial A 24-hour time switch used to switch lighting on and off automatically throughout the year.

solar energy Energy derived from the sun.

solar gain (solar heat gain, passive solar gain) The amount of additional heat gained within a space by the transmission of solar radiation.

solar heating The process of utilizing solar radiation from the sun to heat a building. *See also* PASSIVE SOLAR HEATING.

solar hot-water system A type of hot-water heating system where solar radiation from the sun is used to preheat a glycol mixture as it passes through a *solar collector. The preheated mixture is then passed onto a hot-water storage tank where it may be heated further, usually using a conventional heat source, until it reaches the required temperature.

solar load The amount of cooling that is required to counteract solar gain.

solar protection A device that is used to protect a building against excessive amounts of solar gain, such as louvres or an overhang. External devices are much more effective than internal devices.

solar radiation *See* SOLAR ENERGY.

solder A lead (Pb) and tin (Sn) alloy with a low melting point that is used to join conductors together.

solder balls Spherical parts of *solder.

solder bridging A *solder paste used to make a bridge or path.

soldering A technique used to join conducting objects or metals together using a lead-tin alloy. The solder has a low melting point of 190°C, the joint and solder are heated, and a sound electrical and mechanical joint can be formed.

soldering iron A piece of metal that can be heated then used for *soldering and used to melt and apply *solder.

solder mask A mask used to cover parts not intended for soldering.

solder paste A mixture of minute spherical solder particles, activators, solvent, and a gelling or suspension agent.

soldier A short upstanding member of timber or masonry. Soldier timbers are positioned upright and level, and used for grounds fixing *skirting boards and other board finishes. Bricks on end can also be described as soldiers; *see* SOLDIER COURSE.

soldier arch A flat arch created from a row of bricks on end.

soldier course A row or course of bricks laid on end.

solenoid A coil of wire that surrounds an iron core to form an electromagnet.

solenoid valve An electrically operated valve that can be either opened or closed automatically.

sole plate (footplate, ground sill, mudsill, sole piece, US abutment piece) Large pieces of supporting timber or railway *sleepers laid on the ground to support scaffolding standards, *struts, *props, *shores, or *raking shores.

solid bedding A layer of mortar or adhesive that covers the whole area under the brick or block leaving no gaps or voids.

solid block A concrete block with no perforations or holes that have been deliberately cut into it.

solid brick A brick without any perforations or holes deliberately cut into it.

solid bridging (US block bridging) Solid sections of timber used to add bracing between floor joists. *See also* SOLID STRUTTING.

solid-core door (solid door) A door where a solid material, such as insulating foam, has been used to fill the space between the internal and external face of the door. *See also* HOLLOW-CORE DOOR.

solid floor (solid slab) Concrete floor that rests directly onto compacted hardcore that is in contact with the ground. The term also refers to concrete that rests on insulation, but is still in contact with the ground. A solid floor is different from a suspended floor which has a space between the floor structure and the ground.

solid frame A window or doorframe manufactured from a single piece of wood.

solid-glass door A door constructed completely from glass.

solidification The process whereby a liquid changes to a solid matter on cooling.

solidification range The temperature range between the liquid and solid state for a material on cooling from the liquid state. Many solids (especially metals) do not have a specific or single melting point; indeed, melting occurs over a range, e.g. 20°C, this is the solidification range. When a metal or material is heated and enters this range, it undergoes incipient melting.

solidification shrinkage crack A crack that forms during *solidification.

solid map A map that shows the solid geology (typically bedrock) in reference to geological time with all superficial deposits stripped away. They are often used to assist in the design of site investigations. *See also* DRIFT MAP.

solid moulding (stuck moulding) A moulding that is cut or carved into the timber rather than being fixed or planted onto the surface of the timber.

solid-newel stairs A stone spiral stair, where the inner section of each step is carved so that it forms a round end that is laid directly over the rounded end of the step below, and directly under the rounded end of the step above. As the steps rotate round, a central newel is created from the overlapping stone steps.

solid partition A partition wall without a cavity or gap for insulation.

solid plastering In-situ plastering, not *dry lining. Plaster mixed on-site, placed, and floated against the wall or ceiling where it is desired. This is different from plastered surfaces that use plasterboard.

solid slab A solid concrete floor.

solid stop A rebate with a stop that is cut into the doorframe rather than fixed to it.

solid strutting The insertion of sections of timber between floor or ceiling joists that are the same depth as the joists being braced. *Herringbone strutting uses less timber.

solid timber A unit of timber that is made from one piece being carved, worked, and with rebates cut out to create the desired shape, rather than additional sections of timber planted or fixed to create the unit.

solid wall A wall without a cavity or a gap filled with insulation.

solid-wood floor A floor made out of strips or panels cut from a single element of timber rather than being built up as a *laminated, blockboard, *glued-laminated timber, or *plywood floor.

soluble The ability of a material to be dissolved into a liquid to form a solution. A non-soluble material simply segregates from the liquid and falls to the bottom or rises to the top of the liquid.

solute (solution) A homogeneous mixture composed of two or more substances can be described as a solution. In such a mixture, a solute is dissolved in another substance, which is known as a *solvent. Liquids may dissolve in other liquids becoming one solution. Gases can combine with other gases to form mixtures, rather than solutions. All solutions are characterized by interactions between the solvent phase and solute molecules or ions. Under such a definition, gases typically cannot function as solvents, since in the gas phase interactions between molecules are minimal due to the large distances between the molecules. This lack of interaction is the reason gases can expand freely, and the presence of these interactions is the reason liquids do not expand. Examples of solid solutions include: metal alloys, certain minerals, and polymers containing plasticizers. The ability of one compound to dissolve into another compound is called solubility. The physical properties and characteristics of compounds, such as melting point and boiling point, change when other compounds are added. There are several ways to quantify the amount of one compound dissolved in the other compounds collectively called concentration. Examples include molarity, molality, and parts per million (ppm). The most widely used solvent is water, and another commonly used solvent is ethanol.

solvent A liquid which has the capacity of dissolving another substance and forming a solution; see SOLUTE.

sonar A method of detecting underwater objects by using sound waves.

sonic Relating to sound waves, particularly travelling at the speed of sound in air, which is approximately 1220 km per hour (or 760 miles per hour) at sea level.

sorption To take in something through either *absorption or *adsorption.

sound 1. An audible noise. **2.** Goods or materials that are not damaged. **3.** To measure the depth of a water body using *sonar. **4.** An ocean inlet. **5.** A channel between two larger water bodies or between an island and the mainland.

sound absorption (noise absorption) The process where some of the sound emitted from a source is absorbed as the sound strikes a material. All materials absorb sound to some degree. The **absorption coefficient** is a measure of the level of sound absorption exhibited by a material.

sound boarding *See* PUGGING.

sounder A device used in alarms to make a sound, for instance, a bell, a horn, or a siren.

sound insulation (noise insulation) The main method used to control the transmission of airborne and impact sound from one space to another within a building. Four main factors will have an influence on the level of sound insulation achieved. These are completeness, flexibility, heaviness, and isolation. In addition, the quality of workmanship and detailing is also very important. A construction will only achieve its expected sound insulation if it is constructed without defects.

sound-level meter A device used to measure noise.

soundproofing The process of insulating a space against noise.

sound-reduction factor A measure of the reduction in the intensity of a sound.

southing The distance south in reference to two latitude points.

space heating The process of heating a space with a heater.

spacer Preformed unit of material that is used to maintain a gap, a distance between two objects, or hold one element of construction in a specific position. Spacers and packing *shims are used to temporarily hold the position of window frames, *sole plates, glazing bars, etc. while being fitted. Plastic, concrete, and steel spacers are used to maintain the *cover and position of reinforcement bar in concrete. The spacers are tied to the steel reinforcement and positioned so that a cover of concrete is maintained between the ground and formwork. Spacers are also used to maintain the position and distance between reinforcement within the concrete; and they are used to allow the correct position of parallel sheets of wall formwork.

spacer-lug tile A tile with spacing lugs attached to maintain the correct distance between adjacent tiles.

spacing The distance between two objects, typically in reference to their centre lines or *centres.

spade 1. A hand tool used for digging, composed of a handle, shaft, and rectangular flat blade. The blade or spade can be easily and firmly pushed into stiff soil or clay enabling chunks to be extracted. The smaller the spade cutting end, the easier it is to insert and cut through and extract the soil. **2.** A bulldozer with a large flat blade capable of pushing material over the ground in order to create a flat plane.

spall 1. To break rough edges of stone, masonry, and concrete with a chisel or *spalling hammer. **2.** For concrete or stone to break away from the main structure due to frost attack, aggregate, or material expansion and contraction caused by moisture or thermal movement.

spalling The breaking away or removing by force the edge or face of concrete or masonry; *see* SPALL.

spalling hammer (spall hammer) A heavy hammer with a chiselled head used to break and shape concrete by hitting the surface and causing it to *spall.

span The horizontal distance between two supports.

spandrel 1. The walled area between the arches in an arcade, which is an inverted triangular shape. **2.** The triangular boxing out under a stairway to make a cupboard, store, or simply box out the area. **3.** The area under a window, *spandrel panel.

spandrel panel The solid or opaque panel under a window on one floor which extends down to the window on the floor below.

spandrel step Triangular stone stairs, which, when interlocked together, have a smooth *soffit.

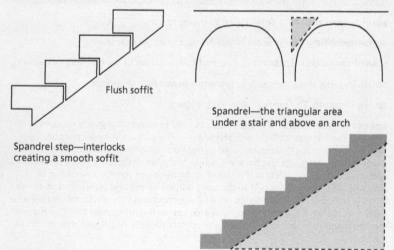

Flush soffit

Spandrel—the triangular area under a stair and above an arch

Spandrel step—interlocks creating a smooth soffit

Spandrel and spandrel step

Spanish tile Half-cylinder-shaped clay roof tile that is wider at the bottom than it is at the top.

span roof A standard pitched roof.

spar angular Limestone and white rock chippings used as ballast to hold down roofing felt and protect it from degradation caused by the sun.

spar dash A render coated in or exposing limestone or white angular pebbles. Dry dash or render pebble dashed with limestone.

sparge pipe A horizontal perforated pipe that produces a spray of water for flushing urinals.

spatterdash (spatter dash) A rich mix 1:1.5 to 1:2 cement and sand, which is flicked onto a surface in small globules by a machine (roughcast machine). It is applied to surfaces that have minimal suction in order to provide a key.

special (special brick) A brick of non-standard dimensions. Special bricks are normally manually cut and expensive, as they are not mass-produced.

special attendance Charge for use of contractors' equipment by a subcontractor for a short period, usually charged at an hourly rate.

specialists 1. Expert or professional who has very specific skills and knowledge, and has a reputation for carrying them out reliably. **2.** Contractor, subcontractor, or consultant selected for their skill and knowledge in a specific area, e.g. structural, mechanical, electrical, and environmental engineers.

special risks *See* FORCE MAJEURE.

specification A description of the works and/or workmanship required. The written description forms part of the *contract documents and normally includes reference to the workmanship, standards, or qualification necessary, tests that may be required to ensure performance, quality, or work, key indicators, and other measures that ensure the work is done properly.

specific gravity Also known as relative density, it is the ratio of a material's density to the density of another material; however, the ratio almost always uses water as the standard comparison. The density of water is 1000 kg/m^3. All other material densities are divided by that of water to give the relative density.

specific heat capacity (specific heat) The amount of heat in joules that is required to raise the temperature of one kg of a material by one degree kelvin. Measured in J/kg/K.

specific leakage area (normalized leakage area) The *effective leakage area divided by the floor area of the building.

specifier (specification writer) A professional employed to write the *specification.

spectrum A range, particularly the distribution of colour light.

speculative building A builder or developer who buys land and builds properties and structures in the anticipation that they will sell on once completed. Speculative building is different from a builder who sells properties 'off-plan', which is selling the building based on a drawing prior to it being completed.

sphere An object that has a circular profile in all directions, e.g. a ball or globe. The pattern is produced when a semicircle rotates about its straight axis tracing a solid object as it turns.

spherical excess The difference between the sum of the angles of a spherical polygon of n sides and $\pi(n-2)$ radians—it may be of significance in surveying regarding the adjustment of the closing error.

spherical polygon A closed geometric shape, such as a triangle, formed on a *sphere.

spigot The end of a pipe that is inserted into the *socket end of another pipe to form a *spigot and socket joint.

spigot-and-socket joint (US bell-and-spigot joint) A type of pipe joint made by inserting the *spigot end of one pipe into the *socket end of another pipe.

spike A sharp pointed end coming from a shaft.

spile A wooden peg or timber supporting post that is driven into the ground.

spillway An overflow structure, in the form of a *bellmouth overflow or a side channel, that excess water is safely discharged through, in order to prevent water from flowing over the top of a *dam.

spindle 1. A piece of wood that has been turned to create a *baluster. **2.** A square section rod that passes through a door from one handle to another, going through the latch mechanism. The rotation of the bar operates the latch. **3.** The axel at the centre of a tap that rotates to open and close the valve.

spine wall A wall inside the building, normally in a central position, that can also offer lateral support.

spiral A continuous curve that increases around a central point.

spiral duct (spiro duct) A tubular duct formed by winding a sheet of metal around in a spiral.

spiral stair A *stair that is helical, not truly spiral, constructed with winders radiating from a central newel.

spire A tall, elongated, and pointed roof structure.

spirit level A straight and flat length of metal with a single bubble captured in a tube of liquid. The position of the bubble reveals whether a surface is level.

spirit stain A dye based in methylated spirits, used for colouring the surface of wood.

splashback (backsplash) A vertical area of tiles, or other easily cleaned material, located behind a sink or cooker to protect the wall from splashes.

splashboard 1. A scaffold board laid on edge on the insider face of the scaffolding platform to protect the building wall from mortar, concrete, and other materials. **2.** An angled moulding placed on the foot of an external door to discharge the rain away from the face of the doorway, helping to ensure water does not enter the building. Also called weather moulding.

splash lap The lap of sheet metal joint that extends onto the flat surface of the sheet metal below.

splat A strip that is used to cover over the joints of wall boards.

splay A sloping cut at an angle to the main surface or edge.

splay brick (cant brick) A special brick that has been cut with either a slant, chamfered, corner edge, or face that is shaped to make a return or form a finished stop at the end of a wall.

splayed coping A stone that is used to provide a cover to a wall that is cut at an angle or chamfered.

splayed grounds Timber grounds that are shaped and angled to provide a key for plaster or render.

splayed heading joint A joint between timber boards where the joint is cut at 45° so that the top of the joint overlaps the bottom.

splayed skirting A bevelled skirting board.

splay knot A timber knot that has been cut along its length, the length of the knot is exposed on the face of the timber.

splice An end-to-end butt joint, where steel plates, splice bars, and crimped couplers are used to strengthen the joint so that it is capable of resisting lateral loads, tension, and compression. The aim is to make the column or beam act as a single material.

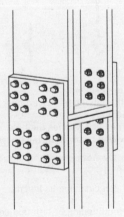

Splice: Column-to-column spliced joint

splice bar A bar used to form a connection between two butt ends, half of the length of the bar penetrates into one half of the joint, with the other half penetrating into the other element to be joined. The bar may be tightly fixed, glued, or mechanically fixed into each adjoining part.

split course A course of split bricks.

split cycle offset optimization technique (SCOOT) A real-time adaptive traffic control system across an urban network. The system will automatically adjust traffic signals from information gained from road sensors to account for the current flow of traffic.

split-face block A block that is made to be split in half, giving one rough face and one fair face. The block is normally made to twice the normal width of a standard block.

split gasket Two strips of rubber (gaskets) used at either side of a pane of glass in *gasket glazing.

split head A prop with a U-shaped head that is used to carry timber or steel beams. The supporting head is formed to allow the beams to simply slot into place and be carried by the prop.

split pin A pin with two legs that is used to secure fixings. The pin is made from a single length of metal that has been doubled over to form a single pin that has a looped head and two legs. The legs sit together so that they can pass through a single hole. The wider part of the looped head prevents the pin passing straight through the hole. Once inserted the pin legs can be splayed out to hold the pin securely in position.

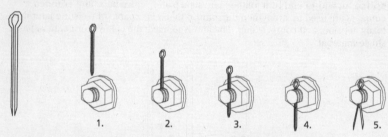

1. The nut is securely fixed in position.
2. The split pin is inserted through the predrilled hole in the bolt.
3. The pin is pushed into the hole.
4. The pin is inserted until the head reaches the bolt.
5. The legs of the pin are splayed apart, ensuring the pin cannot fall out.

Split pin used to secure a nut on a bolt, preventing the bolt from coming undone through vibration

split pipe A pipe that has been cut along its length.

split system A type of air-conditioning system that comprises an indoor *fan coil unit and an outdoor *refrigeration unit.

splittable activity An activity within a programme that is capable of being split into two or more parts and does not have to be completed before the next activity starts. A non-splittable activity has to be delivered in one unit of work and cannot be broken up by other activities.

splitter damper A damper used to divide the air between two ducts.

splitting test (splitting tensile test) A compression test on a concrete cylinder to determine the tensile strength. The cylinder can either be cast or cut, and is placed on its side and loaded through its diameter. The tensile splitting strength is then determined from:

$$2P/\pi dl$$

where P = load, d = diameter and l = length.

spoil (muck) Material and earth from excavation activities.

spokeshave A plane that has handles extending from each side of the cutting blade (like bike handles), enabling the blade to be drawn over curved surfaces to shape them.

sponge-backed rubber flooring (foam-backed rubber flooring) Flooring that is backed with resilient material to reduce impact sound and provide a cushioned effect.

sponging Using a sponge to give a textured finish to paint. The sponge can be used to apply the paint or texture wet paint that is already applied.

spontaneous combustion Combustion that occurs when a material self-ignites.

spore The reproductive seed of fungus.

spot board (mortar board) Square surface with a handle fitted to the underside enabling bricklayers and plasterers to carry plaster or mortar and easily apply it to the surface.

spot gluing Fixing *plasterboard with small *dabs of plaster or adhesive. Small amounts of adhesives are placed at regular intervals around the perimeter and in the centre of the plasterboard panel. The panel is offered to the adhesive and pressed into position so that it is flush and level with surrounding boards.

spot item Small unit of defective work that cannot be linked to just one trade, may require a labourer to break out a hole, a bricklayer to lay blocks, and a plasterer to replaster the area.

spot level 1. A *reduced level. **2.** A reading taken to determine the level, in comparison with an ordinance datum or *site datum.

spotlight (spot) A *luminaire which produces a concentrated beam of light that is used to illuminate a narrow area.

spotting An area of *paint that has a slightly different colour or texture to the rest of the paintwork.

spotting in (spot finishing) The repairing and blending in of small defects; may involve filling in, smoothing off, sanding down, or rubbing up and then painting or finishing, to ensure the repair matches with its surroundings.

spot weld A joint formed by *welding at various points.

spout An outlet on a pipe through which a fluid is poured.

sprag Small nail.

spray A fine jet of water particles moving through the air.

sprayed mineral insulation (firespray) Insulation material that is applied by spraying. Used for thermal insulation purposes or for fire protection.

sprayer (spray nozzle) A shower head or the fixing at the end of a hose used to provide the desired direction or dispersion of the liquid.

spray gun Machine with a pumping unit that is used to disperse and direct paint or another liquid.

spraying The process of projecting and dispersing liquid, such as paint or stiffer liquids, such as plaster and render, through a *spray gun or hose.

spray painting The application of paint using a spray gun rather than brush or roller.

spray plastering The application of plaster using a plaster pump and gun, enables the plaster to be easily applied to the wall. The plaster still needs to be smoothed and levelled by hand using a float.

spray tap A type of very low flow rate tap that discharges the water as a fine spray.

spread To disperse or level out a material over an area.

spread and level (US wasting) To take excess material or soil, or to remove material from a high part of the site, using it to fill in undulations and lower areas of the site. The material is moved around creating a flatter and more level site.

spreader bar 1. A strut used to keep elements of a structure apart, for example, a strut placed at the bottom of a pressed metal doorframe, used to keep the jambs of the door at the correct distance apart. The bar may be positioned below the floor covering or concrete screen, and hidden from view. **2.** A beam used to distribute and spread the loads coming from concentrated or point loads above.

spreading The distribution and levelling out of a material over an area.

spreading rate The area that can be covered by a unit volume of material or paint, applied in one coat or at a specified thickness.

spread of flame *See* FLAMESPREAD.

sprighead roof nail A galvanized steel nail used for fixing down corrugated roofing; the nail has a steel cup under the head to suit the profile of the sheeting.

spring 1. A natural watercourse that arrives at the surface of the ground.
2. A metal helix with elastic properties that when compressed, or stretched, and released will return to its original shape. The coiled metal is used to cushion the impact of vibration and create suspension systems.

springer A stone or brick on which an arch rests or the arch starts.

spring hanger A type of pipe support that uses coiled springs to support the pipe from above, while allowing vertical movement.

spring hinge A hinge with a spring to assist or enable the door to self-close.

springing The intersection of the lower side of the arch and the pier or wall which it abuts.

springing line A horizontal line that joins the two intersections between the arch and the pier or wall carrying the arch.

springing point The point at which stairs are set out, the start of the incline.

sprinkler system A system that is designed to extinguish fires. Comprises a series of water pipes and spray nozzles that automatically operate when a fire is detected.

sprocket A length of timber, often cut diagonally on section and used to lift the edge of the last row of eaves tiles.

sprocket eaves (sprocketed eaves) The eaves tiles that are tilted upwards by a *sprocket.

spun pipe Precast concrete pipes manufactured using a spinning mould.

spur An electrical socket that is connected to the ring main by a single cable. Often used to add additional sockets.

square A triangular or T-shaped engineering tool, set at 90°. The tool allows 90° angles to be set and lines to be scribed at 90° to a surface.

squared log A baulk timber log square sawn.

squared rubble A rubble wall made from stones that have been squared but are different sizes. The courses are lined up every 3 or 4 courses to give the wall a uniform appearance.

square-edged timber (square-sawn timber) Sawn timber that has a cross-section that has been cut at right angles.

square hook An L-shaped hook with a threaded screw.

square joint A joint that is butted up at right angles, 90°.

square-turned baluster (square-turned newel) A newel or baluster that has not been turned but has moulding cut into the faces.

squat To make use of a building or land without permission or ownership.

squatting closet (Asiatic closet, squat pan, squat toilet) A WC that is set into the floor and requires the user to squat.

squint (squint brick) A special brick made with a chamfered face or cut corners. The cut brick allows it to be used to create a wall with changed direction, without parts of the brick projecting from the surface.

squint corner (squint quoin) A corner of a building that is not a right angle, where the bricks or stone project out from the main face.

SS (Special Structural) Grade timber that has been visually graded for situations that demand high-quality timber, capable of sustaining high loads. MSS or Machine grade Special Structural is the equivalent.

stability The ability of an element or unit to resist breakdown or collapse. Stability is a term used in fire resistance as an element's capability to resist collapse during a fire.

S

stabilizer 1. A foot that projects out from a digger, crane, or concrete pump that ensures the vehicle remains safe, does not move, or overturn, also called an *outrigger. **2.** Additive used to prevent a liquid or chemical from breaking down into its separate parts; adds a specific strength, or quality to a material.

stable A building or shed that has been used or is used as a shelter for horses.

stable door (US Dutch door) An external door that is divided into two horizontal leafs, with each leaf being able to be opened independently.

stack 1. A vertical drainage pipe that conveys liquid and solid waste to a drain. **2.** A *chimney stack. **3.** Bonding bricks or blocks simply stacked directly on top of each other, where no attempt is made to stagger the joints. Without ties and brick reinforcement the wall is inherently weak as it is only the mortar that binds the wall together.

stack effect The movement of air into, out of, and within a building or space due to natural buoyancy. In the case of a building, air inside the building is heated by solar gains, heating appliances, people, and equipment. As the internal air increases in temperature, it becomes less dense and rises due to convection. This results in higher pressure air at the top of the building, relative to the external air pressure, leading to exfiltration. As the warm air exits the building, cooler denser air from outside infiltrates the building through cracks and gaps in the construction

stack vent *See* VENT PIPE.

stadia lines (stadia hairs) Two horizontal lines, one above and the other below the cross-hairs of a *theodolite or *level, which are used in *tacheometry.

Staffordshire blue brick A dense clay *engineering brick that is blue in colour, either pressed or wire cut with very high strength and water resisting properties. The brick has one of the highest strength characteristics in the UK. The density of the brick helps it to resist water penetration, enabling it to be used for *damp-proof courses.

staggered joints (break joints) Brick and blocks laid so that the centre of one brick or block sits over the centre of the joint below.

staggered-stud partition A double-skin timber stud wall with a gap between the skins to improve acoustic insulation. Each wall is lined with acoustic insulation and acoustic board. The panels are staggered to reduce the chance of penetration through joints.

staging Low level platform scaffolding or an add-on to a scaffolding platform that takes the surface to a slightly higher level, creating a step up to access higher levels.

stain A solution used for changing the colour of another material. The liquid is commonly used on timber to change the surface colour.

stained-glass window A window that is constructed from pieces of coloured glass **(stained glass)**. The pieces of glass are usually held together with lead.

stainless steel A type of *steel with excellent resistance to corrosion and degradation due to the high chromium content, which is typically 15%. One of the most popular metals; however, due to the relatively high cost of chromium, galvanized steel is a cheaper alternative in the construction industry and is thus more widely used.

stainless-steel plumbing Pipes and fittings made from stainless steel. Used due to their high resistance to corrosion.

stair Steps that are linked together enabling safe vertical movement from one level to another. To enable safe movement, and considering that the steps will be used by people of differing physical ability, the steepness of the stair flight is restricted and landings are used to provide a break point or change direction.

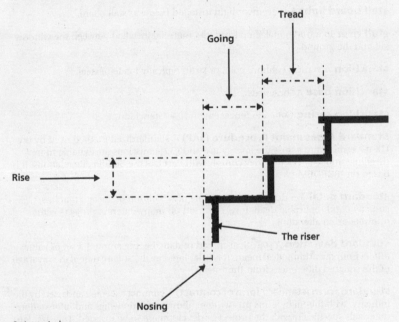

Stair terminology

staircase A unit that contains a series of steps used to gain access from one level to another. For a timber staircase, it is series of *steps that are encased by *strings that carry and contain the steps. The staircase is often prefabricated and brought to site preassembled. The strings are rebated and the *steps and *risers are inserted, glued, and wedged into position. The staircase is then delivered to site where it is fixed to the upper and lower floors and the newel posts and handrails are fitted.

stairlift (inclinator) A power-operated chair that is fixed to the side of a stair, enabling people with disabilities or those who have difficulty using stairs, to be mechanically lifted, by the chair, from one level to another.

stairway A stair that is used as a main circulation route or part of a fire escape.

stairwell The space or gap between flights of stairs that run parallel to each other.

stake A timber post with a pointed end for driving into the ground.

stakeholder A person, group, or organization with an interest, claim, or concern in something, e.g. a business, organization, project, service, or product. Stakeholders can affect or be affected by an organization's activity positively or negatively. They can be internal or external to the organization/project boundaries, and have the power to respond to, negotiate with, and change the strategic direction of the organization/project.

stall board In a traditional shop front, the sill that supports the shop window.

stall board light A *pavement light installed below a *stall board.

stall riser In a traditional shop front, the material installed between the window sill and the ground.

stanchion A vertical column, pole, or strut, typically made of steel.

stanchion base A *baseplate.

stanchion casing Concrete *encasement to a *stanchion.

Standard Assessment Procedure (SAP) A standardized method used by the UK government to assess the energy use and CO_2 emissions attributable to new dwellings and to check that they comply with Part L of the Building Regulations. It is based on *BREDEM.

standard details A drawing, for example of a *column or *beam, that has sufficient and replicable detail it can be used on many different projects with minimal or no alteration.

standard deviation A statistical spread of data, i.e. the amount a set of values differs from the arithmetical mean. It is calculated by the square root of the *average of the squared differences from the *mean.

standard form (standard form of contract) A contract accepted and used by the industry, with the main terms pre-written, allowing for drawings and information that deals specifically with the project or development to be inserted. The standard terms have the advantage that they are known and understood by the industry, have been tested through cases in court, and the implications of the terms, and a breach are largely understood. Typical standard forms of contract include those produced by the Joint Contracting Tribunal: JCT Minor Works Contract, JCT Intermediate Form, JCT Management Contract, JCT Measured Term Contract, JCT Standard Form with Contractors' Design, JCT Prime Cost Contract, JCT Standard Form of Contract. FIDIC (Fédération Internationale des Ingénieurs-Conseils) produces engineering and building contracts for major international projects. The Institute of Civil Engineers publishes standard forms including ICE Conditions of Contract, ICE Minor Works Contract, ICE Design and Construct Contract, and the New Engineering Contract.

standardization To unitize and make components that have regular sizes, fixings, and dimensions, enabling them to fix together. Modules often have regular dimensions to enable them to fit together.

Standard Method of Measurement (SMM) A common method of measuring and describing works. It is accepted as the construction standard and enables multiple parties to describe and price work with a good degree of commonality and ensures consistency.

standard overcast sky A model of a completely cloudy sky. Used when calculating daylight factors.

standard penetration test (SPT) An in-situ test to determine the shear strength of *granular soil. A metal rod is hammered into the ground by dropping a 63 kg mass through 760 mm onto it. The number of blows 'N' is counted for the rod to penetrate the ground by 300 mm. The angle of shear resistance 'Φ' can then be read off a graph defining the relationship between Φ and SPT N value.

standards A set of documents that define a standardized set of methods, procedures, etc. *See* BRITISH STANDARD; CEN; INTERNATIONAL STANDARDS ORGANIZATION.

standard special brick A brick made to a different specification to the standard brick. The brick could be angled, chamfered, skew cut, short, or long. *See also* SPECIAL.

standard wire gauge (swg, imperial swg) An old method of specifying the thickness of metal wire, tube, nails, and sheet material. An SWG of 1 = 7.62 mm, 2 = 7.01, and 3 = 6.401 mm.

standing leaf A folding door leaf that is fixed shut. Often found on double doors, where one leaf is a standing leaf and the other is an *opening leaf.

standing seam A vertical joint in a sheet metal roof that brings together two pieces of metal to form a water-resisting joint. The sheet edges are placed against each other and turned upwards providing a 10 mm upstand. The edges that are now back-to-back are then folded together in the same direction through 180°, making a vertical upstand with the edges both bent in the same direction so that they overlap. The material that has been bent to overlap is turned again through 180°. The joint between the two sheets runs from the ridge to the eaves providing a weather resisting connection.

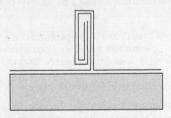

Standing seam: Shape of a standing seam

standpipe (US) *See* FIRE RISER.

stank To stop the flow of water—to prevent something from leaking.

staple 1. A metal U-shaped nail. **2.** A loop for a *hasp and staple gate or door lock.

staple fibres Short-length fibres used in the manufacture of nonwoven needle-punched *geotextiles.

staple gun (stapling machine) A hand-held machine, manually, electrically, or pneumatically powered for firing *staples into materials to fix them together.

starling *See* CUTWATER.

starter (starter bar) A steel reinforcement bar that is used to overlap with and connect to adjoining reinforcement. The bars protrude out from one concrete pour so that they can be fixed to the reinforcement cage in the next pour or bay. The length of the bar protruding is dependent on the length of cover that is required.

start time The commencement of an activity or programme. The commencement of the first activity in a *network diagram or *Gantt chart.

stat An abbreviated version of *thermostat.

statically determinate structure A structure that can be analysed by statics alone, i.e. the resolving forces, taking moments, etc.

statically indeterminate structure A structure that cannot be analysed by statics alone, because there are too many unknowns. Analysis using, for example, *virtual work is required.

static electricity Electrical charges generated from the rubbing together of synthetics, the force fields created from electrical equipment, for example, visual display units. The effect can be reduced and prevented by having floors and equipment earthed.

static head (static pressure) The level or pressure of water above a certain point in a stationary water body.

static penetration test A penetration test in soil which is pushed into the soil (*see* CONE PENETROMETER) rather than hammered (as in the *standard penetration test).

static positioning The determination of a position on the earth using the *global positioning system.

station A position where specific tasks are performed, e.g. a **railway station** is where trains stop and people embark and disembark; a **surveying station** is where a peg, *theodolite, or *level is placed to enable readings to be obtained.

stationary Not moving, is static in one place.

station roof (umbrella roof) An umbrella-shaped roof that is supported on a central column or a row of columns. Often found in railway stations.

statistics Large amounts of numerical data analysed in order to define a population, revealing significant trends or patterns of behaviour.

stay 1. A horizontal bar used in a window to strengthen a mullion.
2. A *casement stay.

stayed-cable bridge A bridge whose deck is supported by cables that are directly hung from masts.

steady-state condition The condition where the internal environment and external conditions are held constant over time.

steam Water that has been boiled and vaporized.

steam stripper (wallpaper stripper) Piece of equipment used to remove wallpaper and paper finishes from walls, using steam to soak the paper and soften the adhesive.

steel A ferrous alloy with a carbon content between 0.1 and 1.0 weight %. Such steels are often referred to as plain carbon steels. Carbon has a very marked effect on the properties of steel. The effect of an addition of a fraction of 1.0 weight % is considerable. Increasing the carbon content of steel increases the tensile strength, the yield strength, and the hardness. However, increasing the carbon content decreases the ductility of the steel. Most constructional steels contain between 0.15% and 0.4% of carbon as this combines *ductility and *tensile strength. This type of metal is used exclusively in structures for civil engineering applications.

steel casement 1. A steel casement window. **2.** The opening part of a steel window.

steel conduit Conduit made from steel.

steel erector A skilled worker who erects the steel frames to structures.

steel-fabric-reinforced concrete Concrete that has the added benefit of steel reinforcement to provide the concrete with tensile as well as compressive strength. The steel is placed where tensile forces are exerted, and the concrete takes the compressive loads.

steel fixer A skilled worker who cuts and bends reinforcing bars to the correct size and shape, and places them in the correct position.

steel fixing (US bar setting) To tie together and assemble steel reinforcement mats and cages for reinforced concrete.

steel frame The support skeleton of certain structures.

steel section A general term used to refer to *universal sections.

steel square (framing square, roofing square) A square which is graduated to enable it to be used for calculating the cutting angles and the length of roof rafters.

steel window A window made from steel.

steelwork A general term used to describe a *steel frame or a number of *steel sections.

steeple A tall tower or *spire.

steeplejack A tradesperson skilled at accessing and maintaining tall buildings, structures, and towers.

stem 1. A threaded rod that is used to open and close the gate of a valve. **2.** A central shaft or column from which other objects or structures extend.

STEM An acronym used to group the following disciplines: science, technology, engineering, and mathematics. It is commonly used in reference to education policy and curriculum choices.

step A platform used by pedestrians to gain access to a higher level. Intermediate platforms may be used to break up different levels to improve the ability of people to move from one level to another with greater ease and safely. A series of steps make up a *stair and is contained within a *staircase.

step flashing A *flashing that is built into raked out joints, providing a weather-resistant junction where a roof abuts a wall. The lead flashing is inserted into the wall and folded so that it laps over and under roof tiles. Individual flashings are cut into the joint or a sheet of lead is cut in steps so that it can be cut into each course of brickwork.

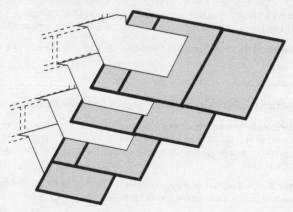

Step flashing

stepladder Foldout ladder with shelf-like platforms to position feet.

stepped Something that has a sudden change.

stepped flashing *See* STEP FLASHING.

stepped footing (stepped foundations) Strip foundation that is stepped to follow the contour of the land. Stepping the foundation reduces the amount of concrete needed to create a stable platform onto which the building loads can be transferred to the ground.

stepped scarf joint (splayed scarf joint) A scarf joint between the ends of veneer or board, formed by lapping the joints diagonally, with a step in the middle enabling the joint to cope with compression and tension.

step-tapered pile (Raymond pile) A cast in-situ pile formed by driving a steel casing, which is wider at the top, into the ground. Concrete is then cast with or without reinforcement—the casing remains in the ground.

stereographic The representation of three-dimensional objects on a two-dimensional plane.

stick slip The rapid movement that occurs between the two sides of a fault when it displaces.

stiff Rigid; something that is not easily deformed.

stiffness The resistance to deformation under load. It is calculated by considering EA/l, where E = *modules of elasticity, A = area, and l = length.

stiffness matrix (K) An array of mathematical elements that generalizes the stiffness of *Hooke's law, used in finite element analysis.

stiffness methods A structural analysis method where forces are expressed as displacements.

Stillson An adjustable grip and wrench in one, used by hand to clamp objects such as pipes and rods, while nuts or other objects can be released.

still water Water with no flow, as would be found in a pond.

stimulation The implementation of a process, technique, or product to increase the yield or output from something else.

stipple Rough-textured finish formed by using a rough-textured brush, *roller or pad that draws the *paint or artex out from the surface creating a undulating pointed finish.

stippler (stippling brush) Tool used to create a *stipple finish.

stippling Creating a stipple finish, using a brush, roller, or pad.

stirrup 1. A rope or hanger to support something e.g. a beam. **2.** A piece of *reinforcing bar to resist a shear force.

stitching The replacement of damaged bricks with new bricks in existing brickwork.

stochastic Random; involving probability or guesswork.

stock 1. Those items that are available and in store. **2.** A major intrusion of a near cylindrical vertical column of intrusive igneous rock formed within the Earth's crust from slowly cooling magma. *See also* BATHOLITH; PLUTON.

stock brick Standard *bricks that are available ready for dispatch.

stockpile To store large quantities of items or equipment, ensuring that sufficient quantity is readily available.

stoichiometry A reference generally to the composition of a material and specifically to the relative atomic proportions of cations and anions.

Stokes' law An expression that defines the settling velocity of spherical particles in liquid. It is used for sedimentation analysis for soil particles, such as silts and clays that have a particle diameter between 0.2 to 0.0002 mm.

stone Rock and other hard naturally occurring, non-metallic material, quarried or collected from the ground. A few examples of rock or stone include: limestone, sandstone, mudstone, slate, shale, granite, and quartz.

stone facing Wall or building that is covered with a veneer of stone or is clad in stone.

stonemason Craftsperson who works with stone. The mason will cut, shape, and lay stones.

stonework Walls that are made or clad in stone.

stop 1. An ornamental end to a decorative moulding. A *special brick that forms the *stop end to a top course or decorative course of bricks.

stop bead (plaster stop) A metal or plastic straight edge that is fixed level with the wall to provide a neat break for a plaster finish, or a break where the finishes change.

stopcock (stop valve) A valve that is used to regulate or stop the flow of a gas or fluid in a pipe.

stop end A moulded or splayed special brick that forms a neat end to a top course of bricks or a decorative course of bricks.

stopped chamfer (stop chamfer) A spayed cut on an edge or a corner, which results in a triangular shape as it spays towards or away from the arris.

stopped mortise A blinded or sub mortise, where the mortise does not penetrate all the way through the timber. One side of the timber has a rebate and the other side of the timber has no sign of the joint.

stopper A plug designed to block the flow through a pipe.

stopping 1. Filling holes in timber, concrete, masonry, or other building materials. **2.** The material used to fill holes.

stopping knife (putty knife) A blunt knife with a rounded edge used by *glaziers for smoothing the putty used to fit windows.

stop valve *See* STOPCOCK.

storage cistern A *cistern.

storage tank A *tank.

storage water heater An appliance that heats water and stores it in a tank until it is required.

storey A full level of a building, the height of which is measured between one floor and the next. An eleven-storey building would mean that there were eleven full floors all enclosed within the building.

storey rod (gauge rod) A length of timber or material that is cut to the full height of a *storey and used to measure off the storey height to ensure all measurements are consistent.

storm cellar (cyclone cellar) A room under the house or dwelling, normally contained below ground, that is built to offer protection during periods of severe weather conditions and storms.

storm clip A clip used to retain glazing in position.

storm door A door installed in front of the main external door into a building that is designed to provide additional protection from the elements.

storm-proof window A window that is designed to resist the wind-driven rain, hail, sleet, and snow experienced in a storm. It usually incorporates double rebates and additional seals.

storm window A window installed in front of the main window of a building that is designed to provide additional protection from the elements.

straddle scaffold A *saddle scaffold that is built over the ridge of a roof, often used to access chimney stacks or abutting walls.

straight arch *See* FLAT ARCH.

straight edge A length of metal or timber that has a straight edge, used as a profile for producing straight and flat surfaces or scribing lines on materials.

straight flight A flight of stairs that goes directly from one level to another without a change in direction.

straight joint 1. Two pieces of material where the ends are simply butted together; a *butt joint. **2.** A defect in brickwork where the joints or one course sit directly above the joints of the course below, rather than being staggered, sitting partway across the top of the brick below, for example, the joints in stretcher bond sit midway above the brick in the course below.

straight-joint tiles *Single-lap tiles designed so that the joints between successive courses are not staggered but straight, running in a straight line from the top of the *ridge to the *eaves.

straight-peen hammer A hammer with a double head where one head has a chisel or wedge-shaped profile.

straight tee (bullhead connector) A T-shaped duct or pipe fitting that has three openings.

straight tongue A straight projection from the edge of a board, such as *tongue-and-groove boarding.

straight tread A regular rectangular tread on a stair (*flyer), where the edges of the tread are all parallel to each other, unlike a *winder, *tapered tread, or *kite winder.

strain A measure of how much a material has extended during deformation when subjected to a tensile force. Strain (ε) has no units and is usually expressed as a percentage; the value is calculated as follows:

$$\varepsilon = \frac{\Delta L}{L}$$

where ε is strain in measured direction,
ΔL is the original length of the material,
L is the current length of the material.

strain energy The energy stored in an elastic element under *strain.

strainer A coarse mesh filter installed on the intake of a pump that is designed to prevent solid particles passing through and damaging the pump.

strain gauge A device such as a *Wheatstone bridge that measures *strain.

strain hardening See COLD ROLLING.

straining beam A horizontal strut.

strain rate The rate at which strain changes in respect to time.

S-trap An S-shaped *trap commonly used on baths, sinks, wash hand basins, and WCs.

strap bolt A bolt with a flat metal bar with pre-drilled holes welded directly to the head of the bolt. The threaded bar is used to firmly secure the panels or unit that the strap is fixed to.

strata The plural of *stratum.

strategic human resource management (SHRM) An approach to managing human resources, supporting long-term business goals and outcomes with a strategic framework. It focuses on longer-term staffing and resourcing issues within the context of an organization's goals and the evolving nature of work, and informs other business and HR strategies in an integrated way (CIPD, 2018).

((⊕)) SEE WEB LINKS
• CIPD strategic human resource factsheet

stratigraphy The study of dating *strata, origin, and composition, represented by a vertical section through the ground.

stratosphere The region of the atmosphere that is 10–60 km (6–30 miles) above the surface of the earth, which contains the *ozone layer.

stratum (bed) A layer or several layers of rock, typically formed from *sedimentary rock.

strawboard A compressed straw slab or panel. The compressed panel is formed through hot-pressing straw. The material is lightweight with good thermal resistance, and can be lined with plasterboard or paper producing a more workable product with a finish. The board can be made with ducts and rebates within it for cables and other services.

straw thatching Thatching made from the yellow straw of red wheat, the most common type of thatching material.

stream 1. A small river. **2.** A constant flow of gas or liquid.

stream gauge A device used to determine the depth of water in respect to a *datum, typically used along a riverbank or at a fjord.

streamline The path followed by particle in a non-turbulent flow of a fluid. Streamline of a particle is determined with respect to a static solid body that the particle passes by.

street furniture *Lights, *lamps, benches, road boards, and signs that are used in *external works.

strength 1. The capacity to withstand a force, pressure, or stress. **2.** The physical power available.

Streptococcus A genus of bacteria that are characterized by being spherical, and used to indicate faecal contamination in drinking water.

stress A measure of the average amount of force exerted per unit area. It depicts the intensity of the total internal forces acting within a body across imaginary internal surfaces, as a reaction to external applied forces and body forces. In general, stress, σ, is expressed as:

$$\sigma = \frac{F}{A}$$

where σ is the average stress, also called engineering or nominal stress, and
    F is the force acting over the cross sectional area A.

The SI unit for stress is the pascal (Pa), which is a shorthand name for one newton (force) per square metre (unit Area). The unit for stress is the same as that of pressure, which is also a measure of force per unit area. Engineering quantities are usually measured in megapascals (MPa), newtons per squared millimetres (N/mm^2), or gigapascals (GPa). In Imperial units, stress is expressed in pounds-force per square inch (psi) or kilopounds-force per square inch (ksi). As with force, stress cannot be measured directly but is usually inferred from measurements of strain and knowledge of elastic properties of the material. Usually, stress is calculated from the results of a tensile test.

stress concentration A localized area where the stress in an element increases; this may be due to a crack or defect or a change in the cross-sectional area.

stress contour plot A multicoloured diagram to represent how the stress changes over the area of something (e.g. the cross-section of a beam); different colours are used to represent different stress values.

stress curve (S-N curve) A graph that indicates failure due to *fatigue, i.e. stress versus the number of fatigue cycles.

stressed skin panel A structural timber panel, with a timber frame and a skin of plywood on one or both sides of the panel. The plywood takes the stress when used as a floor, roof, or wall unit, and transfers them across the timber frame. The unit acts as a whole with the frame and skin, transferring the stress and sustaining the loads.

stress-graded Describes timber that has been classified for its strength capabilities, either through visual stress grading or machine stress grading.

stress-graded timber Timber that has been *stress-graded.

stress grading The process of grading the strength of timber.

stress-grading machine Machine that grades the strength of timber.

stress intensity factor (K) A scale factor used to define and gauge the magnitude of the crack-tip for a stress field, being dependent on the configuration of the cracked component and the way the loads are applied. Such factors are important in the mechanics of metal fracture.

stress relaxation The reduction in tensile stress with time when subjected to constant strain. It occurs in polymeric materials, such as a *geosynthetic used as basal reinforcement under an embankment. It is not to be confused with *creep, which is a time-dependent increase in strain under constant load, as would occur for a geosynthetic in a reinforced wall.

stress strain curve A graph that represents the relationship between stress and strain of a material. It is similar to a load displacement plot, however, with load and displacement, no indication of the size of the sample is given, thus stress (load/area) and strain (change in length/original length) is often used.

stretcher A brick or stone laid lengthways, if laid with other bricks or stones in the same direction, a *stretcher course will be created. Multiple courses of stretchers create a *stretcher bond wall. *Headers and *closers will be used to finish the wall and create other courses.

stretcher bond (common bond, half bond, running bond, stretching bond) Multiple courses of *stretchers laid on top of each other. The centre of the joint on one course sits directly over the centre of the brick or stone on the next course.

stretcher course Stretchers laid end to end throughout the course.

stretcher face The long face of a *stretcher brick which is exposed.

strike To remove the formwork from recently cast concrete. The *formwork should be removed cleanly without pulling concrete away from the structure. A clean strike is achieved when the formwork has been cleaned, properly prepared with release oil, is free from *defects, and the concrete has been properly *vibrated and left to *set.

strike-slip fault A *wrench fault that is characterized by displacement relative to the strike of the fault, such as the San Andreas Fault in California, one of the world's most famous faults.

striking The dismantling of formwork and other temporary supports.

striking off The removal or skimming of any excess, mortar, or concrete from a mould or formwork.

striking plate (keeper, strike, strike plate) A plate fixed to the jamb of a door that has a rectangular hole to receive the door latch.

striking time The time that must be allowed for the concrete to mature and maintain its strength before *striking—when the formwork is removed. The concrete must be sufficiently *set so that it can transfer its own load without fracturing the bonds within the newly set concrete.

striking wedges Wedges used to prise and lever the formwork from the face of the concrete, the triangular wedges are carefully tapped between the concrete and formwork. The wedges should be used only where necessary, since levers and wedges can leave marks on the face of the concrete.

string (stringer) The diagonal boards that carry the *treads and *risers in a staircase. Strings can be closed and rebated out to carry each tread, or can be rough and cut to the profile of the stair with the tread resting on the top of the string. The string is a diagonal beam that supports the staircase, transferring the loads to the upper and lower floors.

string course (band course, belt course) A course in brick that either projects or is indented from the surrounding courses. Courses run in different-coloured brick may also be referred to as a band or belt course.

string line A line pulled between two points that is used for setting out works. String lines are used between the ends of new walls to provide a guide for *courses of brick or blockwork; lines can be pulled between setting out pegs to provide a guide for excavating *trenches and *footings; they are used between steel pins to create a guide when *setting out and laying concrete *kerbs and edgings in road works, and have many uses when marking lines in *formwork *joinery.

string wall A wall that is used to support the treads of a staircase.

strip diffuser (slot diffuser) *See* LINEAR DIFFUSER.

strip flooring Long narrow strips of solid tongue-and-grooved wood flooring.

strip foundation A foundation that is excavated and cast in long lengths, used to carry longitudinal loads such as external walls and walls to houses. The foundations can be wide strip, reinforced to distribute the loads over a greater area, reducing the load per surface area; narrow or deep strip footings are slightly wider than the loads being carried and transfer the loads down to good load-bearing strata.

The foundation runs under continuous loads such as brick walls.
Where the ground has good load-bearing strata, this is the most
economical way to found walls.

Strip foundations used for continuous loads such as walls.

strip lamp A long narrow fluorescent luminaire.

stripper 1. A labourer or formwork joiner who strips formwork. **2.** Chemical paint
remover, used to break down and soften *paint so that it can be scraped away from
the surface it has been applied to.

strip sealant Preformed strip of compressible material used around door and
window frames to exclude draughts; solid rigid material is also used in glazing to
position, space, and hold glass.

strip tie A *twist tie that is stood vertically.

strong back A proprietary metal support used for resisting high loads on concrete
formwork. The beam-type unit can be strapped vertically, or fixed horizontally to
the secondary supports that need to transfer the loads to a strong unit.

Strouhal number A non-dimensional number used to describe oscillation in
flow mechanisms.

struck capacity The amount of material that can be held in an excavator's bucket
which has been levelled off. It is equivalent to the volume of water a bucket would
hold if it was watertight.

struck joint *See* WEATHER-STRUCK JOINT.

structural Relating to a *structure; something that is built such as buildings,
bridges, frameworks that carry loads and resist forces.

structural analysis The mathematical examination of structures and their
behaviour to loads, displacement, stresses, strains, and moments.

structural clay tile *See* HOLLOW CLAY BLOCK.

structural connection The joining of two members that carry loads. The joint
transfers the loads and forces.

structural design The design of structures, particularly in relation to carrying loads and minimizing displacement.

structural drawings The *structural designs and *drawings produced by the *structural engineer or design engineer that show all the *structural members and the structural frame in detail.

structural element *See* STRUCTURAL MEMBER.

structural engineer A professional engineer who specializes in structural design and is a chartered member of the Institution of Structural Engineers (*IstructE).

structural gasket A *gasket used in curtain-walling to hold glass against the mullion.

structural glass Glass, usually coloured, applied to masonry or plastered walls as a decorative finish.

structural glazing Glazing that has some load-bearing capacity. It can either be bonded together using structural sealants such a silicone **(unsupported system)**, or it can be supported on a concealed framework **(supported system)**, which is fixed back to the structure. It is used to create a smooth all-glass façade.

structural integrity The ability of a structure to remain fit for purpose, in respect to carrying loads and stress, and minimizing displacements.

structural member A beam, column, etc., that forms an important part of the structure, such that it carries loads, and provides support or stiffness.

structural steel Steel that has a carbon content of less than 0.25% by weight.

structural steelwork *Structural members manufactured from *structural steel that have been connected to each other to form a *steel frame.

structural temperature gradient The change in the temperature that occurs through a construction.

structural timber Beams and trusses that are manufactured from wood and used in a structure to provide support and carry loads.

structural trades All of the trades that contribute to the structural components and members of the building; includes the ground workers who excavate and cast the footings, bricklayers who erect structural masonry, structural frame erectors who erect either steel, concrete, or timber, and any other trades that contribute to components or elements that are structural.

structure The load-bearing frame of a building, bridge, or other structure; something that is constructed.

structure-borne sound Noise that is transmitted through the fabric of a structure, through a wall or floor, as opposed to an air-borne sound.

strut A structural member inserted to act in compression and is used to hold or brace something; it can be vertical or horizontal.

strutting System of struts used to hold or brace an element; herringbone or solid strutting used to brace floor joists, or struts used to prop up floor formwork.

stub A short element or shorter version of something, for example a *stub tenon.

stub tenon A tenon that is used for a *blind mortise; as the *tenon does not need to penetrate all of the way through the mortise, it is a shorter tenon and described as a *stub.

stucco A material used for render, but has its origins in more classical forms of render. Traditionally, it was smooth with mouldings to represent columns and ornamental structures. Pre-nineteenth century it was made from lime-cement mortar and since that time cement mortars have become much more common. The sustainability movement argues for the reintroduction of lime-cement mortars due to their ability to be reused.

stud Vertical timber (or more recently, pressed steel) supports, used in *stud partition walls. The vertical timbers and horizontal connecting pieces, including *head and *sole plate, and *strutting, act as a structural frame onto which plasterboard can be fixed and supported.

stud partition (stud wall) A *partition wall where the *structure is made of *steel or *timber *studs, creating a *frame onto which *plasterboard finish is applied. Originally the partition wall was just considered a divider of space and either *load-bearing or *non-load-bearing. Today partition walls have various roles to perform in addition to dividing space; they are often required to perform acoustic, thermal, and fire-resisting functions. While in most cases partition walls are not thermally insulated, it is common practice to insert acoustic insulation, and if a fire-rating is required, the walls will have fire protection in the form of fire board, lapped joints, and insulation to delay the passage of fire.

styrene/butadiene copolymer A thermoplastic copolymer comprising styrene and butadiene. One of the outstanding characteristics of styrene/butadiene thermoplastic copolymers is their combination of high transparency and impact resistance. The good miscibility allows adjustment to the desired toughness, while at the same time reducing material costs.

sub-base The lowest layer in the formation of a *pavement, normally constructed of crushed stone or gravel. Its function is to form a working platform and provide strength.

subbie (subby) A slang word used to define a *subcontractor.

sub-board *See* DISTRIBUTION BOARD.

subcontract A contract that is placed by the main contractor to a third party to carry out a proportion of the works. In order to introduce a subcontractor into a contract, agreement should be sought from the client; it is usual for the main contract to state that work cannot be subcontracted without prior agreement.

subcontractor A company or individual who is employed by the main contractor to undertake work. The work is part of the contract that has been awarded by the client to the main contractor under a *subcontract.

subduction When one *tectonic plate moves down under another tectonic plate as they collide in the earth's crust.

sub-floor The smooth floor base onto which the finished material is placed.

sub-frame Any frame or surround used to carry or support finishes or materials to be attached to the main structural component.

subgrade The natural ground below the formation level of a *pavement—*see* FLEXIBLE PAVEMENT.

subletting The subcontracting of work to another party.

sub-level The level below the main level. When discussing the ground it is the strata below that at the top of the ground.

submersible pump A *pump that operates under water, at the bottom of a *borehole to *dewater a site, for instance.

subset A subdivision of a set, e.g. a smaller set of numbers that also belong to a larger set of numbers.

subsidence (settlement) The movement of ground as it compresses, slips, or deforms. Settlement is a problem when buildings are placed on the ground and the structure moves downwards as the ground moves. If the whole building moves downwards at the same rate only the incoming services will have to cope with the movement. If the movement is differential, acting at different rates of fall across the building, then cracks will appear and in severe cases the building may collapse.

subsill A *sill inserted below a *threshold or window sill, designed to ensure that any water that drips down from the threshold or window sill is thrown clear of the wall. A range of materials can be used to form a subsill, including brick, concrete, plain tiles, stone, and timber.

subsoil The compact soil that lies directly beneath the *topsoil.

substance Material matter; from which people and things are made.

substation An electrical facility that contains transformers and other types of electrical equipment that decrease, increase, and regulate the voltage of the electricity that is sent through the power lines.

substrate 1. The surface onto which finishes are applied, such as plaster stuck on a block substrate. **2.** The ground on which the building rests.

substructure The parts of a building that are constructed below ground level or below that which will be the finished ground level once the building is completed. The substructure elements include the foundations and basement. The substructure elements are critical activities, and as they have to be completed before other works can start, they will almost certainly appear on the *critical path of a *Gantt chart.

sub-suite A series of locks that can all be opened using a sub-master key.

subtense bar A bar of standard length typically 2 m, manufactured from *invar, and used in *tacheometry.

suburb Town development that is built against the edge of an existing city or town; it is an *urban area but is not at the heart of the main town.

subway An underground passage for pedestrians to walk from one side of the road to the other. In the US the term is used to refer to an underground railway.

successful tender (success bidder) The bid selected to enter into the contract for the works. Traditionally, the bid would be selected on price alone; today the qualification process and selection has to ensure the capabilities of the contractors, their ability to perform the works, financial stability, level of expertise, and quality of work expected.

succession planning Management activity focused on identifying and growing people to fill leadership and business-critical positions in the future. It is the process of identifying and developing potential future leaders or senior managers as well as individuals to fill other business-critical positions, either in the short- or the long-term. As well as training and development activities, succession planning programmes typically include the provision of practical, tailored work experience relevant for future senior or key roles. The aim is for the organization to be able to fill key roles effectively if the current post holder were to leave the organization (CIPD, 2018).

((⊕)) SEE WEB LINKS
• CIPD succession planning factsheet.

suction The process of moving a fluid from one place to another due to differences in pressure. For instance, a fluid will flow from an area of high pressure to an area of low pressure.

sugar soap A paint remover made from alkaline solution.

suite A set of matching or complementing parts, for example a set of keys matching doors in a hotel, a matching set of furniture, a set of interconnecting rooms making one designated area.

sullage 1. Wastewater from domestic activities. **2.** Sediment or silt deposited by flowing water.

sulphate A type of salt derived from sulphur, the empirical formula of the sulphate ion which is polyatomic is SO_4^{2-}. The tetrahedral arrangement consists of a central sulphur atom surrounded by four oxygen atoms. Many of the sulphates are soluble in water, and exceptions to this rule include: calcium sulphate, strontium sulphate, and barium sulphate.

sulphate attack A reaction between the tricalcium aluminate in concrete and the salts contained in some clays, soils, or flue condensates. The reaction leads to the production of a chemical called hydrated calcium sulphoaluminate (ettringite) which has a greater volume than the cement material which it is replacing. The expanse of material within the concrete or mortar causes it to crack and degrade, losing structural strength, and affecting the finish of the concrete.

sulphate-resisting Portland cement A Portland cement with a chemical composition that is resistant to *sulphate attack.

sump A small pit or well used to drain and collect water.

sump pump A small electrically driven pump used to remove water from a *sump.

sunken fence *See* HA-HA.

sunken gutter A *box gutter.

sunlight Light received from the sun.

sunshade A device designed to reduce *glare and *solar gain.

superconductor A material that has the ability to carry electricity for a prolonged period without dissipating energy in the form of heat. Examples include tin and aluminium. The loops of superconducting wire have the ability to carry electrical currents for many years without experiencing any measurable loss. At absolute zero (−273 °C) to temperatures experienced by liquid nitrogen (−196 °C) there is no electrical resistance.

superelasticity (pseudoelasticity) A phenomenon whereby a material exhibits a large elastic (non-permanent) response to relatively high stress and usually returns to its original shape or dimension on removal of the load.

superheated steam (anhydrous steam, steam gas, surcharged steam) Steam heated to a temperature that is beyond its saturation point at a given pressure. Used in power generation.

superheater A device used to generate superheated steam.

superimposed load *See* LIVE LOAD.

supernatant The clear liquid above settled solids.

superplasticizers A component added to concrete to increase its workability, usually used as an additive or alternative to water.

superposition The summation of a number of effects.

superstructure In general it refers to all the parts of the building that are above ground level, but in specific terms it addresses the structural components that sit on top of the *substructure. The superstructure works include the frame, floors, and walls above the ground, and the roof, finishes, and fittings to all above ground works.

supervision The overseeing and hands-on management of workers or a package of work, offering guidance, checking work done, and coordinating activities.

supplementary angles Two angles (e.g. 130° and 50°) that add up to 180°.

supplier Company that supplies materials, labour, plant, or services for a specific area. The term is more usually associated with companies that provide physical products and materials.

supply air The air that is supplied to a room or space from the air-conditioning, ventilation, or HVAC system.

supply and fix (furnish and install) A contract for the supply and installation of a product and component. The *contractor or *supplier delivering the unit also takes responsibility for ensuring the installation is correct and the product operates correctly.

supply only Designates a contract that is for the supply of a product or material only, and does not include the installation of the unit.

supply pipe The pipe from the building to the site boundary, stop valve, or highway; it is under the consumer's ownership. Beyond the boundary and the stop valve, it is the property of the utility provider. The ownership of the service is important as it designates responsibility for maintenance and repair.

support To bear the weight or hold up something.

supported sheet-metal roofing (flexible metal roofing) Sheets of metal roofing that are supported by *roof deck.

suppressed weir A *weir that is the full width of the channel; *see also* CONTRACTED WEIR.

surcharge Additional loading of the ground, for example, behind a retaining wall or as a form of ground improvement, a temporary loading of the ground to reduce the amount of *consolidation once the structure or embankment has been built.

surface condensation Condensation that occurs on surfaces that are at or below the *dew point temperature of the air immediately adjacent to them. It tends to occur on the internal surface of external elements of a building, and on cold pipes and cisterns within a building.

surface damp-proof membrane (surface damp-proof course) A liquid-based damp-proof membrane or *damp-proof course that is applied to the surface of a material.

surface dry A stage in paint drying, where the surface has a thin film of dried paint and feels dry to touch, although the underlying surface is not dry.

surface filler A fine-grained filling material suitable for creating a smooth finish.

surface-fixed hinge A hinge that is mounted directly onto the surface rather than being recessed or sunk into it.

surface resistance (thermal surface resistance) The conductive/convective coefficient at the surface of materials, with an allowance for variables such as wind speed, emissivity of finishes, and mean radiative temperature. The value of surface resistance is derived from the procedure set out in ISO 6946:2007. It is the conductive/convective thermal resistance at the interface between air and solid material, for example, at the external, internal, and cavity faces of a wall or any other surface of a building. Surface resistances affects the total thermal resistance of the wall. Surface resistance varies; ISO 6946 offers correction formulae for U values—allowing for adjustment of 3% for air voids in insulation, mechanical fasteners penetrating insulation layers, and precipitation on inverted roofs.

surface spread of flame *See* FLAMESPREAD.

surface tension A measure of the ability of the molecules contained within a liquid to stick together. See *also* CAPILLARITY.

surface water Rainwater; also water from streams, rivers, and lakes, as opposed to *groundwater.

surface water drain A *drain that takes the water collected from roofs, the faces of buildings, driveways, and paved areas, and conveys it to the *surface water sewer.

surface water sewer A *sewer that conveys *surface water.

surfacing The treatment or finishing to a surface by smoothing, texturing, or painting.

surfacing material Finishing material, such as paints, bitumens, lining, coverings, chipping, etc., used to protect against water, provide skid resistance, and present an aesthetical finish.

surround A frame, moulding, or trim that is placed around a component or an opening, such as a door.

survey 1. An inspection of a building to determine its structural soundness. **2.** To measure an area of land to enable a scale map or plan to be produced.

surveyor A professional who performs an evaluation or assessment of property or land: a land surveyor describes the nature, level, and lie of the land; a property surveyor determines and evaluates the physical characteristics of a building; or a quantity surveyor assesses the value of land, buildings, work packages, and materials for commercial purposes.

suspended ceiling A ceiling finish that is supported or hung from the structural floor above.

suspended concrete floor slabs Slabs of reinforced concrete held off the surface of the ground so that there is no direct contact with the ground or floor.

suspended floor A floor that is raised and supported above the ground; *see also* SUSPENDED TIMBER GROUND FLOOR.

suspended scaffolding A *flying scaffold, *projecting scaffold, or slung scaffold that is not supported from scaffolding standards or props, but is supported or hung from a building or surrounding structures.

suspended slab Reinforced concrete slab that is raised above the ground, and supported by beams and walls.

suspended timber ground floor A floor constructed from timber joists at ground level, where the joists span across external and intermediate *honeycomb bonds, raising the floor above the ground. The gap below the floor should be ventilated to prevent moisture forming on the timber and causing mould growth.

suspended timber upper floor A floor above ground level that is made from timber joists spanning between walls or beams.

suspension 1. To temporarily stop something. **2.** The dispersion of fine particles in a liquid.

suspension bridge A *bridge, such as the Forth Road Bridge (*see* FORTH RAIL BRIDGE), that is suspended by cables, which are carried by support towers and anchored at either end of the bridge.

suspension cable The main cable, made of steel wire, on a *suspension bridge.

suspension of works Instruction given to halt the flow of works.

Sussex garden wall bond *See* FLEMISH GARDEN WALL BOND.

sustainability The ability to be able to maintain without destroying a natural balance.

sustainable development Gro Harlem Brundtland, the former prime minister of Norway, was selected by the Secretary General of the United Nations (UN) to chair a commission with the remit to unite countries to pursue sustainable development. From this, the most cited (recognized) definition of sustainable development was formed in 1987. Hence, as given in the *Brundtland Report (more correctly entitled *Our Common Future*), sustainable development is defined as: 'development that meets the needs and aspirations of the present without compromising the ability of future generations to meet their own needs'.

Sustainable Development Goals (SDGs) The Sustainable Development Goals are a universal call to action to end poverty, protect the planet, and ensure that all people enjoy peace and prosperity. There are 17 goals aimed at building on the successes of the *Millennium Development Goals (MDGs), while including new areas such as climate change, economic inequality, innovation, sustainable consumption, peace, and justice among other priorities. Following on from the Millennium Development Goals, the SDGs were formed in September 2015, when heads of state from all around the world, from the 193 United Nation's (UN) countries, congregated in New York City. In addition to the 17 SDGs, there are 169 targets to be addressed between 2015 and 2030. The 17 goals are listed below:

SDG 1: End poverty
SDG 2: End hunger
SDG 3: Good health and well-being
SDG 4: Quality education
SDG 5: Gender equality
SDG 6: Clean water and sanitation for all
SDG 7: Affordable and sustainable energy
SDG 8: Decent work for all
SDG 9: Technology to benefit all
SDG 10: Reduce inequality
SDG 11: Safe cities and communities
SDG 12: Responsible consumption
SDG 13: Stop climate change
SDG 14: Protect the ocean
SDG 15: Take care of the Earth
SDG 16: Live in peace
SDG 17: Mechanisms and partnership to reach the goals

The SDGs are relevant to both developed and developing countries, whereas the MDGs were more focused on developing countries. As such, this creates a global platform for *sustainable development to be addressed.

(⊕) SEE WEB LINKS
• UN website with details on the Sustainable Development Goals.

sutro weir A contracted weir that has one of the following cross-sections:

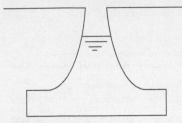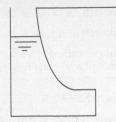

Sutro weirs

swale A swale is a broad, shallow channel that water runs along when it has been raining.

swallow hole A vertical hole or shaft in the ground that connects with an underground passage or tunnel.

swan neck An S-shaped pipe or drain, used to create an air trap in drainage preventing the escape of foul smells from sewers.

swarm A large collection of things, particularly insects.

swash 1. The water that runs up the beach when a wave breaks. **2.** A narrow channel for tides to flow through.

S-wave (secondary wave, shear wave) A type of seismic wave that travels in a perpendicular direction to the direction the wave propagates in.

sweep tee (sweeptee, swept tee) A duct or pipe fitting that has three openings, where a branch joins a slight curve. *See also* STRAIGHT TEE.

swelling An increase in volume, particularly in relation to clays as they absorb water.

swept-sine testing A dynamic loads test.

swept valley *See* LACED VALLEY.

swg The traditional method of describing the thickness of wire, tubes and straws of metal is the *standard wire gauge.

swing To move backwards and forwards.

swing door (double-action door) A door that can open in either direction.

swinging jamb or post A *hanging jamb or post.

swinging scaffold A cradle scaffolding platform that is hung or slung from outriggers that reach out over the edge of the building. The cradle, which can carry the operatives, equipment, and materials, is hung by cables or ropes. Machine operated winches allow the platform to be raised and lowered so that the workers can access the desired area of the building.

swing-up door (up-and-over door, overhead door) A door that opens by lifting up the bottom of the door until it is in a horizontal position. Commonly found on garage doors.

switch An electrical device used to open or close an electrical circuit.

switch and fuse *See* SWITCH-FUSE.

switchboard 1. A panel comprising a number of switches that are used to control power and lighting. **2.** A device that is used to connect telephone callers with one another.

switchbox A box, usually mounted on a wall, that contains an electrical switch.

switch-fuse (switch and fuse) A switch in which a fuse is placed in series with the contact.

switchgear The electrical switches, circuit breakers, and other items of electrical equipment required to control and protect electrical circuits.

switchroom A room that contains *switchgear or a telephone *switchboard.

Sydney Accord An international agreement for accrediting qualifications in the area of engineering technology. It was founded in June 2001 by seven signatories representing Australia, Canada, Hong Kong, Ireland, New Zealand, the United Kingdom, and South Africa. *See also* DUBLIN ACCORD; WASHINGTON ACCORD.

symbols Drawing shapes that represent real manufactured components, materials, or functions. The drawings and diagrams provide a common language that is used to easily interpret the components on a drawing.

syncline A downward *fold in *sedimentary rock, caused by *tectonic movement.

synthetic Not naturally occurring; something that is man-made, such as a *geosynthetic product made from a polymer, polyester, *polyamide, *polypropylene, and *polyethylene, rather than a vegetable fibre such as *coir or *jute.

synthetic fibre (man-made fibre) A strand of material that is not naturally occurring in its final form.

synthetic wastewater The wastewater produced as replica of actual wastewater by the addition of different chemicals into tap water. The term is associated with lab scale studies on wastewater treatment where synthetic wastewater is applied to treatment systems in controlled laboratory conditions. Dextrose (glucose) is added as a substrate for microorganisms, while the addition of macro-nutrients and micro-nutrients is also essential to ensure proper growth of microorganisms.

system building The use of prefabricated, off-site, or factory-built components that are made to fit together as a complete building. The components are made off-site and the assembly designed so that the building becomes a kit of parts that can be quickly bolted and fitted together. The need for wet trades, such as bricklaying, concreting, and plastering are eliminated or drastically reduced, removing drying out time that is required in traditional buildings. Some buildings are volumetric, built off-site as large boxes that contain fully fitted-out bedrooms, hospital rooms, or plant rooms that can be simply bolted together on-site, whereas other buildings are designed to be delivered flat (flat-pack construction) and built up on-site.

table form (table formwork) Horizontal mould used for supporting *in-situ reinforced concrete when casting structural floors. The formwork is delivered partly preassembled in a table shape and is then quickly adjusted to the correct height and length so that it can be craned quickly into position. Large table formwork systems are sometimes called flying forms as they can be easily craned and moved into position.

tacheometry A surveying method that is used to rapidly measure a distance without the need of a chain or tape. The *stadia lines of a theodolite are sighted onto a graduated staff making a fixed *parallactic angle. When the staff is held perpendicular to the collimation line, the collimation distance (from the instrument to the staff) can be computed.

tack A short and very sharp flat-headed nail.

tack-free The point at which a surface that is in the process of drying is no longer tacky. *See also* TACKY.

tack rag Cotton rag or cheese cloth with a small quantity of varnish on the cloth, which is used for dusting or smoothing a surface after it has been rubbed down; it is applied before the main coat. To avoid the cloth hardening, tack rags should be placed in an airtight container.

tacky A term used to describe the stickiness of a surface as it is drying.

taconite A band of low-grade iron ore.

tail 1. A short section of electrical cable that connects the electricity meter to the consumer unit. **2.** The leading edge of a roofing tile or slate.

tailings The waste (fines) remaining from the mining industry after the ore and minerals have been extracted from rock. A **tailing dam** is an embankment that has been constructed using the fines from such material.

tailpiece (US) *See* TRIMMED JOIST.

take-off The compiling of descriptions and recording of quantities of work from drawings and schedules; this is normally the first stage in preparing a *bill of quantities. *Standard Method of Measurement rules should be followed so that the quantities of the work measured are consistent and pricing of the work is comparable.

taking-off The process of compiling the *take-off.

talent management A recruitment and retention strategy to attract, identify, develop, engage, retain, and deploy individuals who are considered particularly valuable to an organization. Individuals may be considered 'high-performing' (those who add value now) and/or 'high-potential' (those who may add value in the future, CIPD, 2018).

SEE WEB LINKS
• CIPD talent management factsheet.

tallboy 1. A long, narrow chimney pot that is designed to prevent down draughts. **2.** A tall cupboard.

tally 1. To check a record or score. **2.** To count something.

tally slates Regular-sized slates that are purchased by the number required, rather than by their weight.

tambour Anything that is cylindrical or drum-shaped; French for drum, it applies to the wall of the structure whether it is situated on the ground or within a ceiling. For example, a cylindrical ceiling with a dome on top.

tamped finish Undulating surface finish formed in concrete by putting a straight edge (*tamper) across the surface of the concrete and taping the surface. By lifting the straight edge off the surface of the concrete, the tension forces the concrete to raise as it attempts to adhere to the tamper. The operation is repeated causing the surface of the concrete to undulate.

tamper Piece of concreting equipment used to create a rough undulating finish on the surface of wet concrete. The straight edge, which sometimes has handles at each end, is raised and lowered onto the surface of the concrete by two *operatives.

tamping The process of creating a *tamped finish.

tang The tapered section of a metal hand-tool that is fixed into the tool's handle.

tangent A straight line that touches a curve at the point where the curve is going in the same direction as the straight line. It is the best straight line approximation to the curve at that point. The straight line would also be at 90° (a right angle) to the radius of the curve.

tangential cut (plain-sawn, slab-cut) A cut made in wood that is at a tangent to the grain.

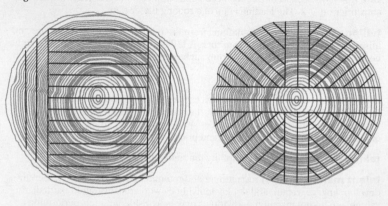

(A) Tangential   (B) Radial

Tangential cut: Wood cut in the tangential and radial direction: (A) tangential and (B) radial

tangent modulus The slope of a line on a non-linear stress–strain curve. Until the proportional limit is reached, the tangent modulus is equivalent to *Young's modulus.

tank A large, usually rectangular container that is used to store liquids or gases.

tanking A waterproof membrane used to prevent groundwater penetrating into a basement. Can be applied to the inside or the outside of the walls and floor.

tap 1. A valve connected to the end of a pipe that is used to control the flow of a liquid or gas. **2.** The process where a connection is made to a water, gas, or electrical main in order to draw from it. **3.** A tool used to produce an internal screw thread on an existing hole.

tape 1. *See* JOINT TAPE. **2.** *See* MEASURING TAPE.

taper 1. To narrow gradually. **2.** A duct or pipe that gradually reduces in size.

taper bend A duct or pipe bend that gradually reduces in size along the bend.

tapered tread A *stair tread that reduces in size along its length.

taper thread A screw thread that gradually reduces in size along its length.

tap holder 1. A bracket used to support a tap. **2.** *See* TAP WRENCH.

taphole A hole in a component for a tap.

tap-off unit An electrical device that is connected to a busway run to provide power to a branch connection.

tap wrench Tool used to cut (tap) a thread. The tool often has a T-shaped handle to ensure even pressure is applied while cutting the thread.

tare weight The weight of something unloaded, for example the unladen weight of a lorry.

target-price contract A contract where the cost of the project is estimated and a price is fixed which the contractor aims for. Any savings made on the 'target price' normally entitle the contractor to a percentage bonus. This contract is different from *fixed-price and reimbursement contracts.

tarmac (tarmacadam) A composite material that is used for surfacing roads. It consists of *macadam (uniform broken stone) and either tar or tar plus *bitumen as the binder.

task An item of work, activity, or an event that has been planned, and normally forms part of a network of events (tasks) that are linked together.

taut Stretched or pulled tight.

T-beam A structural steel section that has been manufactured with a cross-section in the shape of a 'T'.

TBM A *temporary benchmark.

technical assistant An assistant to a quantity surveyor who does *taking-off, or an assistant to an architect who does the technical detailing.

technology The application of tools, methods, and technical knowledge to solve practical problems.

tectonic The large-scale movement and forces that occur within the *lithosphere that are associated with the deformation of the earth's crust; *see* EARTHQUAKE.

tee A T-shaped duct or pipe fitting. *See also* STRAIGHT TEE and SWEEP TEE.

tee bar 1. A T-shaped metal bar. **2.** A type of suspended ceiling that uses T-shaped bars to support ceiling panels.

tee-hinge A T-shaped hinge.

teeming 1. Very heavy rain. **2.** To pour molten metal.

tegula A flat rectangular roof tile that has two upturned edges running along its longest sides.

telemetry 1. To measure or gather data from something remotely. **2.** To measure distance by using a *tellurometer.

telescope An optical device for viewing objects at a distance. Used in surveying instruments such as a *level or a *theodolite.

telescopic boom A hydraulically operated crane *jib that extents outwards and folds back into itself, and as such, is typically used on mobile cranes.

telescopic centring Adjustable *formwork.

telescopic prop An adjustable supporting strut. The adjustable strut can be used to provide support for concrete formwork during casting and curing. Where the vertical support for walls, roofs, and floors has to be removed and replaced, and where it is necessary to provide temporary support, adjustable props can be used. When undertaking remediation, remedial, and other temporary works where it is necessary to use temporary supports, adjustable supports are quick to install to support *pins and beams.

teleseismic Relating to an *earthquake that is at least 1000 km away from the measurement site.

Telford pavement A road pavement that has no *binder and is formed from a layer of large stones laid on the *subgrade, with smaller stones place on top of this, which are hard-rolled to form a smooth surface.

telltale A building survey device used to check for or measure the movement of cracks in existing buildings. The monitoring devices, which are often made out of glass, can be fixed over a crack to check for movement. The ends of the glass strip are securely fixed, using epoxy resin, to the wall; if the crack increases, the glass will break. The movement can be measured by measuring the break in the glass. Modern telltales may be made of two strips of Perspex* with an incremental scale marked on each strip. One strip is fixed to one side of the crack and the other strip is fixed to the other side with the scales overlaying each other. If the crack moves, the degree of movement can be measured on the scales.

tellurometer A surveying device to undertake *electronic distance measurement.

temperature A measure of the how hot or cold a material or element is in comparison to another.

temperature factor A metric used to assess the risk of *surface condensation or mould growth occurring with a particular construction detail.

temperature gradient The difference in temperature across a material. Materials with the highest thermal resistances will have the steepest temperature gradients. Materials with the lowest thermal resistances will have the shallowest temperature gradients. It is used to calculate the risk of interstitial condensation.

temperature movement *See* THERMAL MOVEMENT.

temperature rise The rise in the temperature of a material or component.

tempered glass A type of glass modified to exhibit better toughness. Also known as toughened glass, this type of glass is at least twice as strong as annealed glass. This type of glass is safe as it disintegrates into small fragments upon failure and is commonly utilized for many applications in buildings, i.e. doors, façades, and in bathrooms.

tempering Reducing the brittleness of steel by heating. When steel has been hardened by *annealing (heating and rapid cooling) it is often too hard and brittle for practical use. The steel is then heated to a specific heat, but below its hardening temperature, then it is held at that temperature and slowly allowed to cool. Cooling in the air reduces the stresses in the steel.

template Something that serves as a full-size pattern, for example to allow other things to be cut out to the same size and shape.

temporary Something that is not permanent; *see* TEMPORARY BENCHMARK, TEMPORARY PROTECTION, and TEMPORARY WORKS.

temporary benchmark (TBM) A horizontal platform that is established to provide a known reference point with a set level. Once the level is established, the temporary reference point can be used to check and fix other levels around the site. Using *automatic levels, the temporary benchmark can check ground levels, excavation levels, and heights of structures etc. The height of the temporary benchmark is established by calculating its level in relation to an *ordnance benchmark, which is a permanent establishment with a known level.

temporary protection Resilient materials, such as wood, cardboard, and bubble wrap, that are used to protect the finishes of fittings that have been installed while building work is still ongoing. Staircases are often installed early during the construction programme to allow safe movement between levels; however, the treads must be protected by protective wooden slats to ensure that the impact of safety boots and other traffic does not damage the staircase.

temporary works Additional structural work that is needed to secure, stabilize, make safe, or provide access to the works during the construction or refurbishment stage. Examples include scaffolding, façade retention, earth works support, shoring and propping, hoarding, concrete crane bases, and platforms. Temporary works are priced under the preliminary items in the *bills of quantities.

tenacity The tensile failure strength of a fibre, yarn, textile, or *geosynthetic product. In the textile industry, the cross-section of the material is normally not uniform or solid, for instance a fibre may be hollow and taper toward the ends. Thus, its cross-sectional area is hard to define and its stress value (load/area) is impossible to calculate, so tenacity is often used. Tenacity has a unit of N/tex. Tenacity refers to the value at the breaking load condition and specific stress (N/tex) is used for values up to the breaking load.

tender (bid) An offer submitted by a contractor to undertake the works for a set price or agreed remuneration. The offer is normally competitive where the contractor competes against other contractors for the work. While the tender establishes cost, it is not normally the only factor that is considered when the client evaluates the bid. Selection may be done on company reputation, financial security, quality, and timeframe for completing the works.

tendering (bidding) The process or act of sending out requests for tenders, supported with drawings, bills of quantities, and performance criteria that allow the contractor to estimate and price the works. All of the *tenders are returned before a set date and evaluated in terms of price, time, and quality of work. Company reputation is important for establishing the potential quality of the work that it could deliver.

tendon A high-tensile steel bar, cable, or rope that is stressed. Used to introduce a prestress to concrete, in order to carry and transfer loads.

tenement A dwelling comprising several *apartments or flats.

tenon A rectangular projecting tongue at the end of a component that is shaped to fit into a mortise to form a joint.

tenon saw A small hand saw with a stiffening rib that runs along the top edge of the saw, the strengthening rib makes the saw easier to control enabling more accurate and precise cutting.

tensegrity The property of a structure that contains a collection of continuous tensile members that hold discontinuous compression members together.

tensile Relating to tension; being stretched or pulled.

tensile strength (ultimate tensile strength, UTS) The maximum stress a material can endure under a resultant tensile force/stress. Also usually known as ultimate tensile strength (UTS). If a material is put into tension, this is the stress that must be exceeded to cause tensile fracture. Tensile strength is calculated as:

$$\frac{\text{Tensile Strength (UTS)}}{(\text{N}/\text{mm}^2)} = \frac{\text{Tensile Load to cause Fracture (N)}}{\text{Cross-sectional area (mm}^2)}$$

Units are either N/mm^2 or MPa.

tension A force that pulls or stretches something.

tensor A mathematical entity defined with respect to a particular coordinate system, but capable of undergoing a transformation to other coordinate systems. It is the generalization of a *vector.

terminal 1. A device used to connect electrical cables to an electrical appliance. **2.** An *air terminal unit. **3.** The end of a duct, pipe, flue, or lightning conductor.

terminal velocity In fluid dynamics, the highest velocity of an object falling vertically downwards in a fluid that is achieved when the drag force and buoyancy acting vertically upwards become equal to the force of gravity pulling it down. Once terminal velocity is achieved, the object moves downwards at a constant speed, i.e. zero acceleration.

termination The end of a contract of employment.

termite (white ant) A wood-eating insect found in tropical and warm temperate regions of the world. Properties where termites exist need to be protected with *termite shields.

termite shield (ant cap) Metal sheet that is laid along foundations or within the foundation wall to prevent termites building mud tunnels up to any of the building's timbers. The cap consists of a long metal sheet with edges turned down at 45° preventing *termites climbing the shield. The sheet is laid within the mortar bed of a wall or on top of foundations.

terrace 1. Raised area where the ground has been levelled to create a platform. **2.** Flat roof designed so that it can act as a floor accommodating people and light traffic.

terrace blind *See* AWNING.

terrace house Row of houses positioned side-by-side sharing the intermediate *party walls that separate each dwelling.

terrain A stretch of land viewed in terms of its surface or physical features.

terrestrial Relating to the earth or land rather than from other planets or the sea.

terrigenous sediments *Sediments that were originally formed on land but now form part of the sea bed.

tertiary treatment The final stage in a water-treatment plant, occurring after *secondary treatment, that normally includes a *disinfection stage, such as *chlorination, *ultraviolet light, or *ozone oxidation.

tessellated An area filled with shapes or tiles, with no gaps between the tiles. The tiles need not be square but are normally the same shape, fitting together with no gaps, and filling the area. A mosaic wall or floor may also be considered tessellated.

tessera (*plural* **tesserae)** Cubes used in mosaic work.

test To evaluate something (or someone) by examining its (their) performance under certain conditions; for instance, to determine the tensile strength of a piece of steel, or to assess the knowledge of a person by their ability to answer correctly a series of questions on a particular subject.

testing machine Any machine that is capable of introducing a load to a test sample to determine its physical response, e.g. compression or tensile strength.

tetrahedron A triangular solid that has four faces.

tetrapod Something that has four limbs or legs; for instance, a vertebrate having four legs, a tripod that has four arms at 120° to each other, or a four-legged building block that is used as armour units on breakwaters.

tex The mass in grammes for one km of the fibre/yarn. Also known as linear density. *See also* TENACITY.

textile A fabric structure produced by a woven, nonwoven, or knitted process. When used in intimate association with the earth (*geo), it is referred to as a *geotextile.

thalweg A line connecting the deepest part of a river or valley.

thaumasite A silicate mineral material. Thaumasite has a chemical composition $Ca_3Si(CO_3)(SO_4)(OH)_6 \cdot 12(H_2O)$ and different possible appearances from colourless to white. Thaumasite occurs with zeolites, apophyllite, analcime, calcite, gypsum, and pyrite; and it can also be formed along with other calcium silicate hydrates during cement alteration, especially when sulphate attack develops.

THD (total harmonic distortion) The level differences in harmonic frequencies between the output signal of the amplifier and the input signal. It is the ratio of the sum of the powers of all harmonic frequencies to the power of the fundamental frequency.

theatre A building made to accommodate musical, dramatic, and theatrical productions.

theodolite An optical surveying instrument used for measuring vertical and horizontal angles.

theory The use of scientific principle(s) to explain something.

thermae A Roman spa or public baths.

thermal Relating to or involving heat.

thermal break A material with a low thermal conductivity that is placed within a component to minimize conductive heat loss. For instance, the insertion of polyurethane as a spacer bar between the panes of multiple glazed units.

thermal bridge (cold bridge, heat bridge) An area of the building fabric that has a higher thermal transmission than the surrounding parts of the fabric, resulting in a reduction in the overall thermal insulation of the structure. It occurs when materials that have a much higher thermal conductivity than the surrounding material (i.e. they are poorer thermal insulators) penetrate the thermal envelope or where there are discontinuities in the thermal envelope. Heat then flows through the path created—the path of least resistance—from the warm space (inside) to the cold space (outside). The higher thermal transmission of this part of the fabric results in a reduction in the thermal performance (an increase in U-value) as heat flows through the fabric, and the surfaces of the interior side of the bridge become cooler. The use of the term 'thermal bridge' is somewhat misleading as it implies that the thermal envelope must be 'bridged' in some way for a thermal bridge to occur; this is, in fact, not the case. Thermal bridges can occur in unbridged construction where discontinuities exist in the thermal envelope.

thermal bypass Heat transfer that occurs between the inside and outside of a building that bypasses the *thermal envelope. For example, air that is allowed to move through, around, and between the insulation, in effect bypassing the benefit of the insulation.

thermal camera *See* INFRARED CAMERA.

thermal capacity (heat capacity, thermal inertia, thermal mass) A measure of the ability of a material or structure to store heat. Measured in J/K. All buildings have a thermal capacity that is determined by the quantity and types of materials that have been used to construct the building. The thermal capacity of a building is important as it not only determines how stable the internal temperatures will be, it also determines the time taken for the building to heat up and/or cool down. Buildings with a high thermal capacity will have a high thermal stability, while those with a low thermal capacity will have a low thermal stability. Buildings with a low thermal capacity will respond rapidly to control and will warm up quickly, they will respond more quickly to surrounding temperature changes, and they may overheat in the summer. Buildings with a high thermal capacity will require long preheating times, will respond slowly to surrounding temperature changes, and will cool down slowly.

thermal comfort A subjective state of well-being that exists when the occupants of a building are unaware of their surroundings. Thermal comfort is influenced by a number of individual and environmental factors. Individual factors include activity, age, and clothing while environmental factors include air temperature, mean radiant temperature, humidity, and air movement.

thermal comfort meter A device that measures a range of parameters used to calculate various *thermal comfort indices.

thermal conductivity (heat-transmission value, k-value) The amount of heat flow in watts through a material, which has a thickness of 1 m and a surface area of 1 m^2, for each one degree of temperature difference between the inside and outside surfaces. Measured in W/mK, the thermal conductivity of a material is a property of the material, and is independent of the thickness of the material. Materials with a high thermal conductivity, such as metals, are good conductors of heat and poor insulators, while materials with low thermal conductivities, such as insulation materials, are poor conductors of heat and good insulators.

thermal degradation of timber A degradation mechanism for timber material. When timber undergoes prolonged exposure to elevated temperatures, there is a reduction in both strength and toughness. There is some uncertainty about what minimum temperature will degrade timber, in some cases temperatures as low as 60°C can induce degradation over many years of exposure. The rate of degradation will rise markedly with an increase in temperature and time of exposure; hardwoods appear to be more susceptible to thermal degradation than softwoods. Thermally degraded timber develops a brown colour and breaks easily with a brittle failure.

thermal diode A device where heat can flow more easily in one direction than the other.

thermal envelope Elements of the building *envelope that act as the main barrier to heat loss.

thermal expansion The process where a material or component changes its size (area, length, or volume) due to an increase in temperature.

thermal fusion joint (heat fusion joint) A joint between two components that is created by applying heat.

thermal insulation 1. A material that has a low thermal conductivity, thus reducing the amount of heat that will flow through it. **2.** The process of applying insulation material to a building to reduce its heat loss.

thermal loop The movement of air or liquid within an enclosed space where warm, less dense air or liquid rises at one side and cooler, denser air or liquid descends at the other. This results in a loop of constantly moving air or liquid.

thermally broken A term used to describe a component that incorporates thermal breaks.

thermal movement (temperature movement) The expansion or contraction of a material due to a change in temperature.

thermal resistance (R-value) A measure of a material's resistance to heat transfer. It is calculated by dividing the thickness of the material by its *thermal conductivity. The thicker the material, the greater the thermal resistance. The lower the thermal conductivity of a material, the greater the thermal resistance. All materials present a resistance to the flow of heat through the material. Measured in m^2K/W.

thermal resistivity The inverse of thermal conductivity.

thermal shock Failure of a material due to a sudden increase in temperature. This is highly prevalent in glasses and ceramics, which have a very poor resistance to thermal shock; conversely, porous materials and metals have good resistance. In general materials with low toughness are susceptible to this type of failure.

thermal stress breakage The breakage of glass due to tensile stresses. The stresses result from temperature gradients across the glass, which occur when the glass is non-uniformly heated. If the tensile stresses exceed the edge strength of the glass, thermal breakage occurs.

thermal transmittance (U-value) The rate of heat flow in watts, through 1 m^2 of an element, when there is a temperature difference across the element of 1°C (or K). Measured in W/m^2K.

thermal wheel (heat-recovery wheel) A type of *heat exchanger that comprises a large rotating wheel positioned in between the supply and extract ducts in an air-conditioning or ventilation system.

thermic boring A method that uses an *oxyacetylene flame at the end of a lance to cut through steel or concrete.

thermistor A device for measuring temperature—when the temperature rises the electrical resistance of the device falls.

thermit welding A chemical heat reaction welding process that uses aluminium and iron oxide; used to weld large sections such as railway lines.

thermocouple A device for measuring temperature. Two wires of different metals are joined to give a small potential difference, which is proportional to the temperature of something they touch.

thermodynamics The branch of physics that deals with energy conversion involving heat, pressure, volume, mechanical action, and work.

thermofusion welding The fusing of plastics by heating edges and holding the melted soft edges together.

thermographic camera *See* INFRARED CAMERA.

thermohygrograph A chart recorder that measures and records air temperature and relative humidity simultaneously.

thermometer A device that is used to measure temperature.

thermo-osmosis The passage of water in a porous medium due to a temperature differential.

thermoplastics A category or group of polymers that soften when heated and stiffen when cooled without enduring a physical or chemical change in property (a reversible process). As a result they are very ductile and workable, and they can be easily recycled. Structurally, thermoplastics consist of long carbon chains tangled together and intertwined, without any cross-linking. This accounts for their flexibility and resilience as a group. Examples include polyethylene, PVC, nylon (polyamide), polystyrene, and polycarbonate. Uses in construction include: polyethylene sheet for dpc and vapour barrier systems, uPVC window frames and doors, etc. They can deform elastically, yield, and can deform plastically to a considerable extent (nylon will give over 300% plastic strain). Before yielding they also show non-linear visco-elasticity, a form of behaviour that is partly elastic and partly viscous flow. This accounts for the high levels of *ductility as a group. Thermoplastics can easily be made into complex shapes to make products such as window frame sections, waste pipe systems, water cisterns, and simpler things such as polyethylene sheeting, etc. Thermoplastics are often supplied in the form of small pellets of material, which can be fed into either a moulding or an extrusion process. All polymers can be classified as being either a thermoplastic, a *thermoset, or an *elastomer, according to their properties.

thermosets A group or category of polymers that do not soften when heated to an elevated temperature; instead, they disintegrate with a substantial decrease in the mechanical properties and dimensional stability. The molecular structure of thermosetting polymers is heavily cross-linked. This cross-linking gives them great rigidity, *hardness, and wear resistance. In general, nearly all thermosets are hard, wear-resistant, stiff (high elastic [Young's] modulus, E), and have no ductility. They tend to be strong and *brittle, and thus have to be shaped and polymerized in one operation. Once *polymerization has taken place no further shaping or alteration is possible. Thermosets lend themselves to the production of accurately moulded components such as switch and socket plates, boxes, etc. for electrical products. They are also used for kitchen surfaces. In practice, they are compression moulded, i.e. moulded and polymerized in one. All polymers can be classified as being either a *thermoplastic, a thermoset, or an *elastomer, according to their properties.

thermosetting *See* THERMOSETS.

thermosetting compound A compound that cures by forming a chemical reaction or as a result of the release of a solvent.

thermosiphon *See* GRAVITY CIRCULATION.

thermostat An electronic or mechanical device that is used to control the temperature of an appliance or a system.

thermostatic mixer A thermostatically controlled mixing *valve.

thicknessing machine (panel planer, thicknesser) Machine for reducing the thickness of wood to a set size, producing square, level, and flat surfaces.

thief-resistant lock A security *lock.

Thiessen polygon A graph that represents the rainfall intensity over a particular area from data collected from rainfall gauges over that area.

thimble *See* PIPE SLEEVE.

T-hinge (tee-hinge) A hinge that is T-shaped when opened.

third angle Drawing an object where the view of the top of the object is drawn from above, and the view from the left of the object is drawn to the left, the view from the right is drawn to the right, and the view from the rear is drawn to the extreme of the right perspective.

third fixing (second fixing) Any work carried out after plastering.

third rail An additional rail, either running alongside or between the existing track, which carries an electric current that powers the train that runs on the track.

thixotropy A particular type of viscous behaviour exhibited by some liquids. Thixotropic liquids undergo an increase in viscosity with time; however, upon agitation the viscosity decreases, for example in thin layer mortar and alkyd paints.

Thornthwaite's formula A method that can be used to calculate *evapo-transpiration from vegetation, or used in association with Penman's theory to estimate *evaporation.

thread The spiral-shaped ridge found on the outer face of a screw or bolt or the inner face of a nut.

threaded rod A long cylindrical bar that is threaded at both ends, and may even be threaded along its entire length. Has a variety of uses, such as a fastener, a hanger, or a tie rod.

three-coat work The application of three separate coats of plaster. The first coat is termed the *scratch coat, the second coat the browning coat, and the third the *finishing coat.

three-dimensional (3D) printing The production of a physical model from a computer image by laying down liquid molecules or powder grains which are fused together to form thin layers of material in succession. 3D printing is any of various processes in which material is joined or solidified under computer control to create a three-dimensional object with material being added together (such as liquid molecules or powder grains being fused together). 3D printing is used in both rapid prototyping and additive manufacturing. Objects can be of almost any shape or geometry and typically are produced using digital model data from a 3D model.

three-dimensional (3D) scanning A non-contact 3D laser scanning technology which uses pulse laser light to digitally capture the surface of an object in the form of *point clouds. As light only moves in a straight line, the object (such as a building) is rescanned from a number of different positions; each scan adds more data to the point cloud to improve the resolution of the scan. *See also* LIDAR.

three-hole basin A basin that incorporates three holes; one for the hot-water valve, one for the cold-water valve, and a one for the spout.

three-moment equation An equation that incorporates the internal bending moments in adjacent spans of a continuous elastic beam.

three-phase supply An AC electricity supply that comprises three alternating voltages that are spaced 120° apart. *See also* SINGLE-PHASE.

three pillars of sustainable development (triple bottom line) The formation of a balanced solution that encapsulates economic growth, environmental protection, and social advancement (i.e. profit, plant, and people).

three-pipe system A type of pipework system that has two separate flow supply pipes and a common return.

three-prong plug A plug that has three contacts; one for the live, one for the neutral, and one for the earth.

three-quarter bat A brick that has been reduced in length by a quarter.

three-quarter header A brick that has been reduced in width by a quarter.

three-quarter-round channel An open drainage channel that is three quarters round.

three-way valve (twin-port valve) A valve that is capable of directing the fluid or gas to one of two outlets.

threshold The *sill of an external door.

threshold strip A strip of material, usually metal, positioned under the closed leaf of a door that hides the joint between the floor finishes in different rooms.

throat 1. The narrow portion of a chimney situation between the *gathering and the *flue. **2.** A groove cut along the length of a sill to provide a *drip.

throated sill A sill that incorporates a *throat.

through lintel A lintel that spans the full thickness of an opening.

through stone A stone that penetrates the entire thickness of a wall.

through tenon A *tenon that penetrates through the entire thickness of the timber that contains the mortise.

throw The vertical displacement of strata along a geological *fault. An **up throw** indicates an upwards displacement and **down throw** denotes downwards displacement. *See also* NORMAL FAULT and REVERSE FAULT.

thrust A force causing something to be pushed away.

thumb latch A door latch that is operated by pressing the lever with your thumb.

thumb slide *See* SLIDE BOLT.

tidal Relating to or affected by *tide level.

tide The rise and fall of the sea by the gradational attraction of the moon and sun.

tie 1. Fasten, attach. **2.** A beam, rod, rope, or strap that fastens a material or components together. Designed to operate in tension.

tie beam A horizontal structural member used at the foot of rafters or between two walls to prevent them spreading under load.

tie wire *See* BINDING WIRE.

tight knot A *knot that is firmly fixed in a piece of timber.

tight sheathing (close sheathing, closed sheathing) Horizontal or vertical boards laid tight against one another. Used as a retaining wall in excavations.

tight size (full size, rebate size) The size of the *rebate.

tight tolerances Allowable deviation is less than usual, meaning that setting out, manufacture, and casting must be precise.

tile A thin regularly shaped piece of material used to cover or finish floors, walls, and roofs.

tile-and-a-half tile A roof tile that is the same length but one-and-a-half times the width of the other tiles used on the roof.

tile batten *See* TILING BATTEN.

tile clip An aluminium, plastic, or stainless steel clip used to secure roof tiles to the roof structure. Designed to prevent wind uplift and tile chatter. Three main types are available: eaves clips, tile-to-tile clips, and verge clips.

tile creasing *See* CREASING.

tiled valley A *valley created using *valley tiles.

tile fillet (tile listing) A tile whose upper edge is fully bedded in mortar at an abutment. Used instead of a cover *flashing.

tile hanging (vertical tiling, weather tiling) The process of fixing tiles to a wall.

tile listing *See* TILE FILLET.

tile peg (tile pin) A *peg used to secure a *peg tile.

tiler (tile layer) A tradesperson who lays *tiles.

tiling The process of applying *tiles.

tiling batten (roofing batten, tile batten) A *timber batten used to secure roof tiles to the roof. Usually 25 × 38 mm or 25 × 50 mm in size.

till *See* BOULDER CLAY.

tilt-and-turn window A window that can tilt inwards at the top to provide ventilation, and open inwards or outwards on a hinge.

tilting fillet (doubling piece, skew fillet) A piece of timber inserted at the eaves of a roof to raise the bottom row of tiles.

tilting level An optical surveying instrument used for *levelling. It is characterized by having a tilting telescope on a pivot; *see also* DUMPY LEVEL. The initial horizontal plane is obtained by using the levelling screw; the horizontal line of collimation is then sighted by finely adjusting the tilting screw on the telescope. *See also* AUTOMATIC LEVEL.

tiltmeter An instrument for measuring very small changes from the horizontal level of the ground—it can also be used in structures. Used primarily for monitoring volcanoes, but can be used to monitor changes due to excavation, tunnelling, dewatering, and so on.

TIM *See* TRANSPARENT INSULATION MATERIAL.

timber A natural and extensively utilized material extracted from trees. Timber has been used for centuries and it is still an important construction material today, because of its versatile properties, its diversity, and aesthetic qualities. At least one-third of all timber harvested goes into construction, the rest goes into paper production, is used as fuel, or is wasted during the logging process. The earth's forests, besides being a source of timber, also act as a carbon sink. That is, they absorb carbon dioxide from the atmosphere, and very importantly, breathe out the oxygen. Timber is also a renewable resource, and so can make an important contribution towards the achievement of sustainable building. Because of the way that timber grows, the material has different properties in different directions, thus timber is anisotropic. Some species of timber (often the hardwoods) display good resistance to degradation, while others have little resistance (many softwoods). The agencies of degradation are biological, chemical, photochemical, thermal, fire, and mechanical; however, timber, especially in the UK, is most susceptible to insect and fungal attack.

timber-frame construction The use of timber studs, rails, and sheathing to produce a load-bearing structural frame. Can be site-built or, more commonly, prefabricated in a factory. There are two main types of timber-frame construction; platform frame and balloon frame.

timber-framed building A building that has the main structural support system constructed of timber, rather than the inter-leaf of a cavity wall in conventional UK brick-built houses. In the UK, brickwork is most often used as the external cladding to the timber structure.

timbering 1. The use of timber. **2.** Timber formwork.

Timber Research and Development Association (TRADA) Established in 1934 to protect the use of timber within the construction industry and progress research and development.

((⊕)) SEE WEB LINKS

• Membership-based organization that conducts research and provides information to its members on timber. Students can obtain free membership.

time-and-a-half An overtime rate that pays 50% more for the equivalent period of normal work.

time distribution (time system, master clock system) Network or system that ensures that every clock in the building or organization works and is constantly synchronized with the master clock.

time for completion (construction time, contract time) The period allowed for the construction of the building and site operations. The project usually has a start date and finish date, which are stated in the contract.

timekeeper Sometimes used to describe the clerk of works, especially if this person is monitoring working hours and breaks.

timer An electrical device used to switch something on or off after a period of time, e.g. a switch on lights to delay the lights turning off in stairways and corridors, or a unit on a cooker that sets the start and finish times of cooking.

timescale The x-axis on a programme where the days, weeks, months, or years are represented in a linear scale. Vertical lines corresponding to the dates allow progress of activities to be easily checked and monitored.

tin A non-ferrous metal with symbol Sn. Tin is alloyed with copper to make bronze, which is commonly used in the construction industry; furthermore, it is used in *solders. It has excellent resistance to corrosion, thus can be used to coat other metals to provide corrosion resistance.

tine A prong or pointed projections such as teeth on a excavator's bucket.

tingle A strip of metal used to secure edges of sheet-metal at drips, rolls, and seams.

tin snips Scissors-like tool used for cutting thin metal.

tinted glass (window tint) A thin film applied to the interior part of windows to reduce the amount of ultraviolet (UV), infrared, and visible light passing through.

tinting See TINTED GLASS.

tip 1. The sharpened end of an object, such as a nail. **2.** The process of dumping something, such as rubbish. **3.** To tilt at an angle.

titan crane A large crane mounted on a *portal frame, having a *jib that can rotate around its vertical axis.

titanium A non-ferrous metal, symbol Ti. Titanium is a strong, lightweight metal with excellent resistance to corrosion. It has similar tensile strength to steel yet is 45% lighter. Due to the cost, its application is limited to the aerospace industry, otherwise it could replace steel in nearly all applications in civil engineering.

titanium dioxide Also known as titania or titanium (IV) oxide, chemical formula TiO_2. Titanium dioxide is usually used as a pigment called titanium white. This pigment is particularly used to both colour polymers white, and more importantly, protect polymers from degradation due to UV sunlight.

title block The bordered space within a drawing to add information and text about the drawing. The information contained within the title block usually states the drawing name, the drawing number, date created or updated, scale, and the organization responsible for creating the drawing.

TNT (trinitrotoluene) A compound used as explosives and bombs, particularly for military applications. It has the formula $C_6H_2(NO_2)_3CH_3$, is a solid yellow material with a melting point at 80°C, well below the point at which it will spontaneously detonate, thus the material can be safely poured and combined with other explosives.

toe 1. The portion of a lintel that projects beyond the head of a window or door. **2.** The lowest part of a shutting *stile. *See also* HEEL. **3.** The lower end of something, e.g. the toe of an embankment where the slide slopes meet the ground.

toe board (guard board) A board placed around the outer edge of scaffolding or a roof to prevent equipment, tools, and people from falling over the edge.

toe nailing *See* SKEWNAIL.

toe recess The recess at the base of a kitchen unit.

toggle bolt A type of fastener used to fasten heavy objects to plasterboard walls. Comprises a spring-loaded toggle that is attached to a threaded bolt.

toggle mechanism A linked hinged section of a device that can be used to apply a large pressure onto something from a small amount of force; used in jaw crushers.

toggle switch *See* TUMBLER SWITCH.

toilet A *water closet (WC).

tolerance (permissible deviation) The discrepancy allowed between an exact location or fit, and one that deviates slightly, but is still acceptable and functions. When setting out, cutting, manufacturing, and fitting, it is normal to attempt to obtain total accuracy but, in practice, the process often results in slight variation. As long as the variation is within the acceptable tolerance, then functionality will still be achieved.

toll 1. The fee for driving across a particular road or bridge. **2.** The cost or damage sustained by something, as a consequence of a disaster, for instance, the toll being the number of people killed, or the financial cost to repair property and infrastructure damaged.

tomb A burial chamber or grave, which may include a monument or headstone.

tombolo A narrow strip of sand, spit, or bar that links islands together or to the mainland.

ton (long ton) An imperial unit of weight that is equal to 1016 kg (2,240 lb) in the UK or 907 kg (2000 lb) in the US. Superseded in the UK by the metric equivalent *tonne.

tong tester An instrument for measuring electrical current without the need to disconnect the circuit. The jaws of the instrument are placed around one wire, the instrument then measures the electromagnetic force or field that is created.

tongue A long, narrow piece of timber that projects from the side of a board.

tongue-and-groove (tongued and grooved joint) A type of joint formed by inserting the *tongue of a board into the corresponding *groove on another board.

tonne (t) A metric ton is a measurement of mass equal to 1000 kg or 2204.6 lb. The unit is approximately the weight of 1 cubic metre of water; *see also* TON.

tool (hand tool) Any small tool that can be used for forming, shaping, making holes, texturing, and setting out that is not machine-operated; e.g. a hammer, saw, or screwdriver.

tooled joint A mortar joint that has been finished or shaped by a pointing tool or trowel. The joint is finished and shaped as the work proceeds. Typical joints include: flush, tooled, recessed, keyed, and weathered. All joints, regardless of finish that have been shaped by a pointing tool are tooled joints, but the joint that is specifically a 'tooled' joint has a semicircular finish that can be formed by the rounded handle of a pointing tool, also called a bucket-handle joint.

tooling The process of forming and finishing mortar joints in brick and blockwork.

tool pad (pad, tool holder) A combination tool capable of holding various fittings and tools, such as saws, awls, screwdrivers, and sockets.

toother A stretcher brick that projects at the end of a wall providing a projection that can be used for bonding subsequent walls.

toothing (indenting) Alternate courses are left projecting from the end of a wall so that brick or blockwork can be bonded directly to the wall. Indenting is normally the reverse of toothing, where bricks (half-batts or headers) are left out or cut out to allow new brickwork to be inset and bonded to the wall.

toothplate connector (bulldog plate connector) A steel nail ring, used for bonding or securing two pieces of timber together. The steel ring has sharp teeth that project from both sides of the circular plate. When sandwiched and clamped between two pieces of timber a mechanical bond is formed.

top-course tiles See UNDER-RIDGE TILES.

top-down construction (downwards construction) A method of construction that enables the basement and the superstructure to be constructed simultaneously.

top-hung window A window that comprises a horizontally hinged openable *sash at the top.

top lighting The process of lighting an object from above.

topographic map A map showing the surface profile of the ground with changes in altitude being represented by *contour lines.

topography Relating to the surface characteristics of the ground, particularly mapping both natural and artificial surface features, such as mountains, rivers, roads, and railways.

topping 1. The final layer applied to the top of a screed or concrete floor. **2.** The process of applying the top finish to a screed or concrete floor.

topping out The completion of the building marked by the ceremonial laying of a stone or brick, the planting of a tree, or the unveiling of a plaque. Dignitaries are often invited to help mark the topping out ceremony.

topple To fall forwards or overbalance.

top rail The horizontal highest rail of a door or a *sash.

topsoil The fertile upper (surface) layer of soil that supports vegetation.

torbar A deformed and twisted bar.

torch A gas burner used for heating and working materials, and bringing them to the point of melting. Often used on flat roofs for softening asphalts and bituminous-based materials so that they can be bonded together.

torching (tiering) A combination of mortar and hair that was traditionally troweled into the gaps between the tiles from inside the roof to give additional weather protection. However mortar retains moisture, encouraging rot attack in the battens and can be easily dislodged by movement of the roof timbers. Nowadays, an underlay is used.

tornado A violent column of swirling wind, which is extremely destructive.

torque A rotating force that causes rotation.

torrent A surge of fast-flowing liquid, typically water, or water and debris.

torsion The action of twisting an object by applying an equal and opposite torque at either end.

torsional buckling A mode of behaviour caused by a twisting action about the longitudinal axis in slender compression members that have an unsymmetrical cross-section (e.g. channels and structural tees).

torsion constant (J) A geometrical property of an object's cross-section subject to torsion:

$$J = \frac{T}{G\theta}.$$

where, T = torque, G = modulus of rigidity and θ angle of twist.

total float The allowable slippage or delay of an activity that can occur without affecting subsequent tasks.

total going The sum of the horizontal length of a step; the length of the going is measured from the nose of one step to the nose of the next step.

total potential A concept used in structural analysis, which assumes a structure will displace to a lower position to give the minimal potential energy. The concept can be explained by a ball moving (displacing) to the bottom of a hill when placed halfway up the hill to maintain equilibrium.

Total Productive Maintenance (TPM) A system of maintaining and improving the integrity of production and quality systems through the machines, equipment, processes, and employees that can add business value to an organization. TPM focuses on keeping all equipment in top working condition and on avoiding down time and delays in manufacturing processes.

Total Quality Management (TQM) An organization-wide philosophy which has core values in continual improvement of the quality of its products and services, and the quality of its processes, to meet and exceed customer expectations. This means that everyone in the organization—from top management to employees—plays a role in providing quality products and services to customers. Even suppliers and customers themselves are expected to be part of the TQM process.

total reward A strategic and integrated reward practice that incorporates a holistic consideration of pay and financial benefits together with non-financial benefits, such as recognition, development, and self-actualization, work-life balance, the work environment, and work itself.

total rise The vertical rise from the foot of the stair to the top of the stair, also the sum of all the stairs.

total solar irradiance (G) The total amount of electromagnetic radiation received from the sun per unit area, measured in W/m^2. The product of *direct solar radiation and *diffuse solar radiation.

total stress (σ) The *stress acting on or in a soil mass due to the weight about it (e.g. from the overlying soil, surcharge, etc.). It is the stress carried by the soil skeleton, termed effective stress (σ'), plus the pore pressure (u); $\sigma = \sigma' + u$.

touch dry The point at which a painted surface is sufficiently dry such that it is no longer *tacky to the touch, and the application of light pressure to the surface does not leave any marks. The paint will not be completely dry.

touchless controls Electronic controls that operate automatically using presence detectors. Used to open and close doors, switch lights on and off, switch taps on and off, and in flush urinals.

touch up The process of repainting small missed or damaged areas of paintwork.

toughened *See* TEMPERED GLASS.

toughness A material property indicating the energy required to fail a material; in other words how much energy a material can absorb up to fracture. In general ductile materials tend to be the toughest. The area under a stress–strain plot for a material is a way of measuring toughness.

tough way The hardest direction there is to break out or extract rock; *see also* RIFT.

tourelle A tower that is circular on plan and can have bricks or stone that project out from the main wall to form a corbelled turret.

tower A tall structure, usually slender and higher than surrounding buildings.

tower bolt A large *barrel bolt.

tower crane A tall steel-truss framed structure with a lifting jib mounted on top of the tower. Particularly tall cranes are often tied to the building structure. The lift shaft of a building is often constructed first and provides a rigid stable structure against which the tower crane can be tied. The rotating jib of the crane may be fixed horizontally, or in the case of luffing jib cranes, can rotate up and down assisting lifting. Luffing jib cranes can also be used where the boundary of the site is tight and where permission has not been granted to over-sail surrounding property. Where over-sailing rights have not been granted, the luffing jib is raised to avoid over-sailing the property. On most tower cranes counterweights are used at the back of the jib.

tower scaffold A prefabricated aluminium and steel frame scaffold system. The system is easy to assemble with quick-fit *braces, *ledges, and *outriggers.

town A collection of dwellings larger than a village and smaller than a city.

town gas (coal gas) Gas made by burning coal. Superseded by natural gas.

town house A terraced house close to the centre of a city or large town. Traditionally, townhouses were expensive, large, and tall terraced houses. As land in towns was expensive, the properties were often built with three or more storeys. The house was called the town house as the rich often had a house in the town and one in the country.

town planning Legislation exercised under the Town and Country Planning Act to control and regulate development.

(()) SEE WEB LINKS

• The UK's online planning and building regulations resource. Full building regulations and planning documents can be accessed.

toxaphene An insecticide that is toxic to freshwater and marine aquatic life.

toxic Of or relating to or caused by a *toxin or poison. Any material or object suffering from exposure to toxic substances.

toxin A poisonous substance produced by a living organism.

trace heating (cable heating) The use of an electric element to provide sufficient heat such that the required temperature can be maintained in pipes and vessels.

tracer gas A gas used to undertake a test using the *tracer gas decay method. Common tracer gases include carbon dioxide, freon, helium, nitrous oxide, and sulphur hexafluoride.

tracer gas decay method A non-destructive technique that can be used to measure the in-situ *infiltration or ventilation rate of a space or building. *See also* CONCENTRATION DECAY METHOD; CONSTANT CONCENTRATION DECAY METHOD; CONSTANT INJECTION DECAY METHOD.

tracer gas technique *See* TRACER GAS DECAY METHOD.

tracery (warning tape, marking sand, marker tape) Material laid above buried services used to alert people working in the area that services are positioned under the marker.

track 1. A mark left by a person, animal, or vehicle travelling over soft ground, such as a footprint. **2.** A path or road. **3.** A set of parallel rails upon which a train or tram travels along. **4.** A particular course of action during an investigation.

traction 1. The amount of friction that allows movement to occur. **2.** The action of pulling something along a surface.

tractive force The force available from a vehicle to pull something.

tractor A vehicle used for pulling heavy loads.

TRADA *See* TIMBER RESEARCH AND DEVELOPMENT ASSOCIATION.

trade Work undertaken by craftspeople or professionals rather the general public. A tradesperson describes a skilled person connected to the construction industry.

trade collection A summary of all of the trade work contained within the bills of quantities. The quantity of work contained for each different trade can be summarized.

trade foreman A *Chargehand or group leader who supervises the work of trades such as bricklayers, joiners, and ground workers.

tradeless activity A job that does not require any particular skill and could be undertaken by any tradesperson or worker.

trade-off The relaxation of one regulatory requirement or contractual condition subject to it being met by other agreed means.

trade rubbish Waste materials collected for disposal from developments, construction sites, manufacturing, commercial properties, industry, and other non-domestic activities.

tradesperson (craftsperson) A trained and skilled worker who can undertake work within his or her field with minimum supervision. Recognized trades within construction include electricians, joiners, bricklayers, ground workers, plumbers, decorators, and plasterers.

trade union An organization of people who work in a particular field or industry, that form to represent the common good of the people within the organization. Such organizations gain their power through the number of members signed up and the potential or actual action they take.

traffic The movement of vehicles, goods, and information along a particular route—cars along a road, the illegal trade of drugs from one country to another, the transfer of data along a computer network, and so on.

trafficable roof A roof that is designed to be walked on. *See also* NON-TRAFFICABLE ROOF.

trailing dredge A *dredge that contains a suction pump or cutter to suck up water and sediment as the dredge moves along.

training The provision of opportunities for learning off the job and/or on the job through specific activities that focus on skills provision, or upskilling. Training can also include longer-term and broader development of skills and knowledge, for example through education, work experience, experiential learning, or *apprenticeships.

training wall A wall to divert or contain the flow of a river.

trammel 1. An instruction used to draw ellipses. **2.** A device to check curvature.

tramway A light railway system that uses trams.

transcurrent fault *See* WRENCH FAULT.

transducer A device that converts a physical quantity such as displacement, load, temperature, etc. into another form, such as an electrical signal, and in turn can be recoded by a data logger.

transferred responsibility Duty changed from one person to another by agreement.

transformation piece 1. A connection or adaptor that serves to take one pipe or duct from one size, shape, or fitting to another, for example, from a rectangular duct to a square duct.

transformer A device that changes the voltage, current, phase, or impedance of an alternating current.

transform fault A *fault that is characterized by predominant horizontal displacement occurring where two *tectonic plates slide past each other; typically occurs along the ocean floor presenting a zigzag plate boundary.

transgranular In metals at a microscopic level, any chemical reaction or diffusion that takes place through the grains within the metal. Sometimes also referred to as transcrystalline.

transient actions An are *actions that is applied for a period of time which is much shorter than those of the *design life, such as during a construction or repair phase.

transit 1. US term for a *theodolite. **2.** A change of face on a theodolite. **3.** Travel across something or a route across a particular area. **4.** The passage of a star or planet across the sun or meridian viewed from the earth.

transition 1. The period or activity of change between one operation and another. **2.** Fittings or materials used to change one material or component to another. May also describe the fitting used to change the size or shape of a pipe or duct to another, *see* TRANSFORMATION PIECE.

transition temperature (metals) A temperature for metals where a change in property takes place, e.g. ductile to brittle transition temperature.

transit method A method to distribute the closing error on a closed traverse.

translational The movement along an axis, e.g. a **translational slide** has a failure plane parallel to the surface of the ground at a shallow depth.

translucent An object that allows light to pass through it; however, without optimal clarity, not fully transparent.

transmission The movement of heat, light, or sound from one place to another.

transmission heat-transfer coefficient (heat loss coefficient, HLC) The rate of heat flow that occurs due to the transmission of thermal energy through the building fabric, divided by the temperature difference between the inside and outside of the building envelope. Measured in W/K, it is the total heat-transfer rate from a building resulting from heat transfer through the envelope (UA-value), including the heat transferred or lost from infiltration and/or ventilation (Cv-value) per °C of indoor to outdoor temperature difference in W/°C. If UA-value and Cv value can be considered constant, then HTC can be estimated as HTC = UA + Cv. It can be estimated by means of design data of the building envelope composition and infiltration/ventilation design data. Alternatively, it can be estimated experimentally by means of the co-heating method.

transmissivity 1. A measure of a material's ability to transmit flow, for example the flow of water through an aquifer. **2.** The volumetric flow of water per unit width of a *geosynthetic per unit gradient in a direction parallel to the plane of the product, as defined in BS 6906-7 1990.

transmittance 1. The amount of radiation (visible, UV, total solar, etc.) that passes through a material or construction. In the case of glazing, it is stated as the percentage of radiation **2.** Thermal transmittance.

transmitter A device to generate and transmit radio-frequency waves, for example, broadcasting equipment.

transom The horizontal member that subdivides a *window frame.

transom window (transom light) A small rectangular *window located above a door. May be fixed or horizontally hinged.

transparent See-through, clear; having the property that light passes through it almost undisturbed, such that one can see through it clearly.

transparent insulation material (TIM) A material that has high solar radiation, high light transmission, and high thermal insulation properties. It is often used to improve the performance of windows.

transport Carrying someone or something from one place to another.

Transport Research Laboratory (TRL) An independent company which undertakes consultancy and research work related to transport systems to produce safer, cleaner, and easier transport. For example, the *split cycle offset optimization technique (SCOOT) and the *pedestrian crossing were both developed by the TRL.

transverse Crosswise across something or at right angles to something.

trap A section of pipe located below a drain that is designed to provide a water seal, thus preventing the flow of foul air back up through the drain. Can be J-, P-, S-, or U-shaped.

trapezium A shape with four sides (quadrilateral) that has two parallel sides.

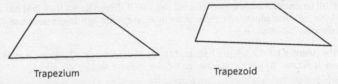

Trapezium   Trapezoid

Trapezium and trapezoid

trapezoid A quadrilateral that has no parallel sides

trapezoidal rule An approximate technique for calculating the definite integral. It is, however, less accurate than *Simpson's rule.

trapezoid tear test A method to determine the tear strength of a *geosynthetic.

trapped waste (waste trap) A trap connected to the waste pipe of an appliance.

trap vent A *branch vent.

travel The allowed movement of an object, such as the distance that a sliding door moves.

traveller 1. In surveying and setting out, a traveller is a T-shaped implement that can be sighted and levelled between two profiles of known height. The profiles are set with horizontal timbers at specific levels to provide a horizontal or sloping sight line. The top of the T-shaped traveller is then sighted so that it is in line with the top of the other profiles. The T-shaped traveller is set at a length so that when it is lined-in, the bottom of the T marks the depth of dig or position of the top of a pipe. **2.** A piece of equipment that moves backwards and forwards along a groove or rod.

traverse A type of survey formed by a series of straight lines and horizontal angles between them. A closed traverse forms a closed loop whereas an open traverse does not.

tray (shower tray) A shallow open container used to collect the water from a shower.

tray aerator A method of aerating water by causing it to fall through a series of perforated trays.

tread The horizontal part of a stair where each foot is placed. The tread is measured from the front of the step, the *nose, to the back of the step, and is the full length of the horizontal surface upon which a foot can be placed.

tread plate Embossed metal sheet with a raised profile to create a slip-resistant surface, often placed on stairs and steps.

treatment A method to alter the physical, chemical or biological character or composition of something, e.g. water treatment removes physical, chemical, and biological impurities from water such that it is fit for human consumption.

tree A perennial woody plant that provides the material *timber.

treenail (trenail) A large dowel-shaped peg that is driven into a hole that has been bored through two pieces of timber to form a joint. This jointing system has considerable historic use.

tremie (tremie tube) A funnel-shaped *hopper containing a large pipe at the bottom. It is used to pour concrete, particularly under water.

trench 1. A long narrow groove cut into a material. **2.** A strip of earth excavated to form a long narrow hole in the ground. The trench is often excavated in order to lay foundations or drainage.

trench duct Tube or pipework laid in the ground to provide a void along which cables and pipes can be threaded through. Once the ducts are in position and the trench is backfilled, any cables housed in the ducts are hidden neatly underground.

trench-fill foundation A long strip of earth is excavated down to a good load-bearing strata and the remaining trench is simply filled with concrete to provide a foundation. The foundation is suited to carrying longitudinal loads such as walls.

trenchless technology A method of installing service, pipes, etc. underground without the need of *open cut.

trench support A supporting structure positioned against the walls of a trench to prevent them collapsing while people are working within the trench.

trepan 1. A tool for boring holes/shafts in rock. **2.** A machine, containing cutting wheels, used for cutting and loading coal from along a coal seam. **3.** To cut a circular groove into a surface.

Tresca yield envelope A criterion to denote the state of stress at yield in a material, with the maximum shear stress being the decisive factor for yield; *see also* VON MISES YIELD.

trestle A supporting tower-like structural framework for bridges, which consists of a horizontal beam held up with raking legs.

trial hole (trial pit, test pit) A small hole excavated either by hand or by an excavator to collect information on the soil, strata, and/or services below the ground. Used during a *site investigation to reveal details of the strata and groundwater condition at shallow depth. For soil investigations samples can be taken from the various levels of strata that will be visible on the sides of the excavation. When attempting to locate the position of services, the exploratory hole is carefully excavated by hand.

triangle A flat, two-dimensional shape with three sides. A triangle has three internal angles and three external angles. The internal angles add up to 180 degrees.

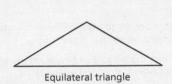
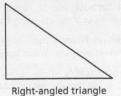

Equilateral triangle   Right-angled triangle

Triangles

triangle of forces *See* POLYGON OF FORCES.

triangulation 1. A method used in surveying that divides an area of land into triangles. The angles of the triangles are measured as well as one side (termed the baseline). The length of the other sides can then be calculated by trigonometric relationships. **2.** The use of different approaches to arrive at the same destination. It is often associated with and used by the armed forces, as well as in the shipping industry as a means of converging on the same point through using different options or routes. **3.** In research methods, a way of using different data collection techniques or research approaches to arrive at convergent views.

triaxial test (triaxial compression test) A geotechnical test to determine the shear strength parameters (*cohesion and *friction) of a cylindrical soil specimen. The specimen has a height-to-diameter ratio of 2:1, typically 76:38 mm or 200:100 mm. The apparatus consists of a cell, which is filled with water under pressure to create the intermediate principal stress (σ_2) and the minor principle stress (σ_3), both being equal in value, termed cell pressure. The specimen, sealed in a rubber membrane within the cell, is loaded vertically via a proving ring or load cell to create a deviator stress (DS). The sum of the DS and σ_3 gives the major principle stress (σ_1). The vertical load on the specimen is increased until failure occurs; the vertical strain being recorded at the same time using a dial gauge or linear displacement transducer. The test is repeated on different specimens from the same soil, using different values of cell pressure. Using the data obtained, a Mohr circle diagram is plotted and the failure envelope established with corresponding values for cohesion and friction being obtained. The triaxial compression test is commonly undertaken on undrained specimens, where a rubber membrane seals the specimen within the cell, and termed **undrained test**. The results are presented in terms of total stress parameters. However, it is possible to measure the pore pressure during the shearing stage, allowing effective stress parameters to be recorded; this is termed **consolidated undrained**. Another method of assessing the effective stress of a specimen is to apply the load at a very slow rate and allow the specimen to drain; this is termed **drained test**.

tribology The science behind the interaction of contacting surfaces, i.e. where friction is encountered.

tribrach The lower part of a theodolite that contains the three levelling screws and the tripod fixing point.

tributary A river or stream flowing into a larger water body, such as a larger stream, river, or lake.

trichloroethane An organic chemical that is monitored in drinking water supplies because it is hazardous to health.

trickling filter A method of treating wastewater, by passing it slowly through a granular filter, where it is subject to aeration and microbial degradation.

trigonometry Mathematics used to calculate the lengths and angles of triangles. Angles and lengths are calculated using the tangent, sine, and cosine ratios.

trihalomethane (THMs) A chemical compound with three halogen atoms, derived from methane; it is formed as a by-product during the chlorination of drinking water and is considered carcinogenic.

trilateration A surveying technique using a network of triangles, where the sides of the triangles are measured using a *tellurometer, as well as the internal angles; *see also* TRIANGULATION.

trim 1. The collective name given to decorative finishes such as architraves and skirting boards. **2.** To cut to the correct size.

trimmed door A door leaf that has been reduced in size once hung.

trimmed joist (tailpiece, US) A short *bridging joist, supported by a *trimmer joist and used to create an opening in a floor, usually for stairs.

trimmer joist A joist that runs perpendicular to the *bridging joist to support the *trimmed joists.

trimming 1. The collective name given to the trimming joist, the trimmer joist, and the trimmed joists used to form an opening. **2.** The process of cutting an object to the correct size.

trimming joist (trimming, Scotland bridling) A floor joist that runs parallel to the *bridging joist to support the end of a *trimmer joist at an opening in the floor.

trimming machine (mitring machine) Guillotine used for trimming mouldings and cutting angles.

triple bottom line *See* THREE PILLARS OF SUSTAINABLE DEVELOPMENT.

tripod A frame or stand with three legs, usually collapsible, used to support a surveying instrument, for example a *theodolite.

triumphal arch Arch or monument built for the return of a victorious army or group.

TRL *See* TRANSPORT RESEARCH LABORATORY.

trochoidal wave A wave that propagates with rotation about a central axis—it approaches the shape of a sine curve at small amplitudes.

trolley (crab) Wheeled frame that runs along the jib of a crane; used to carry and position the lifting hooks.

trolleying The movement inwards and outwards of the *trolley on the jib of a crane. As the trolley comes inwards, the radius of the hook is reduced and as the trolley is positioned further out on the jib the radius is increased.

Trombe wall A thermally massive, usually masonry, south-facing wall that is clad in glazing, with a small air gap between the glazing and the wall. It is designed to maximize the collection of solar radiation during the day and slowly release it at night to the inside of a building. A **mass Trombe wall** is one in which the thermal mass used in the wall is masonry, while a **water Trombe** wall is one where the thermal mass used is water.

tropical storm A storm that develops offshore in the tropics with wind speeds less than 119 km per hour, but has the ability to develop into a *hurricane.

troposphere The lowest layer of the earth's atmosphere, extending 7–20 km (typically 17 km) above the earth's surface. It is the layer that is subjected to the most variation in weather conditions, with decreasing temperature with altitude.

trough A channel, trench, or drain.

troughed sheeting *See* PROFILED SHEETING.

trough gutter *See* BOX GUTTER.

trowelling The final smoothing of plaster, screed, or concrete with the edge of a trowel. The final stage of smoothing takes place after the initial set has taken place. For concrete floors, trowelling takes place after *power-floating.

truck A general term used for a vehicle (such as a *dump truck) for hauling loose material.

true bearing The horizontal angle made between *true north and any line; *see also* MAGNETIC BEARING.

true north The direction north according to the earth's axis.

true strain The value of the actual *strain of a material subjected to a tensile force at any specific point of load. The value of strain between any two specific points.

true stress The value of the actual *stress of a material subjected to a tensile force at any specific point of load. The value of stress between any two specific points.

truncated solid A solid figure, such as a cuboid, with its top cut off making another shape, for example, a frustum

truncated truss A trussed *rafter that has had the top section removed.

trunk The main stem of something e.g. a tree. A **trunk road** is a large main road. A **trunk sewer** is a large main sewer.

trunking A protective duct for cables or pipes that has a removable cover.

trunk lift *See* SERVICE LIFT.

trunnion A cylindrical projection used as a pivoting point on optical instruments, such as theodolites, where they are mounted on either side of the telescope to allow it to rotate in the vertical plane.

truss A structural framework of beams, posts, and struts that is designed to support a structure, such as a roof, or to span an opening.

truss clip A stainless steel clip designed to secure a roof truss to the wall plate.

trussed beam A *beam made in the form of a *truss.

trussed purlin A *purlin made in the form of a *truss.

trussed rafter A prefabricated structural framework, triangular in shape, that is used to support a roof. Commonly used in domestic construction. A wide range of trussed rafters are available including attic, cantilever, fan, fink, Howe, king post, mono, queen post, and scissor.

truss element A two-node member that defines a line, in a finite element model, which can only resist or transmit an axial force.

truth table A method to record values, in rows and columns, for all the possibilities in a logic relation.

try plane A long plane used for truing or final smoothing of wood. Try planes are usually over half a meter in length.

try square A square, used to check that the wood is true; having angles of 90°.

T-square A drawing instrument in the shape of a 'T' used with a drawing board, to draw horizontal straight lines.

tsunami A tidal wave that has been caused by an underwater earthquake or landslide under/on the ocean bed.

tube A pipe, cylinder, hose, or something that has a lateral dimension (typically diameter) far less than its longitudinal dimension, and is typically used to pass liquid through.

tubular In the shape of a tube.

tubular mortise lock *See* BORE LOCK.

tubular scaffolding *Scaffold made using tubular steel tubes.

tubular trap A *trap made using tubular pipe.

tuck A small recess made in the mortar joint prior to *tuck pointing.

tuck pointing 1. A type of decorative *pointing that projects slightly from the face of the masonry courses. **2.** The process of removing and replacing old and damaged mortar between masonry courses; *see* REPOINTING.

Tudor style Architecture from the Tudor period 1485–1603, typified by buildings constructed between the Gothic and Renaissance periods. The Tudor arch was a feature of this period. In domestic construction, wattle-and-daub was often used. External features associated with this period include exposed wood structure with infill white render. Tall decorative chimneys were often constructed on houses of this period.

tumbler The portion of a *lock that enables it to be operated only when the correct key is inserted.

tumbler switch (toggle switch) A spring-loaded electrical switch with a small level.

tumbling bay A manhole or backdrop *inspection chamber.

tumbling in (tumbling courses) A sloping course of brickwork laid across the top of a gable wall. The bricks are laid at 90° to the gable wall course.

tundish An open container used to collect condensate.

tundra A region where tree growth is limited due to very low temperatures and frozen subsoil conditions.

tuned mass damper (TMD) A mechanical system, consisting of an *oscillating *spring and a *dashpot; used to reduce the effects of *motion.

tungsten A non-ferrous metal with the symbol W. Tungsten has a very high melting point (3422°C) and is used mainly in electrical applications, such as lightbulb filaments.

tungsten-filament lamp An incandescent lamp.

tungsten-halogen lamp *See* HALOGEN LAMP.

tungsten-tipped drill (hard metal tungsten drill, masonry drill) Drill bit with a tungsten carbide tip that provides a harder cutting edge; used to drill into masonry.

tunnel An underground channel or passage.

tunnel-boring machine (TBM) A tunnelling machine, which consists of a rotary cutting head that occupies the full face of the tunnel and a system of conveyors (or pumps) to remove the excavated material.

tunnel form Pre-assembled floor and wall formwork allowing the casting of concrete walls and floors in a single pour. The forms are suited to buildings where the room sizes are consistently similar, allowing multiple uses and considerable efficiency in using the forms. Where rooms are the same size, there is minimal setup and adjustment required between pours. Hotels, apartment blocks, and lodges are particularly suited to this type of formwork. The forms are partially constructed with two walls joined by a folding deck. The units can be rolled on wheels, and quickly erected to form the shuttering for the walls and floors. Once the concrete is cast the units are removed by collapsing the foldable deck and rolling the form out on wheels.

tunnelling The process of forming passageways under the ground, through the sides of mountains, below rivers or channels, so that vehicles and trains can pass through.

tupper Local term in some areas in the north of England to describe a bricklayer's labourer.

turbid Describes the cloudiness of water caused by suspended particles. It is measured using a **nephelometer**, which determines how much light is scattered though a column of water to give a **turbidity** value, quoted in **Nephelometric Turbidity Units (NTU)**. The WHO drinking guidelines allow up to 5 NTU to be present in drinking water supplies.

turbidity A geological deposit formed under water, e.g. from a landslide.

turbine A machine that converts fluid energy into mechanical energy. The fluid energy is typically from wind or water, and imparts rotation to the turbine's vane or blades, which in turn drives a generator to create, for example, electricity.

turbulence Disturbed fluid flow, particular in relation to air.

turnbuckle (stretching screw, bottle screw, tension sleeve) A sleeve-like connector used for adjusting tension ropes, cables, etc. The sleeve is threaded and has two threaded nuts or eyelets, which are threaded in the opposite direction to each other, at either end connected in line with the rope or cable to be tensioned. The tension is adjusted by rotating the turnbuckle.

turn cock A *stopcock.

turn-down *See* HOOK INTAKE.

turned bolt A bolt that has been machined to produce a shank dimension of close tolerance.

turning Working or forming an object by rotating and cutting it on a turning lathe. Wood, steel, and other materials can be cut to form more circular sections by rotating it on a lathe and shaping the edges with chisels.

turning piece (centring) Temporary support that provides support to a brick, stone, or other form of segmental arch. The support is often made from plywood with the sheets being cut to the same shape as the arch. The centre or temporary support is then held in position using *telescopic props or temporary adjustable props until the arch is constructed and set.

turning point 1. A crossroads. **2.** A change point in surveying. **3.** The maximum and minimum points of a curve.

turnkey project A contract where the whole design, build, and furnish is undertaken by a contractor. The client signs the contract and expects to be 'turning the key' and opening the door to a fully functional building.

turnpike A toll gate; a place where a *toll is collected on a section of road.

turn-up The adjustment of material where a horizontal surface meets a vertical surface so that the material can be laid flat along the horizontal and angled to travel up the vertical surface. *Flashings, *asphalt to up-stand walls, and *skirtings are used to form turn-ups.

turret A small circular tower, normally quite notably smaller than the main structure.

turret roof Pitched roof to tower or round structure.

turret step Triangular stone steps, traditionally used within *turret towers. The thin end of the triangular stone steps are often rounded, which creates a rounded *newel when the steps are placed on top of each other.

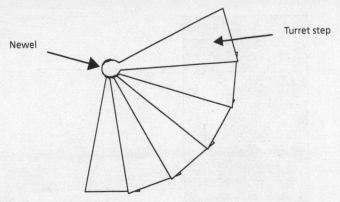

Turret steps

tusk nailing (skew nailing, angle nailing) The practice of installing nails at an angle in timber to provide extra strength to the joint.

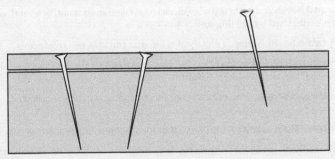

Tusk nailing

tusk tenon Mortise and tenon joint constructed to give additional strength and bearing to the joint. Various forms of tusk tenon joints exist, but the common feature is that the tenon has shoulders cut into the joint to add extra strength.

TV distribution (US master antenna system) The collective name given to all of the components required to provide TV reception in a building, such as the aerial, an amplifier, the wiring, the connectors, and the socket outlets.

twin (metals) A common type of *dislocation which exists in metals at a microscopic level. As the name suggests this type of defect or dislocation comprises two identical dislocations.

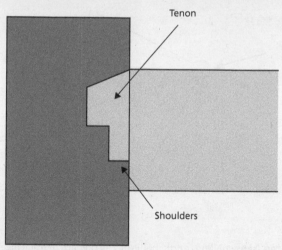

Tenon

Shoulders

Tusk tenon joint

twin and earth A sheathed mains electrical cable that contains two conductors and an earth. Used for lighting and power circuits.

twin cable Sheathed mains electrical cable that contains two conductors.

twin tenon (divided tenon) Two *tenons arranged in a row on the same piece of timber. Often confused with a **double tenon**.

twin-thread screw A *screw that contains two separate, deep interlinked threads.

twin-tube fluorescent A luminaire that incorporates two fluorescent tubes.

twist The warping of timber. If timber is seasoned or dries out too quickly it has a tendency to warp in a helical screw-like manner.

twist drill Hardened steel drill bit with steep helical cutting edge.

twisted fibres Overlapping and interlocking grain found in timber.

twitcher A small trowel cut or shaped to an angle for pointing.

two-bolt lock A lock that incorporates two bolts.

two-brick wall A solid wall two bricks thick.

two-coat work (render and set) Plaster with an undercoat or basecoat, and topcoat or finish. The basecoat helps to level out the surface of the substrate and the finish coat ensures a smooth true surface.

two-peg test *See* COLLIMATION LINE.

two-piece cleat A two-piece *pipe ring.

two-pipe system (dual system, dual-pipe system) An above-ground drainage system that comprises two discharge pipes; one that discharges the soil, and another that discharges the waste. *See also* ONE-PIPE SYSTEM and SINGLE-STACK SYSTEM.

two-stage alarm system A type of fire alarm system that activates an alarm in the area in which a fire is detected, and at the same time activates an alert in the remainder of the building.

two-stage tendering First tender is submitted with approximate quantities; once accepted negotiations commence on the final price and quantities.

two-tailed test A statistical test in which the distribution of the *null hypothesis is considered by looking for any change in the parameter.

two-way slab A slab that spans, and hence has reinforcement, in two directions.

two-way switch A switch that allows lights to be switched on from two separate locations. For instance, at either end of a corridor or stairway.

tying wire *See* BINDING WIRE.

typhoon A *tropical storm that develops in the north-western part of the Pacific Ocean.

typical floor A repetitive floor plan in a multistorey building.

U-bend A U-shaped *trap. *See also* P-TRAP and S-TRAP.

U-bolt U-shaped bar with threads on each end of the bar. The threads enable the U-bolt to be bolted to or fitted with a cross piece, making it suitable for securing pipes, fittings, and appliances.

UCATT (Union of Construction Allied Trades and Technicians) A trade union which formerly pursued the rights of workers in construction and allied industries. UCATT has now merged with Unite, a multisectoral trade union.

⊕ SEE WEB LINKS

U-duct A ducted flue with intake and outlet at roof level.

UF *See* UREA FORMALDEHYDE.

U-gauge Clear glass or plastic U-tube filled with water and used for measuring the effective seal of drainage pipes. Plugs are placed at each end of a drain run. One of the plugs has two ports; one to which a U-gauge is connected, and one to which a pump is attached. Air pressure is applied to the pipe run using the pump until the water in the U-gauge rises to a set level. If the water holds at this level then the drainage run is effectively sealed.

U-liner A high-density polyethylene pipe, which has been deformed during manufacture into a U-shape so that it can be inserted into existing pipes that are damaged. It can then be remoulded into a circle cross-section within the existing pipe. The result is a structurally sound pipe within an old damaged pipe, and the need for excavation is eliminated.

ullage 1. The volume of empty space above a liquid in a container. **2.** The volume of lost liquid from a container due to evaporation or leakage.

ultimate limit state (ULS) A state associated with the collapse, structural failure, excessive deformation, or loss of stability of the whole of a structure or any part of it (e.g. *bearing resistance being exceeded, i.e. its capacity or resistance is in sufficient to support the *actions applied). ULS together with *serviceability limit state (SLS) forms the two subsets which are listed in the *Eurocodes. The six ultimate limit states are:

- EQU – Loss of equilibrium (tilt or rotation)
- STR – Structural failure (e.g. internal failure or excessive structural deformation)
- GEO – Failure of excessive deformation of the ground
- FAT – Fatigue (through time-dependent effects)
- UPL – Uplift or buoyancy (e.g. loss of equilibrium of a structure due to the vertical uplift of water pressure or buoyancy)
- HYD – Hydraulic heave (internal erosion or piping) due to water flow

ultimate tensile strength (UTS) *See* TENSILE STRENGTH.

ultra-filtration A membrane filtration technique in which hydrostatic pressure is used to force a liquid through a very fine filter.

ultra-lightweight aggregate *See* AGGREGATE.

ultrasonic Very high sound frequencies above the human audio range, which are about 20 kilohertz.

ultra very high-strength cement A special type of cement material that imparts very high strength, for example, in high-strength concrete.

ultraviolet (UV) An electromagnetic wave of light that is emitted by the sun and is invisible to the human eye. Though the atmosphere blocks most of the UV radiation, some gets through and it can be potentially harmful for many materials, for example, polymer materials degrade under exposure from ultraviolet radiation thus requiring protection.

ultraviolet radiation (UV radiation) Electromagnetic radiation, with wavelengths from about 5 to 400 nm, that can cause degradation of certain materials.

umber A ferric oxide material usually utilized as a *pigment.

unbonded Something that has failed to stick or has become unstuck.

unbonded screed Improper application of *screed resulting in debonding of the applied layer.

unburnt brick *Brick material that has not been exposed to very high temperatures for a prolonged period during the manufacturing process.

uncased fan A *fan where the rotating blades are not enclosed within a case.

uncased steelwork Structural steelwork that is left exposed and not encased.

unconfined Not restricted, for example an **unconfined triaxial compression test**, where zero lateral pressure is applied.

unconformity A gap in the formation of geological beds, where older rock beds underlie significantly younger rock beds. This is caused by either a period of time in which no sediment is deposited, or an interruption in deposition, or by weathering.

unconsolidated A geology term used to describe loose soil or uncemented sediment.

uncoursed Describes stone walling that is not laid to a set course.

uncoursed rubble (broken range ashlar) Stone of random size that is not laid to set courses.

uncovering The instruction and action to expose work that is suspected of being defective. Where an engineer or *clerk of works responsible for checking the standard of construction suspects that there is defective work, they may give an instruction to remove finishes, coverings, or other materials so that the works can be properly inspected. Where no defects are found, then the costs and delays associated with exposing the structure can normally be reclaimed.

undercloak 1. Tiles or compressed cement board laid at the *verge of a *roof and positioned along the top of the wall so that the ridge tiles slope slightly inwards, thus preventing water dripping over the edge of the roof. **2.** The lower sheet metal of a drip, roll, or seam.

undercoat The penultimate coating applied in the painting process. The undercoat is applied prior to the final top coat. The function of the undercoat is to provide a smooth surface, good covering power, and good adhesion for the finishing coat. Undercoats usually contain a large amount of pigment to give good covering (hiding) power, and most are based on *alkyd resins or *acrylic emulsions. Undercoats and *primers will not survive exposure well without the top coat.

undercooling A procedure whereby a metal is cooled from an elevated temperature (below its melting point) to achieve the desired material chemistry. This procedure is commonly used in the manufacture of many commercial metals, such as steels.

undercroft A cellar or underground storage; it can also be used to describe an open area at the base of a building (street level) that is covered by the first floor of the building.

undercuring A material that has not achieved full curing.

undercutting The process of trimming the bottom of a door leaf while it is still hung.

under-eaves course The lower course of tiles at the eaves of the roof.

underfelt Underlay placed under carpets.

underfired brick A type of brick that has not been subjected to the full *firing process during manufacture. This results in lower strength of the material.

underfloor heating (floor heating, heated floor) A type of central heating system comprising a series of electric cables (dry systems), or water pipes (wet systems) that are embedded in or fixed below the floor. In wet systems, water is circulated through the pipes at between 30–60°C, resulting in a floor surface temperature of around 25–28°C. High surface temperatures should be avoided (≤ 29 °C) to avoid discomfort to the occupants. Surface temperatures must also be compatible with the floor covering. Most types of floor covering can be used with underfloor heating, including ceramic tiles, carpets, vinyl, laminate flooring, and wood flooring.

underfloor space The void that is created between the *suspended floor and the structure that is in contact with the ground.

underground Below the surface of the ground.

underground services Those services that are buried underground.

underground stop valve A *stop cock buried underground that is accessed via a surface box.

underlay Any sheet material that is placed under another material. Resilient sheet material can be placed underneath carpets to offer extra comfort or sheet-felt, such as *sarking felt, can be placed under tiles, to prevent the wind and any moisture penetrating into the roof space. Both could be classed as underlay materials.

underpinning The addition of extra foundations underneath existing foundations to provide extra support. The underpinned foundations are provided where the building loads are to be increased or where the existing foundations have started to settle and are no longer able to provide the support.

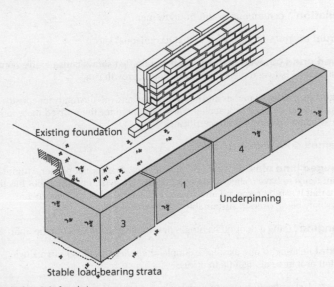

Underpinning: Strip foundations
The numbers in the diagram refer to the sequence in which the foundation is underpinned.

under-reaming A process where the bottom of a *bored pile is made wider than the shaft diameter to increase end-bearing resistance and/or prevent uplift.

under-ridge tile (top course tile) The course of tiles located directly under the *ridge.

under-tile An *eaves tile or an *under-ridge tile.

undertone The presence of another or different colour on a white background that can also be viewed under the presence of transmitted light.

under-vibration Failure to vibrate concrete sufficiently to ensure the concrete compacts, fills the formwork, excludes air bubbles, and achieves its full design strength.

undisturbed sample A sample of soil that represents in-situ conditions, for example, it has not been allowed to be removed or dried out. Such samples are collected during a site investigation and tested using laboratory equipment to determine stress strength parameters, amount of consolidation, and density.

undrained test (immediate test, quick test) A shear strength test where drainage of water is prevented, such that no dissipation of pore pressure is possible. Undrained parameters c_u and ϕ_u are obtained, which are applicable to immediate bearing-capacity issues regarding foundations placed on saturated clays.

undressed timber (unwrought timber) Unplaned timber that is left in its rough cut or sawn state.

undulation A continuous to and fro movement.

unequal angle An *angle section that has unequal legs.

uneven grain (uneven texture) Timber grain that shows considerable contrast and difference between the summer and spring growth rings.

unfixed material Materials and goods ordered and on-site ready for construction that have yet to be used or installed. Partial payment for the unfixed materials is made available within *interim certificates.

unframed door *See* BATTEN DOOR.

ungauged lime plaster Sand and lime plaster. Traditional plaster without gypsum tends to have a slightly lower strength than gypsum plaster, but has the benefit that it can be removed and reused at a later date. Such plasters are increasingly being used in sustainable building projects.

unhanging Lifting a door off its hinges or removing the door from the doorframe.

uniaxial Relating to one axis, for example, the reflection of light in a single direction or a force along a tie in a truss.

uniclass A classification system which allows items to be grouped or arranged under numerical headers, such that they can be classed during the *asset life. It is also used in *BIM models as part of the categorization.

unidirectional Relating to one direction.

uniform Of the same consistency/size throughout.

Uniform Building Code (UBC) Australian equivalent to the *Building Regulations.

uniform corrosion Chemical degradation affecting the whole surface of a material generally. Uniform corrosion is highly prevalent in metals, especially ferrous alloys, and is exemplified by the rusting of steel. Other examples are the formation of a green patina on copper and the tarnishing of silver. When designing a component for a particular environment it is usual and relatively straightforward to adopt measures to prevent uniform corrosion. These can include protective coatings (galvanization), inhibitors, cathodic protection, and appropriate choice of materials.

uniformly distributed load (UDL) A load that is distributed over a portion or the full length of a beam and expressed as kN/m.

uninterruptible power supply (UPS) A battery-operated backup power source that is designed to provide power to specific devices for a short period of time during a power failure.

union A type of threaded pipe joint comprising a male and female end that are joined together and then the joint is sealed using a nut. Enables pipes to be disconnected easily.

union bend A pipe bend that incorporates a *union on one end.

uniqueness theorem A theorem that states that only one entirety can exist under a set of specific conditions or properties, for example, Poisson's equation has a unique gradient for different boundary conditions.

unit 1. A discrete part into which something can be divided. **2.** A standard measurement denoting a quantity, e.g. metre, kilogram, degree, volt, hour. **3.** A single element, component, or module. **4.** A specified proportion of work.

unit air conditioner (packaged unit) An air-conditioning unit where all of the components are housed within a self-contained unit.

unit cell On a microscopic level in metals, this constitutes the whole crystal structure within the metal that is repeated throughout the whole material. The unit cell gives the chemical identity of a metal with regards to the arrangement or packing of atoms. It is the smallest part within a metal that is repeated to form the complete crystal structure.

unit costing A price given per metre, square metre, hour, or item. Listed against standard descriptions in the bill of quantities, prices may be obtained from previous site outputs or from a price index.

unit heater (air heater) A self-contained space heating system that circulates warm air using a fan or fans. Used to heat non-domestic buildings such as factories and warehouses.

unit hydrograph The *hydrograph of a hypothetical unit of *run-off, for example, the effect of rainfall in a specific *basin.

unit of bond The pattern of bricks within a course that repeats itself.

unit price (rate) *See* UNIT COSTING.

universal section A steel section manufactured to a standard size. A **universal beam** is an *I-section and a **universal column** is an *H-section.

universal set A *set that contains all objects, elements, processes, events, or categories.

Universal Transverse Mercator (UTM) A 60-zones grid-based coordinate system, measured in metres, that is used to define locations on the flattened surface of the earth.

unloading The removal of load.

unplasticized polyvinyl chloride (PVC-U (formerly uPVC), rigid PVC) *See* POLYVINYL CHLORIDE.

unreinforced masonry (URM) Buildings or houses whose walls have not been reinforced with steel bars/rods. These walls can be load- or non-load-bearing.

Unsafe theorem *See* UPPER BOUND THEOREM.

unsaturated Not completely full of water, for example, where the pores in the ground have both air and water, as in *vadose water; *see also* SATURATED.

unsocketed pipe (socketless pipe, US hubless joint pipe) A plain-ended pipe that is joined together using a *sleeve coupling.

unsound Structurally unstable or not satisfactory.

unsound knot (rotten knot) A knot that is softer than the surrounding wood.

unvented system (mains pressure system) *See* SEALED SYSTEM.

unwrought timber Undressed lumber.

up-and-over door (overhead door, swing-up door) A large door that swings outwards and upwards to open. Commonly used for garages.

updating (rescheduling) The monitoring of progress against the project programme and making necessary changes to ensure that the delivery comes in on time and within budget.

updrift The opposing direction to which the drift material is being deposited.

uplift Upward movement, for instance, when *quicksand forms due to an upward flow of water that is greater than the weight of the sand acting downwards.

uplift restraint strap Galvanized steel strap attached to the *wall and the *wall plate, or roof *truss to prevent the wind lifting the roof.

uplighter A *luminaire that is designed to project its light upwards.

Upper Bound theorem (Unsafe theorem, plastic collapse) The collapse mechanism cannot be less than the collapse load for a given structure and loading condition.

upper chord The uppermost member in a *truss.

upper floor Floors above ground level.

uprush A sudden upwards movement of something, for instance, the water that rushes onto a beach when a wave breaks.

upset The forming of metals with a swage.

upside-down roof *See* INVERTED ROOF.

upstand (upturn) Upward projection of a roof that is usually designed to prevent rainwater flowing over the edge of the roof. The upward projection may also be used at the edge of floors, roofs, and stairs as a safety barrier, or to offer protection.

upstand beam A floor beam that projects at its ends above floor level.

upstand flashing The under part of lead, copper, or aluminium flashing that is turned up against a wall or an abutment. To fully waterproof the abutment, additional pieces of flashing are embedded in the wall and turned down so that it covers the upturn.

upsurge A rapid rise in something, for instance, an increase in water flow.

uptake 1. An upward air current. **2.** A chimney vent. **3.** The absorption of something into a living organism.

upthrow Upward displacement of a rock adjacent to one side of a *fault.

uptime Time during which plant is available for use, different from downtime where the plant is standing idle because it cannot be used.

uPVC *See* POLYVINYL CHLORIDE.

upward flow filter A filter in which the direction of water flow is upwards so as to reduce the amount of backwashing.

upwelling A process of rising up from a lower depth, for example, where colder water rises from the ocean floor to the surface.

upwind Against the wind, facing the direction in which it is blowing. *See also* WINDWARD.

urban The built-up areas such as towns and cities.

urbanization 1. The changing of an area, developing it into a town or city. **2.** The accommodation of attitude, culture, and other changes that enable city living.

urea formaldehyde (UF) A thermosetting plastic (polymer) (*see* THERMOSETS), made from urea and formaldehyde. Urea formaldehyde is a widely used polymer especially in construction for electrical switch plates and electrical sockets (white plastic plates), and is highly prevalent in dwellings and municipal buildings. UF is characterized by desirable mechanical and physical properties,—*tensile strength, low water absorption, and *hardness. Urea formaldehyde was previously used in cavity walls, however, its use was discontinued in the 1980s due to the emission of toxic formaldehyde. Currently its main use is for electrical sockets, however, UF resin is used in adhesives, finishes, and moulded objects.

urea formaldehyde foam Low-density *urea formaldehyde polymer in foam form typified by the presence of porosity or holes in its internal structure. Urea formaldehyde foam was previously used in cavity wall insulation but was discontinued due to health and safety concerns. *See also* UREA FORMALDEHYDE.

urethane Carbamic acid, polymerized to form *polyurethane.

urinal Bowl or gully used in non-domestic toilets to capture urine. Most commonly used in male toilets although similar female appliances are being developed.

URL (universal resource locator) The address of a web page on the World Wide Web.

usable life Length of time indicative of the durability of a material, object, or component.

use factor The number or diversity of uses an item can facilitate.

U-tie A heavy wall tie.

utile (sipo, *Entandrophragma utile*) A tree from the mahogany family Meliaceae, restricted to tropical Africa. *Entandrophragma utile*.

utilities 1. Services and appliances **2.** Service providers that were traditionally within the public sector, such as water, gas, electricity, telecoms, and sewers.

utilization factor (U) The amount of light given off by a light that reaches the working plane it is designed to illuminate. The surface to be lighted is called the reference plane. The light on the surface is measured in lux.

UV (ultraviolet) *See* ULTRAVIOLET.

U value (air-to-air heat transmission coefficient) *See* THERMAL TRANSMITTANCE.

vacancy (metals) At a microscopic level in metals, this is an unfilled lattice site in a crystal structure. Atoms in a metal occupy positions in an ordered array. Should an atom not be present, this creates a vacancy within the metal. The number and type of vacancies in a metal is very important as it can influence the mechanical properties of the material.

vacuum An enclosed space that is empty of all matter.

vacuum cleaning plant (central vacuum cleaner, central vacuum system) A vacuum cleaning system that is built into a building and comprises a lightweight hose that can be connected into a series of outlets distributed throughout the building. Pipework concealed within the building is then used to transport any picked up material from the hose to a central externally vented vacuum.

vacuum concrete A type of concrete containing a vacuum mat to remove excess water in order to achieve higher compressive strength.

vacuum insulated panel (VIP) An airtight panel where the air within the core of the panel has been evacuated. It is used in construction to provide higher levels of *thermal insulation than would be the case with conventional insulation materials.

vacuum mat A metal screen used to suck out excess water and air from concrete.

vacuum-pressure impregnation Saturation of timber by extracting the air and flooding with preservative.

vacuum pump A device that removes air or vapour molecules from a sealed volume of space in order to maintain a pressure below atmospheric, to cause a partial vacuum.

vadose water (vadose zone) A zone above the *water table in the ground, which is unsaturated, but comprises of: slow-moving water percolating downwards to join the phreatic water (*see* GROUNDWATER) below the water table; and capillary water (*see* CAPILLARITY) held above the water table by surface tension forces (with internal pore pressure less than atmospheric).

valence *See* VALENCY.

valency A measure of the ability of hydrogen atoms to combine with other atoms. For example, helium, neon, and argon have full outer orbits and do not normally form compounds, thus they have valences of zero. Lithium, sodium, and potassium have two electrons missing from their outer orbits and they combine with two hydrogen atoms to have valences of two.

valley The internal angle created when two separate pitched roofs intersect.

valley board A board used to support a *valley gutter.

valley gutter A *gutter used to drain water from a *valley.

valley jack (US) A *jack rafter.

valley rafter A rafter used to form a *valley.

valley tile A specially shaped roof tile used to construct a tiled valley.

value 1. The degree to which an item, product, service, or event meets the desired performance criteria required by the client. **2.** The degree of lightness of a colour.

value added tax (VAT) The government levy added to the cost of goods and services to support governance of the country. VAT rates for new build, demolition, and refurbishment often vary. For details of tax, see HM Customs and Exercise, where guidance, accounting, and information on procedures can be found.

(SEE WEB LINKS)

value cost contract A cost reimbursement contract where the contractor is entitled to a higher fee if the final costs are less than expected.

value engineering A systematic management methodology which aims to improve the value of goods, products, and services through a detailed examination of their function. Positive value engineering is the method to improve the value of goods or products and services which improves both value and/or function and may maintain or reduce the cost. Value is defined as the ratio of function/value to cost. Negative value engineering is the method to improve the value of goods or products and services based on cost alone, specifically attempting to reduce the cost but maintain key functions by virtue of a lack of detailed consideration; however, such approaches can have a detrimental impact on function and performance during operation.

value in exchange The worth or the monetary sums for which a service or product can be traded.

value in use The function of a service or a product that satisfies a need or generates pleasure for its owner. Market prices (or value in exchange) do not necessarily reflect personal notions of value, and individuals' perceptions of value in use vary greatly.

value management The process of determining the function and performance criteria required, and identifying products and services at the lowest cost that meets the criteria.

valve A fitting that is capable of stopping or reducing the flow of gases or liquids through a pipe.

valve boss The raised area around a valve.

vanadium A non-ferrous metal commonly used as an alloying element in steels.

van der Waals bond A non-ionic type of bonding found in various materials.

vane test A simple in-situ test to determine the *shear strength (cohesion) of saturated clay at the bottom of a trial pit or *borehole. The vane is driven or pushed

$$Torque(T) = c\,\frac{\pi d^2}{2}\,(h + d/3)$$

where:

T = Torque
c = Cohesion
d = Diameter
h = Height of vane

Vane test

into the soil and a measured torque applied to it until it rotates. The failure surface is the curved surface plus the flat ends of the cylinder of soil whose diameter and height are that of the vane.

vanity basin A bowl-shaped basin recessed into a *vanity top for washing the face and hands.

vanity cabinet A mirror-fronted cupboard installed in a bathroom.

vanity top A worktop installed in a bathroom.

vanity unit A floor-mounted bathroom cupboard that has a *vanity top and *vanity basin.

vaporizing liquids Liquids that are stored under pressure which vaporize when the pressure is released.

vapour Moisture particles of a substance present in the air.

vapour barrier (vapour control layer) A material that resists water vapour transmission, and is usually made of polythene sheeting. Should always be installed on the warm side of an insulation layer to avoid the risk of *interstitial condensation.

vapour blasting A jet spraying water onto surfaces, normally used to clean or remove debris.

vapour check A material that is similar to a *vapour barrier, but has a higher *vapour resistance.

vapour compression cycle A type of refrigeration cycle where a refrigerant continuously changes state from a liquid to a gas, and back again. In the evaporator, the refrigerant evaporates at low temperature and pressure, absorbing heat from its surroundings resulting in cooling. The refrigerant is then compressed, increasing its pressure and temperature. The refrigerant then passes through a condenser where it condenses, releasing heat and reduces its pressure. It then passes back to the evaporator where the process is repeated.

vapour control layer *See* VAPOUR BARRIER.

vapour diffusion The process where by a vapour moves through a vapour-permeable material.

vapour pressure The pressure associated with a vapour that is in thermodynamic equilibrium with its liquid or solid form.

vapour resistance A measure of the ease with which a given thickness of a material will resist vapour transmission. Calculated by multiplying the *vapour resistivity of a material by the thickness of that material in metres. Measured in meganewton seconds per gram (MNs/g).

vapour resistivity A measure of the ease with which a material will resist vapour transmission. It is a characteristic of the material. Measured in meganewton seconds per gram per metre (MNs/gm).

variable A mathematical symbol (e.g. x_1, x_2, x_3) that represents a value that can change in a *set, rather than constant.

variable air volume box (VAV box) A type of *air terminal unit that uses dampers and thermostats to control the volume of conditioned air entering a space. Reheat coils can be incorporated into the VAV box to provide additional heating if necessary.

variable air volume system (VAV system) A type of centralized all-air air-conditioning system that supplies conditioned air to the space through a variable air volume (VAV) box. These are capable of providing zone control of the temperature from the central air-conditioning plant. However, humidity is still controlled centrally, so humidity conditions may vary widely between different zones. In a VAV system, air is supplied at a constant temperature and relative humidity to all parts of the building. Different cooling requirements are achieved by varying the volume of air supplied to each zone from the central plant. Control of the air volume is performed by a thermostatically controlled VAV box, which is fitted in the air supply ducts. VAV systems are common in new open-plan buildings.

variance A statistical measure of the spread of a *variable around its mean. It is equal to the square of the standard deviation.

variation (change) A modification or alteration to that agreed under the terms of the contract, or that described in the bill of quantities.

variational calculus An area of mathematics that seeks to find a stationary value (normally a minimum or maximum in physical problems) for a given function. Used in continuum mechanics to obtain global matrices.

variation notice Request from the contractor to the client's representative that a written *variation order is required.

variation order A written order from the client's representative authorizing a variation under the terms of the contract.

varnish A transparent, glossy, and hard coating applied to timber and other materials, mainly for gloss and protection.

vault 1. An arched ceiling or roof constructed from masonry. **2.** A room, usually underground, that is used to store valuables. **3.** An underground chamber used to bury the dead.

vault light *See* PAVEMENT LIGHT.

VB consistometer test A laboratory test designed to determine the workability of concrete. The test measures the time in seconds for wet concrete to fill a standard cone.

V-belt (vee-belt) A flexible belt with a cross-section V-shape. The belt sits in V-shaped pulleys helping to ensure good contact between the belt and the pulley.

V-braced frame A structural frame with diagonal braces that provide resistance to lateral forces.

VDC *See* VIRTUAL DESIGN CONSTRUCTION.

vector A quantity (force or velocity) with direction and magnitude.

vee joint A joint made by cutting a groove into the end of the timber resembling the letter 'V', the male joint is also cut to the same profile to fit into the V. The nature of the joint masks the effects of shrinkage on the joint.

velocity The rate at which a body moves: distance divided by time.

velocity head A static head of pressure from a fluid which would produce a certain flow rate. Also defined as:

$$v^2/2g$$

where v = velocity and g = acceleration due to gravity.

vena contracta Where a fluid flows through a contraction, such as an orifice, weir, or nozzle; the stream lines converge before the contraction and turbulent flow occurs after the contraction.

veneer Thin slices of wood glued onto panels.

veneered wall A wall with a facing that is not chemically bonded to it, but is mechanically bonded by wall ties; for example, a brick facing on a timber frame.

veneer finish plaster A coating applied to provide a hard, protective, abrasion-free surface.

veneering The process of affixing two layers of a substance (usually wood), together.

veneer tie A wall tie for holding a skin of masonry onto a timber frame.

Venetian door A central door, often arched, and flanked by flat, square windows on either side.

Venetian window A central window, often arched, flanked by flat, square windows on either side.

Venn diagram A mathematical diagram represented by circles, some that overlap each other, to illustrate relationship between *sets.

vent 1. A pipe or opening that allows gases to enter a space or be discharged to the outside. **2.** The process of discharging gases to the outside.

ventilated lobby A naturally ventilated *fire lobby.

ventilating brick *See* AIR BRICK.

ventilation The process of supplying or removing air, either by natural or mechanical means, to or from an enclosed space.

ventilation branch *See* BRANCH VENT.

ventilation duct A *duct that carries air for *ventilation.

ventilation heat-transfer coefficient The rate of heat flow that occurs due to the movement of air into a building or an enclosed space, divided by the temperature difference between the inside and outside of the building *envelope. The air movement can occur due to *purpose-provided ventilation or *air infiltration. It is measured in W/K.

ventilation pipe *See* VENT PIPE.

ventilation rate The rate at which air within an enclosed space is removed or replaced with air from outside the space. Measured in air changes per hour (ach).

ventilator A device that allows air to enter into or exit an enclosed space. It may take the form of a mechanical device, such as an *extract fan, or may be a passive device, such as an *air brick.

vent light (night vent, ventilator) A small casement window or *fanlight, which sits above the *transom, is top-hung, and opens outwards.

vent pipe (ventilating pipe, US vent stack) 1. The top of a stack drain system that allows the foul drain smells to be discharged at a level higher than any openings (windows and doors) in the house. **2.** An open discharge pipe that sits over the top of a cistern. The pipe is used to discharge steam and hot water, preventing pressure from building up within the system.

vent soaker *See* PIPE FLASHING.

venture A reduction, at a particular section, in a pipe diameter, which is designed to cause a drop in pressure when a fluid or gas flows through it. Used for measuring the flow in closed pipes.

veranda An open-sided roofed structure providing shade to a part of a building.

verdigris The green corrosion—oxidation—that forms on the surface of copper.

verge The sloping section of a pitched roof at the intersection with the *gable wall. The verge may be finished flush or may overhang the gable wall.

verge abutment The portion of a *gable wall that rises above the roof slope at the *verge.

verge board (vergeboard, verge rafter) *See* BARGE BOARD.

verge clip A metal clip attached to the *tiling batten to hold the *verge tile in place.

verge fillet A triangular-shaped piece of timber that is fixed to the end of the *tiling battens at a *gable wall.

verge flashing (verge trim) 1. A flashing used to waterproof the verge of a built-up roof. **2.** The section of a roof that projects beyond the *gable wall.

verge tile A roof tile that is used to form a *verge. The verge tile should project over the gable wall by at least 38 mm to prevent watermarks occurring down the face of the gable wall.

vermicular rustication An extension of the *rustication technique where rough shapes are drilled or chiselled into an external surface or stonework.

vermiculite A hydrated silicate mineral, which expands on heating. It is used in insulation and as a medium for planting in horticulture.

vermiculite-gypsum plaster A plaster with vermiculite used as an aggregate. The vermiculite is a lightweight material that has been expanded at high temperatures, forming a light material, which when used as an aggregate in plaster can improve the fire resistance and thermal performance of the structure.

vertex The highest point of a pyramid, where the intersection of the faces meet.

vertical angle The angle formed from the intersection of a pair of lines in a vertical plane, for example, the angle measured on the vertical circle in a *theodolite.

vertical axis wind turbine A turbine with blades that rotate about a vertical axis, perpendicular to both the direction of the wind and the surface of the ground. The prevailing feature which distinguishes the vertical from horizontal axis turbine, along with its appearance, is that the gearing and generator are normally located at the base of the turbine, which is at ground level.

vertical control A process of using different *benchmarks to check the **vertical** *closing error of a levelling survey.

vertical curve Typically a parabolic curve, in the vertical plane, which is used to relieve gradient changes in roads and railways.

vertical irregularity Any form of discontinuity (with regards to stiffness, strength, geometric, and mass) in one storey of a building in relation to the other storeys.

((()) SEE WEB LINKS
• Vertical irregularities are defined in FEMA 368 Section 5.2.3.3 and Table 5.2.3.3.

vertical lift bridge (lift bridge) A bridge where the deck moves up and down, while the rest remains horizontal.

vertical lift gate A gate that moves in a vertical direction to allow retained water through.

vertical sash A vertically sliding *sash window.

vertical shingling (hanging shingling, weather shingling) Tiles hung on a wall or tile cladding.

vertical shore The props or studs used to support a *needle that provides temporary support to a wall or beam.

vertical sliding window (vertical slider) A vertical *sash window.

vertical tiling The use of vertically hung tiles as an external cladding.

vertical transport Lift, escalator, and other mechanical equipment used to provide movement from one level to another.

vertical twist tie (strip tie) A cavity wall tie, formed by twisting galvanized steel metal. The twist provides the *drip, preventing water passing across the cavity. Steel wall ties with sharp protruding legs are considered a health and safety hazard. As the wall ties are fixed into the internal wall, the legs are left protruding as the external wall is built up; the protruding legs have been responsible for a number of injuries to workers' eyes and faces.

vertical work Bill of quantities description used to describe the application of hot asphalt to vertical surfaces. It is harder and more time-consuming to trowel asphalt onto vertical surfaces than it is to horizontal surfaces.

vestibule 1. An open area or court in front of a building. **2.** A small area within a building, providing a room from which doors open out into other rooms. A larger room is often called a lobby.

vestry An abutting or adjacent room to the church's main area in which clothes for the clergy and choristers are kept. These are called the vestments.

vetting of tenders The reviewing of tenders to ensure all the works have been properly priced, all of those tendering qualify and are capable of performing the works, and finally, the selection of the tender that is most suitable for the work. The work may be awarded on best value or most suitable contractor.

viaduct An arched bridge carrying a road or railway. The bridge is constructed from a series of adjacent arches, and particularly tall viaducts may have arches built on top of each other.

vibrated concrete Concrete that has been subjected to some form of vibration in order to compact it.

vibrating float A vibrating device, typically long and flat, used to finish in-situ concrete.

vibrating pile driver (vibrating pile hammer) A device that is mounted on top of a sheet pile, which vibrates to help drive the pile into the ground or extract it from the ground. Typically used for driving piles into saturated granular soils.

vibrating plate A hand-operated device, which has a flat, square, or rectangular vibrating surface, and is used to compact asphalt, concrete, and earth surfaces over small areas.

vibrating screed board A long horizontal beam that is used to vibrate and level concrete used for roads and floor slabs.

vibrating table A device in the form of a table that vibrates, and is typically used to compact small precast concrete products.

vibration 1. The unwanted shaking of the building and fabric caused by plant and equipment. Flexible and resilient anti-vibration devices may be used to stop the vibration and noise. **2.** Method used to compact concrete and remove air bubbles trapped within the concrete.

vibrator (vibrating poker) A cylindrical probe that is inserted into in-situ concrete, which vibrates to expel entrapped air. Air is trapped into concrete during mixing and placing. If it is not expelled it reduces the cast strength of the concrete. The vibrating part of the poker is typically 450 mm long; with diameters ranging from 20–180 mm. Pokers are either powered by electricity, diesel engine, or compressed air and have cycle speeds between 90 and 250 vibrations per minute.

vibrocompaction *See* VIBROFLOTATION.

vibroflotation An in-situ ground improvement technique used to compact loose granular soils to a depth of 70 m. A crawler crane is used to lower a **Vibroflot**, a cylindrical probe (450 mm diameter, 3–5 m long), into the ground under a vibration/jetting action (using water and/or compressed air). This action causes the surrounding ground to form a denser configuration. The Vibroflot is normally inserted in the ground at regular intervals, on regular grid spacing, over the site. The treated ground will settle less and have improved bearing capacity.

vice A clamp fitted to a workbench used to hold firmly materials that are to be worked. The clamp has a screw fitting, which when tightened secures the material firmly in place.

Vickers hardness test Sometimes known as the Diamond Pyramid Hardness test, it is a popular method of determining the *hardness value of materials, especially metals.

Victorian architecture Buildings and structures constructed during the period from 1837 to 1901. Buildings were constructed on a scale and level that had not been experienced in any previous era. During this period, Paxton's Crystal Palace was constructed in 1851 and back-to-back housing emerged.

Vierendeel truss A structural frame that does not have any diagonal (triangulated) members. Openings between members are therefore rectangular. Fixed joints are required between members to resist vertical shear/bending, rather than pin joints, which are associated with conventional triangulated trusses.

villa 1. A Roman landowner's residence. **2.** Dwelling on the outskirts of the town, a country house, or a holiday home.

village A collection of dwellings in a rural area with a church. A collection of small homes in the country without a church is known as a hamlet.

vinyl Shiny, tough, and flexible plastic, used especially for floor coverings.

violations Intentional failures as a result of human behaviour or actions, where a person deliberately does the wrong thing. Violations of health and safety rules are one of the most common causes of injuries at work.

VIP *See* VACUUM INSULATED PANEL.

virtual design construction (VDC) A technology-based approach for the management of integrated multidisciplinary performance models. It can be viewed as part of the work process within *BIM.

virtual learning environment (VLE) An educational web-based platform in which learning material, activities, interactions, and assessments can be populated.

virtual work The work done on a structure that has resulted from either a virtual force acting through a real displacement, or a real force acting through a virtual displacement.

viscoelastic A material that exhibits both viscous and elastic properties. For example, asphalt could deform under the influence of stress or high temperature to such an extent that it would flow (become a liquid). However once the heat/stress has been removed, it could recover elastically to its original 'solid' shape. If a material has **viscoelasticity** it will exhibit both viscous and elastic characteristics when under deformation.

viscometer An instrument for determining the *viscosity of a liquid.

viscoplasticity A material that exhibits both viscous and plastic properties. Within the plasticity phase, the material would be subjected to a time-dependent stress/strain response, while in the viscous phase the material could flow like a liquid.

viscosity Coefficient of resistance to flow; a measure of how easily a liquid flows.

viscous damping The reduction of energy when the velocity of a particle is resisted by an opposing proportional force.

viscous flow *See* LAMINAR FLOW.

viscous impingement filter *See* IMPINGEMENT FILTER.

visible defect A defect in timber, such as a knot or shake that is identified during *visual grading. The defect may make the timber unsuitable for work that is to have aesthetic qualities. The defects may also affect the structural properties, but the structural properties of timber are now determined by mechanical grading.

vision-proof glass Glass that is obscured preventing one from seeing through it.

visual display unit (VDU) A computer screen.

visual grading The grading of timber by visual examination of the gross features, the defects within the timber.

vitiated air 1. Air in which the oxygen content has been reduced. **2.** Air that is not fresh.

vitreous china A ceramic material manufactured at very high temperatures, thus imparting excellent impermeability properties, making it ideal for bathroom applications since it prevents the ingress of bacteria through the surface.

vitreous enamel A coating material produced by fusing glass that has been ground to a powder. The glass is heated until it melts at temperatures of 750–800°C; in its liquid state it is applied to surfaces and used to provide a protective cover to materials such as metal, providing a smooth vitreous glass finish.

vitreous glass A type of glass that is popular for mosaic tile work. This type of glass is durable, resistant to chemical attack, and comes in an array of colours.

vitrification During firing of a ceramic body, the formation of a liquid phase that upon cooling becomes a glass-bonding matrix. Vitrification is one of the key reactions that occurs during the fabrication of glass.

vitrify To change into glass, usually by heat fusion. **Vitrified clayware** is a ceramic material that has been subjected to a temperature around 1100°C to form a material that has very low water absorption properties; it is used for floor tiles and drain pipes.

Vitruvius Roman architect whose work had a significant influence on Renaissance structures and buildings. Vitruvius defined the Vitruvian Man, as drawn by Leonardo da Vinci, as ideal proportions of the human body defined within geometry.

V-joint A joint formed in a V-shape. *See also* VEE JOINT.

VLE *See* VIRTUAL LEARNING ENVIRONMENT.

V-notched trowel A trowel with a 'V' cut into it for applying even amounts of adhesive to a surface.

void A space or gap; voids occur in concrete due to air bubbles that occur during mixing and hydration and are not vibrated out, or between coarse aggregate where there are no fine particles, such as sand and cement, to fill the gaps.

void ratio (e) The ratio between the volume of voids and the volume of solids in a material, particular soil.

volcanic Relating to a volcano; **volcanic dust**, is fine dust particles suspended in the air from a **volcanic eruption**.

volcanology (vulcanology) The study of volcanoes.

volt (V) The *SI unit of electrical potential difference or electromotive force. One volt is the electromotive force that is required to move one ampere of current through one ohm of electrical resistance.

voltage A measure of the electrical potential difference or electromotive force. Measured in *volts.

voltage drop The drop or difference in voltage that occurs between two points in an electrical circuit.

voltage overload (excess voltage, overvoltage) A voltage load on a component that is in excess of what it was designed to carry. Electrical devices are usually protected from voltage overload by *protection gear.

volume The space that a material or area occupies, measured in cubic meters m^3.

volumetric building (volumetric systems) A preassembled building unit or system. The module can be a component of a building, such as a toilet pod, bathroom pod, chimney stack, or other similar unit, or could be a whole building built in a factory and delivered to site.

volumetric efficiency The actual volume of liquid that is pumped by an engine or piston compared to its theoretical maximum.

volumetric strain The ratio between the change in volume to its original volume, which occurs when a material deforms.

volume yield 1. The volume of concrete that can be determined from the weight of cement, aggregates, and water. **2.** Volume of lime putty obtained from the weight of *quicklime.

volute A spiral that forms part of the uppermost part of a column (column capital).

von Mises yield A criterion to denote that failure will occur when the strain energy reaches the same energy for yield to occur. It is frequently used to calculate the approximate yield of ductile materials and is also known as the **maximum distortion energy criterion, octahedral shear stress theory** or **Maxwell-Huber-Hencky-von Mises theory**.

vortex 1. A circular motion or **forced vortex**, approximating to the flow pattern created by a mechanical rotor or by a body of fluid being whirled around in a container. **2.** A circular motion or **free vortex**, which occurs naturally; for example, a flow down a drain hole or around a river bend. **3.** A forced vortex or **compound vortex**, surrounded by a free vortex; for example, the core of the vortex rotates as a solid body. **4.** A combination of radial flow and a free vortex, or **free spiral vortex**, occurs in the volute of a *centrifugal pump or between the guide vanes of a *reaction turbine.

voussoir V- or wedge-shaped bricks used to form an arch.

voussoir arch An arch in a bridge that contains wedge-shaped blocks, which are placed in such a way as to distribute the load efficiently to maximize the compressive strength of the stone.

vulcanization A chemical process to treat rubber. Sulphur is added to rubber at high temperatures to form crosslinks between individual polymer chains, resulting in a product which is harder, stronger, and more elastic, for example, hoses.

WADGPS (wide-angle differential GPS) A network of ground-based stations that track satellite signals to improve GPS accuracy.

wadi A rocky watercourse that is predominantly dry except after heavy rainfall.

waffle floor A suspended floor slab employing a square grid construction of ribs with deep coffers (recesses). Large spans are possible due to reduced dead load.

wagon drill A pneumatic or percussive-type rock drill, mounted vertically on three or four wheels, used to sink shafts from the surface of the ground.

wailing (US wale) A horizontal timber beam, used to provide support to the side of an excavation or formwork.

wainscot (wainscoting) Timber wall panelling or lining used on an internal wall that extends from the *skirting board up to *dado rail height. Previously used to cover the lower half of internal walls that were prone to damp penetration, nowadays used for decorative purposes.

waiver Voluntarily giving up a right or privilege.

walking line The central line on a stairway where it is expected that people will use the stairs. It is the point at which the going is measured for *kite winders or stairs that are diagonally shaped.

wall A vertical element used to enclose or subdivide a space, it is one of the primary elements of a building.

wall chaser A power tool for cutting a *chase in brickwork, blockwork, and concrete.

wall friction The shear resistance that has been generated at the interface between the back of a retaining wall and the soil that helps to stabilize the wall.

wall-hung Attached to a wall.

wall panel (panel wall) A prefabricated section of wall.

wall plate A structural element (connection) that distributes a load onto a wall, and acts like a mini *lintel.

wall string *See* STRING.

wall tie A fixing element that is used in cavity wall construction. It spans the cavity and is embedded within the mortar joints on both sides to hold the two leafs together. Usually made from stainless steel, galvanized steel, or plastic, and will be designed to prevent water passing from one leaf to the other.

ward The area or grounds of a castle that are surrounded by the castle walls.

warm roof A *pitched roof where the insulation is placed in line with the roof slope. The insulation is placed over and between the roof trusses, creating a warm roof void (*see* figure), which is different from a *cold roof, where the insulation is placed between the joists at the horizontal ceiling level. In most cases where insulation is placed in the upper part of the roof, the roof space is used to create a room in the roof. Where a habitable room in the roof is created, care should be taken to ensure that breather membrane and vapour barriers are in place to prevent moist air entering into the roof structure. Careful attention is needed to ensure airtightness in the roof space. The ceiling structure is often complex, and it is essential that all plane and junction elements are properly formed, interconnected, and sealed.

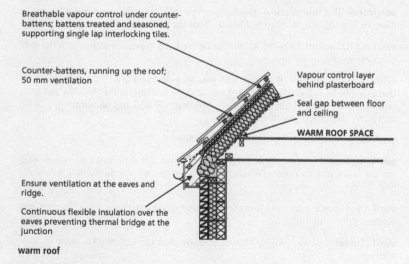

Breathable vapour control under counter-battens; battens treated and seasoned, supporting single lap interlocking tiles.

Counter-battens, running up the roof; 50 mm ventilation

Vapour control layer behind plasterboard

Seal gap between floor and ceiling

WARM ROOF SPACE

Ensure ventilation at the eaves and ridge.

Continuous flexible insulation over the eaves preventing thermal bridge at the junction

warm roof

warning pipe An *overflow connected to a sanitary appliance that discharges to an obvious position outdoors.

warp The distortion or twisting of something in the longitudinal direction, e.g. yarns running in the machine direction of a loom, which become crimped due to the insertion of the *weft (cross-machine direction) yarns.

Warren girder (Warren truss) A triangulated truss that has only sloping members of approximately equal sides between the top and bottom *chords (it has no vertical members).

WASC (water and sewage companies) UK water utilities companies that both supply drinking water and treat sewage.

wash boring A drilling technique that employs a jet of water to displace the soil in front of a pile or casing being installed.

washdown closet (washdown pan) A WC (*water closet) where the contents are removed by a flush of water running down the pan.

washer A thin, usually flat disk that is used to prevent leakage, distribute pressure, reduce friction, or as a spacer.

Washington Accord An international accreditation agreement for professional undergraduate engineering academic degrees between the bodies responsible for accreditation in its signatory countries. Thus, it provides international recognition, quality, standards, and mobility of educational qualifications. From the initial founding six signatories in 1989, it has now grown to include the following countries: Australia, Canada, Taiwan, Pakistan, China, Hong Kong, India, Ireland, Japan, Korea, Malaysia, New Zealand, Russia, Singapore, South Africa, Sri Lanka, Turkey, the United Kingdom, the Philippines, and the United States. Engineering technology and postgraduate programmes are not covered by the Washington Accord. *See also* DUBLIN ACCORD; SYDNEY ACCORD.

washland An area of land between a river and flood embankment, such as a flood plain, where a river can temporarily flood without causing detrimental damage.

wash load Fine sediment that is nearly always in suspension (too fine to settle) in a watercourse such as a stream or river.

washout closet An early type of *water closet where the human waste was deposited in a shallow bowl filled with water. When the toilet was flushed, the contents of the bowl were washed out. Replaced by the *washdown closet.

washout valve A valve used to drain (empty) something completely.

waste 1. Rubbish, something that has been discarded. **2.** To use something carelessly, without effect or purpose.

waste disposal unit A device that flushes away food waste. It is typically fitted beneath a kitchen sink, on the underside of the plughole. Its purpose is to shred food waste into small pieces such that they can be passed through the drainage pipework into the sewage system.

waste pipe A pipe connected to a basin, bath, or sink to carry discharge water to the *soil stack.

waster A mason's chisel characterized by a claw-type cutting head or with a cutting head 19 mm wide.

wastewater Any water that has been used, either domestically or industrially, and now requires water treatment before reuse.

wasteway (waste weir) *See* SPILLWAY.

water bar 1. A strip of material, usually metal, plastic, or rubber, that is designed to prevent water ingress. Can be located in a groove between two components, on the threshold of a door, or can be cast into the construction joints of basement walls to prevent the passage of moisture through the joint **2.** A water barrier

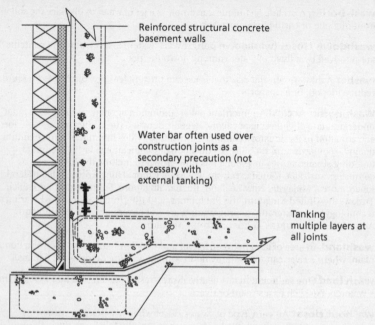

Reinforced structural concrete basement walls

Water bar often used over construction joints as a secondary precaution (not necessary with external tanking)

Tanking multiple layers at all joints

Water bar cast in basement walls preventing ingress of water across construction joints.

(**waterbar**) in the form of a strip or channel. It can be used across a construction joint to prevent the ingress of water or across steep-sloping earth roads to prevent erosion from fast-flowing surface water.

water catchment *See* CATCHMENT AREA.

water/cement ratio The ratio of water and cement/pozzolan (by weight) present in both concrete and mortar. This ratio is very important as it influences both the compressive strength and durability. In most concretes and mortar mixes, an optimum ratio corresponds to the maximum compressive strength.

water closet (WC) A sanitary appliance that collects human waste (excreta and urine), and then uses water to flush the waste to another location. Usually constructed from *vitreous china. *See also* SIPHONIC CLOSET; WASHDOWN CLOSET; WASHOUT CLOSET.

water content *See* MOISTURE CONTENT.

water gauge 1. An instrument that measures the flow of water in a stream or river. **2.** An instrument for measuring the water pressure in a tank or boiler.

water hammer The noise caused by a sudden change in the flow of a fluid in a pipe.

water jet A pressurized stream of fluid forced out of a small nozzle; *see* PELTON WHEEL.

water level The surface level of a body of water.

water main The main underground pipe in a water distribution system.

water meter An instrument that measures the amount of water flowing through it. Used to record for billing purposes how much water has been consumed by a household.

water of capillarity *Groundwater that is held above the *water table due to *capillarity.

waterproof Objects or materials that resist water penetration.

water purification *See* WATER TREATMENT.

water quality The physical, chemical, and biological characteristics of water, which are determined by the end use; *see* DRINKING WATER.

water reducer *See* PLASTICIZER.

water-related diseases Illnesses that stem directly from consuming or indirectly from being in contact with contaminated water. The contaminant can be either biological, or chemical, or a mixture of both. The majority of diseases resulting from microbiological pollution are essentially contracted from water contaminated with human faecal matter and include cholera, typhoid, and other diarrhoeal or non-diarrhoeal infections such as schistosomiasis, skin infections, and yellow fever. Contamination of chemical origin include arsenicosis, which is caused by prolonged low-level exposure to arsenic, and can lead to skin keratosis. Such illnesses are prevalent in developing countries where water treatment is lacking, and are so widespread that they cause more deaths in the world than malaria and HIV/AIDS.

water repellent (waterproofing paints) Special types of paint that are water repellent and can be applied to porous surfaces including brick, concrete, stone, and renderings to prevent damp penetration. Such treatment will not prevent rising damp, but it will allow the continued evaporation of moisture within the masonry.

water-repellent cement *Portland cement that contains a water-repellent agent that has been added during the manufacturing process. Such a cement is used to produce impermeable concrete.

water retention The ability to retain water.

watershed The boundary between two *catchment areas.

water softener A chemical or device that is used to reduce the hardness of water by removing calcium and magnesium salts.

water-soluble The ability to dissolve in water.

water supply The distribution of *drinking water to consumers by a water utility.

water table (phreatic surface) The level of water in the ground below which the pore structure is totally saturated; *see* GROUNDWATER.

water test A hydraulic test used to check the seal in drains. The drain is filled with water to a predetermined head of water and allowed to stand. The level of the water is marked and checked after a period of time to see if the water level has fallen. If the hydrostatic pressure of the water has caused the water level to fall, there is likely to be a leak in the drain.

water tower An elevated tank containing water to be feed into a water distribution system.

water treatment The removal of impurities from raw water in order to make it fit for human consumption. There are a variety of techniques available to treat raw water, which can typically be represented in three main stages: *primary treatment, *secondary treatment, and *tertiary treatment.

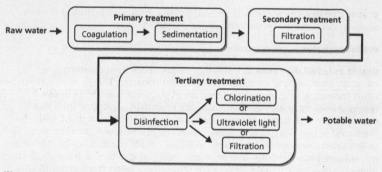

Water treatment process

water turbine A *turbine that is driven by a head of water; *see* KAPLAN TURBINE; REACTION TURBINE.

water utility A company that supplies drinking water to consumers and/or treats wastewater. *See also* WASC.

water vapour Water in a gaseous state.

water wall *See* TROMBE WALL.

waterway 1. A river or canal used by boats and ships. **2.** A channel to drain water away.

waterworks The location where raw water is treated, stored, and distributed into the supply network.

watt (w) The *SI unit of electrical power in terms of one *joule per second.

wattle Sticks or stakes that have been interwoven with branches and twigs to form fences, walls, and roofs.

wattle and daub A type of wall construction comprising interwoven twigs, sticks, and branches (wattle) that are covered with clay, cow dung, or mud (daub). Traditionally used as an infill in post-and-beam timber-frame buildings.

wave A disturbance that propagates through space and time, normally associated with a transfer of energy.

waveband A range of *wavelengths in which transmissions can occur.

wave energy Harnessing energy from sea waves.

wave equation A second-order partial differential equation. Used for modelling sounds, light, and water waves.

waveform The shape or curve of a *wave.

wavelength The distance between two consecutive peaks or troughs on adjacent waves of the same phase.

wavelet A very small *wave such as a ripple.

wave surcharge An increase in water level due to successive grouping of waves.

waybeam A beam that runs beneath the rail to provide support and a place to secure base plates when a bridge section of the rail track is being repaired.

waypoint A signpost on a route noting a change of direction.

WC *See* WATER CLOSET.

weak Not strong; unable to support load, hence carries a high risk of failure.

wear Damage caused by continuous rubbing or friction.

wearables Technology that can be worn to provide various types of information. For example, in a workplace setting a device could be worn by workers to warn them they are getting too close to a dangerous or moving machine.

wear failure A type of degradation mechanism caused by continuous rubbing and friction, especially common in metals. Usually there are three types of wear failures: adhesive, abrasive, and erosive. Most wear failure is characterized by loss of material (from the surface).

weather To allow materials to be exposed to the elements to give them a used and tired look.

weather bar Steel or aluminium strip of metal bar that is rebated in the sill of window and doorframes to prevent water being blown into the property over the sill. The bar provides a physical barrier that prevents water entering the building by either capillary action or driving rain.

weatherboard 1. A piece of *weatherboarding. **2.** A horizontal board fixed to the bottom of an external door to prevent rain penetration.

weatherboarding (clapboard) External timber cladding comprising a series of long thin feather-edged horizontal timber boards.

w

weathercock (weather vane, wind vane) A device, usually located on the top of a building, that is used to indicate the direction of the wind.

weathered joint *See* WEATHER-STRUCK JOINT.

weathering The deterioration due to the exposure to climatic conditions. It is particularly used in reference to the breakdown of rocks, caused by water, ice, chemicals, and changing temperature.

weatherstrip A thin strip of material used around windows and doors to prevent air infiltration through cracks and gaps.

weather-struck joint (struck joint, weathered joint, weathered pointing) A horizontal masonry joint that slopes outwards from the top of the joint to shed rainwater away from the wall.

web The vertical part of a beam or rail, which adjoins the top and bottom horizontal flanges. If the web experiences high compressive stresses and moves out of line **web buckling** can occur.

weber The SI unit of magnetic flux. It measures the amount of magnetic field passing through a conducting surface and is equal to one joule per ampere.

wedge A tapered solid block of wood or steel that is used to separate two objects or to split timber to form two separate pieces.

wedge anchor A *rock anchor that uses a mechanical wedging effect for fixing.

wedge cut A drill hole pattern used in 'drill and blast tunnelling' that forms a 'V' or wedge shape.

wedge theory A method for analysing the force behind a retaining wall, based on the theory that a wall needs to support the weight of a wedge of soil that would move if the wall fails.

weep hole A small water drainage hole. In masonry wall construction, it is used above a *cavity tray and *damp-proof course.

weeping Something that is leaking fluid.

weft The crosswise yarns that pass through the *warp yarns on a loom to produce a woven structure (for example, a *geotextile).

weigh batcher A concrete batching plant where the quality of each material in the mix (excluding water) is measured by weight.

weight The forces exerted on a mass due to gravity. The weight of an object is obtained by multiplying its mass by the acceleration due to *gravity.

weighting To add or put a bias on something so that it prevents or counteracts something occurring, e.g. the use of additional loads to make an object heavier so that it does not float.

weir A small dam or wall built across the full width of a river to regulate flow. It has a horizontal crest or notch to allow water to continue to flow over it but causes the water level to rise upstream. The **weir head** is the depth of water between the weir crest or bottom of the notch, and the upstream water level.

Weisbach triangle A surveying setup used for shafts.

weld *See* WELDING.

weldability The ability of a metal or other material to be welded under specified conditions.

weld decay Corrosion that occurs in some welded stainless steels at regions adjacent to the weld.

weld defects Imperfections occurring during or after the welding process.

welder A tradesperson who specializes in welding materials together.

welding A technique for joining metals in which actual melting of the pieces to be joined occurs in the vicinity of the bond. A filler metal may be used to facilitate the process. In this process both similar and dissimilar metals can be used; several welding techniques exist including arc welding, gas welding, brazing, and soldering.

welding work angle In arc welding, the angle between the electrode and one of the joints.

well A hole or shaft that is dug or drilled into the ground to extract water, oil, gas, and so on.

well-being A holistic model of well-being is based on healthcare, philosophy, psychology, and sociology literatures, which converge on three core dimensions of well-being: psychological (happiness), physical (health), and social (relationships). Happiness refers to the psychological well-being of employees. Key issues in the workplace include satisfaction with work and life in general, subjective experiences and functioning at work, and commitment to the organization. Health refers to the physical and psychological well-being of employees in terms of experiences of strain or work-related stress and outcomes such as cardiovascular disease, hypertension, sleeping problems, mental health issues, and workplace accidents. Relationships concern the interactions and quality of relationships between people, both within the workplace and in their personal life beyond work.

well-conditioned triangle A triangle used in surveying that is, more or less, equilateral, such that any error in the measurement of the angle has minimal effect when computing length.

Welsh arch A small arch spanning less than 12 in. or 300 mm.

welt (welted seam) A raised seam found in sheet-metal roofing.

welted drip A drip used at the eaves of built-up felted roofs.

westing The position of movement westwards in reference to *longitude coordinates.

wet analysis A particle-size sieve analysis that uses water to wash the particles through the various sized sieves. It is normally used for very fine particles that would otherwise agglomerate when dry and would not pass through the aperture.

w

wet and dry bulb thermometer A *hygrometer containing two *thermometers. The bulb on one of the thermometers is wrapped in a muslin cloth soaked in distilled water (wet bulb), while the other thermometer is left unwrapped (dry bulb). The drier the air the quicker the water will evaporate from the wet cloth and in doing so will cool the bulb. Atmospheric humidity can be measured from the difference between the two thermometer readings.

wet bulb temperature The lowest temperature that can be achieved by the air by evaporative cooling. It is measured using a *wet bulb thermometer.

wet bulb thermometer A *thermometer in which the bulb is either covered by a wet cloth or is kept moist.

wet cube strength The strength of a concrete cube after it has been completely saturated with water. The wet strength is lower than the dry strength.

wet deposition The process where by acid gases and particles in the Earth's atmosphere are dissolved in water and deposited on land, water, and surfaces via fog, mist, raindrops, snow, or sleet. *See also* DRY DEPOSITION.

wet dock A *dock in which water levels are kept at high tide by dock gates.

wet drilling The use of water on the drill bit when drilling to reduce dust.

wet edge How easily a painted wet edge can be integrated in areas of overlap.

wet galvanizing A galvanizing technique whereby the metal is introduced into a bath of zinc via molten flux.

wetlands A land area that has a very high *groundwater level and is constantly saturated, including swamps, marshes, and bogs.

wet mix Concrete or mortar containing too much water, immediately evidenced by a runny consistency.

wet rot A decay-affecting timber, caused by alternate wetting and drying.

wetted perimeter The sum of the depth of water at each side of an open channel plus the width of the base of the channel.

wharf A landing area for ships and boats that is sometimes sheltered from the elements. *See also* BERTH or JETTY.

Wheatstone bridge A device used to measure the change in electrical resistance of a flat coil of very fine wire that is glued to the surface of an object.

wheelabrating A form of *shot blasting that utilizes steel grit.

wheeling step (Scotland wheel step) A *winder for a stair.

Whipple-Murphy truss Similar to the *Pratt truss but has diagonals that extend across the base of at least two panels.

Whirley crane A large crane that can rotate 360°.

whirlwind A fast rotating column of air that has a centre of low pressure.

white and coloured Portland cement A type of *Portland cement known for the high degree of whiteness in appearance, although other colours are also possible to obtain.

Whitney stress diagram A plot showing the stress distribution in a concrete reinforced beam based on the theory of ultimate load.

whole-brick wall (one-brick wall) A solid wall one brick thick.

whole-circle bearing A horizontal angle measured in a clockwise direction from a fixed point, which is usually true north.

whole-life cost (WLC) The total of all relevant costs and incomes arising from the purchase and possession of an *asset. It includes the total cost of ownership over the life of an asset, also commonly referred to as cradle-to-grave cost.

Wichert truss A multi-span truss in which vertical members above intermediate supports have been omitted to make the truss statically determinate.

wicket door A small access door located in the leaf of a larger door.

wicking The capillary action (*see* CAPILLARITY) that occurs in fabrics and yarns.

winch A device for hoisting loads using a cable or rope wound round a cylinder, which is turned by hand or by a motor.

wind The natural movement of air across the surface of the earth, caused by pressure differences in the atmosphere.

wind brace A support to a structure that is designed to resist wind loads.

wind effect Air moving around and over a building induces pressure differences across the building and generally results in a small negative pressure internally. Outdoor air will enter the building through cracks and gaps in the construction (*infiltration) on the windward side of the building that is under positive pressure, and indoor air will exit the building (*exfiltration) on the leeward side of the building that is under negative pressure.

wind energy (wind power) Harnessing the energy from the wind through the use of wind turbines.

winder (winding stair) A tapered or triangular stair. The triangular or kite-shaped tread is used to enable a flight of stairs to change direction while still continuing to rise.

wind generator (wind turbine) A device for capturing *wind energy.

windlass A lifting device similar to a *winch used to raise and lower an anchor.

wind load The force exerted on a structure by the *wind.

window An opening formed in a wall or roof to admit daylight through a transparent or translucent material. The most common transparent or translucent material used is glass, although other materials such as Perspex® are used for roof lights. The glass is fixed in a *window frame that may be subdivided by *mullions and *transoms, and may contain a number of *casements. Many types of windows are available, including *casement windows, *pivot windows, *sash windows, and composite action windows.

window back *See* WINDOW BOARD.

window bar *See* GLAZING BAR.

window board (elbow board, elbow lining, Scotland sillboard, window back, US window stool) The panelling on the internal face of an external wall under a window sill.

window frame The members that form the perimeter of a window. Consists of a *sill, a *head, and two *jambs. The frame may be subdivided by *mullions and *transoms.

window sill The horizontal shelf at the bottom of a window frame.

window stool (US) *See* WINDOW BOARD.

window tax A tax system used between 1697 and 1851 where the amount of tax paid on a property was related to the number of its windows. The system resulted in many properties having bricked-up windows to reduce the tax levied.

wind shear A variation in wind velocity or direction over a short distance in the atmosphere.

windshield A screen or wall used to provide protection from the wind.

wind speed The rate at which air moves in the atmosphere.

windtightness A measure of the ability of the external leaf of the building *envelope to prevent air penetration.

wind tunnel A tunnel-shaped testing chamber in which air is blown through at different speeds. Objects or scale models are placed in the tunnel to assess how they disrupt the air flow.

windward Facing the wind, i.e. towards the direction the wind is blowing.

windwashing The process where by the external air penetrates the external leaf of the building *envelope and moves over and through the insulation. This results in additional heat, thus reducing the effectiveness of the insulation.

wings The side projections or outer proportions of a building.

wiped joint A joint prevalent in lead pipes facilitated by the soldering process.

wire A strand of metal used to carry electric current.

wizard stick A battery-operated hand-held smoke-producing wand that can be used to identify the location of *air leakage during a *fan pressurization test.

wobble-wheel roller A roller that has a system of pneumatic tyres suspended on springs.

wood A naturally occurring material derived from trees, also known as timber.

wood adhesive *See* ADHESIVE.

w

woodblock floor Flooring constructed from individual rectangular wooden blocks. Used in areas where heavy traffic is expected.

wood bonding The joining of pieces of wood.

wood-boring insects Species that bore through timber and wood resulting in degradation of the material. In the UK the predominant insects are beetles.

woodworm A beetle larva that bores into wood.

work Can refer to the amount of energy used when an object is moved by force. Work done = strength of force x distance moved, measured in joules (J).

workability The ease of placing and compacting concrete, measured by a *slump test or *compaction factor test.

worker A person working to the terms set within a contract of employment (like an employee*) who will generally have to carry out the work personally. However, some workers may have a limited right to send someone else to carry out the work, such as a subcontractor. Workers could include persons undertaking casual work, agency workers, freelance workers, seasonal workers, and people on zero-hours contracts. Workers are entitled to some employment rights including:
- the National Minimum Wage
- holiday pay
- protection against unlawful discrimination
- the right not to be treated less favourably if they work part-time (ACAS, 2018).

(⊕) SEE WEB LINKS
- ACAS employment status advice.

work hardening (strain hardening, cold work) A procedure used in metallurgy to increase the ultimate tensile strength of a metal by modifying the crystal structure. The process involves the repeated plastic deformation of the material. During work hardening, the dislocation density in metals increases through straining with an applied stress.

working drawings The drawings that are the most current and those which are being used to construct the building. Many drawings may be superseded and would not form the working drawings. The working drawings can often be different from the contract drawings, but would be important in deciding contract variations.

working load (working stress) The maximum safe load or stress that a structure can withstand.

work-life balance The balance that an individual needs between time and other resources allocated for work and aspects of life outside of work, such as personal interests, sports, hobbies, family, and social or leisure activities.

workmanship 1. The skill of a manual worker. **2.** The quality of the work produced.

workmate Person working alongside or within a team; a colleague.

world geodetic system A geographical reference frame for the earth, used in *global positioning systems.

woven The interlacing of two sets of yarns perpendicular to each other. i.e. the *warp and *weft (machine and cross-machine direction respectively). Plain, twill, and satin are the main types of interlacing procedures/patterns. In general, these structures are anisotropic, possess high modulus, have high strength in both the warp/weft directions, but exhibit greater elongation in the machine direction as a result of higher crimp in the warp yarns. *See also* GEOTEXTILE.

wreath A curved section of stair handrail that also inclines or declines.

Wren, Christopher (1632–1723) Architect to Charles II during the Renaissance period, Wren designed St Paul's Cathedral and was responsible for designing many churches and other fine buildings after the Great Fire of London (1666). He is perhaps the most famous English architect.

wrench fault (tear fault, transcurrent fault) A fracture in rock that is characterized by a large lateral displacement and minor vertical displacement.

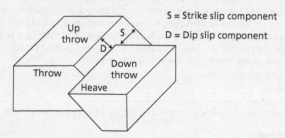

Wrench fault

wrought alloy A metal alloy that is suitable for mechanical forming below melting point temperatures.

wrought iron A ferrous metal that is essentially pure iron, chemical formula Fe. Wrought iron used to be made and utilized in very large quantities for construction and general engineering purposes in Victorian times; applications included bridges and fences. However, currently it is not produced any more, except in small amounts made for demonstration purposes in industrial museums. A distinction is made between cast and wrought metals. Metals that are so brittle that forming or shaping by appreciable deformation is not ordinarily possible, are cast. On the other hand, those that are amenable to mechanical deformation are termed wrought alloys.

w-shape A wide-flange W-shaped *I-section.

WUFI: Wärme Und Feuchte Instationär (WUFI®) A family of software products that allows realistic calculation of the one- and two-dimensional heat and moisture transport in walls and other multilayer building components exposed to natural weather. WUFI software uses the latest findings regarding vapour diffusion and moisture transport in building materials. The software has been validated by detailed comparison with measurements obtained in the laboratory and in outdoor field tests.

wye Something that has a Y-shape, such as a drainage fitting.

X-bracing *See* CROSS-BRACING.

xenolith A rock fragment encapsulated within the formation of the surrounding rock, such as a small foreign piece of rock becoming entrapped in *magma before it cools to form igneous rock.

***Xestobium rufovillosum* (deathwatch beetle)** Wood-boring insect that bores through wood making a ticking sound.

X-rays High-energy electromagnetic radiation, with very short wavelengths (between 0.01 and 10 nanometres) which can penetrate substances such as concrete and metal. The X-ray technique is a very useful method of ascertaining the physical properties of materials, especially physical metallurgy.

X-ray scattering techniques Non-destructive analytical tests and techniques that reveal information on the chemical composition, crystallographic structure, and some of the physical properties of the materials. The techniques observe the scattered intensity of an X-ray beam hitting a material sample.

x-section *See* CROSS-SECTION.

xylonite A *thermoplastic polymer registered in 1870 as Celluloid. It can be easily moulded and shaped and has been used for photo-elastic models. It is also highly flammable and decomposes easily, and is therefore no longer widely used.

yard (yd) Unit of length, defined within the *imperial units system equal to 36 inches or 3 feet, which is equivalent to 0.9144 metres in the *metric system.

yard trap *See* GULLY.

yarn A continuous twisted collection of interlocking fibres, often used for knitting or weaving fabrics, such as *geotextiles.

Y-branch (Y-bend, Y-fitting) *See* WYE.

yield To give way or provide no further resistance.

yielding The start of *plastic deformation when a material is deformed.

yield point (yield stress) The point (stress level) at which a material changes from elastic to plastic behaviour. If a slight increase in loading is applied beyond the elastic limit (where stress is proportional to strain), the material will now start to deform without any increase in loading.

yield strength The stress required to permanently deform a material, the point where *plastic deformation commences. On a stress-strain graph the yield strength corresponds to the maximum value of stress in the linear region of the graph. Yield units of strength are usually measured in MPa or N/mm^2. In some cases the yield point may not be easily discernible from the stress–strain, thus, as a consequence, a convention has been established where a straight line is constructed parallel to the elastic portion of the stress–strain curve at some specified strain offset, usually 0.002. The stress corresponding to the intersection of this line and the stress–strain curve as it bends over in the plastic region will determine the yield strength.

yoke 1. A wooden beam to harness two draught animals. **2.** A clamp to hold/control the movement of parts on a machine. **3.** A strengthening framework which is temporarily fixed around the *perimeter of the *formwork to a square/rectangular column during casting.

Yorkshire bond (Monk bond, flying Flemish bond) The pattern of brickwork where two *stretchers are placed between each *header and successive courses are consistent with equal *lap, to bring alternative headers into regular arrangement.

Yorkshire light A horizontally sliding *sash window where one half moves horizontally and the other half is fixed.

Young's modulus (Young's modulus of elasticity, E) A mechanical property that denotes the initial linear relationship between the stress and strain of a material. It signifies a materials resistance to deformation under load, and when the load is removed, the deformation is recovered.

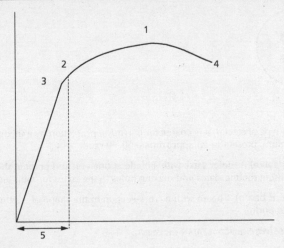

1. Ultimate strength
2. Elastic limit
3. Proportional limit
4. Rupture
5. True elastic limit

Strain = elongation/original length
Stress = force/area

Yield strength: Typical stress–strain behaviour for a typical ductile metal

Ytong A commercial trademark for *aircrete (autoclaved aerated concrete).

yttrium (Y) A silvery metallic element, occurring in nearly all rare-earth minerals, but not a rare-earth element, and used in various metallurgical applications, notably to increase the strength of other metal alloys and materials.

zalutite A type of steel that is coated for corrosion protection by a mixture of zinc and aluminium, present in an approximate 50–50 ratio.

zax (slater's axe) A roofer's axe with a blade at one end and point at the other. Used for cutting roofing slates and making holes in the slates for nails and fixings.

Z-beam (zed beam) A beam with a cross-section in the shape of a letter Z, often used as a *Z-purlin.

zebra crossing *See* PEDESTRIAN CROSSING.

zenith The highest point within the celestial sphere that is directly above the observer.

zeolite One of a large group of silicate minerals, occurring in weathered igneous rocks that have reacted with alkaline groundwater. Zeolite is often used to soften water due to its ion exchange properties.

zero-carbon footprint *See* CARBON-NEUTRAL.

zero harm The notion that no one should experience injuries, diseases, illnesses, or fatalities in a workplace.

zeta potential The amount of *electrostatic repulsive force between two charged particles, with particular reference to the interface between a particle and a liquid; for example, the formation of *floc in water treatment due to charge difference (ions) of the *coagulant.

zinc (Zn) A non–ferrous alloy with a relatively low melting point. Zinc is used to a small extent for roofing and cladding, but its main importance in construction is as corrosion protection for steel as galvanizing. A thin coating of zinc is applied to steel items to protect them against wet corrosion. This process is cheap and it significantly extends the life of steel items, and improves the durability of steel.

zinc chromate primer A *primer of zinc chromate used as a protective coating in metal components.

zinc oxide (Chinese white, zinc white, ZnO) An amorphous white or yellowish powder, used as a pigment in compounding rubber.

zinc phosphate primer A type of primer used for its excellent resistance to the formation of rust.

zinc-rich paint A type of paint with a high concentration of *zinc.

zinc roofing Weather protection on a roof that is formed out of zinc or zinc alloy sheets.

zinc silicate primer A *primer based on zinc silicate used in anti-corrosive paint systems to promote long life of the surface.

zircon ($ZrSiO_4$) A common mineral occurring in small crystals, its appearance is brown to colourless. When heated, cut, and polished, zircon forms a brilliant blue-white gem. Zircon is used as a refractory material when opaque and as a gem when transparent.

zirconium (Zr) A non-ferrous metal relatively abundant in the earth's crust. Zirconium is primarily used as an alloying element due to its excellent resistance to corrosion.

zone Any group of crystal planes that are all parallel to one line, called the **zone axis**.

zoning 1. A system of land-use planning based on boundaries, inside which land can only be used for specific purposes, such as agriculture, dwellings, green belt, industry, or recreation. **2.** The subdivision of the space within a building into separate areas to enable individual control of factors such as air conditioning, building services, fire detection systems, heating, lighting, noise, ventilation, and so on.

zooglea A bacterium used in the treatment of sewage. It is a gram-negative bacillus, with a genus name *Zoogloea*.

Z-purlin (US Z bar) A cold-formed *steel *purlin in the shape of the letter Z. Used as part of a roofing support system or as side rails for vertical cladding.

z transform A mathematical transformation that converts discrete time domain signals into a complex frequency domain, similar to the *Laplace transform.

z

Further Reading

Ahmed, A and Kamau, J (2017) 'Sustainable Construction Using Autoclaved Aerated Concrete (Aircrete) Blocks', *Research & Development in Material Science* 1(4): 1–4.

Blockly, D. (2005) *The Penguin Dictionary of Civil Engineering*, London: Penguin.

Brett, P. (1997) *Illustrated Dictionary of Building*, London: Butterworth-Heineman.

Bucher, W. (1996) *Dictionary of Building Preservation*, New York: Preservation Press, John Wiley.

Chappel, D., Marshall, D., Powell-Smith, V., and Cavender, S. (2004) *Building Contract Dictionary*, Oxford: Blackwell.

CIPD (2018) Chartered Institute of Personnel and Development (CIPD) https://www.cipd.co.uk/.

CIPD (2019) *Workforce Planning*, https://www.cipd.co.uk/knowledge/strategy/organisational-development/workforce-planning-factsheet#8035, accessed 17 October 2019.

Cooke, R. (2007) *Building in the 21st Century*, Oxford: Blackwell.

Chudley, R. and Greeno, R. (2010) *Building Construction Handbook*, Oxford: Elsevier Science and Technology.

Davies, N. and Jokiniemi, E. (2008) *Dictionary of Architecture and Building Construction*, London: Elsevier.

Douglas, J. (2006) *Building Adaptation*, 2nd edn, London: Butterworth-Heineman.

Emmitt, S. and Gorse, C. (2003) *Construction Communication*, Oxford: Blackwell Science.

Emmitt, S. and Gorse, C. (2007) *Communication in Construction Teams*, London: Spon Press, Taylor & Francis.

Emmitt, S. and Gorse, C. (2009) *Barry's Advanced Construction of Buildings*, 2nd edn, Oxford: Wiley-Blackwell.

Fryer, B., Egbu, C., Ellis, R., and Gorse, C. A. (2004) *The Practice of Construction Management*, 3rd edn, Oxford: Blackwell Science.

Garrison, P. (2005) *Basic Structures for Engineers and Architects*, Oxford: Blackwell.

Geller, P. S. (2010) *Built Environment: Design, Management and Applications*, London: Nova Science Publishers.

Gorse, C. and Highfield, D. (2009) *Refurbishment and Upgrading of Buildings*, 2nd edn, London: Spon Press, Taylor & Francis.

Hansen, K. L. and Zenobia, K. E. (2011) *Civil Engineer's Handbook of Professional Practice*, London: John Wiley.

Hicks, T. G. and Hicks, D. (2007) *Handbook of Civil Engineering Calculation*, New York: McGraw Hill.

Lowe, R. J., Bell, M., and Johnston, D. (1996) *Directory of Energy Efficient Housing*. Coventry: UK Chartered Institute of Housing for the Joseph Rowntree Foundation.

Maclean, J. H. and Scott, J. S. (1993) *The Penguin Dictionary of Building*, London: Penguin.

Ranns, R. H. B. and Ranns, E. J. M. (2004) *Practical Construction Management*, London: Taylor & Francis.

Smith, M. (2005) *Using the Building Regulations: Part M: Access*, London: Butterworth-Heinemann.

Stevens, M. (2002) *Project Management Pathways: The Essential Handbook for Project and Programme Managers,* High Wycombe: Association for Project Management.

Open-Access Websites

Planning Portal: The UK Government's online planning and building regulations resource
http://www.planningportal.gov.uk

Virtual Site: Construction resources and information
http://www.leedsmet.ac.uk/teaching/vsite/

Oxford Quick Reference

A Dictionary of Psychology
Andrew M. Colman

Over 9,500 authoritative entries make up the most wide-ranging dictionary of psychology available.

'impressive ... certainly to be recommended'
Times Higher Education Supplement

'probably the best single-volume dictionary of its kind.'
Library Journal

A Dictionary of Economics
John Black, Nigar Hashimzade, and Gareth Myles

Fully up-to-date and jargon-free coverage of economics. Over 3,500 terms on all aspects of economic theory and practice.

'strongly recommended as a handy work of reference.'
Times Higher Education Supplement

A Dictionary of Law

An ideal source of legal terminology for systems based on English law. Over 4,800 clear and concise entries.

'The entries are clearly drafted and succinctly written ... Precision for the professional is combined with a layman's enlightenment.'
Times Literary Supplement

A Dictionary of Education
Susan Wallace

In over 1,000 clear and concise entries, this authoritative dictionary covers all aspects of education, including organizations, qualifications, key figures, major legislation, theory, and curriculum and assessment terminology.

Oxford Quick Reference

A Dictionary of Sociology
John Scott

The most wide-ranging and authoritative dictionary of its kind.

'Readers and especially beginning readers of sociology can scarcely do better ... there is no better single volume compilation for an up-to-date, readable, and authoritative source of definitions, summaries and references in contemporary Sociology.'

A. H. Halsey, Emeritus Professor, Nuffield College,
University of Oxford

The Concise Oxford Dictionary of Politics and International Relations
Garrett Brown, Iain McLean, and Alistair McMillan

The bestselling A–Z of politics with over 1,700 detailed entries.

'A first class work of reference ... probably the most complete as well as the best work of its type available ... Every politics student should have one'

Political Studies Association

A Dictionary of Environment and Conservation
Chris Park and Michael Allaby

An essential guide to all aspects of the environment and conservation containing over 9,000 entries.

'from *aa* to *zygote*, choices are sound and definitions are unspun'
New Scientist

Oxford Quick Reference

A Dictionary of Chemistry

Over 5,000 entries covering all aspects of chemistry, including physical chemistry and biochemistry.

'It should be in every classroom and library ... the reader is drawn inevitably from one entry to the next merely to satisfy curiosity.'

School Science Review

A Dictionary of Physics

Ranging from crystal defects to the solar system, 4,000 clear and concise entries cover all commonly encountered terms and concepts of physics.

A Dictionary of Biology

The perfect guide for those studying biology — with over 5,800 entries on key terms from biology, biochemistry, medicine, and palaeontology.

'lives up to its expectations; the entries are concise, but explanatory'

Biologist

'ideally suited to students of biology, at either secondary or university level, or as a general reference source for anyone with an interest in the life sciences'

Journal of Anatomy